LONDON MATHEMATICAL SOCIETY LECTURE NOTE SERIES

Managing Editor: Professor J.W.S. Cassels, Department of Pure Mathematics and Mathematical Statistics, 16 Mill Lane, Cambridge CB2 1SB, England

The books in the series listed below are available from booksellers, or, in case of difficulty, from Cambridge University Press.

4 Algebraic topology, J.F.ADAMS
5 Commutative algebra, J.T.KNIGHT
8 Integration and harmonic analysis on compact groups, R.E.EDWARDS
11 New developments in topology, G.SEGAL (ed)
12 Symposium on complex analysis, J.CLUNIE & W.K.HAYMAN (eds)
13 Combinatorics, T.P.McDONOUGH & V.C.MAVRON (eds)
16 Topics in finite groups, T.M.GAGEN
17 Differential germs and catastrophes, Th.BROCKER & L.LANDER
18 A geometric approach to homology theory, S.BUONCRISTIANO, C.P.ROURKE & B.J.SANDERSON
20 Sheaf theory, B.R.TENNISON
21 Automatic continuity of linear operators, A.M.SINCLAIR
23 Parallelisms of complete designs, P.J.CAMERON
24 The topology of Stiefel manifolds, I.M.JAMES
25 Lie groups and compact groups, J.F.PRICE
26 Transformation groups, C.KOSNIOWSKI (ed)
27 Skew field constructions, P.M.COHN
29 Pontryagin duality and the structure of LCA groups, S.A.MORRIS
30 Interaction models, N.L.BIGGS
31 Continuous crossed products and type III von Neumann algebras,A.VAN DAELE
32 Uniform algebras and Jensen measures, T.W.GAMELIN
34 Representation theory of Lie groups, M.F. ATIYAH et al.
35 Trace ideals and their applications, B.SIMON
36 Homological group theory, C.T.C.WALL (ed)
37 Partially ordered rings and semi-algebraic geometry, G.W.BRUMFIEL
38 Surveys in combinatorics, B.BOLLOBAS (ed)
39 Affine sets and affine groups, D.G.NORTHCOTT
40 Introduction to Hp spaces, P.J.KOOSIS
41 Theory and applications of Hopf bifurcation, B.D.HASSARD, N.D.KAZARINOFF & Y-H.WAN
42 Topics in the theory of group presentations, D.L.JOHNSON
43 Graphs, codes and designs, P.J.CAMERON & J.H.VAN LINT
44 Z/2-homotopy theory, M.C.CRABB
45 Recursion theory: its generalisations and applications, F.R.DRAKE & S.S.WAINER (eds)
46 p-adic analysis: a short course on recent work, N.KOBLITZ
47 Coding the Universe, A.BELLER, R.JENSEN & P.WELCH
48 Low-dimensional topology, R.BROWN & T.L.THICKSTUN (eds)
49 Finite geometries and designs,P.CAMERON, J.W.P.HIRSCHFELD & D.R.HUGHES
50 Commutator calculus and groups of homotopy classes, H.J.BAUES
51 Synthetic differential geometry, A.KOCK
52 Combinatorics, H.N.V.TEMPERLEY (ed)
54 Markov process and related problems of analysis, E.B.DYNKIN
55 Ordered permutation groups, A.M.W.GLASS
56 Journêes arithmêtiques, J.V.ARMITAGE (ed)
57 Techniques of geometric topology, R.A.FENN
58 Singularities of smooth functions and maps, J.A.MARTINET
59 Applicable differential geometry, M.CRAMPIN & F.A.E.PIRANI
60 Integrable systems, S.P.NOVIKOV et al
61 The core model, A.DODD
62 Economics for mathematicians, J.W.S.CASSELS
63 Continuous semigroups in Banach algebras, A.M.SINCLAIR

64 Basic concepts of enriched category theory, G.M.KELLY
65 Several complex variables and complex manifolds I, M.J.FIELD
66 Several complex variables and complex manifolds II, M.J.FIELD
67 Classification problems in ergodic theory, W.PARRY & S.TUNCEL
68 Complex algebraic surfaces, A.BEAUVILLE
69 Representation theory, I.M.GELFAND *et al.*
70 Stochastic differential equations on manifolds, K.D.ELWORTHY
71 Groups - St Andrews 1981, C.M.CAMPBELL & E.F.ROBERTSON (eds)
72 Commutative algebra: Durham 1981, R.Y.SHARP (ed)
73 Riemann surfaces: a view towards several complex variables,A.T.HUCKLEBERRY
74 Symmetric designs: an algebraic approach, E.S.LANDER
75 New geometric splittings of classical knots, L.SIEBENMANN & F.BONAHON
76 Spectral theory of linear differential operators and comparison algebras, H.O.CORDES
77 Isolated singular points on complete intersections, E.J.N.LOOIJENGA
78 A primer on Riemann surfaces, A.F.BEARDON
79 Probability, statistics and analysis, J.F.C.KINGMAN & G.E.H.REUTER (eds)
80 Introduction to the representation theory of compact and locally compact groups, A.ROBERT
81 Skew fields, P.K.DRAXL
82 Surveys in combinatorics, E.K.LLOYD (ed)
83 Homogeneous structures on Riemannian manifolds, F.TRICERRI & L.VANHECKE
84 Finite group algebras and their modules, P.LANDROCK
85 Solitons, P.G.DRAZIN
86 Topological topics, I.M.JAMES (ed)
87 Surveys in set theory, A.R.D.MATHIAS (ed)
88 FPF ring theory, C.FAITH & S.PAGE
89 An F-space sampler, N.J.KALTON, N.T.PECK & J.W.ROBERTS
90 Polytopes and symmetry, S.A.ROBERTSON
91 Classgroups of group rings, M.J.TAYLOR
92 Representation of rings over skew fields, A.H.SCHOFIELD
93 Aspects of topology, I.M.JAMES & E.H.KRONHEIMER (eds)
94 Representations of general linear groups, G.D.JAMES
95 Low-dimensional topology 1982, R.A.FENN (ed)
96 Diophantine equations over function fields, R.C.MASON
97 Varieties of constructive mathematics, D.S.BRIDGES & F.RICHMAN
98 Localization in Noetherian rings, A.V.JATEGAONKAR
99 Methods of differential geometry in algebraic topology, M.KAROUBI & C.LERUSTE
100 Stopping time techniques for analysts and probabilists, L.EGGHE
101 Groups and geometry, ROGER C.LYNDON
102 Topology of the automorphism group of a free group, S.M.GERSTEN
103 Surveys in combinatorics 1985, I.ANDERSEN (ed)
104 Elliptic structures on 3-manifolds, C.B.THOMAS
105 A local spectral theory for closed operators, I.ERDELYI & WANG SHENGWANG
106 Syzygies, E.G.EVANS & P.GRIFFITH
107 Compactification of Siegel moduli schemes, C-L.CHAI
108 Some topics in graph theory, H.P.YAP
109 Diophantine analysis, J.LOXTON & A.VAN DER POORTEN (eds)
110 An introduction to surreal numbers, H.GONSHOR
111 Analytical and geometric aspects of hyperbolic space, D.B.A.EPSTEIN (ed)
112 Low dimensional topology and Kleinian groups, D.B.A.EPSTEIN (ed)
113 Lectures on the asymptotic theory of ideals, D.REES
114 Lectures on Bochner-Riesz means, K.M.DAVIS & Y-C.CHANG
115 An introduction to independence for analysts, H.G.DALES & W.H.WOODIN
116 Representations of algebras, P.J.WEBB (ed)
117 Homotopy theory, E.REES & J.D.S.JONES (eds)
118 Skew linear groups, M.SHIRVANI & B.WEHRFRITZ

London Mathematical Society Lecture Note Series. 59

Applicable Differential Geometry

M. CRAMPIN
Professor of Mathematics, The Open University

F.A.E. PIRANI
Emeritus Professor of Rational Mechanics
University of London

CAMBRIDGE UNIVERSITY PRESS

Published by the Press Syndicate of the University of Cambridge
The Pitt Building, Trumpington Street, Cambridge CB2 1RP
32 East 57th Street, New York, NY 10022-4211 USA
10 Stamford Road, Oakleigh, Melbourne 3166, Australia

First published 1986
Reprinted 1987 (with corrections), 1988, 1994

Library of Congress cataloging in publication data available

British Library cataloguing in publication data

Crampin, M.
Applicable differential geometry – (London Mathematical Society lecture note series. ISSN 0076-0052; 59)
1. Geometry, Differential
I. Title II. Pirani, F.A.E. III. Series
516.3′6 QA641

ISBN 0 521 23190 6

To John Lighton Synge

and to the memory of Alfred Schild.

Transferred to digital printing 1999

CONTENTS

Preface

This book is based on lecture courses given by the authors over the past decade and a half to various student audiences, most of them final year undergraduates or beginning graduates. It is meant particularly for those who wish to study relativity theory or classical mechanics from a geometrical viewpoint. In each of these subjects one can go quite far without knowing about differentiable manifolds, and the arrangement of the book exploits this. The essential ideas are first introduced in the context of affine space; this is enough for special relativity and vectorial mechanics. Then manifolds are introduced and the essential ideas are suitably adapted; this makes it possible to go on to general relativity and canonical mechanics. The book ends with some chapters on bundles and connections which may be useful in the study of gauge fields and such matters. The "applicability" of the material appears in the choice of examples, and sometimes in the stating of conditions which may not always be the strongest ones and in the omission of some proofs which we feel add little to an understanding of results.

We have included a great many exercises. They range from straightforward verifications to substantial practical calculations. The end of an exercise is marked with the sign □. Exercises are numbered consecutively in each chapter; where we make reference to an exercise, or indeed a section, in a chapter other than the current one we do so in the form "Exercise n of Chapter p" or something of that kind. We conclude each chapter with a brief summary of its most important contents. We hope that beginners will find these summaries useful for review, while those for whom the subject is not entirely new will be able to tell from the summary whether a chapter treats topics they are already familiar with.

We have attempted to make the book as self-contained as is consistent with the level of knowledge assumed of its readers and with the practical limits on its length. We have therefore appended notes to several chapters; these notes summarise background material not covered in the text or give references to fuller treatments of topics not dealt with in detail. The notes should be the first recourse for a reader who meets something unfamiliar and unexplained in the main text.

We are grateful to Elsayed Ahmed, Glauco Amerighi, Marcelo Gleiser, Clive Kilmister, Mikio Nakahara, Tony Solomonides, Brian Sutton and Navin Swaminarayan, who commented on parts of the manuscript, and to Dr. Swaminarayan for working many of the exercises. We are grateful also to Eileen Cadman, who drew the pictures, to Joan Bunn, Carol Lewis and Barbara Robinson, who spent many hours patiently processing the manuscript, and to Mrs Robinson for her careful subediting. Finally we have to thank a number of colleagues for TeXnical help unstintingly given: Sue Brooks, Bob Coates, Steve Daniels, Glen Fulford, Sid Morris and Chris Rowley.

The Norman Foundation generously supported the preparation of the text.

0. THE BACKGROUND: VECTOR CALCULUS

The reader of this book is assumed to have a working knowledge of vector calculus. The book is intended to explain wide generalisations of that subject. In this chapter we identify some aspects of the subject which are not always treated adequately in elementary accounts. These will be the starting points for several later developments.

1. Vectors

The word "vector" is used in slightly different ways in pure mathematics, on the one hand, and in applied mathematics and physics, on the other. The usefulness of the vector concept in applications is enhanced by the convention that vectors may be located at different points of space: thus a force may be represented by a vector located at its point of application. Sometimes a distinction is drawn between "free" vectors and "bound" vectors. By a free vector is meant one which may be moved about in space provided that its length and direction are kept unchanged. A bound vector is one which acts at a definite point.

In the mathematical theory of vector spaces these distinctions are unknown. In that context all vectors, insofar as they are represented by directed line segments, must originate from the same point, namely the zero vector, or origin. Only the parallelogram rule of vector addition makes sense, not the triangle rule.

Closely connected with these distinctions is a difficulty about the representation of the ordinary space of classical physics and the space-time of special relativity theory. On the one hand, one finds it more or less explicitly stated that space is homogeneous: the laws of physics do not prefer any one point of space, or of space-time, over any other. On the other hand, almost any quantitative treatment of a physical problem begins with a particular choice of coordinate axes—a choice which singles out some supposedly unremarkable point for the privileged role of origin. The underlying difficulty here is that the vector space $\mathbf{R}^3$ is not quite appropriate as a model for ordinary physical space. The kind of space which is homogeneous, in which the whereabouts of the origin of coordinates is an act of choice not a dictate of the structure, which retains sufficient vectorial properties to model physical space, and in which a sensible distinction between free and bound vectors can be made, is called an affine space; it will be discussed in detail in Chapter 1. (The concept of a vector space, and the notation $\mathbf{R}^3$, are explained in Note 2 at the end of Chapter 1.)

It is unfortunate that distinctions which appear merely pedantic in the straightforward context of $\mathbf{R}^3$ are sometimes important for generalisations. The scalar product, also called the inner or dot product, is so familiar that it is difficult to keep in mind that $\mathbf{R}^3$ may be given many different scalar products, with similar properties, or no scalar product at all. The scalar product is a secondary structure: if one fails to recognise this one cannot exploit fully the relationship between a vector space and its dual space (the dual space is also defined in Note 2 to Chapter 1).

In other terms, the matrix product of a row and a column vector, resulting in a number, may be constructed without the introduction of any secondary structure, but the scalar product of two column vectors, also resulting in a number, cannot. The first makes use only of vector space notions, combining an element of a vector space, represented as a column vector, and an element of its dual space, represented as a row vector. The product, called a pairing, is represented by matrix multiplication. In tensor calculus this would be expressed by the contraction of a contravariant and a covariant vector. The second requires the additional concept of a scalar product. It is surprising how rich a geometry may be developed without the introduction of a scalar product: after this chapter, we do not introduce scalar products again until Chapter 7. It is also instructive to see which notions of vector algebra and calculus really depend on the scalar product, or on the metrical structure, of Euclidean space.

From the outset we shall distinguish notationally between the underlying n-dimensional vector space of n-tuples, and the same space with the scalar product added, by writing $\mathbf{R}^n$ for the former but $\mathcal{E}^n$ for the latter.

2. Derivatives

Let f be a function on $\mathcal{E}^3$; $\operatorname{grad} f$ is the column vector of its partial derivatives, evaluated at any chosen point. Let $\mathbf{v}$ be a unit vector at that point, with a chosen direction. Then the directional derivative of f in the chosen direction is the scalar product $\mathbf{v} \cdot \operatorname{grad} f$.

A more general directional derivative may be defined by dropping the requirement that $\mathbf{v}$ be a unit vector. This directional derivative may be interpreted as the derivative along any curve which has $\mathbf{v}$ as tangent vector at the point in question, the curve not being necessarily parametrised by its length. If $\mathbf{v}$ is regarded as a velocity vector of a particle then $\mathbf{v} \cdot \operatorname{grad} f$ is the time rate of change of f along the path of the particle. However, the directional derivative may perfectly well be constructed without appealing to the scalar product, by taking the partial derivatives as components of a row vector. This vector is called the differential of f; in these terms the directional derivative is simply the pairing of the tangent vector and the differential. Having no scalar product, one cannot sustain the usual interpretation of the gradient as the normal to a surface $f =$ constant, but the differential may still be used to specify the tangent plane to this surface at the point in question. The main advantage of this point of view is that it is the starting point for a general theory which encompasses non-metrical versions of grad, curl and div, and much more besides.

In vector calculus one sees pretty clear hints of close connections between grad, curl and div, but in the usual treatments they are often not much more than hints. We have in mind for example the relations $\operatorname{curl} \circ \operatorname{grad} = \mathbf{0}$ and $\operatorname{div} \circ \operatorname{curl} = 0$, and the assertions that a vector field is a gradient if and only if its curl vanishes and a curl if and only if its divergence vanishes. These relations all fall into place in the development of the exterior calculus, which is undertaken in Chapters 4 and 5.

We return to consideration of the directional derivative, but from a different point of view. The directional derivative operator associated with a vector field $\mathbf{X}$ will for the time being be denoted by $\mathbf{X}\cdot\mathrm{grad}$, as before, but now we insist on regarding the components of $\mathrm{grad} f$ as the components of the differential, so that there is no need to introduce the scalar product to construct directional derivatives. We list some properties of operators of this type, as applied to functions:

(1) $\mathbf{X}\cdot\mathrm{grad}$ maps functions to functions
(2) $\mathbf{X}\cdot\mathrm{grad}$ is a linear operator, and is linear in $\mathbf{X}$
(3) $(f\mathbf{X})\cdot\mathrm{grad} = f(\mathbf{X}\cdot\mathrm{grad})$
(4) $(\mathbf{X}\cdot\mathrm{grad})(f_1 f_2) = (\mathbf{X}\cdot\mathrm{grad} f_1) f_2 + f_1(\mathbf{X}\cdot\mathrm{grad} f_2)$ (Leibniz's rule).

The composition of directional derivative operators, and their commutation properties, are not often discussed in standard treatments of vector calculus. The composite $(\mathbf{X}\cdot\mathrm{grad})\circ(\mathbf{Y}\cdot\mathrm{grad})$ of two operators is not a directional derivative operator, because it takes second derivatives of any function on which it acts, while directional derivative operators take only first derivatives. However, the commutator

$$(\mathbf{X}\cdot\mathrm{grad})\circ(\mathbf{Y}\cdot\mathrm{grad}) - (\mathbf{Y}\cdot\mathrm{grad})\circ(\mathbf{X}\cdot\mathrm{grad})$$

is a directional derivative operator, which is to say that it is of the form $\mathbf{Z}\cdot\mathrm{grad}$ for some vector field $\mathbf{Z}$. The vector field $\mathbf{Z}$ depends on $\mathbf{X}$ and $\mathbf{Y}$ (and on their derivatives). It is usual to denote the commutator by the use of square brackets, and to extend this notation to the vector fields, writing

$$[\mathbf{X}\cdot\mathrm{grad}, \mathbf{Y}\cdot\mathrm{grad}] = [\mathbf{X},\mathbf{Y}]\cdot\mathrm{grad}.$$

It is not difficult to compute the components of $[\mathbf{X},\mathbf{Y}]$ in terms of the components of $\mathbf{X}$ and $\mathbf{Y}$; this, and the significance and properties of the brackets of vector fields, is discussed at length in Chapter 3.

The directional derivative operator may be applied to a vector field as well as to a function. The Cartesian components of a vector field are functions, and the operator is applied to them one by one: if $\mathbf{E}_1$, $\mathbf{E}_2$, and $\mathbf{E}_3$ are the usual coordinate vector fields and

$$\mathbf{Y} = Y_1\mathbf{E}_1 + Y_2\mathbf{E}_2 + Y_3\mathbf{E}_3$$

then

$$(\mathbf{X}\cdot\mathrm{grad})\mathbf{Y} = (\mathbf{X}\cdot\mathrm{grad}\, Y_1)\,\mathbf{E}_1 + (\mathbf{X}\cdot\mathrm{grad}\, Y_2)\,\mathbf{E}_2 + (\mathbf{X}\cdot\mathrm{grad}\, Y_3)\,\mathbf{E}_3.$$

This operation has properties similar to those of the directional derivative as applied to functions:

(1) $\mathbf{X}\cdot\mathrm{grad}$ maps vector fields to vector fields
(2) $\mathbf{X}\cdot\mathrm{grad}$ is a linear operator and is linear in $\mathbf{X}$
(3) $(f\mathbf{X})\cdot\mathrm{grad} = f(\mathbf{X}\cdot\mathrm{grad})$
(4) $\mathbf{X}\cdot\mathrm{grad}(f\mathbf{Y}) = (\mathbf{X}\cdot\mathrm{grad} f)\mathbf{Y} + f(\mathbf{X}\cdot\mathrm{grad}\,\mathbf{Y})$.

However, the conventional use of the same symbol $\mathbf{X}\cdot\mathrm{grad}$ for what are really two different operators—the directional derivatives of functions and of vector fields—makes the last of these appear more like Leibniz's rule than it really is: on the right hand side each of the two usages of $\mathbf{X}\cdot\mathrm{grad}$ occurs.

The properties of the directional derivative of vector fields listed above are typical of the properties of a covariant derivative; this subject is developed in Chapters 2, 5 and 7, and generalised in Chapters 9, 11, 13 and 15. The application of $\mathbf{X}\cdot\mathrm{grad}$ to vector fields in $\mathbf{R}^3$ is the simplest example of a covariant derivative. The interaction of the covariant derivative with scalar products is exemplified by the formula (in $\mathcal{E}^3$)

$$(\mathbf{X}\cdot\mathrm{grad})(\mathbf{Y}\cdot\mathbf{Z}) = (\mathbf{X}\cdot\mathrm{grad}\,\mathbf{Y})\cdot\mathbf{Z} + \mathbf{Y}\cdot(\mathbf{X}\cdot\mathrm{grad}\,\mathbf{Z}).$$

Note that in this formula the two different meanings of $\mathbf{X}\cdot\mathrm{grad}$ again occur: on the left it acts on a function, on the right, on vector fields. The commutator of two such operators, acting on vector fields, is given by the same formula as for the action on functions:

$$[\mathbf{X}\cdot\mathrm{grad}, \mathbf{Y}\cdot\mathrm{grad}] = [\mathbf{X}, \mathbf{Y}]\cdot\mathrm{grad}.$$

This formula, which is not typical of covariant differentiation formulae, expresses the fact that ordinary Euclidean space is flat, not curved.

We have adopted the usual convention of vector calculus that vectors and vector fields are printed in boldface type. We shall continue to follow this convention, but only for the vectors and vector fields in $\mathcal{E}^3$ with which vector calculus deals: in more general situations vectors and vector fields will be printed in ordinary italic type.

3. Coordinates

One of the byproducts of the approach to be developed here is that the expression in curvilinear coordinates of such constructions as grad, curl and div, which can appear puzzling, becomes relatively straightforward. Coordinate transformations, and the way in which quantities transform in consequence, have an important part to play in the developing argument. However, we do not generally define objects in terms of their transformation properties under change of coordinates, as would be the practice in tensor calculus. We develop the idea that since no one coordinate system is preferable to another, objects of interest should be defined geometrically, without reference to a coordinate system, and their transformation properties deduced from the definitions. In tensor calculus, on the other hand, the transformation law is the primary focus, and generally the basis for the definition of objects.

The arena for most of the geometry described below is (finally) the differentiable manifold, in which coordinates exist locally, but no assumption is made that a single coordinate system may be extended to cover the whole space. The homogeneity of affine space is thus taken many steps further in the definition of a differentiable manifold.

We shall also attempt to give some indications of global matters, which tensor calculus rarely does, it being ill adapted for that purpose. On the other hand, the results we obtain often have tensor calculus equivalents, which will frequently be revealed in the exercises; but our approach is, in a word, geometrical. Our exposition is intended to illustrate Felix Klein's remark that "The geometric properties of any figure must be expressible in formulas which are not changed when one changes the coordinate system ... conversely, any formula which in this sense is invariant under a group of coordinate transformations must represent a geometric property".

4. The Range and Summation Conventions

Throughout this work we shall use certain conventions regarding indices which simplify the representation of sums, and result in considerable savings of space and effort. These are the range and summation conventions, often associated with the name Einstein. The reader who is already familiar with tensor calculus will need no instruction in their use. For other readers, not so prepared, we describe their operation here.

It is simplest to begin with an example. Consider the matrix equation

$$v = \lambda(u).$$

Here u is supposed to be a column vector, of size n say (or $n \times 1$ matrix); λ is an $m \times n$ matrix; and v, therefore, is a column vector of size m ($m \times 1$ matrix). This equation may be interpreted as expressing how each individual component of v is determined from the components of u *via* λ. To write down that expression explicitly one introduces notation for the components of u and v and the elements of λ: say u^a to stand for the ath component of u ($a = 1, 2, \ldots, n$); v^α to stand for the αth component of v ($\alpha = 1, 2, \ldots, m$); and λ^α_a to stand for the (α, a) element of λ, that is, the element in the αth row and ath column. The matrix equation above is then equivalent to the m equations

$$v^\alpha = \sum_{a=1}^{n} \lambda^\alpha_a u^a.$$

The range convention arises from the realisation that it is not necessary to state, at each occurrence of a set of equations like this, that there are m equations involved and that the truth of each is being asserted. This much could be guessed from the appearance of the index α on each side of the equation: for α is a free index, unlike a which is subject to the summation sign. On the other hand, the summation convention follows from the observation that whenever a summation occurs in an expression of this kind it is a summation over an index (here a) which occurs precisely twice in the expression to be summed. Thus summation occurs only where there is a repeated index; and when an index is repeated summation is almost always required. Under these circumstances the summation symbol $\sum_{a=1}^{n}$ serves no useful function, since summation may be recognised by the repetition of an index; it may therefore be omitted.

Thus the component equation above is written, when range and summation conventions are in force, in the simple form

$$v^\alpha = \lambda^\alpha_a u^a.$$

The presence of the repeated index a on the right hand side implies summation over its permitted range of values $1, 2, \ldots, n$ by virtue of the summation convention; while the presence of the free index α on both sides of the equation implies equality for each value $1, 2, \ldots, m$ that it can take, by virtue of the range convention.

In general, the range and summation conventions work as follows. If, in an equation involving indexed quantities, there are free (unrepeated) indices, then the equation holds for all values in the ranges of all free indices, these ranges having been declared previously: this is the *range convention*. Where, in an expression involving indexed quantities, any index is repeated, summation over all possible values in the range of that index is implied, the range again having been declared previously: this is the *summation convention*.

The ranges of indices governed by the range and summation conventions will always be finite: thus only finite sums are involved, and there is no problem of convergence.

Operation of the range and summation conventions in practice is relatively straightforward. One or two rules—often best employed as running checks on the correctness of a calculation—should be mentioned. The number of free indices on the two sides of an equation must be the same; and of course each different free index in an expression must be represented by a different letter. Repeated indices in an expression may occur only in pairs. Replacement of a letter representing an index by another letter is allowed, provided that all occurrences of the letter are changed at the same time and in the same way, and provided that it is understood that the new letter has the same range of values as the one it replaces. The most convenient practice to adopt, where indices with different ranges are involved in a single calculation, is to reserve a small section of a particular alphabet to represent indices with a given range. Thus in the case discussed above one could take a, b, c to range and sum from 1 to n, and α, β, γ to range and sum from 1 to m; then $v^\beta = \lambda^\beta_c u^c$ would mean exactly the same as $v^\alpha = \lambda^\alpha_a u^a$.

From a given expression containing two free indices with the same ranges, a new expression may be formed by making them the same, that is, by taking a sum: this process is known as *contraction*. For example, from the components μ^a_b of a square matrix one may form the number μ^c_c, its trace.

Three points should be made about the way the summation convention is employed in this book. In the first place, we have so arranged matters that the pair of repeated indices implying a summation will (almost always) occur with one index in the upper position and one in the lower. This will already be apparent from the way we have chosen to write the matrix equation above, when some such thing as $v_\alpha = \lambda_{\alpha a} u_a$ might have been expected. The point is related to the importance of distinguishing between a vector space and its dual (column versus row vectors) mentioned several times earlier in this chapter. This distinction is introduced into the notation for components by using an index in the upper position (u^a, v^α) for components of a column vector. For the components of a row vector we shall place the index in the lower position, thus: c_α. Then the multiplication of the matrix λ by a row vector c (of length m), on the left, gives a row vector (of length n) whose components are $c_\alpha \lambda^\alpha_a$. Notice that the type of the resulting vector (row rather than column) is correctly indicated by the position of the free index a.

The pairing of a row and a column vector (in other words, a $1 \times m$ and an $m \times 1$ matrix) by matrix multiplication, mentioned in Section 1, is represented by an expression $c_\alpha v^\alpha$, which conforms to our rule. On the other hand, the scalar product of two column vectors, $\sum_{\alpha=1}^{m} v^\alpha w^\alpha$, cannot be correctly so represented without the introduction of further quantities. What is required is a two-index object, say $\delta_{\alpha\beta}$, with

$$\delta_{11} = \delta_{22} = \ldots = \delta_{mm} = 1, \text{ but } \delta_{\alpha\beta} = 0 \text{ if } \alpha \neq \beta;$$

with the aid of this the expression $\delta_{\alpha\beta} v^\alpha w^\beta$ can be correctly formed. This has the same value as $\sum_{\alpha=1}^{m} v^\alpha w^\alpha$; but the point of the remark is to show again, this time through the application of the summation convention, how the pairing of vectors and duals differs from a scalar product. The extra piece of machinery required in the case of the scalar product, represented by $\delta_{\alpha\beta}$ above, is the Euclidean metric.

The second point we should mention about our use of the range and summation conventions is that, whereas in tensor calculus they are used almost exclusively with indexed quantities which are collections of numbers (or functions), we shall use them with other types of object. For example, basis vectors for an n-dimensional vector space may be written $\{e_a\}$, where a ranges and sums from 1 to n; then any vector u in the space may be written $u = u^a e_a$, where the u^a, its components with respect to the basis, are numbers, but the e_a are vectors.

The third point to watch out for is that an expression such as (x^c) is frequently used to stand for $(x^1, x^2, \ldots, x^n)$. Furthermore, the value of a function of n variables, say f, at (x^c) will be denoted $f(x^c)$. In this situation the index c is subject neither to the summation nor to the range convention. In such a context (x^c) is usually to be thought of as the set of coordinates of a point in some space. Where elements of $\mathbf{R}^n$ are being used as coordinates rather than as the components of velocity vectors or differentials, for example, the distinctions made earlier between vector spaces and their duals, or between column and row vectors, no longer have the same importance.

Note to Chapter 0

Klein's remark is in his splendid *Elementary mathematics from an advanced standpoint, part II, Geometry* (Klein [1939]) p 25.

1. AFFINE SPACES

When one first begins to learn mechanics one is confronted with a space—the "ordinary" space in which mechanical processes take place—which in many ways resembles a vector space, but which lacks a point playing the distinctive role of zero vector. The resemblance lies in the vector addition properties of displacements and of quantities derived from them such as velocities, accelerations and forces. The difference lies in the fact that the mechanical properties of a system are quite independent of its position and orientation in space, so that its behaviour is unaffected by choice of origin. Of course the Sun, or the mass centre of the Solar System, plays a role in the formulation of the Kepler problem of planetary motion, but the relative motion of the planets does not depend on whether displacements are measured from the Sun or from some other point. Nor does it depend on the choice of origin for time measurements.

The same is true in special relativity theory. Here also the behaviour of a physical system is unaffected by the choice of space-time origin.

In neither case can there be ascribed to any point the distinctive position and properties ascribed to the zero vector in a vector space; nor can any meaning be given to the addition of points as if they were vectors. Nevertheless, one learns to manipulate vectors in ordinary Euclidean space or in Minkowski space-time and to give physical significance to these manipulations, without perhaps paying too much attention to the precise nature of the underlying space or space-time. When one wants to be more systematic, however, it is necessary to establish the precise relation between the vectors and the space. A satisfactory construction must allow for vector addition of displacements but may not single out any point with special properties. The result is called an affine space.

It is true that the limitations imposed by formation in an affine mould are too severe for some applications. This became apparent during the course of the nineteenth century, when various generalisations were developed. One line of development culminated in the work of Ricci and Levi-Civita on the tensor calculus, which was exploited by Einstein in the invention of general relativity theory; another line led to the work of Lie in group theory, another to the work of É. Cartan in differential geometry, yet another to the work of Poincaré and Birkhoff in celestial mechanics. The generalisations which were developed include much of the subject matter of the later part of this book (and much else). To a great extent these generalisations may be attained by modifying one or another property of an affine space, so we start with that. Most of the techniques needed in the later work may be explained quite easily in the affine case and extended without much effort. The more general spaces introduced later are called "manifolds". They are defined in Chapter 10. In the first nine chapters we shall develop the differential geometry of affine spaces in a form suitable for applications and adaptable to generalisation.

To start with, the concepts to be explained do not require assumptions of a metrical character—no scalar product or measure of length is required—and so they will be applicable later on in both the Euclidean and the Minkowskian contexts.

1. Affine Spaces

In this section we define affine spaces and introduce coordinates natural to them called affine coordinates.

Affine space defined. We are to define a space $\mathcal{A}$ in which displacements may be represented by vectors. As a model for displacements we shall take a real vector space $\mathcal{V}$ of finite dimension n. We shall not choose any particular basis in the vector space $\mathcal{V}$, so it is not merely a fixed copy of the real number space $\mathbf{R}^n$. From experience in mechanics, one might hope that displacements in $\mathcal{A}$ would enjoy these properties:

(1) a succession of displacements is achieved by the addition of vectors (the triangle law for displacements)

(2) displacement by the zero vector leaves a point where it is

(3) if any two points are chosen, there is a unique displacement from one to the other.

A formal definition which embodies these properties may be given in terms of the effect on any point $x \in \mathcal{A}$ of displacement by the vector $v \in \mathcal{V}$. We shall write $x + v$ to denote the point to which x is displaced, and regard the operation of displacement as a map $\mathcal{A} \times \mathcal{V} \to \mathcal{A}$ by $(x, v) \mapsto x + v$. The definition runs as follows: a set $\mathcal{A}$ is called an *affine space* modelled on the real vector space $\mathcal{V}$ if there is a map, called an *affine structure*, $\mathcal{A} \times \mathcal{V} \to \mathcal{A}$, denoted additively: $(x, v) \mapsto x + v$, with the properties

(1) $(x + v) + w = x + (v + w)$ for all $x \in \mathcal{A}$ and all $v, w \in \mathcal{V}$

(2) $x + 0 = x$ for all $x \in \mathcal{A}$, where $0 \in \mathcal{V}$ is the zero vector

(3) for any pair of points $x, x' \in \mathcal{A}$ there is a unique element of $\mathcal{V}$, denoted $x' - x$, such that $x + (x' - x) = x'$.

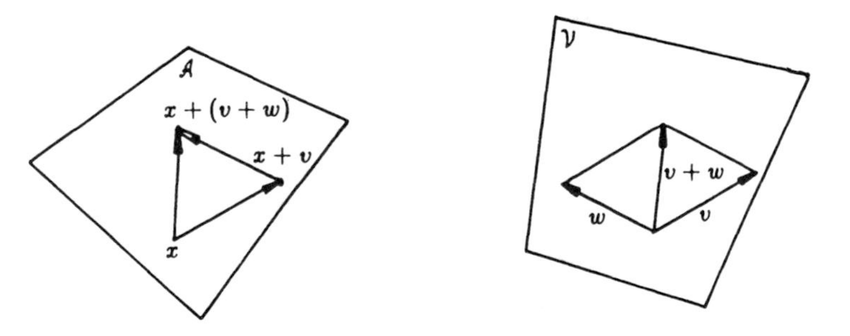

Fig. 1 Points and displacements in an affine space.

Note that the plus sign is given two distinct meanings in this definition: in the expression $x + v$ it denotes the displacement of a point of $\mathcal{A}$ by an element of $\mathcal{V}$, while in the expression $v + w$ it denotes the addition of two elements of $\mathcal{V}$. Moreover, although the displacement from x to x' is denoted $x' - x$, there is no sense in which points of $\mathcal{A}$ may be added together. It is essential to keep these distinctions clearly in mind when working with an affine space.

An affine space $\mathcal{A}$ modelled on a vector space $\mathcal{V}$ of dimension n is also said to be of dimension n. One writes $\dim \mathcal{A} = n$ (and $\dim \mathcal{V} = n$).

Exercise 1. Let $\mathcal{A}$ be an affine space modelled on a vector space $\mathcal{V}$. Show that if, for some $x \in \mathcal{A}$ and some $v \in \mathcal{V}$, $x + v = x$, then $v = 0$. Thus displacement by any vector other than the zero vector leaves no point where it is. □

Exercise 2. Let x_0 be any chosen point of $\mathcal{A}$. Show that the map $\mathcal{A} \to \mathcal{V}$ by $x \mapsto x - x_0$ is bijective (onto and $1:1$). □

Affine coordinates. In order to deal with specific problems one is likely, sooner or later, to want to introduce coordinates into affine spaces. It is true that one can sometimes go a long way in solving mechanics problems without using coordinates, but even so an adept choice of coordinates may much simplify a problem. The same is true in special relativity theory. What is desirable, on the other hand, is to formulate the problem in the first place without using coordinates, so as to be able to recognise whatever symmetry or invariance properties it may have.

Among all the coordinates which may be introduced into an affine space there are those, called affine coordinates, which are especially well-adapted to its structure. These coordinates will be explained here.

A choice of affine coordinates requires only a choice of origin and a choice of axes. The choice of axes is merely a choice of basis for the underlying vector space. If x_0 is any chosen point of $\mathcal{A}$ and $\{e_1, e_2, \ldots, e_n\}$ is any chosen basis for $\mathcal{V}$ then any point x in $\mathcal{A}$ may be written $x = x_0 + (x - x_0)$, and since $x - x_0$ belongs to $\mathcal{V}$ it may in turn be written $x - x_0 = x^a e_a$, where the x^a are its components with respect to the chosen basis $\{e_a\}$. Here for the first time we employ the summation convention explained in Section 4 of Chapter 0: because the index a is repeated, summation from 1 to n ($= \dim \mathcal{A}$) is understood. Thus $x^a e_a$ is short for $\sum_{a=1}^{n} x^a e_a$.

The components $(x^1, x^2, \ldots, x^n)$ are called the *affine coordinates* of x. The point x_0, whose coordinates are evidently $(0, 0, \ldots, 0)$, is called the *origin* of affine coordinates. An assignment of coordinates associates a set of n real numbers $(x^1, x^2, \ldots, x^n)$ to each point of $\mathcal{A}$, and so may be described by a bijective map $\mathcal{A} \to \mathbf{R}^n$ (compare Exercise 2). Thus the dimension of an affine space, like the dimension of the vector space on which it is modelled, is equal to the number of coordinates needed to specify a point of the space.

The notion of dimension and the possibility of using coordinates carry over to manifolds; however, one distinctive property of affine coordinates—that they are valid everywhere in the space at once—will not generalise.

We shall for the time being use $(x^1, x^2, \ldots, x^n)$, often abbreviated to (x^a), to denote the affine coordinates of a general point of $\mathcal{A}$ relative to some system of affine coordinates. Each choice of origin of $\mathcal{A}$ and of basis of $\mathcal{V}$ determines a choice of affine coordinates. If $\{e'_a\}$ is another basis for $\mathcal{V}$, related to the basis $\{e_a\}$ by $e'_b = h^a_b e_a$, where the h^a_b are the elements of a matrix, necessarily non-singular, and if x'_0 is chosen as origin instead of x_0, then the two corresponding sets of affine coordinates $(\acute{x}^a)$, (x^a) of any point x of $\mathcal{A}$ are related by

$$x^a = h^a_b \acute{x}^b + c^a$$

where $x'_0 - x_0 = c^a e_a$, or equivalently by

$$\acute{x}^a = (h^{-1})^a_b x^b + d^a$$

where the $(h^{-1})^a_b$ are the entries in the matrix inverse to the matrix h with entries h^a_b, and $d^a = -(h^{-1})^a_b c^b$. Here we use the range convention as well as the summation convention: the unrepeated index a takes in turn the values $1, 2, \ldots, n$. The transformation from (x^a) to $(\acute{x}^a)$, or *vice versa*, is called an *affine coordinate transformation*.

It should be apparent that the introduction of affine coordinates allows one to identify an affine space $\mathcal{A}$ of dimension n with $\mathbf{R}^n$. The question therefore arises why one should want to consider the more abstract object at all. One reason is that, in a coordinate-based discussion, geometric objects must be defined by giving their transformation properties under change of coordinates. With the more abstract approach one may define geometric objects intrinsically, and deduce the transformation properties from the definition. These alternatives— definition with and without coordinates—represent the points of view of classical tensor calculus and modern differential geometry respectively. We shall repeatedly have occasion to mention the common features of and contrasts between these two points of view. In order to be able to understand the literature of contemporary mathematical physics it is essential to be familiar with both of them.

Exercise 3. Verify that the time of Newtonian mechanics is an affine space of dimension 1. Explain how the process of setting up an affine coordinate system could be distinguished physically. □

Exercise 4. The real number space $\mathbf{R}^n$ may be identified as an affine space modelled on itself as vector space. If the point $(0, 0, \ldots, 0)$ of $\mathbf{R}^n$ is chosen as origin and the vectors $(1, 0, \ldots, 0)$, $(0, 1, \ldots, 0), \ldots (0, 0, \ldots, 1)$ as basis then the point $(\xi^1, \xi^2, \ldots, \xi^n)$ of $\mathbf{R}^n$ has affine coordinates $(\xi^1, \xi^2, \ldots, \xi^n)$ in this affine coordinate system. Show that if the point $(\eta^1, \eta^2, \ldots, \eta^n)$ is chosen as origin and the vectors $e_1 = (e^1_1, e^2_1, \ldots, e^n_1)$, $e_2 = (e^1_2, e^2_2, \ldots, e^n_2)$, $\ldots$, $e_n = (e^1_n, e^2_n, \ldots, e^n_n)$ as basis then the point $(\xi^1, \xi^2, \ldots, \xi^n)$ has coordinates $(x^1, x^2, \ldots, x^n)$ in the new affine coordinate system which are determined uniquely by the equations $e^a_b x^b = \xi^a - \eta^a$. □

Exercise 5. Show that the plane in $\mathbf{R}^3$ through the points $(1,0,0)$, $(0,1,0)$ and $(0,0,1)$ (the set of points $(\xi^1, \xi^2, \xi^3) \in \mathbf{R}^3$ such that $\xi^1 + \xi^2 + \xi^3 = 1$) is an affine space $\mathcal{A}$. Take the point $(0,0,1)$ as origin of affine coordinates in $\mathcal{A}$ and the vectors $e_1 = (1, 0, -1)$ and $e_2 = (0, 1, -1)$ as basis vectors for the vector space on which $\mathcal{A}$ is modelled, and show that the point of $\mathcal{A}$ with affine coordinates (x^1, x^2) is the point $(x^1, x^2, 1 - x^1 - x^2)$ of $\mathbf{R}^3$; next take the point $(\frac{1}{3}, \frac{1}{3}, \frac{1}{3})$ of $\mathcal{A}$ as origin of affine coordinates and the vectors $e'_1 = (\frac{2}{3}, -\frac{1}{3}, -\frac{1}{3})$ and $e'_2 = (-\frac{1}{3}, \frac{2}{3}, -\frac{1}{3})$ as basis vectors, and show that the point of $\mathcal{A}$ with affine coordinates $(\acute{x}^1, \acute{x}^2)$ is the point $(\frac{2}{3}\acute{x}^1 - \frac{1}{3}\acute{x}^2 + \frac{1}{3}, -\frac{1}{3}\acute{x}^1 + \frac{2}{3}\acute{x}^2 + \frac{1}{3}, -\frac{1}{3}\acute{x}^1 - \frac{1}{3}\acute{x}^2 + \frac{1}{3})$ of $\mathbf{R}^3$. □

Exercise 6. Show that the transformation of affine coordinates given in Exercise 5 above is $\acute{x}^1 = 2x^1 + x^2 - 1$, $\acute{x}^2 = x^1 + 2x^2 - 1$. □

Exercise 7. In an affine space, (x^a) are affine coordinates of a point in one chosen coordinate system and $(\acute{x}^a)$ are given by $\acute{x}^a = k_b^a x^b + d^a$, where (k_b^a) is any non-singular matrix and the d^a are any numbers. Show that there is another coordinate system in which $(\acute{x}^a)$ are affine coordinates of the same point. □

Exercise 8. Show that composition (successive application) of two affine coordinate transformations yields another affine coordinate transformation. □

2. Lines and Planes

Let v_0 be a fixed non-zero vector in $\mathcal{V}$ and x_0 a fixed point of $\mathcal{A}$. The map $\mathbf{R} \to \mathcal{A}$ by $t \mapsto x_0 + tv_0$ is called a *line*: it is the line through x_0 determined by v_0. Note that according to this definition a line is a map, not a subset of $\mathcal{A}$. We adopt this approach because in most circumstances in which we have to deal with lines or other curves the parametrisation will be important: we shall want to distinguish between the line through x_0 determined by v_0 and the one through the same point determined by kv_0, $k \neq 0$; these are indistinguishable if one is concerned only with the corresponding subsets of $\mathcal{A}$, but are distinguished by their parametrisations. Using the map to define the line is a convenient way of focussing attention on the parametrisation.

The special nature of lines, by comparison with other kinds of curve one could imagine, may be described as follows. The affine structure map $\mathcal{A} \times \mathcal{V} \to \mathcal{A}$ introduced in Section 1 may be looked at in a slightly different way. Let x_0 be a chosen point of $\mathcal{A}$. Fixing the point x_0 (on the left) in the affine structure map, one obtains a map $a_{x_0}: \mathcal{V} \to \mathcal{A}$, by $v \mapsto x_0 + v$, which takes each vector in $\mathcal{V}$ into the point in $\mathcal{A}$ reached by a displacement from x_0 by that vector. The map a_{x_0} may be thought of as *attaching* the vector space $\mathcal{V}$ to $\mathcal{A}$ at x_0, as space of displacements. The point of this procedure is that it allows one to transfer any figure from a vector space to an affine space modelled on it. Thus, as a subset of $\mathcal{A}$, the image of a line is obtained by attaching a 1-dimensional subspace of $\mathcal{V}$ to $\mathcal{A}$ at x_0.

Any subspace of the vector space $\mathcal{V}$, not only a 1-dimensional one, may in this way be attached to an affine space $\mathcal{A}$ modelled on $\mathcal{V}$. If $\mathcal{W}$ is any subspace of $\mathcal{V}$ then the subspace map, or inclusion, $i: \mathcal{W} \hookrightarrow \mathcal{V}$ takes each vector w, considered as a vector in $\mathcal{W}$, into the same vector w, considered as a vector in $\mathcal{V}$. Following the subspace map by the attachment of $\mathcal{V}$ at x_0 one obtains the map $a_{x_0} \circ i$ which attaches $\mathcal{W}$ to $\mathcal{A}$ at x_0. Attachment of $\mathcal{W}$ at points x_0 and x_1 such that $x_1 - x_0 \in \mathcal{W}$ will result in the same subset of $\mathcal{A}$. Its attachment at points x_0 and x_2 for which $x_2 - x_0 \notin \mathcal{W}$, on the other hand, produces two distinct subsets of $\mathcal{A}$ which are *parallel*.

The set $\{ x_0 + w \mid w \in \mathcal{W} \}$ is called an *affine subspace* of $\mathcal{A}$, or an *affine p-plane* in $\mathcal{A}$, where $p = \dim \mathcal{W}$.

Exercise 9. Let $\mathcal{A}$ be an affine space modelled on a vector space $\mathcal{V}$, and let $\mathcal{B}$ be an affine subspace of $\mathcal{A}$ constructed by attaching the subspace $\mathcal{W}$ of $\mathcal{V}$ to $\mathcal{A}$. Show that $\mathcal{B}$ is in its own right an affine space modelled on $\mathcal{W}$. □

An affine p-plane may be parametrised by p coordinates, say (y^α) (where $\alpha = 1, 2, \ldots, p$), as follows. Let $\{f_\alpha\}$ be a basis for $\mathcal{W}$. Then if the p-plane is attached at x_0, each of its points may be uniqely expressed in the form $x_0 + y^\alpha f_\alpha$. The coordinates of this point with respect to the given basis of $\mathcal{W}$ and origin x_0 are $(y^1, y^2, \ldots, y^p)$, or (y^α). Thus according to our initial definition a line is a parametrised 1-plane.

Exercise 10. Verify that in an affine coordinate system for $\mathcal{A}$ with origin x_0 based on vectors $\{e_a\}$, the affine subspace obtained by attaching a subspace $\mathcal{W}$ at x_1 may be represented by the equations $x^a = c^a + y^\alpha f_\alpha^a$, where (c^a) are the coordinates of x_1, and f_α^a the components of an element of a basis $\{f_\alpha\}$ of $\mathcal{W}$ with respect to the given basis for $\mathcal{V}$. □

In an affine coordinate system with origin x_0 based on vectors $\{e_a\}$, the *coordinate axes* are the 1-planes obtained by attaching at x_0 the 1-dimensional subspaces generated by each of the e_a in turn. Coordinate p-planes and hyperplanes are defined analogously.

Hyperplanes and linear forms. Let $\mathcal{W}$ be a subspace of a vector space $\mathcal{V}$. Then $\dim \mathcal{V} - \dim \mathcal{W}$ is called the codimension of $\mathcal{W}$ in $\mathcal{V}$. Similarly if $\mathcal{B}$ is an affine subspace of an affine space $\mathcal{A}$ then $\dim \mathcal{A} - \dim \mathcal{B}$ is called the *codimension* of $\mathcal{B}$ in $\mathcal{A}$. In particular, an affine subspace of codimension 1 is called a *hyperplane*. Hyperplanes are often described, as point sets, by equations, instead of parametrically, with the help of linear forms. Let α be a linear form on $\mathcal{V}$, that is, a linear function $\mathcal{V} \to \mathbf{R}$. Then provided $\alpha \neq 0$ the set of vectors $v \in \mathcal{V}$ such that $\alpha(v) = 0$ is a subspace of $\mathcal{V}$ of codimension 1; consequently the set of points x in $\mathcal{A}$ such that $\alpha(x - x_0) = 0$ is the hyperplane constructed by attaching this subspace at x_0. Different representations of the same hyperplane are obtained by choosing different points in it at which to attach it to $\mathcal{A}$, and by replacing α by any non-zero multiple of itself. Any one of the possible αs is called a *constraint form* for the hyperplane.

In the usual notation for a pairing $\langle v, \alpha \rangle$ of a vector $v \in \mathcal{V}$ and a linear form or covector $\alpha \in \mathcal{V}^*$, the function $f\colon \mathcal{A} \to \mathbf{R}$ defined by $x \mapsto \langle x - x_0, \alpha \rangle$ determines a hyperplane in $\mathcal{A}$ as the set of points at which this function takes the value zero. (Linear forms and pairings are explained in Note 2 at the end of this chapter.)

A linearly independent set of forms $\{\alpha^1, \alpha^2, \ldots, \alpha^k\}$ of $\mathcal{V}^*$ determines a subspace $\mathcal{W}$ of $\mathcal{V}$, of codimension k, comprising those vectors w for which

$$\langle w, \alpha^1 \rangle = \langle w, \alpha^2 \rangle = \cdots = \langle w, \alpha^k \rangle = 0,$$

and any subspace of codimension k may be specified in this way. Relative to a basis for $\mathcal{V}$, this amounts to a set of k linear equations for the components of w. The affine subspace of $\mathcal{A}$ constructed by attaching $\mathcal{W}$ at x_0 comprises the set of points x for which

$$\langle x - x_0, \alpha^1 \rangle = \langle x - x_0, \alpha^2 \rangle = \cdots = \langle x - x_0, \alpha^k \rangle = 0.$$

Any affine subspace of codimension k may be specified in this way. Different representations of the same subspace (as point set) are obtained by choosing different points in it at which to attach it to $\mathcal{A}$, and by replacing $\{\alpha^1, \alpha^2, \ldots, \alpha^k\}$ by any linearly independent set of k linear combinations of them.

If α is a (non-zero) linear form on $\mathcal{V}$ then the equation $\langle x - x_0, \alpha \rangle = c$ (with c not necessarily zero) also determines a hyperplane in $\mathcal{A}$, because one may always find a vector v such that $\langle v, \alpha \rangle = c$, and then $\langle x - (x_0 + v), \alpha \rangle = 0$. Thus a linear form determines a family of parallel hyperplanes, obtained by giving different values to the constant c. It also determines a spacing between them: if t is a non-zero number, then the linear forms α and $t\alpha$ determine the same family of hyperplanes; but if, for example, $t > 1$ then for any constant c the hyperplane $\langle x - x_0, t\alpha\rangle = c$ lies between the hyperplanes $\langle x - x_0, \alpha\rangle = c$ and $\langle x - x_0, \alpha\rangle = 0$.

Exercise 11. Given an affine coordinate system for $\mathcal{A}$ with origin x_0 and basis $\{e_a\}$ for $\mathcal{V}$, and with dual basis $\{\theta^a\}$ for $\mathcal{V}^*$, show that the equation $\langle x - x_1, \alpha\rangle = 0$ may be written in coordinates $\alpha_a x^a = c$, where (x^a) are the coordinates of x, $\alpha = \alpha_a\theta^a$, and $c = \langle x_1 - x_0, \alpha\rangle$. Show, conversely, that any such linear coordinate equation determines a hyperplane. □

3. Affine Spaces Modelled on Quotients and Direct Sums

The attachment of a subspace is only one of a number of constructions in an affine space which may be derived from the corresponding constructions in a vector space. We now describe two other examples: the fibration of an affine space, which is derived from the quotient of vector spaces; and the construction of a product of affine spaces, from the direct sum of vector spaces. We begin with some essential information about the vector space constructions.

If $\mathcal{V}$ is a vector space and $\mathcal{W}$ a subspace of $\mathcal{V}$, then the *quotient space* $\mathcal{V}/\mathcal{W}$ has as its elements the cosets $v + \mathcal{W} = \{\, v + w \mid w \in \mathcal{W}\,\}$. Sums and scalar products of cosets are defined by

$$(v_1 + \mathcal{W}) + (v_2 + \mathcal{W}) = (v_1 + v_2) + \mathcal{W} \qquad v_1, v_2 \in \mathcal{V}$$
$$k(v + \mathcal{W}) = kv + \mathcal{W} \qquad k \in \mathbf{R},\ v \in \mathcal{V};$$

and with the vector space structure on $\mathcal{V}/\mathcal{W}$ so defined the projection $\pi\colon \mathcal{V} \to \mathcal{V}/\mathcal{W}$, which maps each element of $\mathcal{V}$ to its coset, is a linear map. The dimension of $\mathcal{V}/\mathcal{W}$ is given by

$$\dim(\mathcal{V}/\mathcal{W}) = \dim\mathcal{V} - \dim\mathcal{W},$$

the codimension of $\mathcal{W}$ in $\mathcal{V}$.

Secondly, if $\mathcal{V}$ and $\mathcal{W}$ are vector spaces then their (*external*) *direct sum* $\mathcal{V} \oplus \mathcal{W}$ is the set of all ordered pairs (v, w) of elements $v \in \mathcal{V}$, $w \in \mathcal{W}$, with addition and scalar multiplication defined by

$$(v_1, w_1) + (v_2, w_2) = (v_1 + v_2, w_1 + w_2) \qquad v_1, v_2 \in \mathcal{V},\ w_1, w_2 \in \mathcal{W}$$
$$k(v, w) = (kv, kw) \qquad k \in \mathbf{R},\ v \in \mathcal{V},\ w \in \mathcal{W}.$$

Moreover,

$$\dim(\mathcal{V} \oplus \mathcal{W}) = \dim\mathcal{V} + \dim\mathcal{W}.$$

Projections onto the first and second factors are defined by

$$\Pi_1\colon \mathcal{V} \oplus \mathcal{W} \to \mathcal{V} \text{ by } (v, w) \mapsto v \qquad \Pi_2\colon \mathcal{V} \oplus \mathcal{W} \to \mathcal{W} \text{ by } (v, w) \mapsto w$$

and inclusions by

$$i_1\colon \mathcal{V} \to \mathcal{V} \oplus \mathcal{W} \text{ by } v \mapsto (v, 0) \qquad i_2\colon \mathcal{W} \to \mathcal{V} \oplus \mathcal{W} \text{ by } w \mapsto (0, w).$$

All these are linear maps, and

$$\Pi_1 \circ i_1 = \mathrm{id}_{\mathcal{V}} \qquad \Pi_2 \circ i_2 = \mathrm{id}_{\mathcal{W}}$$

the identities of $\mathcal{V}$ and $\mathcal{W}$. The maps i_1 and i_2 are called sections of the projections Π_1 and Π_2. In general, if $\pi\colon \mathcal{S} \to \mathcal{T}$ is a surjective map of some space $\mathcal{S}$ onto another space $\mathcal{T}$, any map $\sigma\colon \mathcal{T} \to \mathcal{S}$ such that $\pi \circ \sigma = \mathrm{id}_{\mathcal{T}}$ is called a *section* of π.

Exercise 12. Show that the "diagonal" map $\mathcal{V} \to \mathcal{V} \oplus \mathcal{V}$ by $v \mapsto (v, v)$ is a section of the projections on both first and second factors. □

On the other hand, if $\mathcal{V}$ and $\mathcal{W}$ are both subspaces of a vector space $\mathcal{U}$, then their *sum* $\mathcal{V} + \mathcal{W}$, given by

$$\mathcal{V} + \mathcal{W} = \{\, v + w \mid v \in \mathcal{V},\, w \in \mathcal{W} \,\},$$

is the smallest subspace of $\mathcal{U}$ containing both $\mathcal{V}$ and $\mathcal{W}$, while their *intersection* $\mathcal{V} \cap \mathcal{W}$, given by

$$\mathcal{V} \cap \mathcal{W} = \{\, u \in \mathcal{U} \mid u \in \mathcal{V} \text{ and } u \in \mathcal{W} \,\},$$

is the largest subspace contained in both $\mathcal{V}$ and $\mathcal{W}$. The dimensions of these various spaces are related as follows:

$$\dim(\mathcal{V} + \mathcal{W}) + \dim(\mathcal{V} \cap \mathcal{W}) = \dim \mathcal{V} + \dim \mathcal{W}.$$

These constructions are connected with that of the external direct sum as follows. It is easy to see that if $\mathcal{U} = \mathcal{V} \oplus \mathcal{W}$ then $i_1(\mathcal{V}) + i_2(\mathcal{W}) = \mathcal{U}$ and $i_1(\mathcal{V}) \cap i_2(\mathcal{W}) = \{0\}$ (where 0 represents the zero element of $\mathcal{U}$). On the other hand, if $\mathcal{V}$ and $\mathcal{W}$ are subspaces of $\mathcal{U}$ such that $\mathcal{V} + \mathcal{W} = \mathcal{U}$ and $\mathcal{V} \cap \mathcal{W} = \{0\}$ then there is a canonical isomorphism of $\mathcal{U}$ with $\mathcal{V} \oplus \mathcal{W}$ by $v + w \leftrightarrow (v, w)$. In this case $\mathcal{U}$ is said to be the (*internal*) *direct sum* of its subspaces $\mathcal{V}$ and $\mathcal{W}$. The brackets are intended to indicate that the terms "internal" and "external" are used only for emphasis, the type of direct sum under consideration usually being clear from the context. Two subspaces $\mathcal{V}$ and $\mathcal{W}$ of a vector space $\mathcal{U}$, which are such that $\mathcal{V} + \mathcal{W} = \mathcal{U}$ and $\mathcal{V} \cap \mathcal{W} = \{0\}$, are said to be *complementary*; they are then *direct summands* of $\mathcal{U}$. Any subspace of $\mathcal{U}$ is a direct summand: in fact, if $\mathcal{V}$ is a subspace, it has complementary subspaces $\mathcal{W}$, which may be chosen in many different ways, and each complementary subspace is isomorphic to $\mathcal{U}/\mathcal{V}$.

All this may be transferred to an affine space by attaching appropriate subspaces of the space on which it is modelled.

For example, let $\mathcal{A}$ be an affine space modelled on $\mathcal{V}$ and let $\mathcal{B}$ be an affine subspace of $\mathcal{A}$, obtained by attaching a subspace $\mathcal{W}$ of $\mathcal{V}$ to $\mathcal{A}$ at a point x_0. Then the set of all the parallel affine subspaces obtained by attaching $\mathcal{W}$ to $\mathcal{A}$ at every point of it has the structure of an affine space modelled on $\mathcal{V}/\mathcal{W}$, as follows. Consider the set of elements of $\mathcal{V}$ by which one given affine subspace parallel to $\mathcal{B}$ is translated into another. If v belongs to this set then so do all vectors of the form $v + w$ where $w \in \mathcal{W}$, and only these; in short, this set is just the coset $v + \mathcal{W}$. Thus the elements of $\mathcal{V}/\mathcal{W}$ act on the set of parallel affine subspaces; they clearly do so in accordance with the first axiom for an affine space; and given any two of the parallel affine subspaces there is a unique element of $\mathcal{V}/\mathcal{W}$ which maps one to the other, namely the coset of any vector by which some point of the first affine subspace is translated into the second. The set of affine subspaces parallel to $\mathcal{B}$ with this affine structure is called the *quotient affine space* $\mathcal{A}/\mathcal{B}$. The map $\mathcal{A} \to \mathcal{A}/\mathcal{B}$ which associates to each point of $\mathcal{A}$ the subspace parallel to $\mathcal{B}$ in which it lies is called the *projection*. This decomposition of $\mathcal{A}$ into non-intersecting affine subspaces, together with the projection, is an example of a *fibration*, the subspaces being the *fibres*.

Exercise 13. Show that if $\mathcal{V}$ is considered as an affine space modelled on itself, and if $\mathcal{W}$ is considered as an affine subspace of $\mathcal{V}$ attached at the zero vector, then the elements of $\mathcal{V}/\mathcal{W}$ (as an affine space) are just the cosets of $\mathcal{W}$ in $\mathcal{V}$, which may therefore be thought of as the parallels to $\mathcal{W}$. □

Exercise 14. Show that if $\mathcal{A}$ and $\mathcal{B}$ are affine spaces modelled on vector spaces $\mathcal{V}$ and $\mathcal{W}$ then their Cartesian product $\mathcal{A} \times \mathcal{B}$ may be made into an affine space modelled on $\mathcal{V} \oplus \mathcal{W}$ in such a way that displacements in the product space are those of its component parts, carried out simultaneously. □

Thus the external direct sum construction for vector spaces may also be extended to any pair of affine spaces to define their *affine product space*.

Now let $\mathcal{A}$ be an affine space modelled on a vector space $\mathcal{U}$ and let $\mathcal{B}$ and $\mathcal{C}$ be affine subspaces of it modelled on vector subspaces $\mathcal{V}$ and $\mathcal{W}$ of $\mathcal{U}$. Provided it is not empty, $\mathcal{B} \cap \mathcal{C}$ is an affine subspace of $\mathcal{A}$ modelled on $\mathcal{V} \cap \mathcal{W}$; it may consist of a single point, in which case $\mathcal{V} \cap \mathcal{W} = \{0\}$. If $\mathcal{B} \cap \mathcal{C}$ does consist of a single point, say x_0, and if $\mathcal{V}$ and $\mathcal{W}$ are complementary subspaces of $\mathcal{U}$, then for every point $x \in \mathcal{A}$ the vector $x - x_0$ may be uniquely expressed as a sum $v + w$ with $v \in \mathcal{V}$ and $w \in \mathcal{W}$, and the bijective map $x \mapsto (x_0 + v, x_0 + w)$ identifies $\mathcal{A}$ with the affine product space $\mathcal{B} \times \mathcal{C}$. In this case $\mathcal{B}$ and $\mathcal{C}$ are *complementary affine subspaces* of $\mathcal{A}$. If $\mathcal{B}$ is a given affine subspace of $\mathcal{A}$ then each affine subspace $\mathcal{C}$ complementary to $\mathcal{B}$ intersects each subspace parallel to $\mathcal{B}$ just once. The projection map $\mathcal{A} \to \mathcal{A}/\mathcal{B}$ is thus bijective when restricted to a complementary subspace $\mathcal{C}$, and $\mathcal{C}$ provides, in a sense, a concrete realisation of $\mathcal{A}/\mathcal{B}$. The map which sends each element of $\mathcal{A}/\mathcal{B}$ to its intersection with $\mathcal{C}$ is a section of the projection $\mathcal{A} \to \mathcal{A}/\mathcal{B}$. The two figures opposite are intended to illustrate these constructions.

An affine space $\mathcal{A}$ is shown. A 1-dimensional subspace $\mathcal{V}$ of the vector space $\mathcal{U}$ on which $\mathcal{A}$ is modelled is attached to $\mathcal{A}$ as a 1-plane $\mathcal{B}$ through x_0. Parallel 1-planes are shown: $\mathcal{A}$ is to be thought of as filled with 1-planes parallel to $\mathcal{B}$, each one of which constitutes a single element of the quotient space $\mathcal{A}/\mathcal{B}$. The 2-plane $\mathcal{C}$ transverse to the 1-planes, each of which it intersects in just one point, is a subspace complementary to $\mathcal{B}$. Of course the choice of complement is not unique, and there is no reason to prefer one over another: different choices of complement to $\mathcal{V}$ in $\mathcal{U}$, or of the point at which to attach it, may give different complements to $\mathcal{B}$ in $\mathcal{A}$. Figure 3 shows two different 2-planes, each of which is complementary to $\mathcal{B}$.

Space-time as an affine space. These constructions (affine quotient and product) may be exemplified, and the differences between them thereby related to physical considerations, by the different assumptions about space-time underlying different theories of kinematics. We distinguish three views of space-time, as follows. The Aristotelian view, which lasted until the implications of Newton's first law were understood, assumes that there is a state of absolute rest and that all ob-

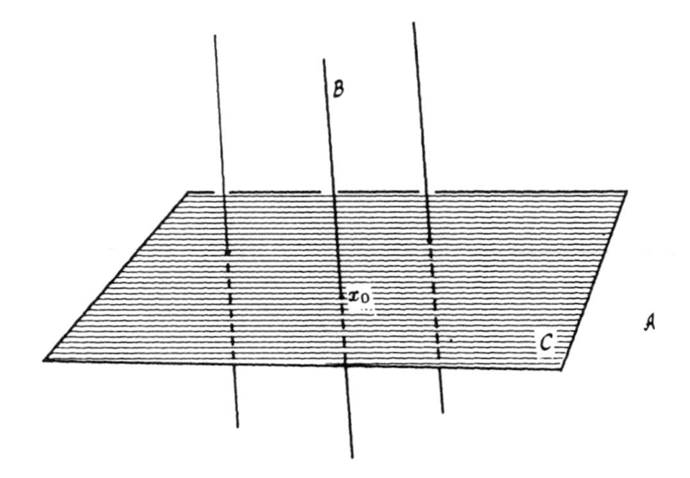

Fig. 2 Complementary subspaces in an affine space.

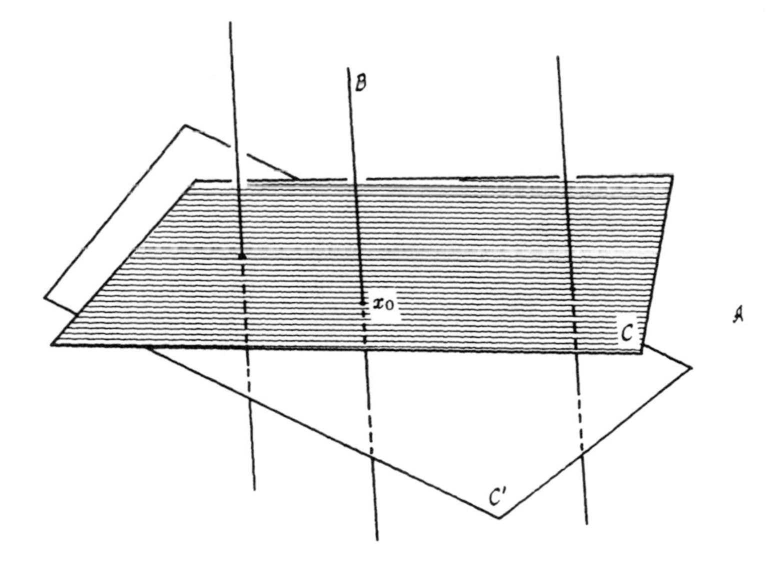

Fig. 3 Two subspaces complementary to a given one.

servers whether at rest or in motion agree whether or not events are simultaneous.

In the Newtonian view there is no state of absolute rest, all inertial observers being regarded as equivalent even though one may be in uniform motion with respect to another; they continue to agree about simultaneity. In the Einsteinian (special relativistic) view, even simultaneity is relative. In each case space-time is considered to be a 4-dimensional affine space, in which the world-lines of unaccelerated observers are affine lines, observers at rest relative to one another having parallel world-lines.

In the Einsteinian case space and time are integrated and nothing further can be said without the introduction of the Minkowskian metric. In Newtonian space-time $\mathcal{N}$, through any point (event) there is a uniquely defined affine 3-plane $\mathcal{S}$ consisting of events simultaneous with that point. The 3-planes parallel to $\mathcal{S}$ define simultaneous events, at different times. The 1-dimensional affine space $\mathcal{N}/\mathcal{S}$ represents Newtonian absolute time. Any line transverse to the 3-planes of simultaneity is the world-line of an inertial observer, but no one such is to be preferred to any other. Thus $\mathcal{N}$ is to be regarded as fibred by the 3-planes of simultaneity, the quotient space representing absolute time. In Aristotelian space-time $\mathcal{A}$, on the other hand, through each event there passes not only the 3-plane $\mathcal{S}$ of simultaneity but also the line $\mathcal{T}$ which is the world-line of an observer at absolute rest; the parallels to $\mathcal{T}$ are the world-lines of all such observers. Thus $\mathcal{A}$ is an affine product $\mathcal{S} \times \mathcal{T}$.

4. Affine Maps

Just as in Newtonian mechanics the translations and rotations of Euclidean space leave its geometrical properties unaltered, and in special relativity translations and Lorentz transformations leave unaltered the geometrical properties of space-time, so in an affine space there are transformations which leave unaffected the defining structure. As one might expect, these transformations are closely related to linear transformations of the underlying vector space on which the affine space is modelled. In this section we shall describe such transformations, and also the more general type of map which maps an affine space to another, possibly different, one in a way which respects the affine structure of the two spaces. Such maps are analogous to linear maps between vector spaces.

Translations. To begin with, we suggest yet another point of view from which to consider the affine structure map $\mathcal{A} \times \mathcal{V} \to \mathcal{A}$ introduced earlier. Instead of considering the effect on a single point x of displacement by the vector v, one may consider the effect on the space $\mathcal{A}$ as a whole, fixing v and displacing every point of $\mathcal{A}$ by it. This action of v on $\mathcal{A}$ is called *translation* by v and will be denoted τ_v. Thus $\tau_v\colon \mathcal{A} \to \mathcal{A}$ by $x \mapsto x+v$ (for all $x \in \mathcal{A}$). Of course, there is no new construction here, but the new point of view suggests further developments. The definition of an affine space may be restated as follows:

(1) composition of translations is a translation: $\tau_w \circ \tau_v = \tau_{v+w}$ for all $v, w \in \mathcal{V}$

(2) $\tau_0 = \mathrm{id}_{\mathcal{A}}$

(3) for any pair of points $x, x' \in \mathcal{A}$ there is a unique translation $\tau_{x'-x}$ taking x to x'.

Notice that translation respects the underlying relation between $\mathcal{A}$ and $\mathcal{V}$, in the sense that

$$\tau_w(x+v) = \tau_w(x) + v$$

for all $x \in \mathcal{A}$ and all $v, w \in \mathcal{V}$.

Affine maps. Besides translations of a single affine space there are other maps of affine spaces which respect the relation of an affine space to the underlying vector space: they are the extensions to affine spaces of linear maps of vector spaces.

Suppose that $\mathcal{A}$ and $\mathcal{B}$ are affine spaces, modelled on vector spaces $\mathcal{V}$ and $\mathcal{W}$ respectively. Let $\lambda: \mathcal{V} \to \mathcal{W}$ be a linear map; it has the property that $\lambda(cv + c'v') = c\lambda(v) + c'\lambda(v')$ for any $v, v' \in \mathcal{A}$ and any numbers c and c'. Now choose two points, $x_0 \in \mathcal{A}$ and $y_0 \in \mathcal{B}$. If x is any point in $\mathcal{A}$ then $x - x_0$ is a vector in $\mathcal{V}$ and $\lambda(x - x_0)$ is a vector in $\mathcal{W}$. Displacing y_0 by this vector, one obtains a map $\Lambda: \mathcal{A} \to \mathcal{B}$ by $x \mapsto y_0 + \lambda(x - x_0)$. This map depends, of course, on the choice of x_0 and y_0 as well as on the choice of the linear map λ. It satisfies

$$\Lambda(x + v) = y_0 + \lambda(x + v - x_0) = y_0 + \lambda(x - x_0) + \lambda(v) = \Lambda(x) + \lambda(v)$$

for all $x \in \mathcal{A}$ and all $v \in \mathcal{V}$; in other words, $\Lambda(x + v) - \Lambda(x)$ depends linearly on v.

The property of maps of affine spaces

$$\Lambda(x + v) = \Lambda(x) + \lambda(v),$$

where λ is linear, generalises the property of translations described at the end of the preceding subsection (where λ was the identity on $\mathcal{V}$). A map of affine spaces with this property is called an *affine map*. Any affine map $\Lambda: \mathcal{A} \to \mathcal{B}$ may be written in the form $\Lambda(x) = y_0 + \lambda(x - x_0)$ by choosing x_0 in $\mathcal{A}$ and then fixing $y_0 = \Lambda(x_0)$. A little more generally, if $\Lambda: \mathcal{A} \to \mathcal{B}$ is an affine map and x_0 and y_0 are any chosen points in $\mathcal{A}$ and $\mathcal{B}$ respectively, then $\Lambda(x) = y_0 + \big(\Lambda(x_0) - y_0\big) + \lambda(x - x_0)$ so that Λ is composed of an affine map taking x_0 to y_0 and a translation of $\mathcal{B}$ by $\Lambda(x_0) - y_0$. The linear map λ which enters the definition of an affine map Λ is called the *linear part* of Λ. If (x^a) are affine coordinates for $\mathcal{A}$ with origin x_0 and (y^α) are affine coordinates for $\mathcal{B}$ with origin y_0 then Λ will be represented in these coordinates by

$$y^\alpha = \lambda^\alpha_a x^a + c^\alpha$$

where the λ^α_a are the entries in the matrix of λ with respect to the bases for $\mathcal{V}$ and $\mathcal{W}$ used to define the coordinates and c^α are the components of the vector $\Lambda(x_0) - y_0$. The choice of coordinates thus allows Λ to be represented as a map $\mathbf{R}^m \to \mathbf{R}^n$, where $m = \dim \mathcal{A}$, $n = \dim \mathcal{B}$.

Exercise 15. Show that if x_1 and y_1 are chosen as origins in place of x_0 and y_0 then $\Lambda(x) = y_1 + \big(\Lambda(x_1) - y_1\big) + \lambda(x - x_1)$, with the same linear part but a different translation. Show also that the difference between the translations is $\lambda(x_1 - x_0) + (y_1 - y_0)$. □

Exercise 16. $\mathcal{A}$ is the plane $\{(\xi^1, \xi^2, \xi^3) \in \mathbf{R}^3 \mid \xi^1 + \xi^2 + \xi^3 = 1\}$ and $\mathcal{B}$ is the plane $\{(\xi^1, \xi^2, \xi^3) \in \mathbf{R}^3 \mid \xi^1 + \xi^2 - \xi^3 = 1\}$; $\Lambda: \mathcal{A} \to \mathcal{B}$ is the map defined by "projection parallel to the ξ^3 axis", that is, $\Lambda(\xi^1, \xi^2, \xi^3)$ is the point(η^1, η^2, η^3) in $\mathcal{B}$ such that $\eta^1 = \xi^1$ and $\eta^2 = \xi^2$. Show that Λ is an affine map. An affine cordinate system is chosen for $\mathcal{A}$ with $(0,0,1)$ as origin and such that the points $(1,0,0)$ and $(0,1,0)$ have coordinates $(1,0)$ and $(0,1)$ respectively, and an affine coordinate system is chosen for $\mathcal{B}$ with $(1,1,1)$ as origin and such that the points $(1,0,0)$ and $(0,1,0)$ have cordinates $(1,0)$ and $(0,1)$ respectively. Find the coordinate representation of Λ. □

Exercise 17. Let $\mathcal{B}$ be the hyperplane $\{ x \in \mathcal{A} \mid \langle x - x_0, \alpha \rangle = 0 \}$ (where α is a non-zero covector) and let x_1 be a point of $\mathcal{A}$ not in $\mathcal{B}$. Show that the map

$$x \mapsto x - \frac{\langle x - x_0, \alpha \rangle}{\langle x_1 - x_0, \alpha \rangle}(x_1 - x_0)$$

is an affine map of $\mathcal{A}$ onto $\mathcal{B}$. □

An affine map which is bijective is called an *affine isomorphism*; two isomorphic affine spaces are, in many circumstances, essentially identical.

Exercise 18. Show that two affine spaces are isomorphic if and only if they have the same dimension. Show that an affine coordinate system is an affine isomorphism with $\mathbf{R}^n$ considered as an affine space modelled on itself. □

Exercise 19. Show that, if $\Lambda: \mathcal{A} \to \mathcal{B}$ is an affine map, the image $\Lambda(\mathcal{A})$ is an affine subspace of $\mathcal{B}$, while for any point $y_0 \in \Lambda(\mathcal{A})$ the set of points of $\mathcal{A}$ which are mapped to y_0, $\Lambda^{-1}(y_0)$, is an affine subspace of $\mathcal{A}$. Show that the dimension of $\Lambda^{-1}(y_0)$ is the same for all y_0, and that $\dim \Lambda(\mathcal{A}) + \dim \Lambda^{-1}(y_0) = \dim \mathcal{A}$. □

Suppose now that $\Lambda: \mathcal{A} \to \mathcal{B}$ and $\mathrm{M}: \mathcal{B} \to \mathcal{C}$ are affine maps with linear parts λ and μ respectively. Then the composition $\mathrm{M} \circ \Lambda$ is an affine map, for if x, $x + v \in \mathcal{A}$ then

$$\mathrm{M}(\Lambda(x+v)) = \mathrm{M}(\Lambda(x) + \lambda(v)) = \mathrm{M}(\Lambda(x)) + \mu(\lambda(v))$$

and $\mu \circ \lambda$ is a linear map; thus the linear part of the composition of affine maps is the composition of linear parts.

Affine transformations. We come now to an important special case: invertible affine maps of an affine space to itself, which are called *affine transformations*. An affine map $\Lambda: \mathcal{A} \to \mathcal{A}$ is invertible if and only if its linear part λ is invertible; if Λ has inverse Λ^{-1} then the linear part of Λ^{-1} is λ^{-1}.

Since the identity transformation is affine, the composition of invertible maps is invertible, and composition is associative, the affine transformations of an affine space form a group, called the *affine group*. We shall now describe the relation between this group and the group $GL(\mathcal{V})$ of non-singular linear transformations of the underlying vector space $\mathcal{V}$.

We have just shown that the map

$$\text{an affine transformation} \mapsto \text{its linear part}$$

preserves multiplication, which is to say that it is a homomorphism from the affine group of $\mathcal{A}$ to $GL(\mathcal{V})$. An affine transformation whose linear part is the identity must be a translation: fix $x_0 \in \mathcal{A}$ and set $v_0 = \Lambda(x_0) - x_0$; then

$$\Lambda(x) = \Lambda(x_0) + (x - x_0) = x + v_0.$$

The identity affine transformation is obtained by setting $v_0 = 0$. The translations constitute a subgroup of the affine group.

Since linear parts compose, any composition of the form $\Lambda \circ \tau_v \circ \Lambda^{-1}$, where Λ is any affine transformation and τ_v is translation by v, must again be a translation. In fact, for any $x \in \mathcal{A}$,

$$\Lambda \circ \tau_v \circ \Lambda^{-1}(x) = \Lambda(\Lambda^{-1}(x) + v) = x + \lambda(v)$$

so that

$$\Lambda \circ \tau_v \circ \Lambda^{-1} = \tau_{\lambda(v)}.$$

Thus the conjugate of τ_v by Λ is $\tau_{\lambda(v)}$. Consequently the translations are invariant under conjugation, and therefore constitute a normal subgroup of the affine group. Moreover, the act of conjugation reproduces the action of the general linear group on $\mathcal{V}$: the conjugated translation vector $\lambda(v)$ is obtained from the original translation vector v by acting with the linear part λ of the conjugating transformation. (The concepts from group theory employed in this subsection are explained in Note 3 at the end of this chapter).

The conclusion that the translations constitute a normal subgroup of the affine group may be reached by another argument: the kernel of the homomorphism in which an affine transformation maps to its linear part consists of those transformations whose linear part is the identity; these are just the translations, whence again it follows that the latter constitute a normal subgroup.

Exercise 20. Show that the only translation which leaves any point fixed is the identity affine transformation. Show that the affine transformations which leave a chosen point fixed constitute a subgroup of the affine group of $\mathcal{A}$ isomorphic to the general linear group $GL(\mathcal{V})$. □

Exercise 21. Show that any affine map $\Lambda: \mathcal{A} \to \mathcal{A}$ may be written in the form $\Lambda(x) = x_0 + \lambda(x - x_0) + v_0$, where $v_0 = \Lambda(x_0) - x_0$; x_0 is a chosen point of $\mathcal{A}$ (and v_0 is fixed once x_0 is chosen). Let $\mathrm{M}: \mathcal{A} \to \mathcal{A}$ be another affine map, with $\mathrm{M}(x) = x_0 + \mu(x - x_0) + w_0$. Show that their composition is given by $\mathrm{M} \circ \Lambda(x) = x_0 + \mu \circ \lambda(x - x_0) + (\mu(v_0) + w_0)$; thus the linear parts compose, but the translation part of the composition depends on μ as well as on v_0 and w_0. Show also that (when it is invertible) the inverse of Λ is given by $\Lambda^{-1}(x) = x_0 + \lambda^{-1}(x - x_0) - \lambda^{-1}(v_0)$. □

Exercise 22. Let ρ_{x_0} denote the affine transformation of $\mathcal{A}$ by $x \mapsto x_0 - (x - x_0)$. Show that ${\rho_{x_0}}^2 = \mathrm{id}_{\mathcal{A}}$ and that the set of two transformations $\{\mathrm{id}_{\mathcal{A}}, \rho_{x_0}\}$ is a normal subgroup of the group of affine transformations leaving x_0 fixed. Show that $\tau_v \circ \rho_{x_0} \circ \tau_{-v} = \rho_{x_0+v}$ for any $v \in \mathcal{V}$. □

The transformation defined in this exercise is called *reflection in the point* x_0.

The group of affine transformations of $\mathcal{A}$ contains two subgroups of importance: one, the translation subgroup, which is normal, and is isomorphic to $\mathcal{V}$; the other, the subgroup of transformations leaving a chosen point x_0 fixed, which is isomorphic to $GL(\mathcal{V})$ (Exercise 20). The first result of Exercise 21 may be interpreted as saying that every affine transformation may be written as the composition of a transformation leaving x_0 fixed and a translation (the translation being performed second); it is clear that these components of an affine transformation are uniquely determined by it. A group G which has a normal subgroup N and another subgroup H such that every element g of G may be written uniquely in the form nh where $n \in N$ and $h \in H$ is called a *semi-direct product* of N and H. Thus the group of affine transformations is a semi-direct product of the translations and the transformations leaving a given point fixed.

This structure of the group of affine transformations may be described in another way. Starting with $\mathcal{V}$ and $GL(\mathcal{V})$, one may construct a new group whose elements are pairs (v, λ) (with $v \in \mathcal{V}$ and $\lambda \in GL(\mathcal{V})$), and whose multiplication rule is

$$(w, \mu) \cdot (v, \lambda) = (w + \mu(v), \mu \circ \lambda).$$

This, too, is called a semi-direct product: the alternative definitions differ in much the same way as do those of the internal and external direct sum of vector spaces. From the expression for the composition of two affine transformations in Exercise 21 it is easy to see that the group of affine transformations is isomorphic to this semi-direct product of $\mathcal{V}$ and $GL(\mathcal{V})$. This makes clear the relation between the group of affine transformations of $\mathcal{A}$, the group of linear transformations of the underlying vector space $GL(\mathcal{V})$, and the vector group $\mathcal{V}$ itself.

Although the formula for the coordinate representation of an affine transformation is indistinguishable, out of context, from the formula for an affine change of coordinates, the two concepts must be kept distinct. A transformation which moves the space around, but leaves the coordinates where they are, is called active, or an alibi transformation ("being somewhere else at the time"): a transformation which leaves the space where it is, but changes the coordinates of the points, is called passive, or an alias transformation ("going under a different name").

Exercise 23. Show that affine coordinate transformations form a group. Show that affine coordinate transformations of the form $\hat{x}^a = x^a + c^a$ constitute a normal subgroup of this group. □

5. Affine Maps of Lines and Hyperplanes

An affine map in general has the property that it maps lines into lines: once again let $\Lambda: \mathcal{A} \to \mathcal{B}$ by $x \mapsto y_0 + \lambda(x - x_0)$ be an affine map, and let $\sigma: \mathbf{R} \to \mathcal{A}$ by $t \mapsto x_1 + tv_0$ be a line in $\mathcal{A}$ through x_1. We shall examine the effect of Λ on the line. Now $\Lambda \circ \sigma: \mathbf{R} \to \mathcal{B}$ is given by

$$\Lambda \circ \sigma(t) = \Lambda(x_1 + tv_0) = \Lambda(x_1) + t\lambda(v_0);$$

thus, provided $\lambda(v_0) \neq 0$, $\Lambda \circ \sigma$ is the line through $\Lambda(x_1)$ determined by $\lambda(v_0)$. If it happens that $\lambda(v_0) = 0$ then the transformed line will degenerate to a single point, but if λ is injective every line is mapped to a line which is not a point.

An affine map in general also maps hyperplanes to hyperplanes, but in the opposite sense to what one might naively expect. If $\Lambda: \mathcal{A} \to \mathcal{B}$ is the affine map given above, and $g: \mathcal{B} \to \mathbf{R}$ by $y \mapsto \langle y - y_0, \alpha \rangle$ is a function determining a hyperplane through y_0, then $g \circ \Lambda$ is a function on $\mathcal{A}$ which will determine a hyperplane unless $\langle \lambda(x - x_0), \alpha \rangle = 0$ for all $x \in \mathcal{A}$. For

$$g \circ \Lambda(x) = \langle \Lambda(x) - y_0, \alpha \rangle = \langle \lambda(x - x_0), \alpha \rangle = \langle x - x_0, \lambda^*(\alpha) \rangle$$

where $\lambda^*(\alpha)$ is the linear form on $\mathcal{V}$ defined by $\langle v, \lambda^*(\alpha) \rangle = \langle \lambda(v), \alpha \rangle$ for all $v \in \mathcal{V}$. Thus $g \circ \Lambda$ determines the hyperplane in $\mathcal{A}$ through x_0 which has $\lambda^*(\alpha)$ as constraint form, provided $\lambda^*(\alpha) \neq 0$. If, however, $\lambda^*(\alpha) = 0$ then $g \circ \Lambda = 0$ identically, and the image of $\mathcal{A}$ is contained in the hyperplane through y_0. But if λ is surjective then $\lambda^*(\alpha) \neq 0$ for $\alpha \neq 0$ and every hyperplane in $\mathcal{B}$ through y_0 determines a hyperplane in $\mathcal{A}$ through x_0.

Exercise 24. Examine the mapping by Λ of hyperplanes in $\mathcal{B}$ not through y_0. □

This reversal of sense of a map is a paradigm of a constantly recurring situation: lines map in the same direction as Λ, or *cogrediently*, whereas hyperplanes map in the opposite direction, or *contragrediently*. It arises because a curve in $\mathcal{A}$ is defined by a map $\sigma\colon \mathbf{R} \to \mathcal{A}$ while a hyperplane in $\mathcal{B}$ is defined by a map $g\colon \mathcal{B} \to \mathbf{R}$; one may compose σ with a map $\Lambda\colon \mathcal{A} \to \mathcal{B}$ on the left to obtain a map $\Lambda \circ \sigma\colon \mathbf{R} \to \mathcal{B}$, but one is forced to compose g with Λ on the right, which gives the map $g \circ \Lambda\colon \mathcal{A} \to \mathbf{R}$. This is much like the situation which arises for a linear map of vector spaces, whose adjoint acts contragrediently on the duals.

Summary of Chapter 1

A set $\mathcal{A}$ is an affine space modelled on a real vector space $\mathcal{V}$ if there is a map (displacement, translation) $\mathcal{A} \times \mathcal{V} \to \mathcal{A}$ by $(x, v) \mapsto x + v$ such that: $(x + v) + w = x + (v + w)$; $x + 0 = x$; there is a unique element $x' - x$ of $\mathcal{V}$ such that $x + (x' - x) = x'$.

Affine coordinates (x^a) of x are defined by $x - x_0 = x^a e_a$, x_0 being a fixed point of $\mathcal{A}$ (the origin of coordinates) and $\{e_a\}$ a basis for $\mathcal{V}$. The dimension of $\mathcal{A}$ is the number of coordinates, that is, $\dim \mathcal{V}$. A change of origin and basis results in a coordinate transformation $\acute{x}^a = k^a_b x^b + c^a$, with (k^a_b) a nonsingular matrix.

The map $\mathcal{V} \to \mathcal{A}$ by $v \mapsto x_0 + v$, with x_0 a fixed point of $\mathcal{A}$, is regarded as attaching $\mathcal{V}$ to $\mathcal{A}$ at x_0. By combining this with the inclusion map one can attach any subspace of $\mathcal{V}$ to $\mathcal{A}$. Attachment of a p-dimensional subspace $\mathcal{W}$ of $\mathcal{V}$ yields a p-plane in $\mathcal{A}$. By choosing a basis for $\mathcal{W}$ one may parametrise the p-plane; in particular, a parametrised 1-plane is a line. Attachment of the 1-dimensional subspaces containing the basis vectors at the origin of affine coordinates produces the coordinate axes. Attachment of the same p-dimensional subspace at all the points of $\mathcal{A}$ gives a family of parallel p-planes. A subspace of $\mathcal{V}$ is attached as a hyperplane in $\mathcal{A}$. A hyperplane may also be defined in terms of a non-zero linear form α on $\mathcal{V}$ by the equation $\langle x - x_0, \alpha\rangle = c$, where c is a constant; α is called a constraint form for the hyperplane. As c varies a family of parallel hyperplanes is obtained.

If $\mathcal{B}$ is an affine subspace of $\mathcal{A}$ (the result of attaching a subspace $\mathcal{W}$ of $\mathcal{V}$) then the set of all affine subspaces parallel to $\mathcal{B}$ is an affine space modelled on $\mathcal{V}/\mathcal{W}$, called the quotient affine space $\mathcal{A}/\mathcal{B}$.

If $\mathcal{A}$ and $\mathcal{B}$ are affine spaces modelled on $\mathcal{V}$ and $\mathcal{W}$ then their Cartesian product is an affine space modelled on $\mathcal{V} \oplus \mathcal{W}$.

An affine map $\Lambda\colon \mathcal{A} \to \mathcal{B}$ satisfies $\Lambda(x + v) = \Lambda(x) + \lambda(v)$ where $\lambda\colon \mathcal{V} \to \mathcal{W}$ is linear. Any affine map may be expressed in the form $\Lambda(x) = y_0 + \lambda(x - x_0)$ with $y_0 = \Lambda(x_0)$. An affine map $\Lambda\colon \mathcal{A} \to \mathcal{A}$ is invertible when its linear part λ is; such affine transformations form a group, with the translations $\{\tau_v\}$ as a normal subgroup; $\Lambda \circ \tau_v \circ \Lambda^{-1} = \tau_{\lambda(v)}$. The group of affine transformations is the semi-direct product of $GL(\mathcal{V})$ (the group of nonsingular linear transformations of $\mathcal{V}$) and the vector group $\mathcal{V}$.

Affine maps in general map lines to lines, and do so cogrediently; and in general they map hyperplanes to hyperplanes, and do so contragrediently.

Notes to Chapter 1

1. Sets and maps. Throughout this book we make use of the notation, and some of the simpler ideas, of the theory of sets.

Any collection of objects, finite or infinite, likely to be encountered here may be called a *set*. The objects are called *members*, *elements*, or *points* of the set. If $\mathcal{S}$ is a set, then $x \in \mathcal{S}$ means that the object x belongs to the set $\mathcal{S}$, and $x \notin \mathcal{S}$ means that x does not belong to $\mathcal{S}$. The elements may be given by enumerating them, usually between braces, separated by commas—thus $\{e_1, e_2, \ldots, e_n\}$ for the basis vectors of an n-dimensional vector space, or $\{e_a\}$ if it is understood that a takes the values 1, 2, ..., n (the range convention; see Chapter 0, Section 4). The elements may be restricted by a condition; thus $\{\, (\xi^1, \xi^2, \xi^3) \in \mathbf{R}^3 \mid \xi^1 + \xi^2 + \xi^3 = 1 \,\}$ (the set which appears in Exercise 5) means the set of triples of real numbers whose sum is 1.

If $\mathcal{S}$ and $\mathcal{T}$ are sets then $\mathcal{S} \subset \mathcal{T}$ or $\mathcal{T} \supset \mathcal{S}$ means that every element of $\mathcal{S}$ is also an element of $\mathcal{T}$; one says that $\mathcal{S}$ is *contained* in $\mathcal{T}$, or that $\mathcal{S}$ is a *subset* of $\mathcal{T}$. If $\mathcal{T}$ is known to contain other elements besides those in $\mathcal{S}$, one says that $\mathcal{S}$ is *properly contained* in $\mathcal{T}$. If $\mathcal{S} \subset \mathcal{T}$ and $\mathcal{T} \subset \mathcal{S}$ then they have the same elements, and one writes $\mathcal{S} = \mathcal{T}$.

The *intersection* $\mathcal{S} \cap \mathcal{T}$ consists of those elements which belong both to $\mathcal{S}$ and to $\mathcal{T}$. The *union* $\mathcal{S} \cup \mathcal{T}$ consists of those elements which belong either to $\mathcal{S}$ or to $\mathcal{T}$ or to both. The *empty set*, which contains no elements, is denoted $\emptyset$; thus if $\mathcal{S}$ and $\mathcal{T}$ have no elements in common, $\mathcal{S} \cap \mathcal{T} = \emptyset$, in which case $\mathcal{S}$ and $\mathcal{T}$ are said to be *disjoint*.

A *map*, or *mapping*, or *function* $\phi\colon \mathcal{S} \to \mathcal{T}$ associates a unique element of $\mathcal{T}$ to each element of $\mathcal{S}$. The set $\mathcal{S}$ is called the *domain* of ϕ and the set $\mathcal{T}$ the *codomain*. If $x \in \mathcal{S}$ the element of $\mathcal{T}$ associated to x by ϕ is called the *image* of x by ϕ and written $\phi(x)$ or ϕx. If $\phi(x) = y$ one writes $\phi\colon x \mapsto y$ to show what happens to this particular element. The set of images is $\operatorname{im}\phi = \{\, \phi(x) \in \mathcal{T} \mid x \in \mathcal{S} \,\}$.

If $\operatorname{im}\phi = \mathcal{T}$ then ϕ is called an *onto* map, or a *surjective map*, or a *surjection*.

If $\mathcal{P} \subset \mathcal{S}$, the map $\mathcal{P} \to \mathcal{T}$ which associates to each element p of $\mathcal{P}$ the element $\phi(p) \in \mathcal{T}$ is called the *restriction* of ϕ to $\mathcal{P}$ and denoted $\phi|_{\mathcal{P}}$.

If $\mathcal{P} \subset \mathcal{S}$, the *inclusion* $i\colon \mathcal{P} \to \mathcal{S}$ assigns to each element of $\mathcal{P}$ the same element, considered as an element of $\mathcal{S}$. Inclusion is often denoted $\mathcal{P} \hookrightarrow \mathcal{S}$.

If $\phi\colon \mathcal{S} \to \mathcal{T}$ and $\psi\colon \mathcal{T} \to \mathcal{U}$ are maps then their composition $\psi \circ \phi$ is the map which results when ϕ and ψ are executed in succession: $\psi \circ \phi(x) = \psi(\phi(x))$. If $\chi\colon \mathcal{U} \to \mathcal{V}$ is another map then $\chi \circ (\psi \circ \phi) = (\chi \circ \psi) \circ \phi$, so one leaves out the brackets and writes $\chi \circ \psi \circ \phi$. By these conventions, maps act on elements written to the right of them, and the right-hand-most map is executed first.

If $\phi\colon \mathcal{S} \to \mathcal{T}$ and $y \in \mathcal{T}$ then the set $\{\, x \in \mathcal{S} \mid \phi(x) = y \,\}$ of elements in $\mathcal{S}$ whose image is y is called the *pre-image* or *inverse image* of y and denoted $\phi^{-1}(y)$. If, for each $y \in \operatorname{im}\phi$, $\phi^{-1}(y)$ consists of a single element, then ϕ is called a *1 : 1* ("one-to-one") map, or an *injective* map, or an *injection*.

A map which is both injective and surjective is called *bijective*, or a *bijection*. A bijection $\phi\colon \mathcal{S} \to \mathcal{T}$ has an inverse $\phi^{-1}\colon \mathcal{T} \to \mathcal{S}$, such that $\phi^{-1} \circ \phi = \mathrm{id}_{\mathcal{S}}$ and

$\phi \circ \phi^{-1} = \mathrm{id}_\mathcal{T}$, where $\mathrm{id}_\mathcal{S}: \mathcal{S} \to \mathcal{S}$ is the identity map, which takes each element to itself.

The *Cartesian product* of two sets, $\mathcal{S} \times \mathcal{T}$, is the set of ordered pairs (x, y) where $x \in \mathcal{S}$ and $y \in \mathcal{T}$. The Cartesian product of n sets $\mathcal{S}_1, \mathcal{S}_2, \ldots, \mathcal{S}_n$ is the set $\mathcal{S}_1 \times \mathcal{S}_2 \times \cdots \times \mathcal{S}_n$ of n-tuples $\{(x_1, x_2, \ldots, x_n) \mid x_k \in \mathcal{S}_k,\ k = 1, 2, \ldots, n\}$. The projection $\Pi_k: \mathcal{S}_1 \times \mathcal{S}_2 \times \cdots \times \mathcal{S}_n \to \mathcal{S}_k$ takes each n-tuple onto its kth entry.

A *partition* of a set $\mathcal{S}$ is a collection of non-empty disjoint subsets of $\mathcal{S}$ such that every element of $\mathcal{S}$ belongs to exactly one of the subsets. It is often convenient to call two elements x and x' *equivalent* and to write $x \sim x'$ if they belong to the same subset; the subsets are then called *equivalence classes*. The equivalence classes may themselves be regarded as the elements of a set, and the map which takes each element into the equivalence class containing it is then called the *canonical projection*.

More complete introductions may be found in the books by Kirillov [1976], Chapter 1, Loomis and Sternberg [1968], Chapter 1, or Porteous [1969], Chapter 1, for example. A very entertaining and readable book is Halmos's *Naive Set Theory* [1960]. A standard text is that by Fraenkel, Bar-Hillel and Levy [1973].

2. Vector spaces. We list the axioms for a vector space, and give some of the basic properties of vector spaces and linear maps between them.

Let $\mathcal{K}$ denote the real numbers $\mathbf{R}$ or the complex numbers $\mathbf{C}$. A *vector space* $\mathcal{V}$ over $\mathcal{K}$ is a set with two composition laws

$+ : \mathcal{V} \times \mathcal{V} \to \mathcal{V}$ (addition)

$\cdot : \mathcal{K} \times \mathcal{V} \to \mathcal{V}$ (multiplication by a scalar)

such that, for all $u, v, w \in \mathcal{V}$ and all $a, b \in \mathcal{K}$,

(1) $v + w = w + v$

(2) $u + (v + w) = (u + v) + w$

(3) $\mathcal{V}$ contains an element 0 such that $v + 0 = v$

(4) $\mathcal{V}$ contains, for each v, an element $-v$ such that $v + (-v) = 0$

(5) $a \cdot (v + w) = a \cdot v + a \cdot w$

(6) $(a + b) \cdot v = a \cdot v + b \cdot v$

(7) $a \cdot (b \cdot v) = (ab) \cdot v$

(8) $1 \cdot v = v$.

The elements of $\mathcal{V}$ are called *vectors*. If $\mathcal{K} = \mathbf{R}$, $\mathcal{V}$ is called a *real* vector space; if $\mathcal{K} = \mathbf{C}$, $\mathcal{V}$ is a *complex* vector space.

Axioms (1) to (4) make $\mathcal{V}$ into an additive Abelian group.

If $\mathcal{K}^n$ denotes the set of ordered n-tuples $(a^1, a^2, \ldots, a^n)$ of elements of $\mathcal{K}$ and $+$ and $\cdot$ are defined by $(a^1, a^2, \ldots, a^n) + (b^1, b^2, \ldots, b^n) = (a^1 + b^1, a^2 + b^2, \ldots, a^n + b^n)$ and $c \cdot (a^1, a^2, \ldots, a^n) = (ca^1, ca^2, \ldots, ca^n)$ then $\mathcal{K}^n$ is a vector space. The real number spaces $\mathbf{R}^n$, in particular, occur frequently in this book.

A subset $\mathcal{W}$ of $\mathcal{V}$ is called a *subspace* if it is itself a vector space with the laws of addition and scalar multiplication it inherits from $\mathcal{V}$.

A finite set of vectors $\{v_1, v_2, \ldots, v_n\}$ is said to be *linearly dependent* if there are numbers $a^1, a^2, \ldots, a^n \in \mathcal{K}$, not all zero, such that $a^1 \cdot v_1 + a^2 \cdot v_2 + \cdots + a^n \cdot v_n = 0$. An infinite set of vectors is called linearly dependent if it contains a linearly dependent finite subset. A set which is not linearly dependent is called *linearly independent*. If, for every positive integer k, $\mathcal{V}$ contains a linearly independent set of k vectors then $\mathcal{V}$ is called *infinite-dimensional*, but if it contains, for some n, a linearly independent set of n vectors but no linearly independent set of $(n+1)$ vectors then it is called *finite-dimensional* and said to be of *dimension* n: one writes $\dim \mathcal{V} = n$.

A subset S of a finite-dimensional vector space $\mathcal{V}$ is a *basis* for $\mathcal{V}$ if it is a linearly independent set and if, for every v not in S, $S \cup \{v\}$ is a linearly dependent set. The number of elements in a basis is equal to the dimension of the space. If $S = \{e_1, e_2, \ldots, e_n\}$ is a basis for $\mathcal{V}$ then every $v \in \mathcal{V}$ may be expressed as a linear combination of elements of S, $v = v^a \cdot e_a$, in a way which is unique except for the order of the terms.

Let $\mathcal{V}$ and $\mathcal{W}$ be vector spaces. A map $\lambda \colon \mathcal{V} \to \mathcal{W}$ is called a *linear map* if $\lambda(c \cdot v + c' \cdot v') = c \cdot \lambda(v) + c' \cdot \lambda(v')$ for all $c, c' \in K$ and all $v, v' \in \mathcal{V}$. A linear map is determined completely by its action on a basis. If $\{e_a\}$ is a basis for $\mathcal{V}$ and $\{f_\alpha\}$ a basis for $\mathcal{W}$, where $\alpha = 1, 2, \ldots, m = \dim \mathcal{W}$, we may write $\lambda(e_a) = \lambda_a^\alpha \cdot f_\alpha$. The λ_a^α are the entries of the *matrix* representing λ with respect to the given bases. The action of λ on an arbitrary vector in $\mathcal{V}$ is given by $\lambda(v) = \lambda_a^\alpha v^a \cdot f_\alpha$, where (v^a) is the n-tuple of components of v with respect to the basis of $\mathcal{V}$. This amounts to the left multiplication of the column vector of components of v by the $n \times m$ matrix (λ_a^α).

If $\lambda \colon \mathcal{V} \to \mathcal{W}$ is a linear map then its *image* $\operatorname{im}\lambda$ is a subspace of $\mathcal{W}$ and its *kernel* $\ker\lambda$, the set of elements of $\mathcal{V}$ mapped to zero by λ, is a subspace of $\mathcal{V}$; $\dim\operatorname{im}\lambda + \dim\ker\lambda = \dim\mathcal{V}$. If $\operatorname{im}\lambda = \mathcal{W}$ then λ is surjective, if $\ker\lambda = \{0\}$ then λ is injective; if both, then λ is bijective, its inverse is also linear, and it is called an *isomorphism*. Two vector spaces which are isomorphic (images of each other by an isomorphism and its inverse) must have the same dimension. An isomorphism whose construction or definition does not depend on a choice of basis in either the domain or codomain is said to be *natural* or *canonical*. Naturally isomorphic spaces may be considered identical for many purposes.

A linear map of a vector space to itself, or linear transformation of the vector space, is said to be *non-singular* if it is invertible, that is, if it is an isomorphism: it is enough for it to be injective to ensure this, by the dimension result above. The set of non-singular linear transformations of $\mathcal{V}$ is a group called the *general linear group* on $\mathcal{V}$, denoted $GL(\mathcal{V})$.

The set of linear maps from $\mathcal{V}$ to $\mathcal{W}$ may itself be made into a vector space by defining

$$+ : (\lambda_1 + \lambda_2)(v) = \lambda_1(v) + \lambda_2(v)$$
$$\cdot : (c \cdot \lambda)(v) = c \cdot (\lambda(v)).$$

An important special case is the vector space $\mathcal{V}^*$ of linear maps from $\mathcal{V}$ to the (1-dimensional) vector space K. Such maps are usually called *linear forms* on $\mathcal{V}$. The space $\mathcal{V}^*$ is called the space *dual* to $\mathcal{V}$. It is of the same dimension as $\mathcal{V}$. Furthermore, $(\mathcal{V}^*)^*$ is canonically isomorphic to $\mathcal{V}$. It is customary to use a notation for the evaluation of linear forms which reflects the symmetry between $\mathcal{V}$ and $\mathcal{V}^*$, namely to write, for $\alpha \in \mathcal{V}^*$ and $v \in \mathcal{V}$, $\langle v, \alpha \rangle$ instead of $\alpha(v)$. The map $\mathcal{V} \times \mathcal{V}^* \to K$ by $(v, \alpha) \mapsto \langle v, \alpha \rangle$ is often called the *pairing* of elements of $\mathcal{V}$ and $\mathcal{V}^*$.

The symmetry between $\mathcal{V}$ and $\mathcal{V}^*$ is also reflected in the use of the term *covariant vector*, or *covector*, instead of linear form, for an element of $\mathcal{V}^*$.

If $\{e_a\}$ is a basis for $\mathcal{V}$, the *dual basis* for $\mathcal{V}^*$ is the set $\{\theta^a\}$ of covectors such that

$$\langle e_a, \theta^b \rangle = \delta_a^b = \begin{cases} 1 & \text{if } a = b, \\ 0 & \text{if } a \neq b; \end{cases}$$

δ_a^b is called the *Kronecker delta*. If the components of a vector $v \in \mathcal{V}$ are written as a column—that is, an $n \times 1$ matrix—and the components of a covector $\alpha \in \mathcal{V}^*$ are written, in the dual basis, as a row—that is, a $1 \times n$ matrix—then the evaluation of $\langle v, \alpha \rangle$ is carried out by (row into column) matrix multiplication.

If $\lambda: \mathcal{V} \to \mathcal{W}$ is a linear map and β is a linear form on $\mathcal{W}$ then $v \mapsto \langle \lambda(v), \beta \rangle$ is a linear form on $\mathcal{V}$ denoted $\lambda^*(\beta)$, so that $\langle v, \lambda^*(\beta) \rangle = \langle \lambda(v), \beta \rangle$ for all $v \in \mathcal{V}$ and any $\beta \in \mathcal{V}^*$. The map $\lambda^*: \mathcal{W}^* \to \mathcal{V}^*$ by $\beta \mapsto \lambda^*(\beta)$ is a linear map called the *adjoint* of λ. If $\lambda: \mathcal{U} \to \mathcal{V}$ and $\mu: \mathcal{V} \to \mathcal{W}$ are linear maps, then $(\mu \circ \lambda)^* = \lambda^* \circ \mu^*$.

The dot denoting scalar multiplication has been used here for emphasis; it is generally omitted.

More extensive discussions may be found in Loomis and Sternberg [1968], Chapter 1, or Bishop and Goldberg [1968], Chapter 2, for example. There is a lovely book by Halmos [1958].

3. Groups. In this note we collect some standard definitions and results from the theory of groups.

A *group* G is a set together with a binary operation $G \times G \to G$ called the *group multiplication*, written $(g_1, g_2) \mapsto g_1 g_2$, such that

(1) multiplication is associative: $(g_1 g_2) g_3 = g_1 (g_2 g_3)$ for all $g_1, g_2, g_3 \in G$

(2) there is an *identity* element e in G such that $ge = eg = g$ for all $g \in G$

(3) each $g \in G$ has an *inverse*, denoted g^{-1}, such that $gg^{-1} = g^{-1}g = e$.

Where more than one group is involved, ambiguity may be avoided by writing e_G for e.

A map of groups $\phi: G \to H$ is called a *homomorphism* if it preserves multiplication: $\phi(g_1 g_2) = \phi(g_1)\phi(g_2)$ for all $g_1, g_2 \in G$. A bijective homomorphism is an *isomorphism*; an isomorphism of a group with itself is an *automorphism*.

A *subgroup* F of G is a subset which is itself a group with the multiplication restricted from G. Equivalently, a group F is a subgroup of G if it is a subset of G and if the inclusion $F \hookrightarrow G$ is a homomorphism.

For any $\acute{g} \in G$ the map $g \mapsto \acute{g} g \acute{g}^{-1}$ is an automorphism of G called *conjugation* by $\acute{g}$; it is also called an *inner automorphism*. If F is a subgroup of G then for each $\acute{g} \in G$, the set $\{ \acute{g} f \acute{g}^{-1} \mid f \in F \}$ is also a subgroup of G; it is called the subgroup *conjugate* to F by $\acute{g}$. A subgroup F is said to be *normal* or *invariant* if it is identical to each of its conjugates, that is, if it is invariant, as a whole, under conjugation.

Let $\phi: G \to H$ be any homomorphism of groups. Its *image* $\operatorname{im}\phi = \{ \phi(g) \mid g \in G \}$ is a subgroup of H, and its *kernel* $\ker\phi = \{ g \in G \mid \phi(g) = e_H \}$ is a normal subgroup of G. Moreover, ϕ is surjective if and only if $\operatorname{im}\phi = H$; injective if and only if $\ker\phi = \{e_G\}$; and therefore bijective if and only if both these conditions hold.

Suitable treatments are to be found in many books, for example MacLane and Birkhoff [1967], Chapter 3, or Kirillov [1976]. Further standard material is introduced in Chapter 11.

2. CURVES, FUNCTIONS AND DERIVATIVES

The ideas introduced in Chapter 1 were all essentially linear—the lines were straight, the subsets were plane, and the maps were affine. In this chapter we drop the restriction to linearity and introduce curves, of which lines are affine special cases, and functions, of which the functions defining affine hyperplanes by constraint are affine special cases. We do not allow curves and functions to be too wild, but impose restrictions which are sufficiently weak to encompass the usual applications but sufficiently strong to allow the usual processes of calculus. These restrictions are embodied in the concept of "smoothness", which is explained in Section 1. We go on to construct tangent vectors to curves, and introduce the idea of the directional derivative, which underlies the idea of a vector field, introduced in Chapter 3, and is central to what follows in the rest of this book. With this additional apparatus to hand, we show how to introduce curvilinear coordinates into an affine space.

1. Curves and Functions

In this section we define curves and functions in an affine space.

Curves. In Section 2 of Chapter 1 a line is defined as a map $\sigma: \mathbf{R} \to \mathcal{A}$ by $t \mapsto x_0 + tv_0$ where $\mathcal{A}$ is an affine space modelled on a vector space $\mathcal{V}$ and v_0 is a non-zero element of $\mathcal{V}$. What distinguishes a line, among other maps $\mathbf{R} \to \mathcal{A}$, is that σ is affine: $\sigma(t+s) = \sigma(t) + \lambda(s)$ where $\lambda: \mathbf{R} \to \mathcal{V}$ is the linear map $s \mapsto sv_0$.

The generalisation which suggests itself, and which one makes use of in applications without giving it any special attention, is to consider any map $\mathbf{R} \to \mathcal{A}$—in other words, to give up the properties of straightness and linearity which distinguish lines. We define a *curve* in $\mathcal{A}$ to be a map $\mathbf{R} \to \mathcal{A}$, or a map $I \to \mathcal{A}$ where I is an open interval of $\mathbf{R}$.

Without further restrictions one could construct some very counter-intuitive examples of curves (for example, space-filling curves). Before making these restrictions, we give the definition of a function, and then impose restrictions on both together.

Functions. In Section 2 of Chapter 1 a hyperplane is defined as the pre-image of 0 by a map $f: \mathcal{A} \to \mathbf{R}$; the construction is then extended to the pre-image of any constant. What distinguishes the hyperplane map, among other maps, is that f is affine: $f(x) = \langle x - x_0, \alpha \rangle$, so that $f(x+v) = f(x) + \langle v, \alpha \rangle$.

We now drop the restriction that the map be affine. A map $f: \mathcal{A} \to \mathbf{R}$ is called a *(real) function* on $\mathcal{A}$.

We shall deal straight away with an awkward problem of notation for functions, which arises repeatedly, and is compounded partly by the cumbersome nature of the usual solutions to this problem, partly by the historical circumstance that mathematicians and physicists usually solve it in different ways. Consider for example a

2-dimensional affine space, with two affine coordinate systems (x^a) and $(\acute{x}^a)$ related by

$$\acute{x}^1 = x^1 + x^2 \qquad \acute{x}^2 = x^1 - x^2;$$

and let f be the function whose value at the point with coordinates (x^1, x^2) (relative to the first coordinate system) is given by $(x^1)^2 - (x^2)^2$. Then a physicist would without hesitation write

$$f(x^1, x^2) = (x^1)^2 - (x^2)^2;$$

and many physicists would write

$$f(\acute{x}^1, \acute{x}^2) = \acute{x}^1\acute{x}^2$$

to mean that the value of this function at the same point is $\acute{x}^1\acute{x}^2$ when its coordinates are given in terms of the second coordinate system. On the other hand, most mathematicians would insist that

$$f(\acute{x}^1, \acute{x}^2) = (\acute{x}^1)^2 - (\acute{x}^2)^2,$$

that is to say, that the symbol f represents the form of the function, not its value, and would introduce another symbol for $\acute{x}^1\acute{x}^2$, say

$$g(\acute{x}^1, \acute{x}^2) = \acute{x}^1\acute{x}^2,$$

so that

$$f(x^1, x^2) = g(\acute{x}^1, \acute{x}^2),$$

the arguments on the right hand side being obtained from those on the left by use of the relations between the two affine systems. Other mathematicians prefer to solve the problem by attaching to f an index which specifies the coordinate system in use.

A related issue concerns the coordinates themselves. An affine coordinate system (x^a) for an n-dimensional affine space $\mathcal{A}$ fixes a set of n functions on $\mathcal{A}$, the ath of which assigns to a point of $\mathcal{A}$ the value of its ath coordinate. These are the *coordinate functions* for the coordinate system. It is natural to denote the ath of these functions by x^a also. But this apparently creates a problem because the same symbol is being used to denote both a function and its value at a particular point. However, in this instance the ambiguity is actually helpful. We shall therefore use $(x^1, x^2, \ldots, x^n)$ to denote either a point of $\mathbf{R}^n$, the coordinates of a point in $\mathcal{A}$, or the map $\mathcal{A} \to \mathbf{R}^n$ which fixes the affine coordinate system, and the context will make it clear which is meant.

No problems arise in either case if one confines oneself to working in one fixed coordinate system, and even if a transformation of coordinates is involved it is usually clear what should be done in any particular instance; but much of what follows is concerned with the effects of changing coordinates in general situations, and then a precise notation is often needed. We shall distinguish between a function, which is a map $\mathcal{A} \to \mathbf{R}$, as in the definition above, and its *coordinate expression* or *coordinate presentation*, which is a map $\mathbf{R}^n \to \mathbf{R}$ obtained by composing the function with the inverse of the map $\mathcal{A} \to \mathbf{R}^n$ which specifies the coordinate system. The coordinate presentation of a function will be distinguished by an index identifying the coordinate system which is being used. When, as in the above instance, there are given two different presentations of the same function f, these will be denoted f^x and $f^{\acute{x}}$ for example. If x^a and $\acute{x}^a$ denote the coordinate functions, then

$$f^x(x^a) = f^{\acute{x}}(\acute{x}^a) = f$$

where composition of a map $\mathcal{A} \to \mathbf{R}^n$ and a map $\mathbf{R}^n \to \mathbf{R}$ is implied in the expressions $f^x(x^a)$ and $f^{\acute{x}}(\acute{x}^a)$; the range convention does not apply here since the free index appears in the argument of a function of n variables (recall the comment in Chapter 0, Section 4). If $\Phi\colon \mathbf{R}^n \to \mathbf{R}^n$ is the affine coordinate transformation which gives $(\acute{x}^a)$ in terms of (x^a) (so that $\acute{x}^a = \Phi^a(x^b) = k^a_b x^b + d^a$ say), then

$$f^x = f^{\acute{x}} \circ \Phi \qquad \text{and} \qquad f^{\acute{x}} = f^x \circ \Phi^{-1}.$$

Thus, in the above example, if $x \in \mathcal{A}$ with $x^1(x) = 3$, $x^2(x) = 2$, then $\acute{x}^1(x) = 5$, $\acute{x}^2(x) = 1$, and $f(x) = f^x(3,2) = f^{\acute{x}}(5,1) = 5$. According to this scheme one should not write $f(3,2)$, since this expects evaluation of a function in a coordinate system which has not been specified. Nor should one write $f(x^1,x^2) = (x^1)^2 - (x^2)^2$, but rather $f^x(x^1,x^2) = (x^1)^2 - (x^2)^2$. However, it is permissible to write $f = (x^1)^2 - (x^2)^2$, where the symbols x^1 and x^2 are now to be interpreted as coordinate functions; and in fact

$$f = (x^1)^2 - (x^2)^2 = \acute{x}^1\acute{x}^2.$$

Exercise 1. Using the coordinate transformation given above, find $(\acute{x}^1)^2 - (\acute{x}^2)^2$ in terms of (x^a), and x^1x^2 in terms of $(\acute{x}^a)$. □

Exercise 2. Let $\mathcal{A}$ be the affine space $\{\, (\xi^1,\xi^2,\xi^3) \in \mathbf{R}^3 \mid \xi^1 + \xi^2 + \xi^3 = 1 \,\}$ and f the function on $\mathcal{A}$ obtained by restricting the function $(\xi^1,\xi^2,\xi^3) \mapsto 2\xi^1 + \xi^2 - 3\xi^3 + 1$. Find the coordinate expressions for f in terms of the two coordinate systems defined in Exercise 5 of Chapter 1, and check the coordinate transformation rule, using the coordinate transformation given in Exercise 6 of that chapter. □

Smoothness. All that has been said so far applies to any curve or function, however counter-intuitive. To preserve the intuition and exploit the calculus one needs to impose some restrictions.

We shall deal only with functions whose coordinate expressions in any (and therefore in every) affine coordinate system have continuous partial derivatives of all orders. This property is unaffected by repeated partial differentiation. Such functions are called *smooth*, or C^∞, which is to say, continuously differentiable "infinitely often". Conditions of differentiability of this kind will occur regularly in this book; they form part of the analytical substratum on which the geometry is built. We shall try to avoid placing more emphasis on analytic technicalities than is absolutely necessary. It would be possible to impose less stringent conditions of differentiability, requiring, for example, only that functions have continuous partial derivatives of all orders up to and including the kth. Such a function is said to be C^k. However, this introduces complications since the derivative of a C^k function is not necessarily C^k, though it will be C^{k-1}. In any case, the functions met with in applications are almost always analytic, when they are differentiable at all, so there would be little practical advantage in relaxing the conditions.

It should be realised, however, that a smooth function is not necessarily analytic: one may certainly construct its Taylor series about any point in its domain of definition, but there is no guarantee that the series will converge to the value of the function at any other point. Again, the only function which is analytic on $\mathbf{R}$ and has the value zero on some open interval is the zero function, while it is possible for a merely smooth function to be identically zero on an open interval but different from zero elsewhere. It is an advantage to be dealing, not just with analytic functions, but with the larger class of smooth functions, precisely because one then has at one's disposal the so-called bump functions: a *bump function* is a smooth function which is positive within a finite interval and zero outside it.

Exercise 3. Show that, for given positive integer k, the function $x^k|x|$ on $\mathbf{R}$ is C^k but not C^{k+1}. □

Exercise 4. The function f on $\mathbf{R}$ defined by

$$f(x) = \begin{cases} e^{-1/x} & \text{if } x > 0 \\ 0 & \text{if } x \leq 0 \end{cases}$$

is smooth. Show that for any $a, b \in \mathbf{R}$ with $a < b$, the function $g_{(a,b)}$ defined by $g_{(a,b)}(x) = f(x-a)f(b-x)$ is smooth, and that $g_{(a,b)}(x) > 0$ for $a < x < b$, while $g_{(a,b)}(x) = 0$ for $x \leq a$ and for $x \geq b$. Show that for any $a, b, c, d \in \mathbf{R}$ with $a < b < c < d$ there is a smooth function h on $\mathbf{R}$ such that $h(x) = 0$ for $x \leq a$ and for $x \geq d$, and $h(x) = 1$ for $b \leq x \leq c$. □

We now define smoothness for curves. We have defined a curve in an affine space $\mathcal{A}$ as a map from the real line (or some open subinterval of it) to $\mathcal{A}$. If affine coordinates are chosen then a curve σ will be represented by n real valued functions $\sigma^a = x^a \circ \sigma$, its *coordinate functions*. A curve σ will be called a *smooth* curve if its coordinate functions are smooth for one, and therefore for every, affine coordinate system. If the domain of definition of the curve is a finite closed interval, as would be appropriate in discussing a curve joining two fixed points of $\mathcal{A}$, then it will be assumed that the curve is the restriction to that interval of a smooth curve defined on a larger, open, interval containing it. Then questions of differentiability at the endpoints of the interval will cause no difficulty, since the curve may be extended beyond them.

Paths, orientations and reparametrisations. As in the case of lines, two curves are counted as different if they are given by different maps, even if their image sets are the same. It is sometimes useful to have a word for the image set of a curve: we call it a *path*.

Curves with the same path may often be distinguished by the sense in which the path is traversed. Two curves which traverse the same path in the same sense are said to have the same *orientation*. An injective curve always fixes an orientation, but it is also possible that a curve will not traverse its path in a unique sense. We shall generally avoid the use of curves which are not injective. It is however convenient to allow constant curves, whose paths are single points of the affine space.

If $h: \mathbf{R} \to \mathbf{R}$ is a smooth function and $\sigma: \mathbf{R} \to \mathcal{A}$ is a smooth curve, then so also is $\sigma \circ h$: it is a *reparametrisation* of σ. One may also consider functions and curves defined on intervals of $\mathbf{R}$. Most reparametrisations of interest are reparametrisations by injective functions of the parameter. A smooth injective function $\mathbf{R} \to \mathbf{R}$ must be either increasing or decreasing; if the curve σ defines an orientation of its path, its reparametrisation by an increasing function defines the same orientation, while its reparametrisation by a decreasing function reverses the orientation.

Exercise 5. Show that the curves in a 3-dimensional affine space with affine coordinate expressions

$$\begin{aligned}
t &\mapsto (a\cos t, a\sin t, bt)\\
t &\mapsto (a\cos t, -a\sin t, -bt)\\
t &\mapsto (a\sin t, a\cos t, b(\pi/2 - t))\\
t &\mapsto (a\cos 2t, a\sin 2t, 2bt)\\
t &\mapsto (a\cos(t^3 - t), a\sin(t^3 - t), b(t^3 - t))
\end{aligned}$$

are all smooth, and all have the same path. Show that all but the last are injective, and distinguish those which have the same orientations. Find the reparametrisations of the first curve which give the others. □

2. Tangent Vectors

The *tangent vector* to a smooth curve σ at the point $\sigma(t_0)$ is the vector

$$\dot\sigma(t_0) = \lim_{\delta \to 0} \frac{1}{\delta}\big(\sigma(t_0 + \delta) - \sigma(t_0)\big).$$

This limit exists, because of the assumed smoothness: if in any affine coordinate system the presentation of σ is $t \mapsto \sigma^a(t)$ then the components of $\dot\sigma(t_0)$ are $\dot\sigma^a(t_0) = d\sigma^a/dt(t_0)$.

Note that the possibility of describing the tangent vector as "the tangent vector at $\sigma(t_0)$" (a point of $\mathcal{A}$) depends on our general assumption that the curves we deal with are injective. Otherwise we should have to say "the tangent vector at $t = t_0$" to avoid ambiguity.

The possibility of making such a definition depends on the fact that the difference $\sigma(t_0 + \delta) - \sigma(t_0)$ is a displacement in $\mathcal{A}$ and hence a vector in $\mathcal{V}$. It is a chord of the curve. The tangent vector is thus an element of $\mathcal{V}$. On the other hand, if x is any point of $\mathcal{A}$ and v is any vector in $\mathcal{V}$, then $t \mapsto x + tv$ is a smooth curve, and its tangent vector at x is v. Thus every vector in $\mathcal{V}$ may occur as tangent vector, and at each point x of $\mathcal{A}$: the set of tangent vectors at a point of $\mathcal{A}$ is a copy of $\mathcal{V}$. The correspondence between vectors in $\mathcal{V}$ and tangent vectors at x is a natural one; in other words, it does not depend on a choice of affine coordinates. Since the spaces of tangent vectors at the different points of $\mathcal{A}$ are all naturally identified with $\mathcal{V}$, they are all naturally identified with each other, and so it makes sense to say whether or not tangent vectors at different points of $\mathcal{A}$ are "equal", or *parallel*.

This construction of a copy of $\mathcal{V}$, as space of tangent vectors, at each point of $\mathcal{A}$ is to be distinguished from the attachment of $\mathcal{V}$ to $\mathcal{A}$ as space of displacement vectors introduced in Chapter 1. The results are similar, but nothing like the displacement vector construction can be achieved in the manifolds to be discussed later, while a development of the tangent vector construction, the directional derivative, can be generalised quite easily. The directional derivative is explained in the next section.

Even though tangent vectors to $\mathcal{A}$ are to be distinguished in concept from elements of $\mathcal{V}$ we shall not make any notational distinction between the two; thus v will denote an element of $\mathcal{V}$ or a tangent vector, it being clear from the context which is intended, and in the latter case, at which point of $\mathcal{A}$ it is tangent.

Exercise 6. Show that if $\acute{x}^a = k^a_b x^b + d^a$ and if (σ^a) and (σ'^a) are the coordinate presentations of a curve σ with respect to the two affine coordinate systems (x^a) and $(\acute{x}^a)$ then $\dot{\sigma}'^a(t) = k^a_b \dot{\sigma}^b(t)$. □

Exercise 7. Show that the tangent vector to a constant curve is the zero vector. □

Exercise 8. Show that if $\rho = \sigma \circ h$ is a reparametrisation then $\dot{\rho}(t) = \dot{h}(t)\dot{\sigma}(h(t))$. □

One very simple reparametrisation which is often useful is a change of origin. Let $\sigma: \mathbf{R} \to \mathcal{A}$ be a smooth curve and let $\tau_c: \mathbf{R} \to \mathbf{R}$ be the function $t \mapsto t + c$. A *change of origin* on σ is a reparametrisation $\sigma \circ \tau_c$ of σ. We denote the reparametrised curve σ_c. A change of origin is the only reparametrisation which does not alter tangent vectors: $\sigma_c(t) = \sigma(t + c)$ and $\dot{\sigma}_c(t) = \dot{\sigma}(t + c)$. Of course all the curves σ_c yield the same path, but they should be regarded as different curves, for different values of c, because of the convention that different maps count as different curves. It is evidently possible to choose c so that the point $\sigma_c(0)$ coincides with any given point of the path of σ. We shall call a set of curves which differ only by change of origin a *congruent* set. The second and third curves in Exercise 5 belong to the same congruent set.

More generally, a reparametrisation induced by an affine map $t \mapsto at + b$, $a \neq 0$ of $\mathbf{R}$ is called an *affine change of parameter*. It has the effect of multiplying tangent vectors by the constant a.

3. Directional Derivatives

In this section we show how a directional derivative may be defined along any tangent vector; this is a generalisation of the operator $\mathbf{v} \cdot \text{grad}$ in elementary vector calculus discussed in Chapter 0 and may be used as an alternative definition of a tangent vector.

Directional derivatives. If f is a smooth function on an affine space $\mathcal{A}$, and σ is a smooth curve in $\mathcal{A}$, then $f \circ \sigma$ is a smooth function on $\mathbf{R}$. The derivative $d/dt(f \circ \sigma)$ measures the rate of change of the function along the curve. In affine coordinates

$$\frac{d}{dt}(f \circ \sigma)(t_0) = \frac{d}{dt}(f^x(\sigma^a))(t_0) = \frac{\partial f^x}{\partial x^b}\dot{\sigma}^b(t_0),$$

the partial derivatives in the last expression being evaluated at $(\sigma^a(t_0))$. The derivative along a curve at a point thus depends only on the point and on the tangent vector to the curve there; it does not depend on the curve in any more complicated way. To put it otherwise: if curves σ and ρ meet at a point $x_0 = \sigma(0) = \rho(0)$ (we may change origins, if necessary, to achieve this agreement of parameters), and if they have the same tangent vectors there, then

$$\frac{d}{dt}(f \circ \sigma)(0) = \frac{d}{dt}(f \circ \rho)(0)$$

for any function f. Thus the derivative of any function along each of two curves at a point is the same whenever the two curves have the same tangent vector at that point.

One may therefore define a directional derivative along tangent vectors, as follows: given a tangent vector v at a point x_0, and a function f, the directional derivative of f along v, written $v(f)$ or simply vf, is the number

$$\frac{d}{dt}(f \circ \sigma)(0),$$

where σ is any curve such that $\sigma(0) = x_0$ and $\dot{\sigma}(0) = v$. One possible choice for σ is $t \mapsto x_0 + tv$. In terms of an affine coordinate system,

$$vf = v^a \frac{\partial f^x}{\partial x^a}$$

where the v^a are the components of v in this coordinate system, and the partial derivatives are evaluated at $(x^a(x_0))$.

Exercise 9. Show that two curves through a point x_0 which yield the same directional derivative for all functions at x_0 have the same tangent vector there. □

In many ways it is more satisfactory to equate a tangent vector with the directional derivative operator it defines than to regard it as the limit of a chord. One reason for this is that the operator interpretation offers the prospect of generalisation to manifolds, on which no affine structure is available and no chords can be constructed. It is therefore desirable to characterise directional derivative operators by their properties, which are

(1) $v(af + bg) = avf + bvg$

(2) $v(fg) = (vf)g(x_0) + f(x_0)(vg)$

for all $a, b \in \mathbf{R}$ and all smooth functions f and g. The first of these says that, as an operator, v is linear, and the second that it obeys the appropriate version of Leibniz's rule. That v, as a directional derivative, does have these properties follows from its definition in terms of ordinary differentiation of a real function. It is also true that, conversely, any operator which maps smooth functions to numbers and satisfies these conditions is a directional derivative operator: we shall show this in detail in Chapter 10. In fact, it can be shown that such an operator may be represented as the derivative along a smooth curve as described above. We formalise these changes of emphasis in a new definition of a tangent vector: a *tangent vector* at a point in an affine space $\mathcal{A}$ is an operator on smooth functions which maps functions to numbers and is linear and satisfies Leibniz's rule as set out above.

We shall denote by $T_{x_0}\mathcal{A}$ the set of tangent vectors at $x_0 \in \mathcal{A}$. As we have remarked above, association of a tangent vector with an element of $\mathcal{V}$ gives a natural identification of $T_{x_0}\mathcal{A}$ with $\mathcal{V}$. As a consequence of this identification we may endow $T_{x_0}\mathcal{A}$ with the structure of a vector space, by defining $av+bw$, where $v, w \in T_{x_0}\mathcal{A}$ and $a, b \in \mathbf{R}$, to be the tangent vector at x_0 corresponding to the element $av + bw$ of $\mathcal{V}$. Alternatively, $av+bw$ is the tangent vector at $t = 0$ to the curve $t \mapsto x_0+t(av+bw)$.

Exercise 10. Show that, as an operator, $(av + bw)f = avf + bwf$ for any smooth function f. □

Not only is $T_{x_0}\mathcal{A}$ naturally identified with $\mathcal{V}$, it is isomorphic to it as a vector space. Nevertheless, the two spaces are conceptually distinct, and each tangent space is distinct from every other. In generalisations to manifolds the naturalness of the isomorphism (its independence of coordinates) gets lost, and it then becomes imperative to regard tangent spaces at different points as distinct.

Given a basis $\{e_a\}$ of $\mathcal{V}$, the tangent vector at a point $x_0 \in \mathcal{A}$ corresponding to the basis vector e_a is the tangent at $t = 0$ to the coordinate line $t \mapsto x_0 + te_a$ of any affine coordinate system based on $\{e_a\}$. This tangent vector has a particularly simple representation as an operator: its action on a function f is given by $f \mapsto \partial f^x/\partial x^a$, the partial derivative being evaluated at $\big(x^a(x_0)\big)$. In accordance with our change of emphasis towards tangent vectors as operators, we shall use a notation for the tangent vectors to coordinate lines which is suggested by this observation: we shall write $\partial_1, \partial_2, \ldots, \partial_n$ for these tangent vectors (the point x_0 being understood); where it is necessary to distinguish the coordinate system we shall use

$$\frac{\partial}{\partial x^1}, \frac{\partial}{\partial x^2}, \ldots, \frac{\partial}{\partial x^n}.$$

These *coordinate tangent vectors* form, at any point x_0, a basis for $T_{x_0}\mathcal{A}$. Any $v \in T_{x_0}\mathcal{A}$ may be uniquely written $v = v^a\partial_a$, where the v^a are the components of v (considered as an element of $\mathcal{V}$) with respect to the basis $\{e_a\}$; and

$$vf = v^a\partial_a f = v^a\frac{\partial f^x}{\partial x^a},$$

the partial derivatives being evaluated at $\big(x^a(x_0)\big)$, as before. Thus the operation of v on a function expressed explicitly in terms of coordinate functions amounts simply to carrying out the indicated partial differentiations, evaluating at the coordinates of x_0, and taking the appropriate linear combination of the results.

Exercise 11. Show that $v^a = v(x^a)$, where x^a is thought of as a (coordinate) function. □

Exercise 12. The point x_0 in a 3-dimensional affine space $\mathcal{A}$ has coordinates $(3, 1, -2)$ with respect to an affine coordinate system (x^a); also $v = \partial_1 + 2\partial_2 + 3\partial_3$ and $f = x^1x^2 + x^2x^3 + x^3x^1$. Show that $vf = 13$. □

Exercise 13. Show that if f is the affine function $x \mapsto \langle x - x_0, \alpha\rangle$ determining a hyperplane and v is a tangent vector then $vf = \langle v, \alpha\rangle$. □

4. Cotangent Vectors

The set of points in an affine space at which a given smooth function takes a particular fixed value is called, if it is not empty, a *level surface* of the function. In general a level surface has, at each point on it, a tangent hyperplane, which contains all the tangent vectors at that point to all the curves lying in the surface and passing through the point. If the function in question is f, and the point is x_0, then for any curve σ in the surface $f \circ \sigma$ is constant and so $d/dt\big(f \circ \sigma\big)(0) = 0$, where $x_0 = \sigma(0)$. Thus the tangent vectors at x_0 to curves in the level surface are those which satisfy $vf = 0$.

Now for a fixed function f the map $T_{x_0}\mathcal{A} \to \mathbf{R}$ by $v \mapsto vf$ is linear and therefore defines a linear form on $T_{x_0}\mathcal{A}$, that is, an element of the space dual to $T_{x_0}\mathcal{A}$. This space is denoted $T^*_{x_0}\mathcal{A}$ and called the *cotangent space* to $\mathcal{A}$ at x_0. The linear forms in the cotangent space are often called *cotangent vectors*, or *covectors* for short. The covector determined by f in this way is denoted df and called the *differential* of f at x_0. Thus

$$\langle v, df\rangle = vf.$$

Provided that df at x_0 is not identically zero, the tangent hyperplane at x_0 to the level surface of f is given by $\langle v, df\rangle = 0$. This defines the tangent hyperplane as a subspace of $T_{x_0}\mathcal{A}$. If the tangent space is identified with $\mathcal{V}$ then df fixes an element of $\mathcal{V}^*$, and thereby a hyperplane in $\mathcal{A}$ attached at x_0. This hyperplane consists of the tangent lines at x_0 to curves in the level surface. If, at x_0, df is zero then it is not possible to define a tangent hyperplane by these means, and in fact there may not even be one.

Thus with each function f and each point $x_0 \in \mathcal{A}$ there is associated an element df of $T^*_{x_0}\mathcal{A}$, a linear form or covector at x_0. (It is important to remember the role of the point in this construction, since it is not evident from the notation df.) From the formula for the coordinate representation of vf, namely $vf = v^a\partial_a f = v^a\partial f^x/\partial x^a$, it will be seen that df is determined by the partial derivatives of f^x evaluated at $\left(x^a(x_0)\right)$. (In future, when the arguments of a particular derivative are evident from the context, we shall not mention them explicitly.) The coordinate functions x^a define linear forms dx^a, the *coordinate differentials*, which constitute the basis of $T^*_{x_0}\mathcal{A}$ dual to the basis $\{\partial_a\}$ of $T_{x_0}\mathcal{A}$. Thus any element of $T^*_{x_0}\mathcal{A}$ may be written uniquely in the form $c_a dx^a$, and in particular

$$df = \langle\partial_a, df\rangle dx^a = (\partial_a f)dx^a = \frac{\partial f^x}{\partial x^a}dx^a.$$

An arbitrary element of $T^*_{x_0}\mathcal{A}$ may be obtained from many different functions on $\mathcal{A}$, and in particular from just one function of the form $x \mapsto \langle x - x_0, \alpha\rangle$, where $\alpha \in \mathcal{V}^*$; this constitutes a natural identification of $T^*_{x_0}\mathcal{A}$ with $\mathcal{V}^*$. The level surface of the function defined by α is a hyperplane in $\mathcal{A}$.

The linear form df determines, when it is not zero, the tangent hyperplane to the level surface of f through x_0. However, any nonzero multiple of df would determine the same hyperplane; thus df contains a little more information about the level surfaces of f: it affords the possibility of comparing the rates at which level surfaces are crossed by any curve transverse to them. The function cf, for constant c, has the same level surfaces as f, though if $c \neq 1$ they are differently labelled; this difference of labelling shows up in the fact that $d(cf) = c\,df$.

The reader will no doubt have noticed that the components of df are the same as those of grad f^x in ordinary vector calculus. However, it makes no sense at this stage to say that df is orthogonal to the level surfaces of f, since no measure of angle or concept of orthogonality has been introduced into the space. If f is a smooth function on an affine space of dimension 4, for example, df will be defined and have the same value regardless of whether that space is Newtonian space-time or Minkowskian space-time or something altogether different. The definition of a gradient involves a metric structure, which will be introduced in Chapter 7.

The reader may also have been reminded, by the notation, of infinitesimals. Infinitesimals in the sense of l'Hôpital—"infinitely small but nonzero quantities"—have long been banished from standard mathematics. However, this particular piece of mathematics does provide a sensible parallel to such a statement as "in a displacement from (x^a) to $(x^a + dx^a)$ the change in f is given by $(\partial f/\partial x^a)dx^a$, and if $dx^a = v^a dt$ then $df = v^a(\partial f/\partial x^a)dt$"; and the notation reflects this.

Exercise 14. Show from the linearity and Leibniz rules that $d(af + bg) = adf + bdg$ and $d(fg) = g(x_0)df + f(x_0)dg$. Show that if $h: \mathbf{R} \to \mathbf{R}$ is a smooth function then, at x_0, $d(h \circ f) = \dot{h}(f(x_0))df$. □

Exercise 15. Compute df, at x_0, in terms of dx^a, for the function $f = x^1x^2 + x^2x^3 + x^3x^1$, where $(x^a(x_0)) = (3, 1, -2)$. Show that the tangent hyperplane through x_0 to the level surface of this function is given by $-x^1 + x^2 + 4x^3 + 10 = 0$. Show that $df = 0$ at the origin of coordinates, that the three coordinate axes all lie in the level surface of the function through the origin, but that (for example) no other line through the origin in the x^1x^2-plane does so, and that therefore the level surface through the origin has no tangent hyperplane there though the function is certainly smooth there. □

This level surface is a cone, with the origin as its vertex.

5. Induced Maps

The defining property of an affine map is that it acts as a linear map of displacement vectors: if $\Lambda: \mathcal{A} \to \mathcal{B}$ by $x \mapsto y_0 + \lambda(x - x_0)$ then $\Lambda(x + v) = \Lambda(x) + \lambda(v)$. An affine map takes lines into lines; it also takes curves into curves, for if $\sigma: \mathbf{R} \to \mathcal{A}$ is a curve, then $\Lambda \circ \sigma: \mathbf{R} \to \mathcal{B}$ is also a curve, which is easily seen to be smooth if σ is. Since tangent vectors (as distinct from displacement vectors) arise in the first place from curves, it should not be surprising that an affine map also takes tangent vectors into tangent vectors, in a way consistent with their definition in terms of curves, and in agreement with the linear map of displacement vectors. In fact the tangent vector to $\Lambda \circ \sigma$ is given, as a limit of chords, by

$$\lim_{\delta \to 0} \frac{1}{\delta}\big(\Lambda(\sigma(t + \delta)) - \Lambda(\sigma(t))\big) = \lim_{\delta \to 0} \frac{1}{\delta}\lambda\big(\sigma(t + \delta) - \sigma(t)\big) = \lambda\big(\dot{\sigma}(t)\big).$$

Thus the linear part λ gives the transformation of tangent vectors, just as it gives the transformation of displacement vectors. The vector $\lambda(\dot{\sigma}(t))$ at $\Lambda(\sigma(t))$ is called the *image* of $\dot{\sigma}(t)$ by Λ.

As a directional derivative operator, the image of a tangent vector v at $x \in \mathcal{A}$ may be defined as the operator $g \mapsto d/dt(g \circ \Lambda \circ \sigma)(0)$ for any function g on $\mathcal{B}$, where σ is any curve such that $\sigma(0) = x$ and $\dot{\sigma}(0) = v$. But $g \circ \Lambda \circ \sigma$ may be constructed by first composing g with Λ, and then composing the result, $g \circ \Lambda$, with σ. Read in this way, $d/dt(g \circ \Lambda \circ \sigma)(0) = v(g \circ \Lambda)$. It may be verified easily that the operator $g \mapsto v(g \circ \Lambda)$ satisfies the linearity condition and Leibniz's rule, and it is therefore a tangent vector at $\Lambda(x) \in \mathcal{B}$. Moreover, the map $T_x\mathcal{A} \to T_{\Lambda(x)}\mathcal{B}$ so defined is evidently a linear one, which we denote Λ_*. Thus $\Lambda_*(v)$ is the element of $T_{\Lambda(x)}\mathcal{B}$ given by

$$(\Lambda_*(v))g = v(g \circ \Lambda).$$

When $T_x\mathcal{A}$ is identified with $\mathcal{V}$ and $T_{\Lambda(x)}\mathcal{B}$ with $\mathcal{W}$, $\Lambda_*(v)$ is identified with $\lambda(v)$ and Λ_* therefore with λ.

The *adjoint* of the linear map $\Lambda_*: T_x\mathcal{A} \to T_{\Lambda(x)}\mathcal{B}$ is a linear map of cotangent spaces $\Lambda^*: T^*_{\Lambda(x)}\mathcal{B} \to T^*_x\mathcal{A}$. It is defined as follows: for $\beta \in T^*_{\Lambda(x)}\mathcal{B}$,

$$\langle v, \Lambda^*(\beta)\rangle = \langle \Lambda_*(v), \beta\rangle \qquad \text{for all } v \in T_x\mathcal{A}.$$

In particular, for any function g on $\mathcal{B}$,

$$\langle v, \Lambda^*(dg)\rangle = \big(\Lambda_*(v)\big)g = v(g \circ \Lambda) = \langle v, d(g \circ \Lambda)\rangle.$$

Thus

$$\Lambda^*(dg) = d(g \circ \Lambda).$$

With respect to affine coordinates (x^a), (y^α), with Λ represented by $y^\alpha \circ \Lambda = \lambda^\alpha_a x^a + c^\alpha$,

$$\frac{\partial}{\partial x^a}(y^\alpha \circ \Lambda) = \lambda^\alpha_a.$$

Using this, one reads off the coordinate expressions for the maps $\Lambda_*: T_x\mathcal{A} \to T_{\Lambda(x)}\mathcal{B}$ and $\Lambda^*: T^*_{\Lambda(x)}\mathcal{B} \to T^*_x\mathcal{A}$ as follows:

$$\left\langle \Lambda_*\left(\frac{\partial}{\partial x^a}\right), dy^\alpha \right\rangle = \lambda^\alpha_a, \quad \text{so that} \quad \Lambda_*\left(\frac{\partial}{\partial x^a}\right) = \lambda^\alpha_a \frac{\partial}{\partial y^\alpha}$$

and

$$\left\langle \frac{\partial}{\partial x^a}, \Lambda^*(dy^\alpha) \right\rangle = \lambda^\alpha_a, \quad \text{so that} \quad \Lambda^*(dy^\alpha) = \lambda^\alpha_a dx^a.$$

The maps Λ_* of tangent spaces and Λ^* of cotangent spaces are said to be *induced* by the affine map Λ. Note that Λ_* is cogredient with Λ while Λ^* is contragredient to it.

Exercise 16. Show that for any affine map Λ

$$\Lambda_*\big(v^a(\partial/\partial x^a)\big) = \lambda^\alpha_a v^a(\partial/\partial y^\alpha) \quad \text{and} \quad \Lambda^*(c_\alpha dy^\alpha) = c_\alpha \lambda^\alpha_a dx^a. \qquad \square$$

Exercise 17. Show that if $\Lambda: \mathcal{A} \to \mathcal{B}$ and $\mathrm{M}: \mathcal{B} \to \mathcal{C}$ are affine maps then

$$(\mathrm{M} \circ \Lambda)_* = \mathrm{M}_* \circ \Lambda_* \quad \text{and} \quad (\mathrm{M} \circ \Lambda)^* = \Lambda^* \circ \mathrm{M}^*. \qquad \square$$

6. Curvilinear Coordinates

We have so far found it unnecessary to use any but affine coordinates. The reader will be aware of the possibility, indeed the advantage under certain circumstances, of using other kinds of coordinates: polar, spherical polar, cylindrical or whatever. In the sequel we shall often use curvilinear coordinates—not any specific kind, but in a rather general way. We shall devote this section to defining curvilinear coordinates and describing the modifications required to the matters so far discussed as a result of introducing them.

Before attempting a definition we must point out one possible difficulty with curvilinear coordinates, which arises even in such a simple case as that of polar coordinates for the plane. An affine coordinate system has the desirable property that each point of the affine space has unique coordinates. In polar coordinates this is not so, the origin being the exceptional point. Moreover, points which have nearby affine coordinates need not necessarily also have nearby polar coordinates, since their angular coordinates may differ by almost 2π. Of course, in the particular case of polar coordinates one adopts various *ad hoc* methods for dealing with the consequences of these defects: but this will not be possible in general. Another way of getting over the difficulty with polar coordinates is to restrict their domain so that single-valuedness and continuity are restored, by deleting the non-positive x-axis. This is the lead which we shall follow in the general case. We shall allow for a curvilinear coordinate system to be *local*, that is to say, defined, single-valued, and smooth with respect to affine coordinates only on some open subset of the space, not necessarily on the whole of it.

We noted in Section 1 of Chapter 1 that an affine coordinate system on an n-dimensional affine space may be described as a bijective map from the space to $\mathbf{R}^n$, namely the map which assigns to each point the n-tuple of its coordinates. Two different affine coordinate systems are related by a coordinate transformation, which is a map from $\mathbf{R}^n$ to itself. These are also essential features of our definition of curvilinear coordinates, which follows.

A local curvilinear coordinate system, or *local coordinate chart*, for an n-dimensional affine space $\mathcal{A}$ is a bijective map ψ from an open subset $\mathcal{P}$ of $\mathcal{A}$, called the *coordinate patch*, to an open subset of $\mathbf{R}^n$; this map is to be smooth with respect to affine coordinates in the following sense: if $\phi\colon \mathcal{A} \to \mathbf{R}^n$ is the bijective map defining an affine coordinate system on $\mathcal{A}$ then the map $\psi \circ \phi^{-1}$, which takes affine coordinates into curvilinear coordinates, and which is a bijective map between two open subsets of $\mathbf{R}^n$, is to be smooth and have a smooth inverse. The map $\psi \circ \phi^{-1}$, which is called the *coordinate transformation* from the affine to the curvilinear coordinates, may be thought of as a vector-valued function of n variables; it will be smooth if all its component functions have continuous partial derivatives of all orders. Since affine coordinate transformations are clearly smooth, a local coordinate chart which is smooth with respect to one affine coordinate system is smooth with respect to all.

For any differentiable map Φ of an open subset of $\mathbf{R}^n$ into $\mathbf{R}^n$ we shall denote by Φ' the matrix of partial derivatives, or *Jacobian matrix*, of Φ. It is a smooth $n \times n$ matrix-valued function on the domain of Φ. If one writes

$$\Phi(\xi^a) = (\Phi^1(\xi^a), \Phi^2(\xi^a), \ldots, \Phi^n(\xi^a))$$

then

$$\left(\frac{\partial \Phi^b}{\partial \xi^1}, \frac{\partial \Phi^b}{\partial \xi^2}, \ldots, \frac{\partial \Phi^b}{\partial \xi^n}\right)$$

is the bth row of Φ'. There are important connections between the invertibility of the map Φ and the invertibility of the matrix Φ'. In the first place, if Φ is invertible then Φ' is non-singular and $(\Phi^{-1})' = (\Phi' \circ \Phi^{-1})^{-1}$. Furthermore, the inverse function theorem states that if Φ is smooth on an open set containing a point ξ and $\Phi'(\xi)$ is non-singular then there is an open set $\mathcal{O}$ containing ξ and an open set $\tilde{\mathcal{O}}$ containing $\Phi(\xi)$ such that $\Phi\colon \mathcal{O} \to \tilde{\mathcal{O}}$ has a smooth inverse $\Phi^{-1}\colon \tilde{\mathcal{O}} \to \mathcal{O}$. It is also known that if $\Phi\colon \mathcal{O} \to \mathbf{R}^n$ is injective and Φ' is non-singular at all points of the open set $\mathcal{O}$ then $\Phi(\mathcal{O})$ is open and $\Phi^{-1}\colon \Phi(\mathcal{O}) \to \mathcal{O}$ is smooth. These results sometimes allow one to infer the existence of a local coordinate chart from the invertibility of a Jacobian matrix. In practice, a specific curvilinear coordinate system is usually given by coordinate transformations from some affine coordinate system; to check the validity of the curvilinear coordinates it is therefore necessary merely to find where the Jacobian of the coordinate transformation $\Phi = \psi \circ \phi^{-1}$ is non-singular and confirm that the transformation is injective there. Alternatively, it may be more convenient to work with the inverse of the coordinate transformation.

Exercise 18. Let $\mathcal{O}$ be the open subset of $\mathbf{R}^2$ consisting of all points other than those on the non-positive ξ^1-axis. The function $\vartheta\colon \mathcal{O} \to (-\pi, \pi)$ is defined by

$$\vartheta(\xi^1, \xi^2) = \begin{cases} \arctan(\xi^2/\xi^1) & \text{if } \xi^1 > 0 \\ \pi + \arctan(\xi^2/\xi^1) & \text{if } \xi^1 < 0, \xi^2 > 0 \\ -\pi + \arctan(\xi^2/\xi^1) & \text{if } \xi^1 < 0, \xi^2 < 0 \\ \pi/2 & \text{if } \xi^1 = 0, \xi^2 > 0 \\ -\pi/2 & \text{if } \xi^1 = 0, \xi^2 < 0. \end{cases}$$

Show that the map $\mathcal{O} \to \mathbf{R}^2$ by $(\xi^1, \xi^2) \mapsto \left(\sqrt{(\xi^1)^2 + (\xi^2)^2}, \vartheta(\xi^1, \xi^2)\right)$ defines a coordinate transformation from any affine coordinates on a 2-dimensional affine space to curvilinear coordinates ("polar coordinates"). □

Exercise 19. Let x be an affine coordinate on a 1-dimensional affine space $\mathcal{A}$. Show that, although the function $\mathcal{A} \to \mathbf{R}$ by $x \mapsto x^3$ is bijective, it does not define a local coordinate chart on $\mathcal{A}$. □

If $\psi\colon \mathcal{P} \to \mathbf{R}^n$ and $\chi\colon \mathcal{Q} \to \mathbf{R}^n$ are two local coordinate charts such that $\mathcal{P} \cap \mathcal{Q}$ is non-empty, then $\psi \circ \chi^{-1}$ and $\chi \circ \psi^{-1}$, which are the *coordinate transformations* between the charts, are smooth maps of open subsets of $\mathbf{R}^n$.

The *coordinate functions* for a local coordinate chart are defined in the same way as for affine coordinates: the ath coordinate function assigns to each point in the coordinate patch the value of its ath coordinate. In other words, $x^a = \Pi^a \circ \psi$, where ψ is the chart and $\Pi^a\colon \mathbf{R}^n \to \mathbf{R}$ is projection onto the ath component. The coordinate functions are *local functions*, that is, not necessarily defined on the whole of the space; they must however be smooth on their domain.

Exercise 20. Let $(\acute{x}^a)$ be the coordinate functions for a local coordinate chart and (x^a) those for an affine coordinate system. Show that $\partial_a \acute{x}^b$, the function obtained by applying the coordinate tangent vector $\partial_a = \partial/\partial x^a$ to the function $\acute{x}^a$, has for its coordinate expression with respect to the affine coordinates the (b, a) element of the Jacobian matrix of the coordinate transformation from affine coordinates to curvilinear ones. □

The differentials of the curvilinear coordinate functions $(\acute{x}^a)$ are given in terms of those of the affine coordinate functions (x^a) by

$$d\acute{x}^b = (\partial_a \acute{x}^b) dx^a.$$

The coefficient matrix is non-singular, by Exercise 20; the linear forms $\{d\acute{x}^a\}$ therefore constitute a basis for the cotangent space, at each point of the coordinate patch. They will be called the *coordinate differentials* for the curvilinear coordinate system.

The ath *coordinate curve* is the curve given in terms of the curvilinear coordinates by $t \mapsto (\acute{x}^1, \acute{x}^2, \ldots, \acute{x}^a + t, \ldots, \acute{x}^n)$. The tangent vector to the ath coordinate curve is denoted by $\acute{\partial}_a$, or $\partial/\partial \acute{x}^a$, just as in the case of affine coordinates, and for the same reason. These *coordinate tangent vectors* form a basis for the tangent space, at each point of the coordinate patch, which is dual to the basis of the cotangent space at the same point given by the coordinate differentials:

$$\left\langle \acute{\partial}_a, d\acute{x}^b \right\rangle = \acute{\partial}_a \acute{x}^b = \frac{\partial}{\partial \acute{x}^a} \acute{x}^b = \delta_a^b.$$

The components of $\acute{\partial}_a$ are given in terms of the coordinate vectors for the affine coordinate system by

$$\left\langle \acute{\partial}_a, dx^b \right\rangle = \acute{\partial}_a x^b, \quad \text{so that} \quad \acute{\partial}_a = \left(\acute{\partial}_a x^b \right) \partial_b.$$

We define the coordinate expressions for curves and functions in terms of curvilinear coordinates just as we did for affine coordinates, making allowance if necessary for the local nature of the curvilinear coordinates.

Exercise 21. Show that the matrices $(\acute{\partial}_a x^b)$ and $(\partial_a \acute{x}^b)$ are inverses of each other. □

Exercise 22. Show that the coordinate differentials and vector fields of any two coordinate systems are related in the same way as those of a curvilinear and an affine coordinate system. Let v^a be the components of a tangent vector v in one coordinate system (x^a) (curvilinear or affine) and let $\acute{v}^a$ be the components of the same tangent vector in any other coordinate system $(\acute{x}^a)$. Show that $\acute{v}^a = (\partial_b \acute{x}^a) v^b$. Show that the components c_a and $\acute{c}_a$ of a linear form are related by $\acute{c}_a = (\acute{\partial}_a x^b) c_b$. □

Exercise 23. Show that the differential of a function f takes the form $df = (\partial_a f)\, dx^a$ with respect to any coordinate system. □

Exercise 24. Let (x^1, x^2, x^3) be affine coordinates in a 3-dimensional affine space $\mathcal{A}$ and let (r, ϑ, φ) be the curvilinear coordinates ("spherical polars") given by

$$x^1 = r \sin\vartheta \cos\varphi \qquad x^2 = r \sin\vartheta \sin\varphi \qquad x^3 = r \cos\vartheta.$$

Show that the open subset of $\mathcal{A}$ obtained by deleting the half-plane on which $x^2 = 0$, $x^1 \leq 0$ is a suitable domain for (r, ϑ, φ), and that no larger open subset of $\mathcal{A}$ will do. Verify that these functions do define a coordinate chart; identify the corresponding coordinate patch (in terms of the affine coordinates). Compute the components of the affine coordinate differentials and vectors in terms of the curvilinear coordinates, and *vice-versa*. □

The great majority of coordinate formulae carry over to the case of curvilinear coordinates without change of appearance, but it must be remembered that in general they hold only locally, that is on the coordinate patch. Where in the sequel we have occasion to derive a result that is true only for affine coordinates, or some other special coordinates, we shall draw the reader's attention to this; otherwise it may be safely assumed that any coordinate expression is valid in any coordinate system.

7. Smooth Maps

So far in this chapter we have shown how various affine objects—lines, hyperplanes, affine coordinate systems—may be generalised by relinquishing the conditions of global linearity. By retaining the requirement of smoothness, however, one ensures that a measure of linearity is preserved, albeit only on an infinitesimal scale. We now make a similar generalisation, from affine maps to smooth maps. The process of inducing linear maps of vectors and covectors from an affine map will be generalised at the same time, to give a way of constructing, from a smooth map, linear maps of tangent and cotangent spaces.

An affine map is represented, in terms of affine coordinates, by inhomogeneous linear functions; but the functions representing the same affine map in terms of curvilinear coordinates will not be linear, though they will be smooth. The map's affine property, in other words, will not be very apparent from its representation in curvilinear coordinates. Nevertheless, the construction of the corresponding induced map of vectors (for example) must still be possible, since its definition does not depend on any particular choice of coordinates. The representation of this induced map, with respect to the coordinate vectors of the curvilinear coordinate system, will be a matrix; but, again in contrast to the case of affine coordinates, this matrix will vary from point to point. These observations give a clear guide as to how to proceed in general, and what to expect.

Smooth maps defined. Let $\phi\colon \mathcal{A} \to \mathcal{B}$ be a map of affine spaces. Such a map may be represented with respect to any coordinates (x^a) on $\mathcal{A}$ and (y^α) on $\mathcal{B}$ by $n = \dim \mathcal{B}$ functions ϕ^α of $m = \dim \mathcal{A}$ variables, as follows: for each $x \in \mathcal{A}$,

$$\phi^\alpha(x^a(x)) = y^\alpha(\phi(x)).$$

Here x^a and y^α are to be interpreted as coordinate functions. The functions ϕ^α may be considered as the components of a map

$$(\xi^1, \xi^2, \dots, \xi^m) \mapsto (\phi^1(\xi^a), \phi^2(\xi^a), \dots, \phi^n(\xi^a))$$

from $\mathbf{R}^m$ (or some open subset of it) to $\mathbf{R}^n$: thus if (x^a) are the coordinates of a point $x \in \mathcal{A}$ then $(\phi^\alpha(x^a))$ are the coordinates of the image point $\phi(x) \in \mathcal{B}$. We may also write the defining relation in the form $\phi^\alpha(x^a) = y^\alpha \circ \phi$, or describe $y^\alpha = \phi^\alpha(x^a)$ as the *coordinate presentation* of ϕ. It will frequently be convenient to define a map ϕ between affine spaces by giving its coordinate presentation, that is, by specifying the functions ϕ^α which represent it with respect to some given coordinate systems on $\mathcal{A}$ and $\mathcal{B}$. Of course, in order for the map to be globally defined (that is, defined all over $\mathcal{A}$) it is necessary that the coordinates used for $\mathcal{A}$ should cover $\mathcal{A}$; and correspondingly, use of a coordinate system for $\mathcal{B}$ which does not cover $\mathcal{B}$ restricts the possible range of the image set. These difficulties can arise only when the coordinates chosen for $\mathcal{A}$ or $\mathcal{B}$ are non-affine (and not necessarily even then): for affine coordinates no such problems arise.

Exercise 25. Explain how the coordinate presentation of a map $\mathcal{A} \to \mathcal{B}$ is affected by a change of coordinates in $\mathcal{A}$ and $\mathcal{B}$. □

A map $\phi\colon \mathcal{A} \to \mathcal{B}$ is *smooth* if the functions ϕ^α which represent it with respect to affine coordinate systems on $\mathcal{A}$ and $\mathcal{B}$ are smooth.

Exercise 26. Show that if $\phi\colon \mathcal{A} \to \mathcal{B}$ is smooth then the functions which represent it with respect to any coordinate systems on $\mathcal{A}$ and $\mathcal{B}$, affine or not, are smooth (on their domain). □

If ϕ is an affine map of affine spaces then the functions ϕ^α which represent it with respect to affine coordinates are inhomogeneous linear: $\phi^\alpha(x^a) = \lambda^\alpha_a x^a + c^\alpha$;

and so ϕ is certainly smooth. The definition of a smooth map is clearly also consistent in concept with the definitions of a smooth curve and a smooth function (though a smooth curve in $\mathcal{B}$ is not quite the same thing as a smooth map from a 1-dimensional affine space to $\mathcal{B}$, nor is a smooth function on $\mathcal{A}$ quite the same thing as a smooth map from $\mathcal{A}$ to a 1-dimensional affine space).

Sometimes we shall have to deal with the case of a map between affine spaces $\mathcal{A}$ and $\mathcal{B}$ whose domain is not the whole of $\mathcal{A}$; but provided that the domain is an open subset of $\mathcal{A}$ the definition of smoothness carries over without essential change (this is analogous to the situation that occurs when a curve is defined only on an open interval in $\mathbf{R}$).

Induced maps of vectors and covectors. We have already described maps of tangent vectors and covectors induced by affine maps (Section 5). We have implicitly introduced them again in defining the coordinate tangent vectors and differentials for curvilinear coordinates (Section 6). We now repeat the argument in a more general context, where the map is no longer assumed to be affine, nor between spaces of the same dimension.

Let $\mathcal{A}$ and $\mathcal{B}$ be affine spaces, and $\mathcal{O}$ an open subset of $\mathcal{A}$ (which may be the whole of $\mathcal{A}$). Let $\phi\colon \mathcal{O} \to \mathcal{B}$ be a smooth map. We shall first construct the map of tangent vectors induced by ϕ. This construction depends on little more than that ϕ takes curves into curves.

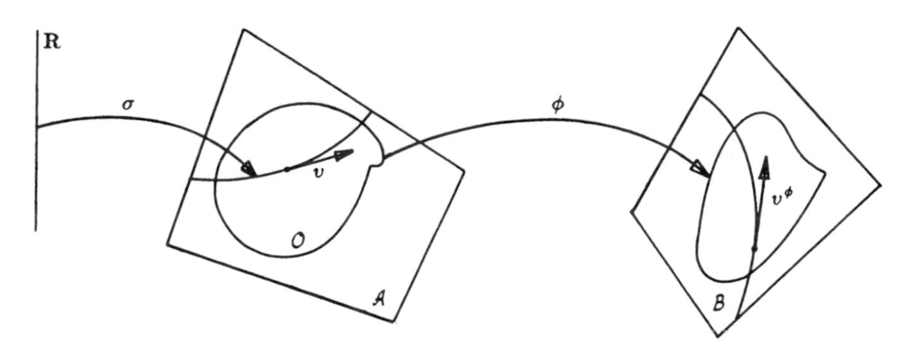

Fig. 1 Induced map of a curve and a tangent vector.

Let v be any tangent vector at a point x in $\mathcal{O}$, and let σ be any curve which has v as tangent vector at x. The map ϕ takes σ to a curve $\sigma^\phi = \phi \circ \sigma$ through $\phi(x)$, and σ^ϕ has a tangent vector v^ϕ there, which may be constructed either as a limit of chords or, better for this purpose, through its directional derivative. Let f be any function on $\mathcal{A}$. Then

$$v^\phi f = \frac{d}{dt}(f \circ \sigma^\phi)(0)$$

but

$$f \circ \sigma^\phi = f \circ (\phi \circ \sigma) = (f \circ \phi) \circ \sigma$$

so that

$$v^\phi f = \frac{d}{dt}((f \circ \phi) \circ \sigma)(0)$$

or

$$v^\phi f = v(f \circ \phi).$$

This expression reveals that the tangent vector to σ^ϕ at $\phi(x)$ depends only on the tangent vector to σ at x, as the notation has anticipated. The construction works in essentially the same way as in the affine case. The fact that ϕ may be non-affine, and not necessarily defined all over $\mathcal{A}$, makes no significant difference to the construction. Note that v^ϕ is a tangent vector at $\phi(x) \in \mathcal{B}$.

Exercise 27. Show that the alternative approach, defining the tangent vector as a limit of chords, leads to the expression

$$v^\phi = \lim_{t\to 0} \frac{1}{t}(\phi(\sigma(t)) - \phi(x)).$$ □

The map $\phi_*: T_x\mathcal{A} \to T_{\phi(x)}\mathcal{B}$ by $v \mapsto v^\phi$ with $v^\phi f = v(f \circ \phi)$ is called the *induced map* of tangent spaces.

Exercise 28. Verify that $\phi_*: T_x\mathcal{A} \to T_{\phi(x)}\mathcal{B}$ is a linear map. □

The important difference between the affine case and the general one is that in the affine case the induced map is a fixed map of the underlying vector spaces $\mathcal{V} \to \mathcal{W}$ whereas here the map ϕ_* depends on x. If it is necessary to specify where ϕ_* acts we shall write ϕ_{*x}, but we avoid this as far as possible. We write $\phi_* v$ or $\phi_{*x} v$ for the image of $v \in T_x\mathcal{A}$.

The computation of $\phi_{*x} v$ is often most conveniently carried out by choosing a curve through x which has v as its tangent vector there, finding the image of the curve under ϕ, and computing the tangent vector to the image curve. The line $t \mapsto x + tv$ is an obvious choice of curve for this computation.

The adjoint map is defined essentially as in the affine case. It is the linear map $\phi^*: T^*_{\phi(x)}\mathcal{B} \to T^*_x\mathcal{A}$ defined by

$$\langle v, \phi^*\beta\rangle = \langle \phi_* v, \beta\rangle$$

for any β in $T^*_{\phi(x)}\mathcal{B}$ and all v in $T_x\mathcal{A}$. Note that for any function f on $\mathcal{B}$

$$\langle v, \phi^*(df)\rangle = \langle \phi_* v, df\rangle = v^\phi f = v(f \circ \phi) = \langle v, d(f \circ \phi)\rangle$$

from which follows the important formula

$$\phi^* df = d(f \circ \phi).$$

As in the affine case, one may read off the coordinate expressions of ϕ_* and ϕ^*. Recalling that the components of a vector are given by its action on the coordinate functions, and introducing local, possibly curvilinear, coordinates (x^a) around x and (y^α) around $\phi(x)$, one obtains at once

$$(v^\phi)^\alpha = v^\phi(y^\alpha) = v\phi^\alpha = v^a \partial_a \phi^\alpha,$$

so that

$$(\phi_* v)^\alpha = v^a \frac{\partial \phi^\alpha}{\partial x^a},$$

giving the components $(\phi_* v)^\alpha$ of the induced vector in terms of the components v^a of the original vector, the Jacobian matrix (matrix of partial derivatives) $(\partial\phi^\alpha/\partial x^a)$ being evaluated at $(x^b(x))$, and $\phi^\alpha(x^a) = y^\alpha \circ \phi$ being the coordinate presentation of ϕ.

Exercise 29. Obtain the same expression for $(\phi_*)^\alpha$ from the result of Exercise 27. □

Exercise 30. Show that

$$\phi_*(\partial_a) = \frac{\partial \phi^\alpha}{\partial x^a}\partial_\alpha \quad \text{and that} \quad \phi^*(dy^\alpha) = \frac{\partial \phi^\alpha}{\partial x^a} dx^a.$$ □

Exercise 31. Let $\phi\colon \mathcal{A} \to \mathcal{B}$ and let $v \in \mathcal{V}$, the space on which $\mathcal{A}$ is modelled. Show that

$$\phi(x + tv) = \phi(x) + t\phi_{*x}(v) + O_2,$$

where O_2 represents a vector of order 2 in t. Thus ϕ_* determines the best affine approximation to ϕ near x, in the sense of Taylor's series. □

Exercise 32. Show that if $\phi\colon \mathcal{A} \to \mathcal{B}$ and $\psi\colon \mathcal{B} \to \mathcal{C}$ are smooth maps then

$$(\psi \circ \phi)_* = \psi_* \circ \phi_* \quad \text{and} \quad (\psi \circ \phi)^* = \phi^* \circ \psi^*.$$ □

Exercise 33. Let $\phi\colon \mathcal{A} \to \mathcal{A}$ be bijective and have a smooth inverse. Show that ϕ_* and ϕ^* are isomorphisms and that $(\phi_*)^{-1} = (\phi^{-1})_*$ and $(\phi^*)^{-1} = (\phi^{-1})^*$, paying due attention to the domains and codomains of these linear maps. □

Exercise 34. Let ϕ be a map of a 2-dimensional affine space to itself given in terms of some coordinates by $(x^1, x^2) \mapsto ((x^1)^2 - (x^2)^2, 2x^1x^2)$. Compute $\phi_*(\partial_1)$ and $\phi_*(\partial_2)$, $\phi^*(dx^1)$ and $\phi^*(dx^2)$. □

8. Parallelism

In this section we exploit the natural identification of tangent spaces at different points of an affine space, described in Section 2, to establish the idea of parallelism of vectors at different points. We go on to introduce a more restricted idea, of parallelism of vectors along a curve, which is easily generalised to other spaces.

Complete parallelism and parallelism along a curve. As we explained earlier, the tangent spaces at different points of an affine space may be naturally identified with the vector space on which it is modelled, and therefore with each other. Thus given a tangent vector at a point of the affine space, one may draw a vector equal to it, in the sense of this identification, at any other point. The two vectors are said to be *parallel*, and this property of affine spaces, that they admit a criterion of equality of vectors at different points, is called *complete*, or *absolute*, *parallelism*. This property could have been inferred, from the definition of an affine space, for displacement vectors, but we prefer to point it out for tangent vectors, which will continue to be important throughout the book.

Except in special cases, manifolds are not endowed with complete parallelism, and a restricted idea of parallelism—parallelism along a curve—has turned out to be more appropriate to them. We introduce this idea next.

A *vector field V along a curve σ* in an affine space $\mathcal{A}$ is an assignment of an element $V(t)$ of the tangent space $T_{\sigma(t)}\mathcal{A}$ at each point $\sigma(t)$ (we use uppercase italic letters V, W and so on to denote vector fields, here and subsequently; they should be easily distinguished from the script letters $\mathcal{V}$, $\mathcal{W}$ and so on used to denote vector spaces). The components of $V(t)$, with respect to an affine or a curvilinear coordinate system, will be functions of t, which will be assumed to be smooth functions.

If a vector v is given at a point $\sigma(t_0)$ on the curve then by the natural identification mentioned above one may construct the parallel vector at every other point of the curve. By this construction v is said to be *parallelly transported* along σ, and the resulting vector field V along σ is called a *parallel vector field* (on σ). The construction depends only on the affine structure of $\mathcal{A}$, and does not rely on the existence of a metric; nor does it depend on the parametrisation of σ.

Exercise 35. Show that V is a parallel vector field along σ if and only if its components in an affine coordinate system are constants. □

Exercise 36. Show that if σ is an affine line $t \mapsto x_0 + tv$ then the field of tangent vectors to σ is a parallel vector field along σ. □

Equations of parallel transport. The components of a parallel vector field will not, in general, be constants in a curvilinear coordinate system. However, it is easy to calculate the condition which they must satisfy. Let σ be a curve in $\mathcal{A}$, V a parallel vector field along σ, $(\acute{x}^a)$ curvilinear coordinates in a patch which σ crosses, and (x^a) any affine coordinates. Let $\acute{V}^a(t)$ and $V^a(t)$ be the curvilinear and affine components of $V(t)$ respectively so that (Exercise 22)

$$V^b(t) = \frac{\partial x^b}{\partial \acute{x}^c}\acute{V}^c(t)$$

(the derivatives being evaluated at $\sigma(t)$). Then the V^b are constants. Differentiating with respect to t, and writing $(\acute{\sigma}^a)$ for the (curvilinear) coordinate presentation of σ, one obtains

$$0 = \frac{\partial x^d}{\partial \acute{x}^c}\frac{d\acute{V}^c}{dt} + \frac{\partial^2 x^d}{\partial \acute{x}^b \partial \acute{x}^c}\acute{V}^b \frac{d\acute{\sigma}^c}{dt}$$

where $d\acute{\sigma}^c/dt$ are the (curvilinear) components of the tangent vector to σ. On multiplying by $(\partial \acute{x}^a/\partial x^d)$ one obtains the *equations of parallel transport*

$$\frac{d\acute{V}^a}{dt} + \Gamma^a_{bc}\acute{V}^b\frac{d\acute{\sigma}^c}{dt} = 0$$

where

$$\Gamma^a_{bc} = \frac{\partial \acute{x}^a}{\partial x^d}\frac{\partial^2 x^d}{\partial \acute{x}^b \partial \acute{x}^c} = (\partial_d \acute{x}^a)(\acute{\partial}_b \acute{\partial}_c x^d)$$

are the *connection coefficients* (for the given system of curvilinear coordinates). The equations of parallel transport hold in any system of coordinates if we define the connection coefficients for affine coordinates to be zero.

Exercise 37. Show that the Γ^a_{bc} are unchanged if the chosen system of affine coordinates is replaced by another one. □

Exercise 38. Show that $\Gamma^a_{bc} = \Gamma^a_{cb}$. □

Exercise 39. Compute the equations of parallel transport for the spherical polar coordinates given in Exercise 24, and show that they are satisfied by the affine coordinate vector fields (expressed in spherical polar coordinates) along any curve. □

9. Covariant Derivatives

Covariant derivative of a vector field. The idea of parallelism along a curve may be exploited to define a derivative, along a curve, of any vector field given along the curve. This derivative is called the ***absolute***, or ***covariant***, ***derivative.*** The covariant derivative of a vector field V along σ is the vector field DV/Dt along σ defined by

$$\frac{DV}{Dt}(t) = \lim_{\delta \to 0} \frac{1}{\delta}\left(V(t+\delta)_{\parallel} - V(t)\right)$$

where $V(t+\delta)_{\parallel}$ is the vector at $\sigma(t)$ parallel to $V(t+\delta)$ (which is a vector at $\sigma(t+\delta)$). The limit process is carried out in $T_{\sigma(t)}\mathcal{A}$ and so the result is again an element of the same space.

Exercise 40. Let U, V and W be vector fields along σ and f a function on $\mathcal{A}$. Show that

$$\frac{D}{Dt}(U+V) = \frac{DU}{Dt} + \frac{DV}{Dt} \quad \text{and} \quad \frac{D}{Dt}(fW) = f\frac{DW}{Dt} + \frac{df}{dt}W$$

(equality at each point of σ being implied). □

Exercise 41. Show that if V is parallel along σ then $DV/Dt = 0$. □

Exercise 42. Show that the components DV^a/Dt of DV/Dt in any affine coordinate system are simply

$$\frac{DV^a}{Dt} = \frac{dV^a}{dt},$$

while the components $D\acute{V}^a/Dt$ of DV/Dt in a curvilinear coordinate system are given by

$$\frac{D\acute{V}^a}{Dt} = \frac{d\acute{V}^a}{dt} + \Gamma^a_{bc}\acute{V}^b\frac{d\acute{\sigma}^c}{dt}.$$ □

Exercise 43. Let $\tilde{\sigma} = \sigma \circ h$ be a reparametrisation of σ and let $\tilde{V} = V \circ h$ be the vector field along $\tilde{\sigma}$ obtained by reparametrising V. Show that

$$\frac{D\tilde{V}}{Dt} = \dot{h}\frac{DV}{Dt}.$$ □

Covariant derivative of a covector field. The natural identification with $\mathcal{V}^*$ of cotangent spaces at different points of $\mathcal{A}$ allows one to define parallel covectors in exactly the same way as parallel vectors were defined: two covectors at different points of $\mathcal{A}$ are said to be *parallel* covectors if they are identified with the same element of $\mathcal{V}^*$. Further, a *covector field* α along a curve σ is an assignment of an element $\alpha(t)$ of $T^*_{\sigma(t)}\mathcal{A}$ at each point $\sigma(t)$, and if the covectors assigned at different points are parallel the covector field is called *parallel* along σ. The components of a covector field along σ will be assumed to be smooth functions of t in any (affine or curvilinear) coordinate system.

Exercise 44. Show that α is a parallel covector field along σ if and only if its components in an affine coordinate system are constants. Show that in an arbitrary curvilinear coordinate system the components $\acute{\alpha}_a(t)$ of a parallel covector field α satisfy the equations

$$\frac{d\acute{\alpha}_a}{dt} - \Gamma^b_{ac}\acute{\alpha}_b\frac{d\acute{\sigma}^c}{dt} = 0.$$ □

These are the *equations of parallel transport* for a covector field.

The *covariant derivative* $D\alpha/Dt$ of a covector field is

$$\frac{D\alpha}{Dt}(t) = \lim_{\delta \to 0} \frac{1}{\delta}\left(\alpha(t+\delta)_{\parallel} - \alpha(t)\right)$$

where $\alpha(t+\delta)_{\parallel}$ is the covector at $\sigma(t)$ parallel to $\alpha(t+\delta)$ (which is a covector at $\sigma(t+\delta)$).

Exercise 45. Let α, β and γ be covector fields along σ and f a function on $\mathcal{A}$. Show that

$$\frac{D}{Dt}(\alpha+\beta) = \frac{D\alpha}{Dt} + \frac{D\beta}{Dt} \quad \text{and} \quad \frac{D}{Dt}(f\gamma) = f\frac{D\gamma}{Dt} + \frac{df}{dt}\gamma.$$

Show that if α is parallel along σ then $D\alpha/Dt = 0$. □

Exercise 46. Show that the components $D\acute{\alpha}_a/Dt$ of $D\alpha/Dt$ in any coordinate system are given by

$$\frac{D\acute{\alpha}_a}{Dt} = \frac{d\acute{\alpha}_a}{dt} - \Gamma^b_{ac}\acute{\alpha}_b\frac{d\acute{\sigma}^c}{dt},$$

where the connection coefficients are zero if the coordinates are affine. □

Exercise 47. Let V be a vector field and α a covector field along σ. Show that

$$\frac{d}{dt}\langle V, \alpha\rangle = \left\langle \frac{DV}{Dt}, \alpha \right\rangle + \left\langle V, \frac{D\alpha}{Dt} \right\rangle.$$ □

Summary of Chapter 2

A coordinate expression for a function $f\colon \mathcal{A} \to \mathbf{R}$ is the function f^x on $\mathbf{R}^n$ such that $f^x(x^a) = f$; $f^{\acute{x}} = f^x \circ \Phi^{-1}$, where Φ is the coordinate transformation from (x^a) to $(\acute{x}^a)$. A function is smooth if its coordinate expression is smooth, that is, has continuous partial derivatives of all orders, with respect to any affine coordinates.

A curve σ in $\mathcal{A}$ is a map $\mathbf{R} \to \mathcal{A}$ (or a map $I \to \mathcal{A}$, where I is an open interval of $\mathbf{R}$); it is smooth if its coordinate functions $\sigma^a = x^a \circ \sigma$ are.

The tangent space $T_x\mathcal{A}$ is the vector space of directional derivative operators at x, that is, maps v of functions to numbers satisfying: $v(af + bg) = avf + bvg$ (linearity); $v(fg) = (vf)g(x) + f(x)(vg)$ (Leibniz). Each curve σ defines a directional derivative $\dot{\sigma}(t)$, its tangent vector at $\sigma(t) = x$, by $\dot{\sigma}(t)f = d/dt(f \circ \sigma)(t)$. Tangent spaces are naturally isomorphic to $\mathcal{V}$ and to each other. For any affine coordinates the operators $\{\partial_a\}$ are a basis for the tangent space at any point.

Each smooth function f defines a covector df, an element of $T_x^*\mathcal{A}$, the cotangent space at x, by $\langle v, df\rangle = vf$. The cotangent space is the vector space dual to $T_x\mathcal{A}$. The cotangent spaces are naturally isomorphic to $\mathcal{V}^*$.

An affine map $\Lambda\colon \mathcal{A} \to \mathcal{B}$ induces linear maps $\Lambda_*\colon T_x\mathcal{A} \to T_{\Lambda(x)}\mathcal{B}$ by $(\Lambda_* v)f = v(f \circ \Lambda)$. The dual map $\Lambda^*\colon T^*_{\Lambda(x)}\mathcal{B} \to T_x^*\mathcal{A}$ satisfies $\Lambda^* df = d(f \circ \Lambda)$. The linear map Λ_* is essentially the linear part of Λ. Tangent spaces map in the same direction as Λ, cotangent spaces oppositely.

Curvilinear (non-affine) coordinates are in general defined only locally, on a coordinate patch. A local curvilinear coordinate system, or coordinate chart, for an affine space $\mathcal{A}$ is a bijective map ψ from the coordinate patch (an open subset of $\mathcal{A}$) to an open subset of $\mathbf{R}^n$ which is smooth with respect to any affine coordinate system. Coordinate functions, differentials and vectors are defined just as for affine coordinate systems, and have much the same properties. The basic transformation rules are: $\acute{v}^a = (\partial_b \acute{x}^a) v^b$ for the components of a vector; $\acute{c}_a = (\acute{\partial}_a x^b) c_b$ for the components of a covector.

Elements of $T_x\mathcal{A}$ and $T_{\hat{x}}\mathcal{A}$ which correspond to the same element of $\mathcal{V}$ are parallel. The absolute derivative of a vector field V along a curve σ is the vector field defined by $DV/Dt = \lim_{\delta \to 0}(V(t+\delta)_{\|} - V(t))/\delta$, where $V(t+\delta)_{\|}$ is the vector at $\sigma(t)$ parallel to the vector $V(t+\delta)$ at $\sigma(t+\delta)$. The components of DV/Dt in affine coordinates (x^a) are obtained by differentiating the components of V with respect to t. In terms of curvilinear coordinates $(\acute{x}^a)$, $D\acute{V}^a/Dt = d\acute{V}^a/dt + \Gamma^a_{bc}\acute{V}^b d\acute{\sigma}^c/dt$, where $\Gamma^a_{bc} = (\partial_d \acute{x}^a)(\acute{\partial}_b \acute{\partial}_c x^d)$ are the connection coefficients.

Notes to Chapter 2

1. Topology. From everyday perception of ordinary space one acquires an intuitive idea of nearness. In elementary calculus, and again in vector calculus, this idea is made more precise. It is then exploited in definitions of continuity and differentiability. In the more general context of this book a more general formulation of the same ideas is needed. This formulation makes it easy to exclude from consideration various awkward cases of little interest.

We begin the formulation by recalling from elementary calculus some ideas about subsets of the set $\mathbf{R}$ of real numbers. An *open interval* of $\mathbf{R}$ is a set $\{x \mid a < x < b\}$ where a and b are real numbers (and $a < b$). This interval will be denoted (a, b). In other words, an open interval is an unbroken segment of the real line, deprived of its endpoints.

An *open set* of $\mathbf{R}$ is a union of any number (not necessarily a finite number) of open intervals. For example the half-infinite intervals $\{x \mid a < x\}$, denoted (a, ∞), and $\{x \mid x < b\}$, denoted $(-\infty, b)$, are open sets. The whole of $\mathbf{R}$ is also an open set, and it is convenient to count the empty set $\emptyset$ as open.

It is not difficult to see that the intersection of a finite number of open sets is an open set. On the other hand, the intersection of $(-1, 1)$, $(-\frac{1}{2}, \frac{1}{2})$, $(-\frac{1}{3}, \frac{1}{3})$, $\ldots, (-\frac{1}{n}, \frac{1}{n}), \ldots$ comprises only the single point 0, which is not an open set.

The complement of an open set is called a *closed set*. For example, the *closed interval* $\{x \mid a \leq x \leq b\}$, which is the complement of $(-\infty, a) \cup (b, \infty)$, is a closed set, denoted by $[a, b]$. In other words, a closed interval is an unbroken segment of the real line, including its end points. It is very often the case that open sets are defined by inequalities, closed sets by equalities. In particular, a single point is a closed set.

Abstraction from these ideas about subsets of $\mathbf{R}$ leads to the definition of a topological space and a topology, as follows. A set $\mathcal{S}$ is called a *topological space* if there is given a collection T of subsets of $\mathcal{S}$, called a *topology* for $\mathcal{S}$, with the properties

(1) $\mathcal{S}$ is in T and the empty set $\emptyset$ is in T

(2) the union of any number of elements of T is also in T

(3) the intersection of any finite number of elements of T is also in T.

It follows at once from the preceding discussion that, with the collection of open sets for T, the real line $\mathbf{R}$ is a topological space. This choice of T is called the *usual topology* for $\mathbf{R}$. In the general case the sets in the collection T are also called *open sets* of S. Other topologies for $\mathbf{R}$ are possible besides the usual one. For example, if every subset of $\mathbf{R}$ is included in T, the conditions for a topology are satisfied, and in particular, each point is an open set. This is called the *discrete topology*. At the other extreme, if there are no open sets but the empty set $\emptyset$ and the whole of $\mathbf{R}$, the conditions are again satisfied. This is called the *trivial topology*. These constructions may evidently be applied to any set S.

In general, an arbitrary subset of a topological space need be neither open nor closed. For example, in the case of $\mathbf{R}$ with the usual topology, the union of an open interval with a point outside it is neither open nor closed. An instance is the half-open interval $(a, b] = \{x \mid a < x \leq b\}$, with one end-point included but not the other.

A subcollection T_0 of T is called a *basis* for the topology if every open set is a union of members of T_0. In the case of $\mathbf{R}$ with the usual topology, the original construction shows that the collection of open intervals constitutes a basis.

The power of the idea of a topological space begins to become apparent when one considers maps between such spaces. Continuous maps, which generalise the idea of a continuous function, are usually the only maps of interest. Let S and T be topological spaces and let a be a point of S. A map $f: S \to T$ is *continuous at* a if for every neighbourhood Q of $f(a)$ there is a neighbourhood P of a such that $f(P) \subset Q$. A map is called *continuous* if it is continuous at every point of its domain. In order for a map to be continuous it must have the property that the pre-image of each open subset of its codomain is an open subset of its domain.

Two topological spaces are indistinguishable, as far as their topological properties are concerned, if there is a bijective continuous map from one to the other, with a continuous inverse. Such a map is called a *homeomorphism*, and spaces connected by a homeomorphism are said to be *homeomorphic*.

There are certain topologies which are generally appropriate to subsets and product sets. If S is a topological space and T is a subset of S, the *induced topology* on T is the collection of sets $Q \cap T$, where Q is any open set of S. For example, if $\mathbf{R}$ has the usual topology and $[a, b]$ is a closed interval, the induced topology on $[a, b]$ has a basis consisting of

(1) open subintervals of $[a, b]$
(2) half-open intervals $[a, x)$, with $x \leq b$
(3) half-open intervals $(x, b]$, with $a \leq x$.

If S_1 and S_2 are topological spaces with topologies T_1 and T_2 respectively, the *product topology* on their Cartesian product $S_1 \times S_2$ is the topology with basis $\{Q_1 \times Q_2 \mid Q_1 \in \mathsf{T}_1, Q_2 \in \mathsf{T}_2\}$. This definition generalises to a product with any number of factors.

The product topology on $\mathbf{R}^m = \mathbf{R} \times \mathbf{R} \times \cdots \times \mathbf{R}$ (m factors) is called the *usual topology* for $\mathbf{R}^m$. It has a basis consisting of hypercubes without their boundaries

$$\{(\xi^1, \xi^2, \ldots, \xi^m) \mid \xi^a \in I_a \quad a = 1, 2, \ldots, m\},$$

where the I_a are open intervals of $\mathbf{R}$. Each *open ball*

$$\{(\xi^1, \xi^2, \ldots, \xi^m) \mid (\xi^1 - \xi_0^1)^2 + (\xi^2 - \xi_0^2)^2 + \cdots + (\xi^m - \xi_0^m)^2 < r^2\}$$

is an open set in $\mathbf{R}^m$. The $(m-1)$-*sphere*

$$\{(\xi^1, \xi^2, \ldots, \xi^m) \mid (\xi^1 - \xi_0^1)^2 + (\xi^2 - \xi_0^2)^2 + \cdots + (\xi^m - \xi_0^m)^2 = r^2\}$$

is an example of a closed set. If $f: \mathbf{R}^m \to \mathbf{R}$ is continuous then for any $c \in \mathbf{R}$ the set $\{(\xi^1, \xi^2, \ldots, \xi^m) \mid f(\xi^1, \xi^2, \ldots, \xi^m) < c\}$ is open in $\mathbf{R}^m$; open sets are often defined as the solution sets of strict inequalities involving continuous functions. The definition of a closed set, on the other hand, usually involves weak inequalities or equalities. Another example of a closed set in $\mathbf{R}^m$ is the hyperplane

$$\{(\xi^1, \xi^2, \ldots, \xi^m) \mid a_1\xi^1 + a_2\xi^2 + \cdots + a_m\xi^m = 1\}.$$

No lower-dimensional subset of $\mathbf{R}^m$ can be open: thus an interval (a, b) of the ξ^m-axis, with $\xi^1 = \xi^2 = \cdots \xi^{m-1} = 0$, is open in the induced topology of the ξ^m-axis, but is neither open nor closed in the usual topology of $\mathbf{R}^m$.

A standard reference is Kelley [1955].

2. The inverse function theorem. Let Φ be a map from $\mathbf{R}^n$ to $\mathbf{R}^n$ which is smooth in an open neighbourhood of a point ξ of its domain at which the Jacobian matrix $\Phi'(\xi)$ is non-singular. The inverse function theorem asserts that Φ is invertible in some neighbourhood of ξ, with a smooth inverse. The size of the neighbourhood depends on the detailed form of Φ. Proofs will be found in many books on advanced calculus. There is one near to the point of view of this book in Spivak [1965].

3. VECTOR FIELDS AND FLOWS

The steady flow of a fluid in a Euclidean space is an appropriate model for the ideas developed in this chapter. The essential ideas are

(1) that the fluid is supposed to fill the space, so that there is a streamline through each point

(2) that the velocity of the fluid at each point specifies a vector field in the space

(3) that the movement of the fluid along the streamlines for a fixed interval of time specifies a transformation of the space into itself.

The fluid flow is thus considered both passively, as a collection of streamlines, and actively, as a collection of transformations of the space. Besides these integral appearances it also appears differentially, through its velocity field.

Let ϕ_t denote the transformation of the space into itself by movement along the streamlines during a time interval of length t. To be specific, given any point x of the space, $\phi_t(x)$ is the point reached by a particle of the fluid, initially at x and flowing along the streamline of the fluid through x, after the lapse of a time t. The set of such transformations has the almost self-evident properties

(1) ϕ_0 is the identity transformation

(2) $\phi_s \circ \phi_t = \phi_{s+t}$.

A set of transformations with these two properties (for all s and t) is called a *one-parameter group of transformations*. The study of such transformations, and of the streamlines and vector fields associated with them, forms the subject matter of this chapter.

We begin in Section 1 with a special case, in which the transformations ϕ_t are all affine transformations. The general case is developed in Sections 2 to 4. In Sections 5 to 7 we introduce a new and powerful construction, the Lie derivative, which measures the deformation of a moving object relative to one which is moved along the streamlines. In Section 8 we develop the idea of vector fields as differential operators, and exhibit some of their properties from this point of view.

1. One-parameter Affine Groups

In this section we develop the ideas introduced above for the case in which all the transformations involved are affine transformations. We begin with a simple example. Let $\mathcal{A}$ denote an affine space of dimension n modelled on a vector space $\mathcal{V}$ and, as in Section 4 of Chapter 1, let $\tau_v: x \mapsto x + v$ denote the translation of $\mathcal{A}$ by v, where v is any vector in $\mathcal{V}$. Then $\phi_t = \tau_{tv}$ is a one-parameter group of transformations; it is easily seen that ϕ_0 is the identity transformation, and that $\phi_s \circ \phi_t = \phi_{s+t}$, as the definition requires. Moreover, the transformations are smooth, in the following sense: in any affine coordinate system the coordinates of $\phi_t(x)$ are $(x^a + tv^a)$, where (x^a) are the affine coordinates of x and v^a the components of v; so the coordinates of $\phi_t(x)$ are certainly smooth functions of the x^a and t jointly. This joint smoothness is the smoothness condition which will be imposed later on one-parameter groups in general, although a transformation of a general one-parameter group will not have as simple a coordinate representation as a translation does, of course.

Let x be any point of $\mathcal{A}$. The set of points into which x is transformed by the translations ϕ_t, as t varies, is described by a curve (actually a line) denoted σ_x and called the *orbit* of x under ϕ_t. Thus

$$\sigma_x(t) = \phi_t(x) = x + vt.$$

Note that we write σ_x for the orbit, which is a map $\mathbf{R} \to \mathcal{A}$, with x fixed, and ϕ_t for the transformation, which is a map $\mathcal{A} \to \mathcal{A}$, with t fixed.

Every point of $\mathcal{A}$ lies on (the image set of) an orbit; moreover if y lies on the orbit of x then σ_x and σ_y are congruent (Chapter 2, Section 2), because $y = x+vs = \sigma_x(s)$, for some s; and then for all t, $\sigma_y(t) = y + tv = x + (s+t)v = \sigma_x(s+t)$. The orbits may be partitioned into disjoint sets, any two orbits in the same set being congruent; every point lies on the orbits of a congruent set, and no two orbits from distinct congruent sets intersect.

The definition of an orbit extends in an obvious manner to any one-parameter group of transformations, and the property of orbits just described continues to hold. A collection of curves on $\mathcal{A}$, such that each point of $\mathcal{A}$ lies on the curves of a congruent set, and no two curves from distinct congruent sets intersect, is called a *congruence* of curves. With any congruence one may associate a unique tangent vector at each point of $\mathcal{A}$, namely the tangent at that point to any one of the set of congruent curves through it; in particular this is true for the congruence of orbits of a one-parameter group.

The congruence is the geometrical abstraction of the collection of streamlines introduced in the context of fluid flow at the beginning of this chapter. Notice that, in this context, the congruence property of streamlines is a consequence (indeed, more or less a definition) of the steadiness of the flow: in effect, a particle initially at x, and one which arrives there a time t later, follow the same streamline, but separated always by a length of time t. Equally, the one-parameter group property is a consequence of the assumed steadiness of the flow.

The abstraction of the velocity field of the fluid is a vector field. A choice of tangent vector at each point of $\mathcal{A}$ is called a *vector field* on $\mathcal{A}$. Associated with any one-parameter group there is, as we have seen, a vector field, namely the field of tangent vectors to its orbits. This is often called the *generator*, *infinitesimal generator*, or *generating vector field* of the one-parameter group.

Suppose now that an affine coordinate system has been chosen for $\mathcal{A}$, and consider the generator of the one-parameter group of translations of $\mathcal{A}$ by te_a where e_a is one of the basis vectors of the underlying vector space from which the coordinates are built. This generator is obtained by choosing, at each point of $\mathcal{A}$, the coordinate vector ∂_a; so, naturally, we denote it by the same symbol: in future, ∂_a (or $\partial/\partial x^a$) may denote either a coordinate vector at a point, or a coordinate vector field; which of the two is meant will generally be clear from the context.

An arbitrary vector field V may be expressed in terms of the coordinate vector fields of an affine coordinate system in the form $V = V^a\partial_a$, where the V^a are functions on $\mathcal{A}$. For any $x \in \mathcal{A}$, $V^a(x)$ are the components of V_x, the tangent vector at x determined by V. The functions V^a are therefore called the *components* of V with respect to the (affine) coordinate system.

Exercise 1. Show that if $\acute{x}^a = k^a_b x^b + c^a$ is a new affine coordinate system then $\acute{\partial}_a = (k^{-1})^b_a \partial_b$, and if $V = V^a\partial_a = \acute{V}^a\acute{\partial}_a$ then $\acute{V}^a = k^a_b V^b$. □

We shall have to deal only with vector fields whose components with respect to one (and thus, by the exercise, to any) affine coordinate system are smooth functions. Indeed, for our immediate concerns we shall need vector fields whose affine components are very simple functions.

A vector field with constant affine components is a field of parallel vectors. If such a field is given, one can reconstruct the congruence to which it is tangent—a congruence of parallel lines—and the one-parameter group which it generates—a one-parameter group of translations. It is taken for granted in fluid dynamics that this reconstruction is possible in general, not only for a parallel field—that if one knows the velocity field then one can determine the path of each fluid particle and the motion of the fluid, at least in principle. This presumption is in fact justified, as we shall explain later.

Before going on to describe one-parameter groups of affine transformations in general, we give two more examples.

(1) A family of affine transformations of a 2-dimensional affine space $\mathcal{A}$ is given in affine coordinates by

$$\phi^1_t(x^1,x^2) = x^1\cos t - x^2\sin t \qquad \phi^2_t(x^1,x^2) = x^1\sin t + x^2\cos t.$$

If $\mathcal{A}$ has a Euclidean structure these are just rotations about the origin, t is the angle of rotation, and the orbits are circles. It is not necessary to invoke a Euclidean structure in order to define these transformations, however. Observe that $\phi^a_0(x^1,x^2) = x^a$, $a = 1,2$, and that

$$\begin{aligned}
&\phi^1_s(\phi^1_t(x^1,x^2),\phi^2_t(x^1,x^2))\\
&\qquad = (x^1\cos t - x^2\sin t)\cos s - (x^1\sin t + x^2\cos t)\sin s\\
&\qquad = x^1\cos(s+t) - x^2\sin(s+t) = \phi^1_{s+t}(x^1,x^2),
\end{aligned}$$

and similarly

$$\begin{aligned}
&\phi^2_s(\phi^1_t(x^1,x^2),\phi^2_t(x^1,x^2))\\
&\qquad = (x^1\cos t - x^2\sin t)\sin s + (x^1\sin t + x^2\cos t)\cos s\\
&\qquad = x^1\sin(s+t) + x^2\cos(s+t) = \phi^2_{s+t}(x^1,x^2);
\end{aligned}$$

moreover, ϕ^1_t and ϕ^2_t are smooth functions of t, x^1 and x^2. The given family of transformations is therefore a one-parameter group of affine transformations. To find its generating vector field V, observe that the orbit of the point whose coordinates are (x^1,x^2) is given in coordinates by $t \mapsto (x^1\cos t - x^2\sin t, x^1\sin t +$

$x^2 \cos t)$, and its tangent vector at $t = 0$ has components

$$\frac{d}{dt}\left(x^1 \cos t - x^2 \sin t\right)_{t=0} = -x^2 \qquad \text{and} \qquad \frac{d}{dt}\left(x^1 \sin t + x^2 \cos t\right)_{t=0} = x^1.$$

Thus V is the vector field whose value at a point with coordinates (x^1, x^2) is given by $-x^2\partial_1 + x^1\partial_2$. In fact, this expression defines the vector field, if one interprets x^1 and x^2 as coordinate functions.

(2) A family of affine transformations of $\mathcal{A}$ is given in affine coordinates by

$$\phi_t^1(x^1, x^2) = x^1 + tx^2 \qquad \phi_t^2(x^1, x^2) = x^2.$$

Observe that $\phi_0^a(x^1, x^2) = x^a$, and that

$$\phi_s^1(\phi_t^1(x^1, x^2), \phi_t^2(x^1, x^2)) = (x^1 + tx^2) + sx^2 = x^1 + (s+t)x^2 = \phi_{s+t}^1(x^1, x^2)$$

and

$$\phi_s^2(\phi_t^1(x^1, x^2), \phi_t^2(x^1, x^2)) = x^2 = \phi_{s+t}^2(x^1, x^2);$$

moreover ϕ_t^1 and ϕ_t^2 are smooth functions of t, x^1 and x^2. Therefore this family is also a one-parameter group of affine transformations. To find its generator V, observe that the orbit of the point whose coordinates are (x^1, x^2) is given by $t \mapsto (x^1 + tx^2, x^2)$ and its tangent vector at $t = 0$ is therefore $x^2\partial_1$; again, this is the required expression for the vector field V if one interprets x^2 as a coordinate function, rather than just a coordinate.

One-parameter affine groups defined. A set $\{\phi_t \mid t \in \mathbf{R}\}$ of affine transformations, such that ϕ_0 is the identity and $\phi_s \circ \phi_t = \phi_{s+t}$ for every $s, t \in \mathbf{R}$, is called a *one-parameter group of affine transformations*, or *one-parameter affine group* for short. We require that the functions $(t, x^a) \mapsto \phi_t^b(x^a)$ representing the transformations of a one-parameter affine group with respect to affine coordinates (x^a) be smooth functions of all the variables (t, x^a). Clearly if this condition is satisfied for one affine coordinate system it is satisfied for all. In fact, all that is required in practice is that if $\phi_t^b(x^a) = \lambda_c^b(t)x^c + d^b(t)$ then each λ_c^b and d^b should be a smooth real function; we give the smoothness condition in its more general form to make it obviously consistent with what comes later.

We have already furnished several examples of one-parameter affine groups. We shall frequently use ϕ_t, or some similar expression, to denote a one-parameter affine group, though strictly speaking it should represent one specific transformation drawn from the group.

Exercise 2. Show that the transformations given in affine coordinates by $\phi_t^1(x^1, x^2) = e^{k_1 t}x^1$, $\phi_t^2(x^1, x^2) = e^{k_2 t}x^2$ form a one-parameter affine group. Show that the transformations $(x^1, x^2) \mapsto (k_1 t x^1, k_2 t x^2)$, on the other hand, do not form a one-parameter affine group. □

Exercise 3. Suppose that $\phi_t: x \mapsto x_0 + \lambda_t(x - x_0) + v_t$ defines a one-parameter group of affine transformations of $\mathcal{A}$. Deduce that λ_t must be a one-parameter group of linear transformations of the underlying vector space $\mathcal{V}$, and that $v_{s+t} = \lambda_s(v_t) + v_s = \lambda_t(v_s) + v_t$. Observe that if ϕ_t leaves x_0 fixed for all t then $v_t = 0$ for all t. □

Exercise 4. Show that any transformation of a one-parameter group has an inverse, which is obtained by changing the sign of the parameter: $(\phi_t)^{-1} = \phi_{-t}$. □

Exercise 5. Let ϕ_t be a one-parameter group of affine transformations of $\mathcal{A}$. Define $\phi\colon \mathbf{R}\times\mathcal{A}\to\mathcal{A}$ by $\phi(t,x)=\phi_t(x)$. Show that $\phi(0,x)=x$ and that $\phi(s,\phi(t,x))=\phi(t,\phi(s,x))=\phi(s+t,x)$. □

Exercise 6. Show that the following constitutes an alternative definition of a one-parameter affine group: a family $\{\phi_t \mid t\in\mathbf{R}\}$ of affine transformations is a one-parameter affine group if the map $t\mapsto\phi_t$ is a homomorphism of $\mathbf{R}$ (the additive group of real numbers) into the group of affine transformations. □

Exercise 7. Let v be a given vector in $\mathcal{V}$, α a given covector in $\mathcal{V}^*$, and x_0 a given point of $\mathcal{A}$. Show that the one-parameter family of transformations given by $x\mapsto x+t\langle x-x_0,\alpha\rangle v$ is a one-parameter group if and only if $\langle v,\alpha\rangle=0$. □

The transformations defined in this exercise, when they form a one-parameter affine group, are called *shears* along the hyperplanes $\langle x-x_0,\alpha\rangle$ = constant in the v direction. The one-parameter group described in Example (2) is a special case.

Exercise 8. Let ϕ_t be a one-parameter group of affine transformations of $\mathcal{A}$, and let $V=V^a\partial_a$ be its generating vector field. Show that in affine coordinates $V^a(x)=d/dt(\phi_t^a(x))_{t=0}$ and that $V^a(\phi_s(x))=d/dt(\phi_t^a(x))_{t=s}$. □

What vector fields can be the generators of one-parameter affine groups? We have shown that one-parameter groups of translations are generated by parallel vector fields. Translations move all points equally: a transformation which leaves a point fixed might be considered the opposite extreme. Consider therefore a one-parameter group of affine transformations ϕ_t which leaves fixed the point x_0: $\phi_t(x_0)=x_0$ for all t. Each such transformation is determined by its linear part λ_t (Exercise 3): $\phi_t(x)=x_0+\lambda_t(x-x_0)$, with $\lambda_s\circ\lambda_t=\lambda_{s+t}$. In affine coordinates, with x_0 as origin, each λ_t is represented by a matrix L_t with $L_sL_t=L_{s+t}$. To find the generator, one has only to find the tangent vector to an orbit: $V^a(x)\partial_a=d/dt\big((L_t)^a_bx^b\big)_{t=0}$, so that in affine coordinates the generator has the form

$$V=A^a_bx^b\partial_a,$$

where the matrix A, which is constant, is given by $A=d/dt(L_t)(0)$. Although the matrices L_t must be non-singular, A may be singular. Therefore the generator V must be a linear homogenous vector field, which means to say that in affine coordinates, with the fixed point as origin, its components are linear homogeneous functions of the coordinates. Note that the orbit of the fixed point is a constant curve, and that the generator vanishes at the fixed point.

Exercise 9. The set of vectors $\{e_1,e_2\}$ is a basis for $\mathcal{V}$, $\{\theta^1,\theta^2\}$ the dual basis for $\mathcal{V}^*$. A set of transformations of $\mathcal{A}$ is given by $\phi_t(x)=x_0+e^{k_1t}\langle x-x_0,\theta^1\rangle e_1+e^{k_2t}\langle x-x_0,\theta^2\rangle e_2$, where k_1 and k_2 are real constants and x_0 is a chosen point of $\mathcal{A}$. Show that ϕ_t is a one-parameter affine group, and that, in affine coordinates with x_0 as origin and $\{e_1,e_2\}$ as coordinate basis, ϕ_t is represented by the matrix $\operatorname{diag}(e^{k_1t},e^{k_2t})$. Show also that the generator of ϕ_t is $k_1x^1\partial_1+k_2x^2\partial_2$. □

This is an example of a one-parameter group of *dilations*. The same group is given in coordinate form in Exercise 2.

Exercise 10. Verify that, under a transformation from one affine coordinate system to another with the same origin, the vector field $V=A^a_bx^b\partial_a$ remains linear homogeneous, but that if the origin is changed then the components of V, though still linear, are no longer homogeneous. □

Exercise 11. A one-parameter group of affine transformations of a 1-dimensional affine space, which leaves a point fixed, is given in affine coordinates with the fixed point as origin by $x^1 \mapsto \lambda(t)x^1$, where λ is a real function with $\lambda(0) = 1$, $\lambda(s)\lambda(t) = \lambda(s+t)$. Show that its generator is $Ax^1\partial_1$, where $A = \dot\lambda(0)$. From the one-parameter group property, infer that $\dot\lambda(t) = A\lambda(t)$, and deduce that $\lambda(t) = e^{At}$, so that the one-parameter group must have the form $x^1 \mapsto e^{At}x^1$. □

The result of the preceding exercise generalises to any number n of dimensions, in terms of the matrix exponential. Let ϕ_t be a one-parameter affine group of $\mathcal{A}$ which leaves fixed the point x_0. Let the matrix L_t represent ϕ_t in affine coordinates with x_0 as origin. Then the generator of ϕ_t is $A^a_b x^b \partial_a$ where the matrix A is given by $A = d/dt(L_t)(0)$. By the one-parameter group property, $d/dt(L_t) = AL_t$. It is known that this matrix differential equation has a unique solution such that $L_0 = I_n$ (the $n \times n$ identity matrix); it is called the *matrix exponential* of A, written e^{tA} or $\exp(tA)$.

Exercise 12. Show that for each t, the matrix $\exp(tA)$ is non-singular, that for each s and t, $\exp(sA)\exp(tA) = \exp((s+t)A)$, and that the matrix exponential has Taylor expansion
$$\exp(tA) = I_n + tA + \frac{1}{2!}t^2A^2 + \frac{1}{3!}t^3A^3 + \cdots$$ □

Exercise 13. Show, by means of the Taylor expansion described in Exercise 12, that the exponentials of the matrices
$$\begin{pmatrix} 0 & 1 \\ 0 & 0 \end{pmatrix}, \quad \begin{pmatrix} 0 & 1 \\ -1 & 0 \end{pmatrix} \quad \text{and} \quad \begin{pmatrix} k_1 & 0 \\ 0 & k_2 \end{pmatrix}$$
are
$$\begin{pmatrix} 1 & t \\ 0 & 1 \end{pmatrix}, \quad \begin{pmatrix} \cos t & \sin t \\ -\sin t & \cos t \end{pmatrix} \quad \text{and} \quad \begin{pmatrix} e^{k_1 t} & 0 \\ 0 & e^{k_2 t} \end{pmatrix}$$
respectively. □

Exercise 14. Let $\phi_t(x) = x_0 + \lambda_t(x - x_0) + v_t$ be a one-parameter group of affine transformations which do not necessarily leave any point fixed. Show that in affine coordinates with x_0 as origin the generator of ϕ_t has the form $V = (A^a_b x^b + B^a)\partial_a$, where $A = d/dt(L_t)(0)$ and $B^a = d/dt(v^a_t)(0)$. Show also that if V vanishes at some point x_1, then in affine coordinates with x_1 as origin each component V^a is linear homogeneous, but that if V does not vanish anywhere then no choice of affine coordinates will make its components homogeneous. Verify that the translations correspond to the case $\lambda_t = \mathrm{id}_\mathcal{V}$, $A = 0$. □

Exercise 14 answers the question raised above: the generator of a one-parameter affine group is a linear vector field (when expressed in terms of affine coordinates), and in general an inhomogeneous one. It may be shown that, conversely, every linear vector field generates a one-parameter affine group.

2. One-parameter Groups: the General Case

In this section we discuss one-parameter groups of transformations which are not necessarily affine.

We begin with a simple example of a non-affine one-parameter group of transformations of a 2-dimensional affine space. Let ϕ_t be the transformation given in affine coordinates (x^1, x^2) by

$$\phi_t^1(x^1, x^2) = x^1 + t \qquad \phi_t^2(x^1, x^2) = x^2 - \sin x^1 + \sin(x^1 + t).$$

It is easily checked that ϕ_0 is the identity map and that $\phi_s \circ \phi_t = \phi_{s+t}$ for all s and t. Moreover, the functions ϕ_t^a are smooth (in fact analytic) functions of t, x^1 and x^2. Thus ϕ_t is a one-parameter group of transformations. Though ϕ_t is not a one-parameter affine group one may still define its orbits and its generating vector field. The orbit σ_x of a point x is the curve given by $\sigma_x(t) = \phi_t(x)$ as before; and the generating vector field V is the field of tangents to the orbits. To compute V, observe that the orbit of a point x is given in coordinates by $t \mapsto (x^1+t, x^2-\sin x^1+\sin(x^1+t))$, a curve whose tangent vector at $t = 0$ is $\partial_1+\cos x^1\partial_2$, so that

$$V = \partial_1 + \cos x^1\partial_2.$$

Exercise 15. Show that the set of transformations given in coordinates by

$$\phi_t^1(x^1, x^2, x^3) = (x^1 + t\sin x^3)\cos t + (x^2 - \sin x^3 + t\cos x^3)\sin t$$
$$\phi_t^2(x^1, x^2, x^3) = -(x^1 - \cos x^3 + t\sin x^3)\sin t + (x^2 + t\cos x^3)\cos t$$
$$\phi_t^3(x^1, x^2, x^3) = x^3 + t$$

is a one-parameter group of transformations whose generator is

$$x^2\partial_1 - (x^1 - 2\cos x^3)\partial_2 + \partial_3. \qquad \square$$

Generating a one-parameter group from a vector field. We have shown by examples how to derive from a one-parameter group, affine or not, a vector field, its generator. As the name implies, the vector field may be used, on the other hand, to generate the one-parameter group. Suppose that one is given a smooth vector field V (one whose components with respect to any affine coordinates are smooth functions): then by turning the calculation of the generator on its head one obtains a one-parameter group of which V is the generating vector field. (Actually this process may not be completely successful for technical reasons which will be explained below, but to begin with we wish to describe the general principles.) The first step is to find the congruence of curves to which V is tangent; these will be the orbits of points under the action of the one-parameter group. These curves are called, in this context, *integral curves* of V: a curve σ is an integral curve of a vector field V if, for each t in its domain, $\dot\sigma(t) = V_{\sigma(t)}$. One can find the integral curves in terms of the coordinate presentation: in order for σ to be an integral curve of V its components σ^a must satisfy the differential equation

$$\frac{d\sigma^a}{dt} = V^a(\sigma^b) \qquad \text{where } V = V^a\partial_a.$$

To find the integral curves of V, therefore, one solves this system of first order differential equations.

We shall illustrate the process of generating a one-parameter group from a vector field by taking as vector field $\partial_1 + \cos x^1\partial_2$ and reconstructing its one-parameter group, given at the beginning of this section. The conditions that σ be an integral curve of this vector field are

$$\frac{d\sigma^1}{dt} = 1 \qquad \frac{d\sigma^2}{dt} = \cos\sigma^1.$$

Integrating the first equation one obtains $\sigma^1(t) = t + c^1$, and, substituting in the second equation and integrating again, $\sigma^2(t) = \sin(t + c^1) + c^2$. The constants of integration c^1 and c^2 are constant on a particular integral curve, but serve to distinguish different integral curves. Note that $\sigma^1(0) = c^1$, $\sigma^2(0) = \sin c^1 + c^2$. To obtain the integral curve which passes through the point (x^1, x^2) when $t = 0$, one evidently has to take $c^1 = x^1$, $c^2 = x^2 - \sin x^1$. The curve is then $t \mapsto (t + x^1, \sin(t + x^1) + x^2 - \sin x^1)$, which is indeed an orbit of the one-parameter group we started with.

Exercise 16. A vector field on a 2-dimensional affine space is given in affine coordinates (x^1, x^2) by $V = \operatorname{sech}^2 x^2 \partial_1 + \partial_2$. Find its integral curves, verify that they form a congruence, and construct the one-parameter group whose orbits are the curves of this congruence. Describe the orbits. □

Exercise 17. Let (x^1, x^2) be affine coordinates. Find the curves with tangent vector field $\cos^2 x^1 \partial_1$ and determine whether they are the orbits of a one-parameter group. If so, determine the transformations in the group. □

Exercise 18. Let (x^1, x^2) be affine coordinates on a 2-dimensional affine space. Define (global) non-affine coordinates $(\acute{x}^1, \acute{x}^2)$ by $\acute{x}^1 = x^1$, $\acute{x}^2 = x^2 + \sin x^1$. Show that the set of transformations given by

$$\acute{\phi}_t^1(\acute{x}^1, \acute{x}^2) = \acute{x}^1 + t, \qquad \acute{\phi}_t^2(\acute{x}^1, \acute{x}^2) = \acute{x}^2 - \sin \acute{x}^1 + \sin(\acute{x}^1 + t)$$

is a one-parameter affine group, and identify these transformations. □

3. Flows

In the above example, we showed first how one may pass from a one-parameter group to the associated congruence of orbits and tangent vector field, and then how one may go back from the vector field to the congruence and to the one-parameter group. But, as we mentioned there, this last process is not always possible, even for smooth vector fields; the following exercise illustrates the difficulty that may occur. In order to deal with this difficulty we shall have to widen the definition of one-parameter group: the resulting object is called a flow.

Exercise 19. Let $V = (x^1)^2 \partial_1$ (a vector field on a 1-dimensional affine space, with affine coordinate x^1). Show that the integral curve of V which passes through the point with coordinate x_0^1 when $t = 0$ is given by $\sigma^1(t) = x_0^1(1 - t x_0^1)^{-1}$, where
 if $x_0^1 > 0$, t lies in the interval $(-\infty, 1/x_0^1)$
 if $x_0^1 = 0$, t may take any value, and
 if $x_0^1 < 0$, t lies in the interval $(1/x_0^1, \infty)$.
For each x^1, and for each t for which it makes sense, set $\phi_t^1(x^1) = x^1(1 - t x^1)^{-1}$. Show that these "transformations" ϕ_t have the properties that ϕ_0 is the identity transformation and that $\phi_s \circ \phi_t = \phi_{s+t}$ whenever both sides make sense; and show that V is tangent to the "orbits" $t \mapsto \phi_t(x)$. □

In this exercise, $\phi_t(x)$ is well-defined for all t only if $x^1(x) = 0$. But there is nothing pathological about the vector field or the orbits. The vector field is smooth and, where they are defined, the transformations ϕ_t are smooth in t and x^1. But the orbits "get to infinity in a finite time". As a consequence, ϕ_t is not a one-parameter group of transformations. However, this situation arises so easily and so frequently that it cannot be excluded from consideration. We must therefore introduce modifications of the idea of one-parameter group which will make it possible to deal with it.

Exercise 20. A vector field on a 2-dimensional affine space has in affine coordinates the expression $V = (x^1)^2\partial_1 + (x^2)^2\partial_2$. Find coordinate expressions for its integral curves, and find their domains. □

Flows and their congruences. Examples like the ones in Exercises 19 and 20 are incompatible with the idea of a one-parameter group of transformations unless the requirement that the transformations be defined for all values of the parameter is given up. The set of transformations then ceases to be a group in the usual sense; it is called a "local group" or "flow". The idea of the modified definition is that each orbit should be specified for some range of values of the parameter, but not necessarily for all values. It is natural to specify values around 0, since this corresponds to the always possible identity transformation. The definition is framed along lines suggested by the construction in Exercise 5, which makes it easy to impose a condition of smoothness in the coordinates and the parameter jointly. A *flow*, or *local one-parameter group of local transformations*, on an affine space $\mathcal{A}$ is a smooth map $\phi: \mathcal{D} \to \mathcal{A}$, where $\mathcal{D}$ is an open set in $\mathbf{R} \times \mathcal{A}$ which contains $\{0\} \times \mathcal{A}$ and is such that for each $x \in \mathcal{A}$ the set $\{\, t \in \mathbf{R} \mid (t,x) \in \mathcal{D}\,\}$ is an open interval (possibly infinite) in $\mathbf{R}$, and ϕ satisfies the conditions

(1) $\phi(0,x) = x$ for each $x \in \mathcal{A}$

(2) $\phi(s,\phi(t,x)) = \phi(s+t,x)$ whenever both sides are meaningful.

For each $x \in \mathcal{A}$ a smooth curve σ_x, with domain an open interval containing 0, may be defined, by $\sigma_x(t) = \phi(t,x)$. For each t a smooth map ϕ_t of an open subset of $\mathcal{A}$ into $\mathcal{A}$ may be defined, by $\phi_t(x) = \phi(t,x)$. However, there need be no x for which $\sigma_x(t)$ is specified for all t, and no t except 0 for which $\phi_t(x)$ is specified for all x. In the special case that $\mathcal{D} = \mathbf{R} \times \mathcal{A}$, $\phi_t(x)$ is specified for all t and x, and then ϕ_t is a *one-parameter group of transformations* of $\mathcal{A}$, the provisions about smoothness being added to the definition originally given at the beginning of the chapter. Thus the idea of a flow includes the idea of a one-parameter group as a special case. We shall denote a flow ϕ or something similar.

Exercise 21. Show that the set of transformations constructed in Exercise 19 constitutes a flow. □

The curve σ_x is called the *orbit* of x under the flow ϕ. Thus $\sigma_x(t) = \phi_t(x)$, as in the case of a one-parameter group, but now it may be that σ_x is specified only on an open interval, not on the whole of $\mathbf{R}$. The idea of change of origin (Chapter 2, Section 2) has to be modified to take account of this. Let τ_s denote the translation $t \mapsto t+s$ of $\mathbf{R}$. If I is an open interval of $\mathbf{R}$, say $I = (a,b)$, and σ is a curve defined on I, then $\sigma \circ \tau_s$ is defined on $\tau_{-s}(I) = (a-s, b-s)$. A change of origin on σ is defined to be a reparametrisation $\sigma \circ \tau_s$, defined on $\tau_{-s}(I)$. The parameter value at a given image point is decreased by s, and the endpoints of the interval on which the curve is defined are also decreased by s. The tangent vector at any image point is unaffected.

Exercise 22. Devise a definition of change of origin for a curve defined on (a,∞) and for one defined on $(-\infty,b)$. □

As before, a set of curves which differ only by change of origin is called a *congruent* set. It would be possible, again as before, to call a collection of curves on $\mathcal{A}$, such that each point of $\mathcal{A}$ lies on a congruent set, and no two congruent sets intersect, a congruence of curves. However, it is more convenient to modify the definition in such a way that a flow may be associated with every congruence. Accordingly, a *congruence of curves* on an affine space $\mathcal{A}$ is defined to be a set of curves $\sigma_x: I_x \to \mathcal{A}$, one for each $x \in \mathcal{A}$, where for every x, I_x is an open interval containing 0, such that

(1) $\sigma_x(0) = x$

(2) each x lies on exactly one congruent set

(3) the set of points $\mathcal{D} = \bigcup_{x\in\mathcal{A}} I_x \times \{x\}$ is an open subset of $\mathbf{R} \times \mathcal{A}$ and the map $\sigma: \mathcal{D} \to \mathcal{A}$ by $(t,x) \mapsto \sigma_x(t)$ is represented by smooth functions when the curves are presented in any (affine or curvilinear) coordinate system.

The third of these conditions expresses the requirement that the curves be smooth and vary smoothly from point to point. In coordinates, σ will be represented by n functions σ^a of $n+1$ variables t and x^a, and interchange of the order of differentiation with respect to t and any x^a, as well as with respect to any two x^a, will be permissible.

Since the curves in each congruent set all have the same tangent vector at any point, a vector field may still be associated with any congruence, and therefore with any flow. It is often called the *generator* of the flow, by analogy with the case of a one-parameter group. The smoothness condition for a congruence ensures that the tangent vector field will be smooth.

Exercise 23. Show that $\phi: \mathcal{D} \to \mathcal{A}$ by $\phi(t,x) = \sigma_x(t)$ associates a flow ϕ with the congruence σ. □

Exercise 24. Show that in any coordinate system (x^a) the tangent vector field V to a congruence is given by $V^a\partial_a$ where $V^a(x) = (d\sigma_x^a/dt)_{t=0}$. □

Exercise 25. Let x^1 be an affine coordinate on a 1-dimensional affine space. Show that, for any given $k > 1$,

$$\phi(t,x^1) = x^1\left(1-(k-1)t(x^1)^{k-1}\right)^{-1/(k-1)}$$

is a flow whose generator is $(x^1)^k\partial_1$. □

Exercise 26. A collection of maps ϕ_t is given in affine coordinates (x^1,x^2) by

$$\phi_t^1(x^1,x^2) = \log(e^{x^1}+t) \qquad \phi_t^2(x^1,x^2) = x^2+t$$

(where these make sense). Show that it is a flow. □

4. Flows Associated with Vector Fields

The three related concepts—flow, congruence and vector field—may be exhibited in the following diagram:

$$\begin{array}{ccc} & \text{vector field} & \\ \nearrow & & \nwarrow \\ \text{flow} & \longleftrightarrow & \text{congruence} \end{array}$$

The arrows denote implication of existence: with every flow there is associated a congruence, and *vice versa*, and with every congruence a vector field. The remaining question, which we raised earlier but have not yet completely answered, is whether a congruence, and thus a flow, may be associated with every vector field. This amounts to the question, whether a certain system of differential equations has a solution, because if V is a given vector field, with coordinate expression $V = V^a\partial_a$, the condition that a curve σ be an integral curve of V is that its coordinate functions satisfy

$$\frac{d\sigma^a}{dt} = V^a(\sigma^b).$$

Such systems of differential equations are known to have solutions, which are uniquely determined by their initial values. To be precise, for each (x^a) there is a solution (σ^a), defined on some open interval of $\mathbf{R}$ containing 0, such that $\sigma^a(0) = x^a$; and any two solutions satisfying the same initial conditions are identical on the intersection of their domains. This result of differential equation theory guarantees, for each point x, the local existence and uniqueness of an integral curve σ_x of V such that $\sigma_x(0) = x$. An integral curve σ_x such that $\sigma_x(0) = x$ is called a *maximal integral curve through* x if every other integral curve through x is the restriction of the maximal one to an open subinterval of its domain. By piecing together local integral curves it is possible to construct, for each point x, a unique maximal integral curve through x. The key result in generating a flow from a vector field V is that because of the uniqueness property the collection of maximal integral curves of a vector field forms a congruence. To establish this we have to show that the maximal integral curves through two points lying on the same maximal integral curve are congruent, that is, differ only by a change of origin. We denote by I_x the domain of the maximal integral curve σ_x through x; I_x is an open interval, and may very well be a proper subset of $\mathbf{R}$, as Exercise 19 shows. Suppose that y lies on the path of σ_x, so that $y = \sigma_x(s)$ for some $s \in I_x$. Then the curve $\sigma_x \circ \tau_s$ is certainly an integral curve of V (since a change of origin does not affect tangent vectors), and its initial point, $(\sigma_x \circ \tau_s)(0)$, is just $\sigma_x(s) = y$. Thus $\sigma_x \circ \tau_s$ is at worst the restriction of the maximal integral curve through y to some subinterval of its domain. The domain of $\sigma_x \circ \tau_s$ is $\tau_{-s}(I_x)$, so it follows that $\tau_{-s}(I_x) \subset I_y$. But the same argument, with the roles of x and y interchanged, gives $\tau_s(I_y) \subset I_x$, from which it follows that I_x and I_y are just translates of each other. Thus maximal integral curves may be partitioned into congruent sets, and each point lies on precisely one congruent set.

The smoothness requirement and the requirement of openness on the set $\mathcal{D}$ for maximal integral curves to form a congruence may also be deduced from the theory of systems of first order differential equations. We shall not go into the details. The conclusion of this argument is that given a smooth vector field V on an affine space $\mathcal{A}$ there is a congruence of curves on $\mathcal{A}$ such that V is the generator of the corresponding flow.

The diagram may thus be extended to

$$\begin{array}{ccc} & \text{vector field} & \\ \swarrow\!\!\nearrow & & \searrow\!\!\nwarrow \\ \text{flow} & \longleftrightarrow & \text{congruence} \end{array}$$

The implications expressed in this diagram, that whenever one of the three constructions is given then the existence of the other two is assured, will be exploited frequently throughout this book.

Exercise 27. Find the flow whose infinitesimal generator, in affine coordinates (x^1, x^2), is $((x^1)^2 + (x^2)^2)(x^2\partial_1 - x^1\partial_2)$. Describe the orbits. □

Exercise 28. Let V be a vector field on an affine space $\mathcal{A}$ generating a flow ϕ, let $\Psi\colon \mathcal{A} \to \mathcal{A}$ be any smooth invertible map with smooth inverse, and let $\Phi(t,x) = \Psi(\phi(t, \Psi^{-1}(x)))$. Show that Φ is also a flow on $\mathcal{A}$, and that its generator V^Ψ is given by $V_x^\Psi = \Psi_*(V_{\Psi^{-1}(x)})$. □

5. Lie Transport

The constructions to be described in the following sections are among the most useful and most elegant in differential geometry. Lie transport is a process for displacing a given geometric object along a flow. The object may be a vector, a covector, or something more complicated. Lie transport might also be called "convective transport"; it is quite distinct from parallel transport, and in many ways more fundamental.

The Lie derivative, to be described in the next section, is a process using Lie transport to measure the rate of change of a field of objects along a flow. It is a directional derivative operator constructed from the flow and expressed in terms of the generator.

In this section we shall discuss the Lie transport of vector fields and covector fields. We begin with a simple example, the Lie transport of a displacement vector along a one-parameter affine group. The general case, which follows, entails the use of induced maps of tangent vectors and covectors since the transformations are no longer affine.

Lie transport along a one-parameter affine group. Let ϕ_t be a one-parameter group of affine transformations of an affine space $\mathcal{A}$, and let w be a displacement vector attached to $\mathcal{A}$ at a point x. Under the action of ϕ_t the points x and $x + w$ will be moved along their orbits to $\phi_t(x)$ and $\phi_t(x) + \lambda_t(w)$ where λ_t is the linear part of ϕ_t. The result of this process, for each t, is thus to transform the displacement vector w, attached at x, into $\lambda_t(w)$, attached at $\phi_t(x)$. In this way a displacement vector may be constructed at each point of the orbit σ_x of x. An assignment of a displacement vector at each point of a curve, like this, is called a ***field of displacement vectors*** along the curve. We denote by W the field of displacement vectors, and by $W(t)$ the vector it defines at $\sigma_x(t) = \phi_t(x)$, so that $W(t) = \lambda_t(w)$. The process of construction of W from w is called the ***Lie transport of the displacement vector*** w along σ_x by ϕ_t (or just "along ϕ_t").

The significance of W, so far as the action of ϕ_t is concerned, is that for each t, $W(t)$ connects corresponding points of the orbits of x and $x + w$.

Suppose for example that ϕ_t is the one-parameter group of affine transformations of a 2-dimensional affine space $\mathcal{A}$ given in affine coordinates by

$$(x^1, x^2) \mapsto (x^1 \cos t - x^2 \sin t, x^1 \sin t + x^2 \cos t).$$

(This example was treated in Section 1.) Let w be the displacement vector from

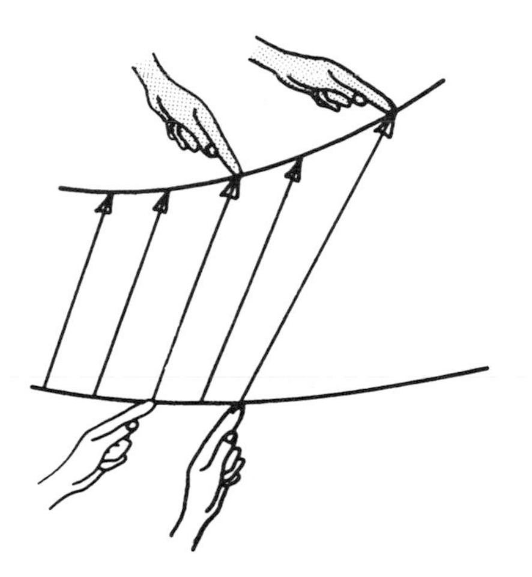

Fig. 1 Lie transport of a displacement vector along a one-parameter affine group.

(x^1, x^2) with components (w^1, w^2). Then ϕ_t takes $(x^1 + w^1, x^2 + w^2)$ to

$$\begin{aligned}&\left((x^1+w^1)\cos t - (x^2+w^2)\sin t, (x^1+w^1)\sin t + (x^2+w^2)\cos t\right)\\&\quad = \left(\phi_t^1(x^1,x^2), \phi_t^2(x^1,x^2)\right) + \left(w^1\cos t - w^2\sin t, w^1\sin t + w^2\cos t\right)\end{aligned}$$

Thus $W(t)$ has components $(w^1\cos t - w^2\sin t, w^1\sin t + w^2\cos t)$. In the Euclidean interpretation, the effect of ϕ_t is to rotate w through angle t, as well as moving its point of attachment around the origin through the same angle.

Exercise 29. Let ϕ_t be a one-parameter group of translations. Show that any displacement vector is Lie transported into a field of parallel vectors by ϕ_t. □

Exercise 30. Let ϕ_t be the one-parameter group of affine transformations given in affine coordinates by $(x^1, x^2) \mapsto (x^1 + tx^2, x^2)$ (another example treated in Section 1). Describe the Lie transport of a displacement vector, and draw a sketch. □

Exercise 31. Let ϕ_t be the one-parameter affine group of a 2-dimensional affine space given by $x \mapsto x_0 + e^{kt}(x - x_0)$ (a special case, with $k_1 = k_2$, of Exercises 2 and 9). Describe the Lie transport of a vector attached at x_0, and of one attached anywhere else. □

Exercise 32. Let x_0, v and α be a point in an affine space $\mathcal{A}$, a vector, and a covector, respectively, such that $\langle v, \alpha\rangle = 0$. Describe the Lie transport of an arbitrary vector w by the one-parameter affine group $\phi_t: x \mapsto x + t\langle x - x_0, \alpha\rangle v$. Distinguish between the cases $\langle w, \alpha\rangle = 0$ and $\langle w, \alpha\rangle \neq 0$. □

Exercise 33. Let ϕ_t be the one-parameter affine group of a 3-dimensional affine space given in affine coordinates by

$$(x^1, x^2, x^3) \mapsto (x^1\cos kt - x^2\sin kt, x^1\sin kt + x^2\cos kt, x^3 + t).$$

Show that the orbit of the origin is a straight line, whatever the value of k, but the Lie transport of a vector specified at the origin yields a parallel field if $k = 0$, while for $k \neq 0$ the Lie transported vector spirals round the x^3-axis. □

Lie transport may be applied to other figures besides displacement vectors. In the first place, the whole of the line joining the point x to $x + w$ will be transformed by ϕ_t into a new line, the one joining $\phi_t(x)$ to $\phi_t(x+w)$, since affine transformations take lines into lines. Thus Lie transport along a one-parameter affine group may be extended from displacement vectors to lines. Moreover, affine transformations map hyperplanes to hyperplanes, and this may be used to define Lie transport of hyperplanes along a one-parameter affine group. Consider the hyperplane through x consisting of all points $x + w$ such that $\langle w, \alpha \rangle = 0$, where α is some nonzero covector. Then the points $\phi_t(x + w) = \phi_t(x) + \lambda_t(w)$ lie in the hyperplane through $\phi_t(x)$ determined by the covector $\lambda_{-t}{}^*(\alpha)$, since $\langle \lambda_t(w), \lambda_{-t}{}^*(\alpha) \rangle = \langle w, \alpha \rangle = 0$. The field of hyperplanes along the orbit σ_x of x constructed in this way is said to be Lie transported; the Lie transported hyperplane at $\sigma_x(t)$ is the one defined by the covector $\lambda_{-t}{}^*(\alpha)$. The minus sign arises because linear forms map contragrediently under affine transformations.

The result of the process of Lie transport, in each case, is the construction of copies of a chosen object along the orbit of a one-parameter affine group through the point where the object was specified originally. The form of the copies is determined by the configuration of neighbouring orbits, not by a single one.

The special feature of the affine one-parameter groups in these examples is that they take lines into lines and hyperplanes into hyperplanes, so that it makes sense to speak of the transport of an extended figure into one of the same kind. A general one-parameter group, or a flow, does not preserve these objects, but nevertheless a process similar to the one already described may be carried out in tangent spaces. We next describe this general process.

Lie transport of a tangent vector along any flow. By the use of induced maps one can generalise Lie transport from one-parameter affine groups to arbitrary flows. Suppose there to be given a flow ϕ on an affine space $\mathcal{A}$ and an element w of the tangent space to $\mathcal{A}$ at some point x. We shall explain how to construct from w a vector field (field of tangent vectors) along σ_x, the orbit of x, by application of the flow ϕ. (Vector fields along curves were defined in Chapter 2, Section 8.)

As before, we denote by I_x the maximal interval on which σ_x is defined. For each fixed $t \in I_x$ the domain of the map $y \mapsto \phi(t, y)$ contains some open neighbourhood of x. We denote this map ϕ_t. The corresponding induced map $\phi_{t*}: T_x\mathcal{A} \to T_{\phi_t(x)}\mathcal{A} = T_{\sigma_x(t)}\mathcal{A}$ is thereby defined. This enables one to construct, from the vector $w \in T_x\mathcal{A}$, a succession of induced vectors $\phi_{t*}w \in T_{\sigma_x(t)}\mathcal{A}$ along the orbit σ_x, that is, a vector field along σ. The process is best imagined as one in which t varies continuously; the construction is called the *Lie transport* of w by ϕ.

In contrast to the affine case it is not possible to interpret the Lie transported vector as a displacement vector joining corresponding points on different orbits. However it does relate neighbouring orbits in a certain infinitesimal sense, which may be described as follows. Consider the line through x determined by w, which is given by $s \mapsto x + sw$. It will no longer be the case that the transform of this line, namely $s \mapsto \phi_t(x + sw)$, will be a line; it will however be a smooth curve (for each fixed t), and the Lie transported vector $\phi_{t*}w$ is the tangent vector to this curve at $s = 0$. (Of course, $\phi_t(x + sw)$ will not necessarily be defined for all $s \in \mathbf{R}$ if ϕ is not a one-parameter group, but it will be defined for s in some open interval containing 0, which is sufficient for our purposes). So one could say that the displacement vector from $\phi_t(x)$ to the corresponding point $\phi_t(x + sw)$ on a neighbouring orbit is approximated by $\phi_{t*}sw$, and this the more accurately the closer s is to zero.

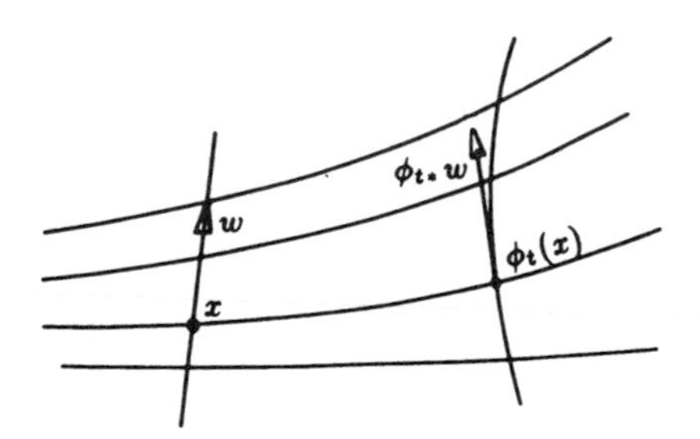

Fig. 2 Lie transport along a flow.

Exercise 34. Show that if ϕ_t is a one-parameter affine group then the Lie transport of displacement vectors defined above agrees with the Lie transport of tangent vectors just defined. □

In the next two exercises, affine coordinates on a 2-dimensional affine space are denoted as usual by (x^1, x^2).

Exercise 35. Carry out the Lie transport of ∂_1 and of ∂_2 from the origin of affine coordinates along the orbits of the one-parameter group

$$(x^1, x^2) \mapsto (x^1 + t, x^2 - \sin x^1 + \sin(x^1 + t)).$$ □

Exercise 36. Carry out the Lie transport of ∂_1 from any point of the affine space along the flow described in Exercise 26. □

The same construction may be applied to covectors, paying due account to contragredience. In the interval of definition of σ_x one may construct, from a covector $\alpha \in T_x^*\mathcal{A}$, a succession of induced covectors $\phi_{-t}{}^*\alpha \in T^*_{\sigma_x(t)}\mathcal{A}$ along the orbit σ_x. Again the process should be seen as a continuous one, giving rise to a covector field along σ_x, and the construction is called the *Lie transport* of α by ϕ.

Exercise 37. Carry out the Lie transport of dx^1 and of dx^2 along the flows given in Exercises 26 and 35. □

Exercise 38. Show that if V and α are respectively a vector field and a covector field obtained by Lie transport along an orbit of some flow then $\langle V, \alpha \rangle$ is constant along the orbit. □

6. Lie Difference and Lie Derivative

Suppose that W is a vector field along the orbit of a point x under a flow ϕ on an affine space $\mathcal{A}$. If W is defined by Lie transport then for each s, $W(s) = \phi_{s*}W(0)$, which one may equally well write $\phi_{-s*}W(s) - W(0) = 0$. Even if W is not defined by Lie transport one may form $\phi_{-s*}W(s) - W(0)$, which is a vector at x; it is called a *Lie difference*. In general a Lie difference will be nonzero, and in fact it will provide some measure of the departure of W from being defined by Lie transport.

The Lie difference may be constructed for each s in some interval containing 0. A more useful quantity will be obtained if this dependence on s is eliminated, as may be done by taking a suitable limit as $s \to 0$. Now $\lim_{s\to 0}(\phi_{-s*}W(s) - W(0)) = 0$; but it happens that $\lim_{s\to 0} \frac{1}{s}(\phi_{-s*}W(s) - W(0))$ always exists, is not necessarily zero, and gives useful information about W: it is called the *Lie derivative* of W along ϕ at x. Since it is constructed from vectors at x the Lie derivative is also a vector at x. If W is defined by Lie transport then its Lie derivative is zero; otherwise its Lie derivative measures its rate of change along the orbit of x, in a sense in which a field defined by Lie transport is to be regarded as unchanging.

To see why the limit in the definition of the Lie derivative exists it is advantageous to regard $s \mapsto \phi_{-s*}W(s) - W(0)$ as defining a curve in the tangent space at x; this curve is evidently smooth, and passes through the origin of $T_x\mathcal{A}$ when $s = 0$. The Lie derivative is simply the tangent vector to this curve at $s = 0$, regarded as an element of $T_x\mathcal{A}$.

There is nothing particularly special about the role of x in this definition, which may be easily modified to apply to any point on the orbit: the Lie derivative of W along ϕ at $\sigma_x(t)$ is defined to be

$$\lim_{s\to 0} \frac{1}{s}(\phi_{-s*}W(s+t) - W(t)).$$

In this way one may construct from W a new vector field along the orbit, whose value at any point of the orbit is the value of the Lie derivative of W there. This new field is again called the *Lie derivative* of W. It will become clear, from a formula derived below, that it is a smooth field.

Given the equivalence of a flow and its generator, one may as well regard the Lie derivative as being defined in terms of the vector field V which generates ϕ as in terms of ϕ itself. In fact it is usual to include V rather than ϕ in the notation for a Lie derivative: one writes $\mathcal{L}_V W$ for the Lie derivative of W along the flow generated by V, and calls it the Lie derivative of W with respect to V. Here W is assumed to be a vector field along an integral curve σ_x of V, and $\mathcal{L}_V W$ is then a vector field along the same integral curve. We denote by $\mathcal{L}_V W(t)$ the Lie derivative of W at $\sigma_x(t)$: thus

$$\mathcal{L}_V W(t) = \lim_{s\to 0} \frac{1}{s}(\phi_{-s*}W(s+t) - W(t)).$$

It is suggestive and convenient also to write this

$$\mathcal{L}_V W(t) = \frac{d}{ds}(\phi_{-s*}W(s+t))_{s=0},$$

taking advantage of the fact that ϕ_{0*} is the identity.

It is often the case that V generates a one-parameter group, not only a flow, and that the domain of W is more than a single orbit, but the construction of $\mathcal{L}_V W$ is unaffected by this.

Before giving an example of the calculation of a Lie derivative we shall conclude the story of the relationship between Lie derivative and Lie transport. If W is defined by Lie transport along σ_x then $\mathcal{L}_V W = 0$: for in this case,

$$W(s+t) = \phi_{(s+t)*}W(0) = (\phi_s \circ \phi_t)_* W(0) = \phi_{s*}\phi_{t*}W(0) = \phi_{s*}W(t).$$

Conversely, suppose that W is a vector field along σ_x such that $\mathcal{L}_V W = 0$. We consider the curve $r \mapsto \phi_{-r*}W(r)$ in $T_x\mathcal{A}$, and show that it is in fact a constant curve. We have, for each fixed t,

$$\frac{d}{dr}\left(\phi_{-r*}W(r)\right)_{r=t} = \frac{d}{ds}\left(\phi_{-(s+t)*}W(s+t)\right)_{s=0}$$
$$= \phi_{-t*}\frac{d}{ds}\left(\phi_{-s*}W(s+t)\right)_{s=0} = \phi_{-t*}\mathcal{L}_V W(t) = 0$$

(where the variable has been changed from r to $s = r - t$). Thus $\phi_{-r*}W(r)$ is constant and equal to its value when $r = 0$, which is $W(0)$; and so W is defined by Lie transport of $W(0)$. Thus $\mathcal{L}_V W = 0$ is a necessary and sufficient condition for W to be defined by Lie transport.

Computing the Lie derivative. We turn now to the computation of Lie derivatives.

As an example, let $V = -x^2\partial_1 + x^1\partial_2$, which generates the one-parameter affine group $(x^1, x^2) \mapsto (x^1\cos t - x^2\sin t, x^1\sin t + x^2\cos t)$; and let W be the parallel vector field $\partial_1 + \partial_2$. We compute $\mathcal{L}_V W(0)$ along the integral curve of V through the point with coordinates $(1,0)$. The result will therefore be a vector at that point.

The integral curve is $t \mapsto (\cos t, \sin t)$. We have therefore to compute $\phi_{-s*}W(s)$, where $W(s)$ is the vector $\partial_1 + \partial_2$ at the point $(\cos s, \sin s)$. One simple way of carrying out this computation is to choose a curve through $(\cos s, \sin s)$ to which $W(s)$ is the tangent, compute the image of this curve under ϕ_{-s}, and find its tangent vector. A suitable choice of curve is the line $r \mapsto (r + \cos s, r + \sin s)$; its image is the curve $r \mapsto (1 + r(\cos s + \sin s), r(\cos s - \sin s))$. (The image curve is also a line, because ϕ_{-s} is affine, though this is incidental.) The tangent vector to the image curve at $r = 0$ is $(\cos s + \sin s)\partial_1 + (\cos s - \sin s)\partial_2 = \phi_{-s*}W(s)$. This is a vector at $(1,0)$: ∂_1 and ∂_2 are the coordinate vectors at that point. To compute the Lie derivative we have merely to evaluate the derivatives of the components with respect to s, at $s = 0$; we obtain $\mathcal{L}_V W(0) = \partial_1 - \partial_2$.

Exercise 39. Let ϕ_t be the one-parameter affine group given in affine coordinates by $(x^1, x^2) \mapsto (x^1 + tx^2, x^2)$ and let W be the vector field given along the orbit of $(0,1)$ by $W(t) = \cos t\partial_1 + \sin t\partial_2$. Find $\mathcal{L}_V W$ as a function of t, where V is the generator of ϕ_t. □

Exercise 40. Let ϕ_t be the one-parameter affine group given in affine coordinates by $(x^1, x^2, x^3) \mapsto (e^t x^1, e^{2t}x^2, e^{3t}x^3)$ and let W be the parallel vector field with components $(3,2,1)$. Find $\mathcal{L}_V W$ on the orbit of $(1,1,1)$ as a function of t, where V is the generator of ϕ_t. Also compute V. □

Exercise 41. Let ψ_t be the one-parameter affine group given in affine coordinates by $(x^1, x^2, x^3) \mapsto (x^1 + 3t, x^2 + 2t, x^3 + t)$ and let $V = x^1\partial_1 + 2x^2\partial_2 + 3x^3\partial_3$. Find $\mathcal{L}_W V$ on the orbit of $(1,1,1)$, where W is the generator of ψ_t. Determine W. Describe the connections between this exercise and the previous one. Compare $\mathcal{L}_V W$ with $\mathcal{L}_W V$ at $(1,1,1)$. □

Exercise 42. Let $\phi_t: x \mapsto x_0 + \lambda_t(x - x_0) + v_t$ be a one-parameter affine group with generator V. Let W be a vector field given along the orbit of x. Show that in an affine coordinate system the components of $\mathcal{L}_V W(0)$ are $\dot W^a(0) - A^a_b W^b(0)$, where $W^a(t)$ are the components of W and the A^a_b are the entries in the matrix of $d/dt(\lambda_t)(0)$. □

Exercise 43. Show that for any vector field V, $\mathcal{L}_V V = 0$. □

Exercise 44. Show that if V generates the flow ϕ and W is a vector field given along an orbit of ϕ then $\mathcal{L}_V(\phi_{t*}W) = \phi_{t*}\mathcal{L}_V W$. □

Exercise 45. Let W_1 and W_2 be vector fields defined along an orbit of a flow ϕ. Show that $\mathcal{L}_V(c_1W_1 + c_2W_2) = c_1\mathcal{L}_V W_1 + c_2\mathcal{L}_V W_2$ where V is the generator of ϕ and c_1 and c_2 are any constants. □

Exercise 46. Let f be a function defined on a neighbourhood of x. Show that for any vector field W defined along the orbit of x

$$\mathcal{L}_V(fW)(0) = (V_x f)W(0) + f(x)(\mathcal{L}_V W)(0). \quad \square$$

The Lie derivative of a covector field. The Lie derivative of a covector field corresponds to the Lie transport of a covector in the same way as the Lie derivative of a vector field corresponds to the Lie transport of a vector. Let ϕ be a flow on $\mathcal{A}$, σ_x the orbit of the point x, and α a covector field specified on σ_x. One may construct a curve in $T_x^*\mathcal{A}$ by Lie transporting to that point covectors specified at other points, obtaining $\phi_s{}^*\alpha(s)$ from $\alpha(s)$ given at $\sigma_x(s)$. The change in sign (compared with the case of a vector field) arises from the contragredience: ϕ_s maps x to $\phi_s(x) = \sigma_x(s)$, so $\phi_s{}^*$ pulls $\alpha(s)$ back from $\sigma_x(s)$ to x. The Lie difference is $\phi_s{}^*\alpha(s) - \alpha(0)$, and if V denotes the generator of ϕ, then the Lie derivative of α along σ_x, with respect to V, at $t = 0$, is

$$\mathcal{L}_V\alpha(0) = \lim_{s\to 0}\frac{1}{s}\left(\phi_s{}^*\alpha(s) - \alpha(0)\right) = \frac{d}{ds}\left(\phi_s{}^*\alpha(s)\right)(0).$$

In this way one can define a new covector field with the same domain as the original one. The Lie derivative measures the rate of change of α along the flow, and is zero if and only if α is defined by Lie transport, just as is the case for vectors.

Exercise 47. Show from the definition that if W is a vector field and α a covector field defined along an orbit of a vector field V then at each point of the orbit $V\langle W,\alpha\rangle = \langle\mathcal{L}_V W,\alpha\rangle + \langle W,\mathcal{L}_V\alpha\rangle$ where $\langle W,\alpha\rangle$ is regarded as a function defined along the orbit. (The relative signs in the definitions are chosen so that this Leibniz formula will hold). □

Exercise 48. Let α_1 and α_2 be covector fields specified along an orbit of V. Show that for any constants c_1 and c_2, $\mathcal{L}_V(c_1\alpha_1 + c_2\alpha_2) = c_1\mathcal{L}_V\alpha_1 + c_2\mathcal{L}_V\alpha_2$. □

Exercise 49. Show that $\mathcal{L}_V(f\alpha)(0) = (V_x f)\alpha(0) + f(x)(\mathcal{L}_V\alpha)(0)$, f being a function defined on a neighbourhood of x, and α a covector field specified along the orbit of x. □

7. The Lie Derivative of a Vector Field as a Directional Derivative

We now exhibit an explicit representation for the Lie derivative of a vector field as a directional derivative operator, acting on functions. Before giving the relevant expression we point out that from a vector field U and a function h one can derive a new function Uh by setting $(Uh)(x) = U_x h$. In coordinates, if $U = U^a\partial_a$ then $Uh = U^a\partial_a h = U^a\partial h^x/\partial x^a$. If U is defined only on a curve then Uh is a function on the same curve, while if h is a function on an integral curve of U then Uh makes sense and is again a function on the same curve.

The formula in question is

$$(\mathcal{L}_V W)f = V(Wf) - W(Vf).$$

Here W is specified along an orbit of V, and f in an open neighbourhood of a point of the orbit, so that Wf is determined along the orbit, and V, which differentiates along the orbit, can sensibly act on it.

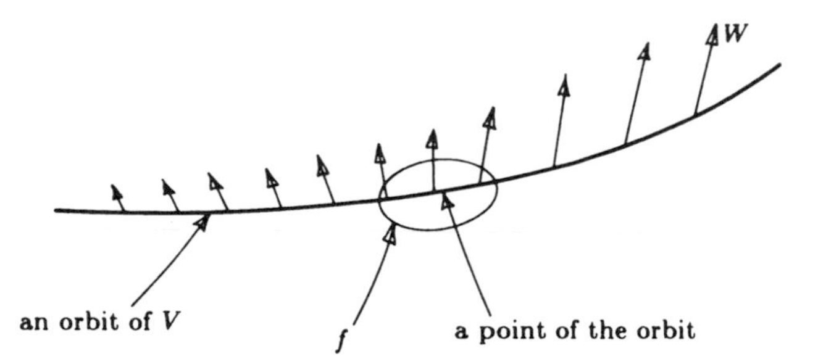

Fig. 3 Specification of W and f.

The calculation entails combination of an induced map formula with the definition of Lie derivative. For any smooth map of affine spaces ψ and any function g on the codomain of ψ, $\psi^*dg = d(g \circ \psi)$, where ψ^* denotes the induced map which pulls covectors back from $\psi(x)$ to x (Chapter 2, Section 7). Here dg is considered as a covector at $\psi(x)$, and $d(g \circ \psi)$ as a covector at x. On the other hand, the Lie derivative of a covector field α is given by $d/ds(\phi_s{}^*\alpha(s))(0)$. We use this expression with α the field of covectors along σ_x defined as follows: $\alpha(t)$ is the covector determined by df at $\sigma_x(t)$. It follows from the cotangent map formula just stated that $\phi_s{}^*\alpha(s) = d(f \circ \phi_s)$, as covectors at x. Thus with this choice of α

$$\mathcal{L}_V\alpha(0) = \frac{d}{ds}\big(d(f \circ \phi_s)\big)(0).$$

The calculation of each component of this covector involves first the partial differentiation of the function and evaluation at x, and second the differentiation of the result with respect to the parameter s and substitution of 0 for s. It follows from our assumption about the smoothness of a flow that these steps are interchangeable. Moreover, $d/ds(f \circ \phi_s)_{s=0} = Vf$ since, for any point y, $s \mapsto \phi_s(y)$ is the orbit of y, to which V_y is tangent. On interchanging the differentiations we therefore obtain

$$\mathcal{L}_V\alpha(0) = d(Vf)$$

(as covectors at x). We combine this result with the Leibniz formula (Exercise 47)

$$V\langle W, \alpha\rangle = \langle \mathcal{L}_V W, \alpha\rangle + \langle W, \mathcal{L}_V\alpha\rangle;$$

recognising that $\langle W, \alpha\rangle = Wf$, one obtains

$$V(Wf) = (\mathcal{L}_V W)f + \langle W, d(Vf)\rangle = (\mathcal{L}_V W)f + W(Vf)$$

so that

$$(\mathcal{L}_V W)f = V(Wf) - W(Vf),$$

as asserted. This formula holds at any point where the operations are defined.

By choosing $f = x^a$ in the formula, one obtains $(\mathcal{L}_V W)^a = V(W^a) - W(V^a)$. The component W^a of W is to be regarded a function of the parameter along an integral curve of V, so that $V(W^a) = \dot{W}^a$; thus

$$\mathcal{L}_V W(t) = (\dot{W}^a(t) - W(t)(V^a))\partial_a = (\dot{W}^a(t) - W^b(t)\partial_b V^a)\partial_a.$$

The components of $\mathcal{L}_V W$ are clearly smooth functions of t, assuming V and W to be smooth; thus $\mathcal{L}_V W$ is a smooth vector field.

Exercise 50. Compute $\mathcal{L}_V W$ again from the data given in Exercises 40 and 41 using this formula. □

Exercise 51. Show that for any covector field α

$$\mathcal{L}_V \alpha(t) = (\dot{\alpha}_a(t) + \alpha_b(t)\partial_a V^b)dx^a. \qquad \square$$

8. Vector Fields as Differential Operators

A vector field, like a tangent vector, is a directional derivative operator. If V is a vector field and f a function then Vf is a function, whose value at x is $V_x f$. These ideas have arisen already, in the previous section, but there the vector fields in question need be specified only along a single curve. We now consider vector fields specified all over an affine space, or at least over an open subset of it, and describe some of their operator properties.

The module of vector fields. The smooth vector fields on an affine space $\mathcal{A}$ constitute an Abelian group (axioms 1 to 4, Note 2 of Chapter 1). Moreover, if U and V are vector fields and f and g are functions then fU is a vector field, defined by $(fU)_x = f(x)U_x$, and

(1) $f(U + V) = fU + fV$
(2) $(f + g)U = fU + gU$
(3) $f(gU) = (fg)U$
(4) $1U = U$

(where 1 denotes the constant function which takes the value 1 at every point). These properties are formally similar to those which define a vector space. However, the set of smooth functions, which here plays the role of scalars (the role played by $\mathbf{R}$ in the definition of a real vector space) differs from $\mathbf{R}$ in one important respect: the product of two functions may be zero without either factor being zero (vanishing identically). Nevertheless, the set of smooth functions has many of the other properties of $\mathbf{R}$ (commutativity of addition and multiplication, distributivity, and so on), and includes $\mathbf{R}$ in the form of the constant functions. It is an example of an algebra over $\mathbf{R}$. A set with the vector-space-like properties enjoyed by the set of smooth vector fields is called a *module* over the algebra in question (in this case the algebra of smooth functions on $\mathcal{A}$). The algebra of smooth functions on $\mathcal{A}$ will be denoted $\mathcal{F}(\mathcal{A})$, and the module of vector fields $\mathcal{X}(\mathcal{A})$. The coordinate vector fields $\{\partial_a\}$ in any affine coordinate system constitute a basis for the module $\mathcal{X}(\mathcal{A})$: every vector field may be written as a linear combination of these, with coefficients from the underlying algebra of functions. The existence of a basis is a property of all affine spaces but does not extend to manifolds. Other bases for $\mathcal{X}(\mathcal{A})$ may be found, not necessarily consisting of coordinate vector fields: indeed, if functions U_a^b are such as to form a non-singular matrix at each point of $\mathcal{A}$, then the vector fields $U_a = U_a^b \partial_b$ constitute a basis for $\mathcal{X}(\mathcal{A})$.

A vector field V acts on functions as a linear operator which satisfies Leibniz's rule:

(1) $V(af + bg) = aVf + bVg \qquad a, b \in \mathbf{R},\ f, g \in \mathcal{F}(\mathcal{A})$
(2) $V(fg) = (Vf)g + fVg$.

An operator with these two properties is called a *derivation*. Thus the vector fields on $\mathcal{A}$ are derivations of the algebra of functions $\mathcal{F}(\mathcal{A})$. It may be shown that, conversely, any derivation of $\mathcal{F}(\mathcal{A})$ corresponds to a smooth vector field on $\mathcal{A}$. Besides the module and derivation properties we mention one other, the *chain rule*, which is useful for computational purposes: if $h: \mathbf{R} \to \mathbf{R}$ is a smooth real function and V and f are any vector field and any function on $\mathcal{A}$, then

$$V(h \circ f) = (\dot{h} \circ f)Vf.$$

With these properties to hand it is easy to compute the action of a vector field V on any function given explicitly in terms of the coordinate functions, when the components of V with respect to the coordinate basis are known: for if $V = V^a \partial_a$ then $V^a = V(x^a)$. As an example, suppose that $f = x^1 x^2 + \sin^2 x^3$: then for any vector field V

$$Vf = V(x^1)x^2 + x^1 V(x^2) + 2\sin x^3 \cos x^3\, V(x^3) = x^2 V^1 + x^1 V^2 + 2\sin x^3 \cos x^3 V^3;$$

and if, for example, $V = \partial_1 + x^3 \partial_2 - x^2 \partial_3$ then

$$Vf = x^2 + x^1 x^3 - 2x^2 \sin x^3 \cos x^3.$$

Exercise 52. Let $V = x^1 \partial_1 + x^2 \partial_2 + x^3 \partial_3$ and $f = x^2 x^3 + x^3 x^1 + x^1 x^2$. Compute Vf. □

9. Brackets and Commutators

The composite of two vector fields V and W—that is, the operator $f \mapsto V(Wf)$—is linear but does not satisfy Leibniz's rule:

$$\begin{aligned} V\big(W(fg)\big) &= V\big((Wf)g + f(Wg)\big) \\ &= \big(V(Wf)\big)g + (Wf)(Vg) + (Vf)(Wg) + f\big(V(Wg)\big). \end{aligned}$$

However, the symmetry in V and W of the unwanted terms reveals that the commutator $V \circ W - W \circ V$ is a derivation of $\mathcal{F}(\mathcal{A})$ and is therefore a vector field. The commutator of two vector field operators is usually written between square brackets and is therefore called their *bracket*:

$$[V, W]f = V(Wf) - W(Vf).$$

The geometrical significance of the bracket becomes immediately apparent if this formula is compared with the one derived in the previous section for the Lie derivative: formally,

$$[V, W] = \mathcal{L}_V W.$$

However, the present point of view differs from that of the preceding section: there V was considered throughout in its role of generator of a flow, while W was supposed specified only along an integral curve of V, which was enough for the definition of the Lie derivative of W. Here, on the other hand, V and W have equal status in their roles as directional derivative operators, specified everywhere on $\mathcal{A}$, or at least on an open subset. Nevertheless it should be clear that the point of view of the preceding section can be regained by considering the restriction of W to a single integral curve of V.

Substituting $V = V^a\partial_a$ and $W = W^a\partial_a$, where (x^a) is any coordinate system, affine or curvilinear, in the expression for the bracket, one obtains immediately its coordinate expression

$$[V,W] = (V^b\partial_b W^a - W^b\partial_b V^a)\partial_a.$$

Notice that on restriction to an integral curve of V the first term becomes the derivative of W^a along the integral curve.

The following properties of the bracket are simple consequences of its definition; they reveal some properties of the Lie derivative not so far apparent:

(1) the bracket is skew-symmetric: $[V,W] = -[W,V]$
(2) the bracket is bilinear, that is linear (over $\mathbf{R}$) in each argument
(3) $[V,fW] = f[V,W] + (Vf)W$ for any $f \in \mathcal{F}(\mathcal{A})$
(4) $[U,[V,W]] + [V,[W,U]] + [W,[U,V]] = 0$.

The last of these, which is known as *Jacobi's identity*, is a general property of commutators of linear operators. Its proof is a simple computation:

$$\begin{aligned}
&[U,[V,W]] + [V,[W,U]] + [W,[U,V]] \\
&\quad = U\circ V\circ W - U\circ W\circ V - V\circ W\circ U + W\circ V\circ U \\
&\quad + V\circ W\circ U - V\circ U\circ W - W\circ U\circ V + U\circ W\circ V \\
&\quad + W\circ U\circ V - W\circ V\circ U - U\circ V\circ W + V\circ U\circ W = 0.
\end{aligned}$$

Exercise 53. Infer that $\mathcal{L}_W V = -\mathcal{L}_V W$ when V and W are globally defined vector fields. □

Exercise 54. Deduce from property (3) that for $f \in \mathcal{F}(\mathcal{A})$

$$\mathcal{L}_V(fW) = f\mathcal{L}_V W + (Vf)W \quad \text{while} \quad \mathcal{L}_{fV}W = f\mathcal{L}_V W - (Wf)V.$$ □

Exercise 55. Show that Jacobi's identity may be written in the form $[U,[V,W]] = [[U,V],W] + [V,[U,W]]$, which bears some resemblance to Leibniz's rule. □

Exercise 56. Show that $\mathcal{L}_V\mathcal{L}_W U - \mathcal{L}_W\mathcal{L}_V U = \mathcal{L}_{[V,W]}U$. □

Exercise 57. Show that if $V = (A^a_b x^b + K^a)\partial_a$ and $W = (B^a_b x^b + L^a)\partial_a$ are linear vector fields then

$$[V,W] = ((B^a_c A^c_b - A^a_c B^c_b)x^b + (B^a_b K^b - A^a_b L^b))\partial_a.$$

In particular, any two constant vector fields commute; the bracket of a constant with a homogeneous linear vector field reproduces the action of a matrix on a column vector; and the bracket of two homogeneous linear vector fields reproduces (except for sign) the commutator of matrices, $[A,B] = AB - BA$. □

Exercise 58. Show that the commutator of matrices satisfies conditions (1), (2) and (4) above, and that so does the "bracket" of vectors in 3-dimensional Euclidean space defined by $[\mathbf{a},\mathbf{b}] = \mathbf{a}\times\mathbf{b}$. □

Exercise 59. Compute the commutator of each pair of vector fields that may be formed from the three vector fields ∂_1, $x^1\partial_1$, and $(x^1)^2\partial_1$ on a 1-dimensional affine space. □

Exercise 60. Show that if one of the vector fields U, V and W is a linear combination of the other two, with constant coefficients, then Jacobi's identity is a consequence of skew-symmetry and bilinearity of the bracket. □

Vector fields related by a smooth map. Let $\Psi\colon \mathcal{A} \to \mathcal{B}$ be a smooth map. It induces a linear map Ψ_* of vectors tangent to $\mathcal{A}$ at any point. However, it is not necessarily the case that a vector field V given on $\mathcal{A}$ will map under Ψ_* to a vector field on $\mathcal{B}$. We may certainly form $\Psi_{*x}V_x$ for all $x \in \mathcal{A}$: but if Ψ is not injective there is no reason to suppose that $\Psi_{*\acute{x}}V_{\acute{x}} = \Psi_{*x}V_x$ when $\Psi(\acute{x}) = \Psi(x)$; even if Ψ is injective this procedure will not define a vector at each point of $\mathcal{B}$ if Ψ is not also surjective. Thus only when Ψ is bijective can one be sure that Ψ_* maps vector fields on $\mathcal{A}$ to vector fields on $\mathcal{B}$. When this is the case, and Ψ has a smooth inverse, the image vector field has the flow Φ given by $\Phi(t,y) = \Psi\big(\phi(t,\Psi^{-1}(y))\big)$ for $y \in \mathcal{B}$, where ϕ is the flow of V. (This result was obtained in Exercise 28 for a smooth invertible map of $\mathcal{A}$ onto itself; the more general case is a simple extension.)

Thus for a general smooth map no theory of induced maps is possible which applies to all vector fields. However, there are many occasions when one wishes to consider vector fields V on $\mathcal{A}$ and W on $\mathcal{B}$ which happen to be related *via* the induced maps of a smooth map Ψ in the sense that, for all $x \in \mathcal{A}$,

$$W_{\Psi(x)} = \Psi_{*x}V_x$$

The vector fields W is then said to be *Ψ-related* to the vector field V. We now describe some properties of Ψ-related vector fields, and in particular show that if W_1 is Ψ-related to V_1 and W_2 is Ψ-related to V_2, then $[W_1,W_2]$ is Ψ-related to $[V_1,V_2]$.

Let W be a vector field on $\mathcal{B}$, Ψ-related to the vector field V on $\mathcal{A}$. We note first of all that for any function f on $\mathcal{B}$

$$(Wf)\big(\Psi(x)\big) = W_{\Psi(x)}f = (\Psi_{*x}V_x)f = V_x(f \circ \Psi)$$

and thus

$$(Wf) \circ \Psi = V(f \circ \Psi).$$

Conversely, if V and W satisfy this relation for every function f on $\mathcal{B}$ then W is Ψ-related to V. This gives an alternative criterion for Ψ-relatedness in terms of the vector fields as operators.

The property of the brackets of Ψ-related vector fields is now almost immediate. Let W_1 be Ψ-related to V_1 and W_2 to W_2. To show that $[W_1,W_2]$ is Ψ-related to $[V_1,V_2]$ we have to show that for every function f on $\mathcal{B}$

$$\big([W_1,W_2]f\big) \circ \Psi = [V_1,V_2](f \circ \Psi).$$

But

$$\begin{aligned}
\big([W_1,W_2]f\big) \circ \Psi &= \big(W_1(W_2 f)\big) \circ \Psi - \big(W_2(W_1 f)\big) \circ \Psi \\
&= V_1\big((W_2 f) \circ \Psi\big) - V_2\big((W_2 f) \circ \Psi\big) \\
&= V_1\big(V_2(f \circ \Psi)\big) - V_2\big(V_1(f \circ \Psi)\big) \\
&= [V_1,V_2](f \circ \Psi)
\end{aligned}$$

as required.

Exercise 61. Show that if W is Ψ-related to V then the flow ψ of W is related to the flow ϕ of V by $\psi(t, \Psi(x)) = \Psi(\phi(t, x))$. □

10. Covector Fields and the Lie Derivative

Just as the concept of a vector field has been enlarged, from an object defined along a curve to one defined all over the affine space, so also may the concept of a covector field. A *covector field* on an affine space $\mathcal{A}$ is a choice of element of each cotangent space to $\mathcal{A}$. A covector field α may be expressed in the form $\alpha = \alpha_a dx^a$ in terms of a coordinate system (x^a), affine or not; the dx^a are the *coordinate covector fields*. The α_a are functions on the coordinate chart; the covector field is smooth if these component functions, for an affine coordinate system, are smooth.

A function f on $\mathcal{A}$ may be used to define a covector field, its *differential*, whose value at x is just df at x. We denote by df the differential of f as a field also. Given any covector at a point there is a function (indeed an affine one) whose differential agrees with the covector at that point. However, it is not necessarily the case that given a covector field there is a function whose differential agrees with the covector field everywhere. The conditions for this to be so are related to the conditions for a vector field to be a gradient which are discussed in vector calculus. We shall return to this point in Chapter 5.

The Lie derivative may be adapted to apply to covector fields, in a pretty well self-evident way. Properties of the Lie derivative of a covector field along an integral curve extend to the new situation in analogy with what happens for vector fields.

Exercise 62. From the formula $V\langle W, \alpha\rangle = \langle \mathcal{L}_V W, \alpha\rangle + \langle W, \mathcal{L}_V \alpha\rangle$ deduce that $\mathcal{L}_V \mathcal{L}_W \alpha - \mathcal{L}_W \mathcal{L}_V \alpha = \mathcal{L}_{[V,W]}\alpha$ for any covector field α. □

The definition of the bracket and the results of Exercises 56 and 62 may be given a coherent formulation if first of all the Lie derivative on functions is defined to be the directional derivative:

$$\mathcal{L}_V f = Vf,$$

and secondly the bracket of Lie derivative operators is defined to be their commutator:

$$[\mathcal{L}_V, \mathcal{L}_W] = \mathcal{L}_V \circ \mathcal{L}_W - \mathcal{L}_W \circ \mathcal{L}_V;$$

for then the Lie derivative on functions, vector fields and covector fields satisfies

$$[\mathcal{L}_V, \mathcal{L}_W] = \mathcal{L}_{[V,W]}.$$

Thus the whole structure of Lie derivative operators is closely related to the bracket structure of vector fields.

Exercise 63. Show that, so far as operation on functions is concerned, $\mathcal{L}_V \circ d = d \circ \mathcal{L}_V$. □

11. Lie Derivative and Covariant Derivative Compared

In Section 9 of Chapter 2 the covariant derivative of a vector field along a curve was defined by exploiting the parallelism of affine space to identify tangent spaces at different points. Like the Lie derivative, the covariant derivative may be extended to an operation of a vector field on a vector field, which results in a further vector field. It may also be extended, again like the Lie derivative, to an operation of a vector field on a covector field, leading to a further covector field. In this section we shall first of all explain these constructions, and then compare the resulting operation with the Lie derivative.

Suppose that W is a vector field defined along a curve σ in an affine space $\mathcal{A}$; then in affine coordinates the components DW^a/Dt of DW/Dt are given simply by $DW^a/Dt = dW^a/dt$. At any chosen point $\sigma(t)$ of the curve, therefore, the covariant derivative of W along the curve may be expressed in terms of the tangent vector to the curve at that point, $\dot{\sigma}(t)$, as follows:

$$\frac{DW}{Dt} = \dot{\sigma}(t)(W^a)\partial_a.$$

(Differentiation of W^a along the tangent vector $\dot{\sigma}(t)$ is intended on the right hand side.) Note that, W being given, it is enough to know the tangent vector to σ at any point in order to compute DW/Dt at that point.

Suppose now that W is no longer a vector field defined along just one curve, but is instead a vector field defined all over $\mathcal{A}$, or at least on some open neighbourhood of a point x in it. Then we may define, for each non-zero $v \in T_x\mathcal{A}$, the *covariant derivative* of W along v as the value of DW/Dt at x along any curve through x which has v as its tangent there. If $v = 0$ then we define the covariant derivative to be zero also. We shall denote this newly defined object $\nabla_v W$; it is an element of $T_x\mathcal{A}$.

Exercise 64. Show that in terms of not necessarily affine coordinates

$$\nabla_v W = v^c(\partial_c W^a + \Gamma^a_{bc} W^b)\partial_a,$$

where the Γ^a_{bc} are the appropriate connection coefficients for the given coordinates. □

Exercise 65. Devise a corresponding definition for the covariant derivative of a covector field along a tangent vector at a point. □

This construction may be extended to a definition of the *covariant derivative* $\nabla_V W$ *of a vector field* W by a vector field V, as follows: $\nabla_V W$ is the vector field whose value at x is $\nabla_{V_x} W$. By adapting the result of the Exercise 64 above we obtain the following expression for the covariant derivative in coordinates:

$$\nabla_V W = V^c(\partial_c W^a + \Gamma^a_{bc} W^b)\partial_a.$$

Exercise 66. Find the corresponding expression for the covariant derivative of a covector field. □

In terms of affine coordinates the covariant derivative takes the simple form

$$\nabla_V W = V^c(\partial_c W^a)\partial_a.$$

However, in the more general situations to be treated later, the non-affine version is the safer guide.

From these expressions, or from the properties of D/Dt exhibited in Chapter 2, Section 9, it is easy to see that the covariant derivative has the following properties:

(1) $\nabla_{U+V}W = \nabla_U W + \nabla_V W$

(2) $\nabla_{fV}W = f\nabla_V W$ $\qquad f \in \mathcal{F}(\mathcal{A})$

(3) $\nabla_U(aV + bW) = a\nabla_U V + b\nabla_U W$ $\qquad a, b \in \mathbf{R}$

(4) $\nabla_U(fV) = f\nabla_U V + (Uf)V$.

We consider next the relation between the covariant and the Lie derivative, *via* the bracket. It is clear from the formula $[V,W] = \big(V(W^a) - W(V^a)\big)\partial_a$ that $\nabla_V W$ provides just the first half of $[V,W]$, and that in fact

$$[V,W] = \nabla_V W - \nabla_W V.$$

We call this the *first order commutation relation* of covariant differentiation. Furthermore, using the expression for the covariant derivative in affine coordinates,

$$\nabla_U(\nabla_V W) = \nabla_U\big(V(W^a)\partial_a\big) = U\big(V(W^a)\big)\partial_a,$$

and so

$$\nabla_U(\nabla_V W) - \nabla_V(\nabla_U W) = \nabla_{[U,V]}W.$$

This is the *second order commutation relation.*

The formula in Exercise 47 of Chapter 2 shows that the covariant derivative operator acts on pairings in the same way as the Lie derivative: that formula may be written immediately in terms of vector fields

$$V\langle W,\alpha\rangle = \langle \nabla_V W, \alpha\rangle + \langle W, \nabla_V \alpha\rangle.$$

Thus although covariant derivatives are defined in terms of parallelism and Lie derivatives in terms of flows, they share many properties. We sum up by listing first their similarities, and then their differences. We write D to stand equally for ∇ or for $\mathcal{L}$ when a statement is true for both:

(1) each is an operator depending on a vector field, which sends vector fields to vector fields and covector fields to covector fields

(2) each is linear in both arguments over $\mathbf{R}$

(3) each satisfies the following version of Leibniz's rule in the second variable: $D_V(fW) = fD_V W + (Vf)W$

(4) for each, its operations on vector and on covector fields are related as follows: $V\langle W,\alpha\rangle = \langle D_V W,\alpha\rangle + \langle W, D_V\alpha\rangle$

(5) for each, the commutator of operators corresponds to the bracket of vector fields: $[D_V, D_W] = D_{[V,W]}$.

The reader should be warned, however, that so far as the covariant derivative is concerned, this last property is specific to the absolute parallelism which one finds in affine space, and does not generalise.

There are the following differences between covariant and Lie derivatives:

(1) $(\nabla_V W)_x$ depends only on the value of V at x, whereas $(\mathcal{L}_V W)_x$ depends both on the value of V at x and on the value of the partial derivatives of its components at x

(2) $\nabla_{fV}W = f\nabla_V W$, but $\mathcal{L}_{fV}W = f\mathcal{L}_V W - (Wf)V$ (Exercise 54)

(3) for affine coordinate fields ∂_a, $\nabla_V\partial_a = 0$ but in general $\mathcal{L}_V\partial_a \neq 0$

(4) $\mathcal{L}_W V = -\mathcal{L}_V W$, but in general $\nabla_W V \neq -\nabla_V W$

(5) $\nabla_V W - \nabla_W V = [V,W]$, whereas $\mathcal{L}_V W = [V,W]$ (and so $\mathcal{L}_V W - \mathcal{L}_W V = 2[V,W]$).

Other expressions for the covariant derivative. We have shown that with respect to non-affine coordinates (x^a) the covariant derivative operator takes the form

$$\nabla_V W = V^c(\partial_c W^a + \Gamma^a_{bc} W^b)\partial_a$$

where the connection coefficients Γ^a_{bc} are defined in terms of affine coordinates $(\acute{x}^a)$ by

$$\Gamma^a_{bc} = \frac{\partial x^a}{\partial \acute{x}^d}\frac{\partial^2 \acute{x}^d}{\partial x^b \partial x^c} \quad \text{or} \quad \nabla_{\partial_c}\partial_b = \Gamma^a_{bc}\partial_a.$$

More generally, the covariant derivative may be referred to any basis of vector fields $\{U_a\}$, not necessarily coordinate fields. If we define the *connection coefficients* γ^a_{bc} with respect to $\{U_a\}$ by

$$\nabla_{U_c} U_b = \gamma^a_{bc} U_a,$$

then for any vector fields $V = V^a U_a$ and $W = W^a U_a$ (care is needed here: the V^a and W^a are functions, the U_a are vector fields)

$$\nabla_V W = V^c(U_c(W^a) + \gamma^a_{bc} W^b)U_a.$$

Exercise 67. State what type of object each symbol occurring in this equation represents. □

If we express the new basis with respect to affine coordinate fields, say $U_a = U^b_a \partial_b$, then

$$\nabla_{U_c} U_b = U^d_c(\partial_d U^e_b)\partial_e = U^d_c(\partial_d U^e_b)(U^{-1})^a_e U_a$$

where the $(U^{-1})^b_a$ are the components of the matrix inverse to (U^b_a) (non-singular because the vector fields U_a are linearly independent); thus

$$\gamma^a_{bc} = U^d_c(\partial_d U^e_b)(U^{-1})^a_e.$$

It is not necessarily the case that $\gamma^a_{cb} = \gamma^a_{bc}$, since $[U_b, U_c]$ is not necessarily zero; in fact

$$[U_b, U_c] = (\gamma^a_{cb} - \gamma^a_{bc})U_a.$$

Exercise 68. Confirm, from the definition of γ^a_{bc}. □

Exercise 69. Show that if $\{\hat{U}_a\}$ is another basis, and $\hat{U}_a = A^b_a U_b$, then

$$A^a_f \hat{\gamma}^f_{bc} = \hat{U}_c(A^a_b) + \gamma^a_{de} A^d_b A^e_c.$$ □

12. The Geometrical Significance of the Bracket

In this section we tease out some of the geometrical consequences of the identification of the Lie derivative with the bracket of vector fields, and show the relation of the bracket to the corresponding flows.

First of all we show that if the bracket vanishes then the flows commute. Let V and W be vector fields on an affine space $\mathcal{A}$ such that $[V,W] = 0$, and let ϕ and ψ be the flows generated by V and W respectively. We assume throughout this section that the parameters labelling the flows are confined to those domains for which the equations make sense. If the flows are one-parameter groups, the equations will make sense everywhere on $\mathcal{A}$. We showed in Section 6 that $[V,W] = \mathcal{L}_V W$ vanishes on an orbit of V if and only if W is Lie transported by ϕ, that is,

$$W_{\phi_t(x)} = \phi_{t*}W_x.$$

On the other hand, from the result of Exercise 28, one may conclude, interchanging the roles of V and W, that for each t, $\tilde{\psi}_s = \phi_t \circ \psi_s \circ \phi_{-t}$ is a flow on $\mathcal{A}$ with generator W^{ϕ_t} given by

$$(W^{\phi_t})_x = \phi_{t*}W_{\phi_{-t}(x)}.$$

Therefore $\mathcal{L}_V W = 0$ if and only if $W^{\phi_t} = W$. It then follows from the uniqueness of integral curves that the flows of W^{ϕ_t} and W must coincide: $\tilde{\psi}_s = \phi_t \circ \psi_s \circ \phi_{-t} = \psi_s$, whence

$$\phi_t \circ \psi_s = \psi_s \circ \phi_t.$$

This proves that if the bracket vanishes then the flows commute.

Exercise 70. Show that if the flows commute then the bracket vanishes. □

The next two exercises are concerned with a particular type of one-parameter group of interest, the matrix exponential, which we introduced in Section 1. Results about general one-parameter groups or flows have interesting consequences for matrices; conversely, matrix exponentials can give useful pointers to the general theory.

Exercise 71. Show that if A and B are square matrices, then $\exp(tA)$ and $\exp(sB)$ commute if and only if A and B commute. Infer that $\exp(tA)$ and $\exp(sA)$ always commute and that $\exp(tA)$ commutes with A. □

Exercise 72. Show, using the Taylor expansion given in Exercise 12, that for a commutator of matrix exponentials

$$\exp(-sB)\exp(-tA)\exp(sB)\exp(tA) = I_n - st[A,B]$$

correct to second order terms in the Taylor expansion. □

The result of Exercise 72 suggests that it may be possible to interpret the bracket of arbitrary vector fields in terms of the commutator of their flows $\psi_{-s} \circ \phi_{-t} \circ \psi_s \circ \phi_t$. This is indeed the case, as we shall show. The following exercise in coordinates paves the way.

Exercise 73. Let ϕ and ψ be flows on $\mathcal{A}$ generated by vector fields V and W respectively. Show that in affine coordinates (x^a)

$$\begin{aligned}\phi_t^a(x^e) &= x^a + tV^a + \tfrac{1}{2}t^2V^b\partial_b V^a + O_3 \\ \psi_s^a(\phi_t^c(x^e)) &= x^a + tV^a + sW^a + \tfrac{1}{2}t^2V^b\partial_b V^a \\ &\quad + stV^b\partial_b W^a + \tfrac{1}{2}s^2W^b\partial_b W^a + O_3\end{aligned}$$

where each of the expressions on the right is evaluated at (x^e), and O_3 denotes terms of third order in s and t. Infer that

$$\psi_{-s}^a(\phi_{-t}^b(\psi_s^c(\phi_t^d(x^e)))) = x^a + st[V,W]^a + O_3,$$

where the bracket is evaluated at (x^e). □

It appears from this result that when $[V,W]_x \neq 0$ the "square" obtained by transforming the point x successively by ϕ_t, ψ_t, ϕ_{-t} and ψ_{-t} does not close, and that $t^2[V,W]_x$ is an approximation to the displacement vector between its ends.

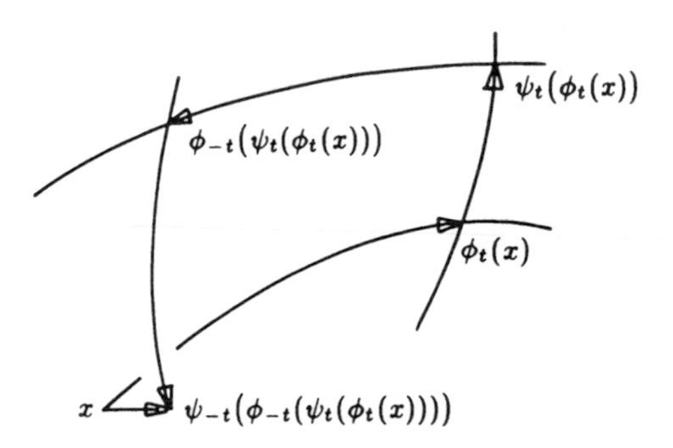

Fig. 4 Non-commuting flows.

A more precise interpretation of $[V,W]_x$ may be given in terms of the curve $t \mapsto \psi_{-t}(\phi_{-t}(\psi_t(\phi_t(x))))$. It would be pleasant to be able to say that $[V,W]_x$ is the tangent vector to that curve at $t = 0$, but a moment's reflection will reveal that the curve has zero tangent vector there. However, when a curve has zero tangent vector at a point one may define a *second-order tangent vector* there, as in the following exercise.

Exercise 74. Suppose that the curve σ has zero tangent vector at $t = 0$. Show that the map $\mathcal{F}(\mathcal{A}) \to \mathcal{F}(\mathcal{A})$ by $f \mapsto d^2/dt^2(f \circ \sigma(t))(0)$ has the properties of a tangent vector (that is, the appropriate linearity and Leibniz properties). It is the second-order tangent vector to σ at $t = 0$. □

We shall show that $[V,W]_x$ is one half of the second-order tangent vector to the curve $t \mapsto \psi_{-t}(\phi_{-t}(\psi_t(\phi_t(x))))$ at $t = 0$. To do so we consider, for any $f \in \mathcal{F}(\mathcal{A})$, the real function F given by $F(t) = f(\psi_{-t}(\phi_{-t}(\psi_t(\phi_t(x)))))$. We have to show that $\dot{F}(0) = 0$ and that $\ddot{F}(0) = 2[V,W]_x f$. In order to compute derivatives of F it is convenient to introduce another function G, defined on a neighbourhood of $(0,0)$ in $\mathbf{R}^2$ by

$$G(r,s) = f(\psi_{-r}(\phi_{-s}(\psi_r(\phi_s(x))))).$$

We denote the derivatives of G with respect to its first and second arguments by G_r and G_s respectively. Then

$$F(t) = G(t,t)$$
$$\dot{F}(0) = G_r(0,0) + G_s(0,0)$$
$$\ddot{F}(0) = G_{rr}(0,0) + 2G_{rs}(0,0) + G_{ss}(0,0).$$

We shall compute $G_{rs}(0,0)$ from $G_r(0,s)$, and we therefore require to know only $G_r(r,0)$, $G_r(0,s)$ and $G_s(0,s)$. Now $G(r,0) = G(0,s) = f(x)$ and so $G_r(r,0) = dG(r,0)/dr = 0$, and similarly $G_s(0,s) = 0$. Thus $\dot{F}(0) = 0$; and $G_{rr}(0,0) = G_{ss}(0,0) = 0$. It remains to compute $G_r(0,s)$ and $G_{rs}(0,0)$. To do so, we note that for fixed s we may write $G(r,s) = f(\psi_{-r}(\chi_r(x)))$ where $\chi_r = \phi_{-s} \circ \psi_r \circ \phi_s$ is a flow, whose generator is $W^{\phi_{-s}}$. It is necessary, therefore, to evaluate $d/dr\big(f(\psi_{-r}(\chi_r(x)))\big)(0)$ where ψ and χ are two flows.

Once again it is convenient to introduce a function of two variables, say H, by $H(u,v) = f\big(\psi_u(\chi_v(x))\big)$, whose domain again contains a neighbourhood of $(0,0)$ in $\mathbf{R}^2$. Then

$$\frac{d}{dr}\big(f(\psi_{-r}(\chi_r(x)))\big)(0) = -H_u(0,0) + H_v(0,0)$$

But
$$H_u(0,0) = \frac{d}{du}H(u,0)(0) = W_x f,$$

while
$$H_v(0,0) = \frac{d}{dv}H(0,v)(0) = (W^{\phi_{-s}})_x f,$$

and so
$$G_r(0,s) = -W_x f + (W^{\phi_{-s}})_x f.$$

Thus, recalling that $(W^{\phi_{-s}})_x = \phi_{-s*}W_{\phi_s(x)}$, we see that

$$G_{rs}(0,0) = \frac{d}{ds}\big(\phi_{-s*}W_{\phi_s(x)}\big)(0)f = (\mathcal{L}_V W)(0)f = [V,W]_x f$$

as required.

Exercise 75. Show, using these methods, that for vector fields V, W which generate flows ϕ, ψ, the tangent vector at $t = 0$ to the curve $t \mapsto \phi_t(\psi_t(x))$ is $V_x + W_x$. Show that $\phi_t \circ \psi_t$ defines a flow if and only if ϕ_s and ψ_t commute for all relevant s and t. □

Notice that if $[V,W]_x \neq 0$ then it is possible, by tracking round a "square" built from orbits of V and W, to travel from x in a direction transverse to the 2-dimensional subspace of $T_x\mathcal{A}$ spanned by V_x and W_x. We develop this idea in Chapter 6.

Summary of Chapter 3

A vector field V on an affine space $\mathcal{A}$ is a choice of element V_x of $T_x\mathcal{A}$ for each $x \in \mathcal{A}$. In coordinates, $V = V^a\partial_a$, the V^a being functions on $\mathcal{A}$; V is smooth if the V^a are. The collection of smooth vector fields, $\mathcal{X}(\mathcal{A})$, is a module over the algebra $\mathcal{F}(\mathcal{A})$ of smooth functions on $\mathcal{A}$. Vector fields act linearly, as directional derivative operators, on $\mathcal{F}(\mathcal{A})$; they also satisfy Leibniz's rule. The bracket $[V,W]$ of vector fields, which is to say, their commutator as operators, $[V,W]f = V(Wf) - W(Vf)$, is again a vector field; $[V,W] = (V^b\partial_b W^a - W^b\partial_b V^a)\partial_a$.

A flow on $\mathcal{A}$ is a smooth map ϕ of a suitable open subset of $\mathbf{R} \times \mathcal{A}$, containing $\{0\} \times \mathcal{A}$, to $\mathcal{A}$ such that $\phi(0,x) = x$, and $\phi(s,\phi(t,x)) = \phi(s+t,x)$ whenever both sides make sense. Fixing x defines a curve σ_x, the orbit of x under the flow; its domain may not be the whole of $\mathbf{R}$; $\sigma_x(0) = x$; and if $y = \sigma_x(s)$ then $\sigma_y(t) = \sigma_x(s+t)$ (this is change of origin of the parameter). A collection of curves with these properties is a congruence. Fixing t defines a transformation ϕ_t; its domain may not be the whole of $\mathcal{A}$; $\phi_0 = \mathrm{id}_{\mathcal{A}}$; $\phi_s \circ \phi_t = \phi_{s+t}$. When the domain of a flow is the whole of $\mathbf{R} \times \mathcal{A}$ the corresponding transformations form a one-parameter group. If these transformations are always elements of a particular group (translation group, affine transformation group) the one-parameter group (of translations, of affine transformations) is the image of a homomorphism of $\mathbf{R}$ into that transformation group.

Vector fields, congruences of curves and flows are equivalent to each other. Each vector field defines a congruence, of its integral curves, which are given in coordinates as the solutions of a system of ordinary differential equations; the congruence property is a consequence of the theorem on the uniqueness of solutions of such a system. Each congruence defines a flow, whose orbits are its curves, in which $\phi_t(x)$ is the point a parameter distance t along the curve of the congruence through x. Each flow defines a vector field, its generator, whose value at x is the tangent vector at $t = 0$ to the orbit of x.

The generators of affine transformations are linear but, in general, inhomogeneous vector fields; of translations, constant vector fields. The one-parameter group generated by a linear homogeneous vector field is determined by exponentiation of a matrix.

The Lie derivative $\mathcal{L}_V W$ ($\mathcal{L}_V \alpha$) of a vector field V (covector field α) along an orbit of the flow ϕ generated by V is the vector (covector) field along the orbit whose value at t is $d/ds(\phi_{-s*}W(s+t))_{s=0}$ (for a vector field) $d/ds(\phi_{s*}\alpha(s+t))_{s=0}$ (for a covector field). In coordinates,

$$\mathcal{L}_V W = (\dot{W}^a - W^b \partial_b V^a)\partial_a \qquad \mathcal{L}_V \alpha = (\dot{\alpha}_a + \alpha_b \partial_a V^b) dx^a.$$

If W is a vector field on $\mathcal{A}$ then $\mathcal{L}_V W = [V, W]$. The Lie derivative is then skew symmetric and linear in both arguments, and

$$\mathcal{L}_V(fW) = f\mathcal{L}_V W + (Vf)W \qquad \mathcal{L}_{fV} W = f\mathcal{L}_V W - (Wf)V$$
$$\langle \mathcal{L}_V W, \alpha \rangle + \langle W, \mathcal{L}_V \alpha \rangle = V\langle W, \alpha \rangle \qquad [\mathcal{L}_V, \mathcal{L}_W] = \mathcal{L}_{[V,W]}.$$

The last property is related to Jacobi's identity

$$[U,[V,W]] + [V,[W,U]] + [W,[U,V]] = 0.$$

If $\Psi\colon \mathcal{A} \to \mathcal{B}$ is a smooth map, a vector field W on $\mathcal{B}$ is said to be Ψ-related to a vector field V on $\mathcal{A}$ if $W_{\Psi(x)} = \Psi_{*x} V_x$ for all $x \in \mathcal{A}$. If W_1, W_2 are Ψ-related to V_1, V_2 respectively then $[W_1, W_2]$ is Ψ-related to $[V_1, V_2]$.

The tangent vector at $t = 0$ to the curve $t \mapsto \phi_t(\psi_t(x))$ is $V_x + W_x$, where V and W are the generators of ϕ and ψ. The tangent vector at $t = 0$ to the curve $t \mapsto \psi_{-t}(\phi_{-t}(\psi_t(\phi_t(x))))$ is zero, but its second-order tangent vector is $2[V,W]_x$. Flows commute if and only if their generators do.

The covariant derivative operator ∇ is defined by $(\nabla_V W)_x = \nabla_{V_x} W = DW/Dt$, where DW/Dt is the covariant derivative of the restriction of W to the integral curve of V through x (or any other curve through x to which V_x is tangent) evaluated at x. It is distinct from the Lie derivative, depending on parallelism for its definition. It is linear in both arguments, and

$$\nabla_V(fW) = f\nabla_V W + (Vf)W \qquad \nabla_{fV} W = f\nabla_V W$$
$$\langle \nabla_V W, \alpha \rangle + \langle W, \nabla_V \alpha \rangle = V\langle W, \alpha \rangle \qquad \nabla_V W - \nabla_W V = [V, W]$$
$$[\nabla_V, \nabla_W] = \nabla_{[V,W]}$$

(this last property being special to affine space).

In the terminology of vector calculus, the operation of directional differentiation of a function f by a vector field $\mathbf{V}$ would be written $\mathbf{V} \cdot \operatorname{grad} f$. The covariant derivative would be written $(\mathbf{V} \cdot \operatorname{grad})\mathbf{W}$. The Lie derivative is practically unknown in vector calculus.

Note to Chapter 3

Solution of systems of ordinary differential equations. Let V^a, $a = 1, 2, \ldots, n$, be smooth functions defined on some open connected subset $\mathcal{O}$ of $\mathbf{R}^n$. Then for every point $(x^a) \in \mathcal{O}$ there are smooth functions σ^a, defined on some open interval of $\mathbf{R}$ containing 0, such that

$$\frac{d\sigma^a}{dt} = V^a(\sigma^b) \qquad \text{and} \qquad \sigma^a(0) = x^a.$$

Moreover, the functions σ^a are unique, in the sense that any other functions with the same properties coincide with the σ^a on the intersection of their domains, which is an open interval about 0. In other words, the system of ordinary first-order differential equations $\dot{x}^a = V^a(x^b)$ has a unique solution with given initial conditions.

This theorem is proved in many books. Very often the more general situation in which the functions V^a depend also on the variable t is considered. However, the result is most often proved with the assumption that the V^a are C^1 only. See for example Sanchez [1968] Chapter 6, where the problem of piecing together local solutions to obtain a maximal solution is also discussed. The proof of the theorem under the smoothness conditions stated above is more difficult: a proof may be found in Lang [1969], pp 126*ff*.

In case the map $(x^a) \mapsto (V^a(x^b))$ is linear the equations always admit the solution $\sigma^a = 0$; thus, by uniqueness, a solution of the equations which is zero anywhere is zero everywhere.

The books by Arnold [1973] and by Coddington and Levinson [1955] are standard.

4. VOLUMES AND SUBSPACES: EXTERIOR ALGEBRA

In ordinary Euclidean space the volume of a parallelepiped whose edges are vectors $\mathbf{e}_1, \mathbf{e}_2$ and $\mathbf{e}_3$ is $\det(e_a^b)$ where e_1^a, e_2^a and e_3^a are the orthogonal Cartesian components of $\mathbf{e}_1, \mathbf{e}_2$ and $\mathbf{e}_3$. In an affine space without additional structure, on the other hand, the idea of volume is without intrinsic significance for, like length, volume is not preserved by general affine transformations. However, as we shall show in this chapter, it is possible, exploiting the properties of determinants, to introduce an idea of volume into an affine space without introducing a Euclidean measure of length. Thus the availability of a measure of length is sufficient for the definition of volume, but it is not necessary.

The statement that in Euclidean space the volume of a parallelepiped is given by a determinant requires some qualification: the value of a determinant may turn out to be zero or negative, both somewhat unlikely "volumes", as the word is commonly used. However, it is convenient, in a systematic treatment, to give up the common usage which expects volumes to be positive numbers. The value zero is obtained when the vectors along the edges of the parallepiped are linearly dependent, so that it collapses into a plane figure. Whether a non-zero value is positive or negative depends on a convention. To explain the convention we distinguish between right-handed and left-handed sets of vectors: a set of mutually perpendicular vectors $\mathbf{e}_1$, $\mathbf{e}_2$ and $\mathbf{e}_3$ in ordinary Euclidean space is called *right-handed* if when the vector $\mathbf{e}_3$ is grasped by the right hand, thumb extended in the sense of that vector, the fingers wrap around the vector in the sense of rotation from $\mathbf{e}_1$ to $\mathbf{e}_2$. The set is called *left-handed* if the same is true when the vector $\mathbf{e}_3$ is grasped by the left hand.

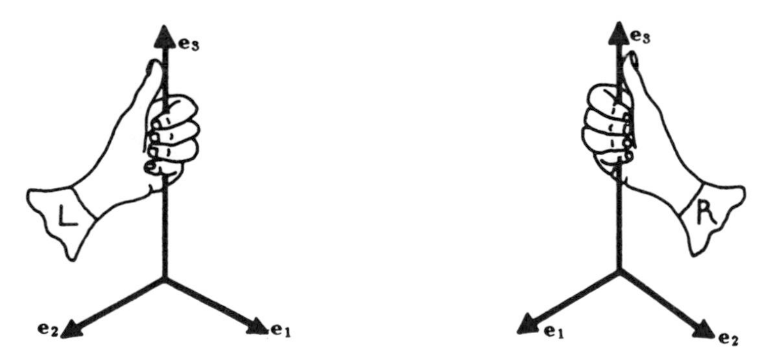

Fig. 1 Left- and right-handed sets of vectors.

The usual convention is to assign positive volume to a parallelepiped whose edges can be obtained from a right-handed set by a transformation with positive determinant, and negative volume to a parallelepiped whose edges can be obtained from a right-handed set by a transformation with negative determinant. Since a left-handed set is obtained from a right-handed set by a transformation with negative determinant, a parallelepiped whose edges can be obtained from a left-handed set by a transformation with positive determinant will have negative volume, according to this convention. An ordered set of vectors which comprises the edges of a parallelepiped with positive volume is said to have *positive orientation*, with negative volume, *negative orientation*. However, it should be emphasised that an initial choice is necessary, and that it would be perfectly consistent to assign positive volume and orientation to left-handed sets of axes instead of right-handed ones.

To give substance to the idea of volume, one has not only to specify it, but also to pick out those affine transformations which preserve it. As we shall show, such transformations form a subgroup of the affine group.

Another application of determinants is to the characterisation of subspaces of an affine space. This is so because one can identify a subspace by a determinant of pairings formed from vectors which span it. The familiar algebraic properties of determinants may be summed up by saying that a determinant is an alternating multilinear function on its rows or columns, and in studying subspaces one is confronted by such functions at every turn, which leads one to investigate them more carefully for the sake of their geometrical interpretations. The study of alternating multilinear functions, which is called *exterior algebra*, underlies all of the developments in this chapter.

In the first seven sections we develop first the idea of volume and then the ideas about subspaces going, as far as possible, from geometric property to formula. In the subsequent sections we give an introduction to exterior algebra, and go from some of the algebraic formulae to the corresponding geometric properties. (Determinants are defined, and some of their properties listed, in Note 2 to this chapter.)

1. Volume of a Parallelepiped

A parallelepiped is the many-dimensional generalisation of a parallelogram. The area of a parallelogram may be written as a determinant. In this section we define the volume of any parallelepiped in an affine space and show how it, too, may be written as a determinant.

The argument takes place in an affine space $\mathcal{A}$ modelled on a real n-dimensional vector space $\mathcal{V}$. A *parallelepiped* $(x; v_1, v_2, \ldots, v_n)$ in $\mathcal{A}$ is specified by giving a point x, the *principal vertex* of the parallelepiped, and a set of vectors $v_1, v_2, \ldots, v_n$ in a definite order, its *edges*. The parallelepiped consists of the set of points $\{ x + t^a v_a \mid 0 \leq t^a \leq 1 \}$ (range and summation conventions for a). The *vertices* are the points $x + t^a v_a$ for which each t^a is either 0 or 1, and the *faces* are portions of hyperplanes obtained by setting one of the t^a equal to 0 or 1 and allowing the others to vary in their domain. The $2n$ faces are thus divided into n pairs, the faces in each pair lying in parallel hyperplanes. If the vectors v_a are not linearly independent then the parallelepiped will be degenerate, and some of these assertions will need modification: some faces may be lower dimensional and some pairs of faces may lie in the same hyperplane.

Two parallelepipeds are counted as different, even though they comprise the same point sets, if a different vertex is chosen to be the principal one or if the edges are given in a different order.

An affine transformation of $\mathcal{A}$ takes parallelepipeds into parallelepipeds. The affine transformation A with linear part λ takes the parallelepiped $(x; v_1, v_2, \ldots, v_n)$ into $(A(x); \lambda(v_1), \lambda(v_2), \ldots, \lambda(v_n))$. In particular, a translation τ_w takes the parallelepiped $(x; v_1, v_2, \ldots, v_n)$ into $(x + w; v_1, v_2, \ldots, v_n)$.

Exercise 1. Show that the parallelepiped $\tau_v(x; -v, v_2, \ldots, v_n)$ encompasses the same point set as $(x; v, v_2, \ldots, v_n)$. □

Volume functions. It should be emphasised that, while a parallelepiped may be defined entirely in terms of the affine structure on an affine space, a measure of volume of parallelepipeds is something which is introduced as an additional structure, and which entails not only a conventional choice of sign, as in the case of ordinary Euclidean space described above, but also a choice of scale for volumes. In an affine space there are, to start with, no orthogonal unit vectors in terms of which to define unit volume, so that the unit of volume, as well as the orientation, has to be chosen. It is not to be expected, therefore, that there will be a single function with the properties of a measure of volume on an affine space, even allowing for the ambiguity of sign.

We shall approach the investigation of volume functions by setting out axioms which any measure of volume might be expected to satisfy. We use Ω to stand for a volume function, and $\Omega(x; v_1, v_2, \ldots, v_n)$ to denote the volume it ascribes to the parallelepiped $(x; v_1, v_2, \ldots, v_n)$. The axioms are as follows:

(1) the volume of a parallelepiped is a real number, and there is at least one parallelepiped for which it is not zero

(2) volume is unaltered by translation:

$$\Omega(x + w; v_1, v_2, \ldots, v_n) = \Omega(x; v_1, v_2, \ldots, v_n) \qquad \text{for all } w \in \mathcal{V}$$

(3) if an edge of a parallelepiped is scaled by factor k, its volume is scaled by the same factor:

$$\Omega(x; v_1, v_2, \ldots, kv_c, \ldots, v_n) = k\Omega(x; v_1, v_2, \ldots, v_c, \ldots, v_n)$$

for all real k

(4) if a multiple of one edge is added to another, the volume is unaltered:

$$\Omega(x; v_1, v_2, \ldots, v_b + kv_c, \ldots, v_c, \ldots, v_n)$$
$$= \Omega(x; v_1, v_2, \ldots, v_b, \ldots, v_c, \ldots, v_n) \qquad \text{for } b \neq c.$$

Assumption (1) implies that Ω is a function $\mathcal{A} \times \mathcal{V}^n \to \mathbf{R}$ (where $\mathcal{V}^n$ means $\mathcal{V} \times \mathcal{V} \times \cdots \times \mathcal{V}$ with n factors $\mathcal{V}$). It prohibits the assignment of volume 0 to all parallelepipeds—an assignment which would be consistent with practically any other plausible assumptions about volume, but is neither useful nor interesting. Assumption (2) requires that volumes respect the homogeneity of affine space. It asserts that the volume of a parallelepiped depends only on its edges and not on the position of the principal vertex, and so reduces the study of volume to considerations about the vector space $\mathcal{V}$ underlying the affine space $\mathcal{A}$: Ω is independent of its first argument, and accordingly one may write $\Omega(v_1, v_2, \ldots, v_n)$ instead of $\Omega(x; v_1, v_2, \ldots, v_n)$ and regard Ω as a map $\mathcal{V}^n \to \mathbf{R}$. Assumption (3) asserts, for example, that doubling an edge doubles the volume, and that if any edge is 0 then the volume is 0. It asserts also that if an edge is reversed then the sign of the volume is changed, the magnitude remaining the same. In view of Exercise 1 and assumption (2) this is consistent with what has been said already about orientation, and will allow us to define the orientation of any linearly independent ordered set of vectors in an affine space with a volume function: the ordered set of vectors $(v_1, v_2, \ldots, v_n)$ will be said to have positive orientation if the volume $\Omega(v_1, v_2, \ldots, v_n)$ is positive, negative orientation if $\Omega(v_1, v_2, \ldots, v_n)$ is negative. Assumption (4) generalises the rule that figures with the same base, lying between the same parallels, have the same volume. Consider, for example, two parallelepipeds with the same principal vertex x and with edges $(v_1, v_2, \ldots, v_n)$ and $(v_1, v_2, \ldots, v_n + kv_1)$ respectively, where $\{v_a\}$ are linearly independent. Each parallelepiped lies between the same two hyperplanes, the one through x spanned by $v_1, v_2, \ldots, v_{n-1}$ and the parallel one through $x + v_n$. The faces lying in the hyperplane through x coincide, while in the hyperplane through $x + v_n$ the face of the second parallelepiped is obtained from the face of the first by translating it through kv_1. These two parallelepipeds with the same base, lying between the same parallel hyperplanes, are to have the same volume.

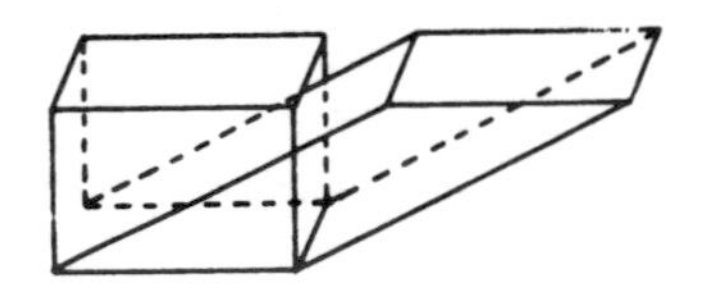

Fig. 2 The significance of axiom (4).

Exercise 2. Show that there is a parallelepiped whose volume is 1. □

Exercise 3. Let $w = w^a v_a \in \mathcal{V}$. Replace the cth edge of $(x; v_1, v_2, \ldots, v_n)$ by w. Show that

$$\Omega(v_1, v_2, \ldots, w, \ldots, v_n) = w^c \Omega(v_1, v_2, \ldots, v_n).$$

Show, in particular, that if the vectors $v_1, v_2, \ldots, v_n$ are linearly dependent then $\Omega(v_1, v_2, \ldots, v_n) = 0$. Deduce that the edges of any parallelepiped whose volume is non-zero constitute a basis for $\mathcal{V}$. □

Exercise 4. Show that the function Δ on triples of vectors in 3-dimensional Euclidean space defined by $\Delta(\mathbf{v}_1, \mathbf{v}_2, \mathbf{v}_3) = \mathbf{v}_1 \cdot \mathbf{v}_2 \times \mathbf{v}_3$ satisfies the axioms for a volume function. □

Exercise 5. Let $\{e_a\}$ be a basis for $\mathcal{V}$. Let Δ be the function which maps $(v_1, v_2, \ldots, v_n)$ to the determinant whose rows are the components of the v_a relative to the e_a as basis: if $v_a = v_a^b e_b$ then $\Delta(v_1, v_2, \ldots, v_n) = \det(v_a^b)$. Show that Δ satisfies the assumptions for a volume function. Let $\{\theta^a\}$ be the basis for $\mathcal{V}^*$ dual to $\{e_a\}$. Show that $\Delta(v_1, v_2, \ldots, v_n) = \det(\langle v_a, \theta^b \rangle)$. Show that any parallelepiped whose edges are $(e_1, e_2, \ldots, e_n)$ has volume 1 (as measured by Δ). □

Exercises 4 and 5 exhibit volume functions, and show, therefore, that our axioms for volume functions make sense. In the next section we show that the construction in Exercise 5 is essentially the only possible construction of a volume function.

2. Volume as an Alternating Multilinear Function

A function on a vector space with two or more arguments, which is linear in each argument, is said to be multilinear; a function which changes sign if any two of its arguments are interchanged is called alternating. To be explicit, suppose that $T: \mathcal{V}^p \to \mathbf{R}$ is a function with p arguments, each taken from $\mathcal{V}$. Then T is linear in its ith argument ($1 \leq i \leq p$) if, for every $v_1, v_2, \ldots, v_i, v_i', \ldots, v_n \in \mathcal{V}$ and every $k, k' \in \mathbf{R}$,

$$T(v_1, v_2, \ldots, kv_i + k'v_i', \ldots, v_p) = kT(v_1, v_2, \ldots, v_i, \ldots, v_p) + k'T(v_1, v_2, \ldots, v_i', \ldots, v_p).$$

The function T is *p-fold multilinear* if it is linear in each argument. It is *alternating* if

$$T(v_1, v_2, \ldots, v_j, \ldots, v_i, \ldots, v_p) = -T(v_1, v_2, \ldots, v_i, \ldots, v_j, \ldots, v_p)$$

for every collection of vector arguments, and for interchange of any pair of arguments. These two properties are of course quite independent; we shall have occasion later to deal with functions which are multilinear without being alternating.

We show now, from the assumptions about volume, that every volume function must be alternating multilinear. We show first, from the results of Exercise 3, that any volume function Ω is multilinear. We have to show that

$$\Omega(v_1, v_2, \ldots, v_c + v_c', \ldots, v_n) = \Omega(v_1, v_2, \ldots, v_c, \ldots, v_n) + \Omega(v_1, v_2, \ldots, v_c', \ldots, v_n);$$

since by assumption $\Omega(v_1, v_2, \ldots, kv_c, \ldots, v_n) = k\Omega(v_1, v_2, \ldots, v_c, \ldots, v_n)$, this is enough to show that Ω is linear in its cth argument, and so that it is multilinear (since the same result will hold for every c).

Suppose first that $\{v_1, v_2, \ldots, v_c, \ldots, v_n\}$ and $\{v_1, v_2, \ldots, v_c', \ldots, v_n\}$ are both linearly dependent. Then $\{v_1, v_2, \ldots, v_c + v_c', \ldots, v_n\}$ is also a linearly dependent set, and so

$$\Omega(v_1, v_2, \ldots, v_c + v_c', \ldots, v_n) = \Omega(v_1, v_2, \ldots, v_c, \ldots, v_n) + \Omega(v_1, v_2, \ldots, v_c', \ldots, v_n)$$

since each term is zero. On the other hand, suppose that at least one of these sets of vectors, say $\{v_1, v_2, \ldots, v_c, \ldots, v_n\}$, is linearly independent. Then $v_c' = kv_c + \sum_{d \neq c} k_d v_d$ say, and

$$\begin{aligned}\Omega(v_1, v_2, \ldots, v_c + v_c', \ldots, v_n) &= \Omega(v_1, v_2, \ldots, (1+k)v_c + \sum_{d \neq c} k_d v_d, \ldots, v_n) \\ &= (1+k)\Omega(v_1, v_2, \ldots, v_n)\end{aligned}$$

by Exercise 3, while

$$\Omega(v_1, v_2, \ldots, v_c', \ldots, v_n) = k\Omega(v_1, v_2, \ldots, v_c, \ldots, v_n)$$

so that again

$$\Omega(v_1, v_2, \ldots, v_c + v_c', \ldots, v_n)$$
$$= \Omega(v_1, v_2, \ldots, v_c, \ldots, v_n) + \Omega(v_1, v_2, \ldots, v_c', \ldots, v_n).$$

Thus Ω is linear in each argument; since it has n arguments it is an n-fold multilinear form on $\mathcal{V}$.

If two of the arguments of Ω are the same, then its arguments are certainly linearly dependent, so that, again by Exercise 3, its value is zero. From the multilinearity it now follows, inserting $v + v'$ in any two places, that

$$\begin{aligned}\Omega(v_1, v_2, \ldots, v + v', \ldots, v + v', \ldots, v_n) &= 0 \\ = \Omega(v_1, v_2, \ldots, v, \ldots, v, \ldots, v_n) &+ \Omega(v_1, v_2, \ldots, v, \ldots, v', \ldots, v_n) \\ + \Omega(v_1, v_2, \ldots, v', \ldots, v, \ldots, v_n) &+ \Omega(v_1, v_2, \ldots, v', \ldots, v', \ldots, v_n)\end{aligned}$$

leaving only

$$\Omega(v_1, v_2, \ldots, v, \ldots, v', \ldots, v_n) + \Omega(v_1, v_2, \ldots, v', \ldots, v, \ldots, v_n) = 0,$$

which is to say, interchange of any pair of arguments changes the sign. Hence Ω is alternating: it is an n-fold alternating multilinear form on $\mathcal{V}$. An alternating multilinear form is also called an exterior form; thus Ω is called an *n-fold exterior form*, or *exterior n-form*, or even simply *n-form*. These latter expressions, although less expressive of the nature of the object, are preferred for their brevity and established by common usage. The word "skew" or "skew-symmetric" is also used as a synonym for "alternating"; however, we shall reserve skew-symmetric to refer to the components of an exterior form, to be introduced shortly.

If $\mathcal{A}$ is an affine space modelled on $\mathcal{V}$, then an exterior form on $\mathcal{V}$ determines an exterior form at each point of $\mathcal{A}$. The same language is used to describe forms on $\mathcal{A}$ and on $\mathcal{V}$.

Volumes and determinants. We now make explicit the connections between volumes, exterior forms, and determinants. Let $(x; e_1, e_2, \ldots, e_n)$ be a parallelepiped whose volume is not zero. The e_a must be linearly independent, and hence a basis for $\mathcal{V}$. Now let π be any permutation of $(1, 2, \ldots, n)$. Then

$$\Omega(e_{\pi(1)}, e_{\pi(2)}, \ldots, e_{\pi(n)}) = \epsilon(\pi)\Omega(e_1, e_2, \ldots, e_n),$$

where $\epsilon(\pi)$ is the parity of π; because if π is written as a product of transpositions, each transposition will effect a change of sign, by the alternating property of Ω, and $\epsilon(\pi)$ is just (-1) to the power of the number of transpositions. We shall express the volume of any other parallelepiped in terms of the volume of this one by writing the vectors which specify its edges relative to the e_a as basis. For this purpose it is convenient to introduce the Levi-Civita *alternating symbol*, which is a tensor-algebraic device for constructing determinants: for each ordered set of n integers $(a_1, a_2, \ldots, a_n)$ with $1 \leq a_i \leq n$ let

$$\epsilon_{a_1 a_2 \ldots a_n} = \epsilon^{a_1 a_2 \ldots a_n} = \begin{cases} 0 & \text{if } a_1, a_2, \ldots, a_n \text{ are not all different} \\ \epsilon(\pi) & \text{otherwise} \end{cases}$$

where, in the second case, π is the permutation which brings $(1,2,\ldots,n)$ to $(a_1,a_2,\ldots,a_n)$.

Now choose any $(a_1,a_2,\ldots,a_n)$, possibly with repetitions, and define

$$\Omega_{a_1a_2\ldots a_n} = \Omega(e_{a_1},e_{a_2},\ldots,e_{a_n}).$$

These numbers (for all possible $(a_1,a_2,\ldots,a_n)$) are called the *components* of Ω relative to the basis $\{e_a\}$. Notice that the array of components is *skew-symmetric*: if any two of its indices are interchanged a component changes sign, and if any two are equal it is zero. Comparing the definition of components with the expression for $\Omega(e_{\pi(1)},e_{\pi(2)},\ldots,e_{\pi(n)})$ given above, one sees that

$$\Omega_{a_1a_2\ldots a_n} = \epsilon_{a_1a_2\ldots a_n}\Omega(e_1,e_2,\ldots,e_n)$$

with each index a_i allowed to range over $1,2,\ldots,n$.

Now let $(x;v_1,v_2,\ldots,v_n)$ be any other parallelepiped, and express the v_a in terms of the e_a, say $v_a = v_a^b e_b$. By the multilinearity of Ω,

$$\begin{aligned}\Omega(v_1,v_2,\ldots,v_n) &= v_1^{a_1}v_2^{a_2}\cdots v_n^{a_n}\Omega(e_{a_1},e_{a_2},\ldots,e_{a_n})\\ &= \epsilon_{a_1a_2\ldots a_n}v_1^{a_1}v_2^{a_2}\cdots v_n^{a_n}\Omega(e_1,e_2,\ldots,e_n);\end{aligned}$$

but from the definition of $\epsilon_{a_1a_2\ldots a_n}$ and the definition of a determinant,

$$\epsilon_{a_1a_2\ldots a_n}v_1^{a_1}v_2^{a_2}\cdots v_n^{a_n} = \sum \epsilon(\pi)v_1^{\pi(1)}v_2^{\pi(2)}\cdots v_n^{\pi(n)} = \det(v_a^b),$$

(the sum on the right being over all permutations π of $(1,2,\ldots,n)$) since the n summations on the left contribute only when $(a_1,a_2,\ldots,a_n)$ is a permutation of $(1,2,\ldots,n)$. Substituting in the preceding formula, one is left with

$$\Omega(v_1,v_2,\ldots,v_n) = \det(v_a^b)\Omega(e_1,e_2,\ldots,e_n).$$

Compare this with the result of Exercise 5, which is a special case. What has been shown here is that any volume, because it is an exterior n-form, may be written as a determinant of its edge vector components times a standard volume.

Exercise 6. Let $\{\acute{e}_a\}$ be another basis for $\mathcal{V}$, related to the basis already introduced by $\acute{e}_a = h_a^b e_b$. Show that

$$\acute{\Omega}_{a_1a_2\ldots a_n} = \det(h_a^b)\Omega_{a_1a_2\ldots a_n},$$

where $\acute{\Omega}_{a_1a_2\ldots a_n}$ are the components of Ω relative to the new basis. □

Exercise 7. Let $\{\theta^1,\theta^2,\ldots,\theta^n\}$ be a set of covectors, given in order, and let $\Omega\colon \mathcal{V}^n \to \mathbf{R}$ be the function defined by $\Omega(v_1,v_2,\ldots,v_n) = \det(\langle v_a,\theta^b\rangle)$. Show that Ω is an exterior n-form, that is, that it is multilinear and alternating, and that every non-zero exterior n-form may be so obtained, by a suitable choice of $\{\theta^a\}$. Show that Ω takes the value zero on every set of vectors if and only if the θ^a are linearly dependent, and that if the θ^a are linearly independent then Ω takes the value zero only when its arguments are linearly dependent. Let $\{\acute{\theta}^a\}$ be another set of covectors, related to the given one by $\acute{\theta}^a = k_b^a\theta^b$. Show that $\{\acute{\theta}^a\}$ yields the same exterior n-form, by $\det(\langle v_a,\acute{\theta}^b\rangle)$, if and only if $\det(k_b^a) = 1$. □

The exterior n-form defined by $\Omega(v_1,v_2,\ldots,v_n) = \det(\langle v_a,\theta^b\rangle)$, the covectors θ^a being given, is called the *exterior product* of the θ^a and denoted

$$\Omega = \theta^1 \wedge \theta^2 \wedge \cdots \wedge \theta^n$$

(the mark $\wedge$ is read "wedge"). The order of factors in the product is important. In fact, this construction is not limited to n-forms. If $\{\theta^1, \theta^2, \ldots, \theta^r\}$ are any r covectors, given in order, and $\omega: \mathcal{V}^r \to \mathbf{R}$ is the function defined by

$$\omega(v_1, v_2, \ldots, v_r) = \det(\langle v_\alpha, \theta^\beta \rangle)$$

then ω is an r-fold multilinear alternating form. Its geometrical significance will become apparent in Section 4.

Exercise 8. Show that

$$\theta^{\pi(1)} \wedge \theta^{\pi(2)} \wedge \cdots \wedge \theta^{\pi(n)} = \epsilon(\pi)\theta^1 \wedge \theta^2 \wedge \cdots \wedge \theta^n$$

for any permutation π. □

Exercise 9. Let Ω and $\hat{\Omega}$ be two volume functions on $\mathcal{V}$, let $\{e_a\}$ be a basis for $\mathcal{V}$, and let $\{v_a\}$ be any n vectors in $\mathcal{V}$. Show that if $\Omega(v_1, v_2, \ldots, v_n) = 0$ then $\hat{\Omega}(v_1, v_2, \ldots, v_n) = 0$; and that

$$\frac{\hat{\Omega}(v_1, v_2, \ldots, v_n)}{\Omega(v_1, v_2, \ldots, v_n)} = \frac{\hat{\Omega}(e_1, e_2, \ldots, e_n)}{\Omega(e_1, e_2, \ldots, e_n)}$$

when $\Omega(v_1, v_2, \ldots, v_n) \neq 0$. □

It follows from the result of this exercise that $\hat{\Omega}$ differs from Ω by the same constant factor for the volumes of all parallelepipeds, and that any volume function is determined completely by its value on one ordered basis of $\mathcal{V}$. The only n-form which does not correspond to a volume is the zero form, which takes the value zero whatever its arguments. Consequently, if $\{\theta^a\}$ is any basis for covectors, every exterior n-form may be written $\Omega = k\theta^1 \wedge \theta^2 \wedge \cdots \wedge \theta^n$ for some k, and Ω is a volume function if and only if $k \neq 0$. The non-zero n-forms may be divided into two disjoint sets, the forms in each set differing from one another by a positive factor and from the forms in the other set by a negative factor. This allows one to generalise the idea of an orientation, described in the introduction to this chapter. An *orientation* on $\mathcal{V}$ is a choice of one of these two sets of n-forms, and two forms in the same set are said to define the same orientation. Choice of an orientation for $\mathcal{V}$ amounts to choice of a volume function, up to a positive factor. An orientation having been chosen, an ordered basis $\{e_1, e_2, \ldots, e_n\}$ for $\mathcal{V}$ is said to be *positively* or *negatively oriented* according as $\Omega(e_1, e_2, \ldots, e_n)$ is positive or negative, for some, and hence for any, volume function defining that orientation.

If $\mathcal{A}$ is an affine space modelled on $\mathcal{V}$, then an orientation on $\mathcal{V}$ determines an orientation on $\mathcal{A}$. The other definitions just stated may be repeated word for word for $\mathcal{A}$. As has been pointed out before, the choice of an orientation, like the choice of a volume form, is conventional.

3. Transformation of Volumes

We have already described the effect of an affine transformation on a parallelepiped. We show now what it does to a volume function Ω. Let $(x; v_1, v_2, \ldots, v_n)$ be a parallelepiped in $\mathcal{A}$, and let Λ be an affine transformation of $\mathcal{A}$ with linear part λ. The transformed parallelepiped is $(\Lambda(x); \lambda(v_1), \lambda(v_2), \ldots, \lambda(v_n))$ and its volume is $\Omega(\lambda(v_1), \lambda(v_2), \ldots, \lambda(v_n))$. Notice that the function

$$(v_1, v_2, \ldots, v_n) \mapsto \Omega(\lambda(v_1), \lambda(v_2), \ldots, \lambda(v_n))$$

is alternating and multilinear, so that it is an n-form on $\mathcal{V}$. It must therefore be a multiple of the n-form Ω; call the multiplying factor $D(\lambda)$. We show that $D(\lambda)$ is just the determinant of the matrix representing λ, with respect to any basis. Let $\{e_a\}$ be any basis for $\mathcal{V}$, and let $\lambda(e_a) = \lambda_a^b e_b$. Then

$$\begin{aligned} D(\lambda)\Omega(e_1, e_2, \dots, e_n) &= \Omega(\lambda(e_1), \lambda(e_2), \dots, \lambda(e_n)) \\ &= \lambda_1^{a_1}\lambda_2^{a_2}\cdots\lambda_n^{a_n}\Omega(e_{a_1}, e_{a_2}, \dots, e_{a_n}) \\ &= \epsilon_{a_1 a_2 \dots a_n}\lambda_1^{a_1}\lambda_2^{a_2}\cdots\lambda_n^{a_n}\Omega(e_1, e_2, \dots, e_n) \\ &= \det(\lambda_b^a)\Omega(e_1, e_2, \dots, e_n). \end{aligned}$$

But Ω is non-zero and $\{e_a\}$ a basis, so $\Omega(e_1, e_2, \dots, e_n) \neq 0$. Thus $D(\lambda) = \det(\lambda_b^a)$ as asserted. Since it is independent of the choice of basis, we can write $\det\lambda$ for $\det(\lambda_b^a)$.

Exercise 10. Show that det does not depend on the choice of Ω. □

Exercise 11. Let A and M be affine transformations with linear parts λ and μ. By considering the effect of $\mathrm{M}\circ\mathrm{A}$ on a volume form Ω show that $\det(\mu\circ\lambda) = (\det\mu)(\det\lambda)$. □

Exercise 12. Show that $\det\lambda = 0$ if and only if λ is singular. □

Exercise 13. Let λ be a linear map and Ω a non-zero n-form on $\mathcal{V}$. Show that the function $\mathcal{V}^n \to \mathbf{R}$ given by

$$(v_1, v_2, \dots, v_n) \mapsto \sum_{c=1}^{n} \Omega(v_1, v_2, \dots, \lambda(v_c), \dots, v_n)$$

is an n-form on $\mathcal{V}$, and is therefore a constant multiple of Ω. Show that the factor is the sum of the diagonal entries of the matrix representing λ with respect to any basis for $\mathcal{V}$, and that it is independent of choice of Ω. □

The factor is called the *trace* of λ and denoted $\operatorname{tr}\lambda$.

Whether or not an affine transformation changes orientation is determined by the sign of the determinant of its linear part: if $\det\lambda > 0$ then orientation is preserved, if $\det\lambda < 0$ it is reversed. The orientation-preserving affine transformations form a subgroup of the group of all affine transformations. The linear parts λ with $\det\lambda > 0$ form a subgroup of the general linear group $GL(\mathcal{V})$. The affine transformations with $\det\lambda = 1$ preserve not only orientation but volume itself, and no other affine transformations preserve volume. A linear transformation λ with $\det\lambda = 1$ is called *unimodular*, and an affine transformation with unimodular linear part may also be called unimodular.

Exercise 14. Show that the unimodular linear transformations of a vector space $\mathcal{V}$ form a group, and that this group is a normal subgroup of $GL(\mathcal{V})$. Show that the unimodular affine transformation of an affine space form a group, and that this group is a normal subgroup of the group of all affine transformations. □

The groups of unimodular transformations of $\mathbf{R}^n$ and of $\mathcal{V}$ respectively are denoted $SL(n)$ or $SL(n, \mathbf{R})$ and $SL(\mathcal{V})$ and called *special linear*, "special" being in this context a synonym for "unimodular".

The assignment of an orientation and the assignment of a volume function are two examples of the addition, to the affine structure of an affine space, of a further structure whose preservation entails the restriction of transformations of the space to a subgroup of the affine group. This is a frequently occurring situation; further examples will appear in Chapter 7.

It is instructive to compute the rate of change of volume under the transformations of a one-parameter affine group. Let ϕ_t be a one-parameter affine group, with linear part λ_t. Then

$$\frac{d}{dt}\left(\Omega(\lambda_t(v_1),\lambda_t(v_2),\ldots,\lambda_t(v_n))\right)_{t=0} = \sum_{c=1}^{n} \Omega(v_1,v_2,\ldots,A(v_c),\ldots,v_n)$$
$$= (\operatorname{tr} A)\Omega(v_1,v_2,\ldots,v_n)$$

where $A = d/dt(\lambda_t)(0)$. The vector field V which generates ϕ_t is given with respect to affine coordinates by $(A^a_b x^b + B^a)\partial_a$, where $A = (A^a_b)$, and $\operatorname{tr} A$ may be suggestively expressed in the form $\partial_a V^a$. We shall use this in the next chapter as the basis for a definition of the divergence of a vector field. Notice that if ϕ_t is a one-parameter group of unimodular affine transformations (which preserve volume) then $\operatorname{tr} A = 0$.

4. Subspaces

In this section we show how alternating multilinear forms may be used to characterise affine subspaces. The setting is again an affine space $\mathcal{A}$ modelled on a vector space $\mathcal{V}$ of dimension n. An affine subspace of $\mathcal{A}$ of dimension p, or p-plane, denoted $\mathcal{B}$, is constructed in $\mathcal{A}$ by attaching at the point x_0 a p-dimensional subspace $\mathcal{W}$ of $\mathcal{V}$. In Chapter 1, Section 2 we described two different ways of characterising the p-plane $\mathcal{B}$:

(1) parametrically, choosing a basis $\{w_\alpha\}$ for $\mathcal{W}$ $(\alpha = 1,2,\ldots,p)$: any point y of $\mathcal{B}$ may be written $y = x_0 + y^\alpha w_\alpha$

(2) by constraints, choosing a basis $\{\theta^\rho\}$ for constraint forms of hyperplanes which intersect in the p-plane $(\rho = p+1, p+2,\ldots,p+n)$: any point y of $\mathcal{B}$ must satisfy the equations $\langle y - x_0, \theta^\rho\rangle = 0$.

These two descriptions are related by the fact that the constraint forms vanish on vectors which lie in the p-plane, so that $\langle w_\alpha, \theta^\rho\rangle = 0$ for $\alpha = 1,2,\ldots,p$ and $\rho = p+1, p+2,\ldots,n$. They may be regarded as dual descriptions, in the sense that one is in terms of elements of $\mathcal{V}$, the other in terms of elements of the dual space $\mathcal{V}^*$. This duality pervades the developments which follow. The notation will be adapted to it: throughout this section, indices α, β from the beginning of the Greek alphabet will range and sum over $1,2,\ldots,p$, and indices ρ, σ from the middle of the alphabet will range and sum over the complementary values $p+1, p+2,\ldots,n$. Latin indices a, b will range and sum over $1,2,\ldots,n$ as hitherto.

Hyperplanes and multivectors. Each of the descriptions of a p-plane mentioned above is highly redundant; the p-plane may be determined by any set of p independent vectors which lie in it or any set of $n-p$ independent constraint forms for hyperplanes which intersect in it, and in choosing any particular set of either one is giving unnecessary information. Experience shows that descriptions which include the minimum of unnecessary information are likely to be the most revealing, and so it is worth seeking a less redundant description. A description in terms of determinants is suggested by the facts that one could change the order of the vectors, or constraint forms, in the independent set, or add a multiple of one of them to another, without changing the p-plane—compare the properties of volumes set out in previous sections.

A determinantal description turns out to be convenient, and to have other important applications. In formulating it we suppose the point x_0 at which the subspace $\mathcal{B}$ is attached to be fixed once for all, so that we may as well deal with the position of the subspace $\mathcal{W}$ in the vector space $\mathcal{V}$ as with the position of the affine subspace $\mathcal{B}$ in the affine space $\mathcal{A}$.

Out of the basis $\{w_\alpha\}$ for $\mathcal{W}$ we construct a p-fold alternating multilinear function W on the space $\mathcal{V}^*$ of all covectors, as follows: for any $\eta^1, \eta^2, \dots, \eta^p \in \mathcal{V}^*$,

$$W(\eta^1, \eta^2, \dots, \eta^p) = \det\big(\langle w_\alpha, \eta^\beta\rangle\big).$$

Writing this determinant out with the help of the Levi-Civita symbol one finds

$$\begin{aligned} W(\eta^1, \eta^2, \dots, \eta^p) &= \epsilon^{\alpha_1 \alpha_2 \dots \alpha_p} \langle w_{\alpha_1}, \eta^1\rangle \langle w_{\alpha_2}, \eta^2\rangle \cdots \langle w_{\alpha_p}, \eta^p\rangle \\ &= \epsilon^{\alpha_1 \alpha_2 \dots \alpha_p} w^{a_1}_{\alpha_1} w^{a_2}_{\alpha_2} \cdots w^{a_p}_{\alpha_p} \eta^1_{a_1} \eta^2_{a_2} \cdots \eta^p_{a_p} \\ &= W^{a_1 a_2 \dots a_p} \eta^1_{a_1} \eta^2_{a_2} \cdots \eta^p_{a_p} \end{aligned}$$

where

$$W^{a_1 a_2 \dots a_p} = \epsilon^{\alpha_1 \alpha_2 \dots \alpha_p} w^{a_1}_{\alpha_1} w^{a_2}_{\alpha_2} \cdots w^{a_p}_{\alpha_p}$$

are the *components* of W with respect to a basis for $\mathcal{V}$, in which the w_α have components w^a_α, while the η^α have components η^α_a with respect to the dual basis for $\mathcal{V}^*$.

Exercise 15. Verify that this function W is alternating and multilinear. □

Exercise 16. Show that if $\acute{w}_\alpha = k^\beta_\alpha w_\beta$, where $\{\acute{w}_\alpha\}$ is another basis for $\mathcal{W}$, and if $\acute{W}(\eta^1, \eta^2, \dots, \eta^p) = \det(\langle \acute{w}_\alpha, \eta^\beta\rangle)$, then $\acute{W} = \det(k^\beta_\alpha) W$. □

A multilinear alternating function on $\mathcal{V}^*$ is called a *multivector*, and a p-fold multivector is called a *p-vector*. A 2-vector is usually called a *bivector*. It follows from the last exercise that each p-plane determines a p-vector, up to a non-zero scalar factor. Any one of these p-vectors will be called a *characterising p-vector* for the p-plane. The p-vector W defined above is denoted

$$W = w_1 \wedge w_2 \wedge \cdots \wedge w_p$$

and called the *exterior product* of the w_α (and $\wedge$ is again read "wedge"). Any multivector which may be written in this way as an exterior product of vectors is called *decomposable*, but not all multivectors are decomposable.

One can easily retrieve from a decomposable p-vector W a basis for the vector subspace which it characterises, but one cannot, for $p > 1$, reconstruct the individual vectors of which it was formed as exterior product—the whole idea, after all, was to find a description of the p-plane avoiding any particular choice of basis for it. The retrieval can be carried out by acting with W on $p - 1$ covectors, leaving one argument to be filled: $W(\,\cdot\,, \eta^2, \dots, \eta^p)$, with first argument left empty, is a linear form on $\mathcal{V}^*$, to be evaluated by filling in the first argument, and is thus an element of $\mathcal{V}$. Explicitly,

$$W(\,\cdot\,,\eta^2,\ldots,\eta^p) = \epsilon^{\alpha_1\alpha_2\ldots\alpha_p}\langle w_{\alpha_2},\eta^2\rangle\cdots\langle w_{\alpha_p},\eta^p\rangle w_{\alpha_1};$$

this is a linear combination of the original vectors, hence lies in the subspace $\mathcal{W}$. Choosing the η^α $p-1$ at a time from a basis for $\mathcal{V}$, one recovers a spanning set for $\mathcal{W}$. This shows that the p-vector W determines the subspace $\mathcal{W}$, and hence also, given a point x_0 on it, the p-plane $\mathcal{B}$.

Exercise 17. By completing $\{w_\alpha\}$ in any manner to a basis for $\mathcal{V}$, and choosing the η^α from the dual basis for $\mathcal{V}^*$, show that $W(\,\cdot\,,\eta^2,\ldots,\eta^p)$ spans the subspace $\mathcal{W}$ as the η^α are varied. □

Exercise 18. Let π be a permutation of $(1,2,\ldots,p)$. Show that

$$w_{\pi(1)}\wedge w_{\pi(2)}\wedge\cdots\wedge w_{\pi(p)} = \epsilon(\pi)w_1\wedge w_2\wedge\cdots\wedge w_p. \quad □$$

Exercise 19. Let $\{e_a\}$ be a basis for $\mathcal{V}$, relative to which a p-vector W has components $W^{a_1a_2\cdots a_p}$; let $\{\acute{e}_a\}$ be another basis for $\mathcal{V}$, related to $\{e_a\}$ by $\acute{e}_a = h_a^b e_b$; and let $\acute{W}^{a_1a_2\cdots a_p}$ be the components of W relative to $\{\acute{e}_a\}$. Show that $\acute{W}^{a_1a_2\cdots a_p} = h_{b_1}^{a_1}h_{b_2}^{a_2}\cdots h_{b_p}^{a_p}W^{b_1b_2\cdots b_p}$. Thus each index of W transforms in the same way as a vector index under change of basis. □

Hyperplanes and exterior forms. We now develop the dual description of a p-plane, in terms of constraint forms for the hyperplanes which intersect in it. The process is very similar to that just carried out. The notation is as before: $\{\theta^\rho\}$ is a basis for constraint forms determining the affine subspace $\mathcal{B}$ constructed by attaching at the chosen point x_0 of $\mathcal{A}$ the vector subspace $\mathcal{W}$ of $\mathcal{V}$. Therefore $\langle w,\theta^\rho\rangle = 0$ for every $w\in\mathcal{W}$ and for $\rho = p+1,p+2,\ldots,n$. Out of the basis $\{\theta^\rho\}$ for covectors annihilating all vectors in $\mathcal{W}$ we construct an $(n-p)$-fold alternating multilinear function ω on $\mathcal{V}$, as follows:

$$\omega(v_{p+1},v_{p+2},\ldots,v_n) = \det\big(\langle v_\rho,\theta^\sigma\rangle\big)$$

for any $v_{p+1},v_{p+2},\ldots,v_n\in\mathcal{V}$. Writing this determinant out with the help of the Levi-Civita symbol one finds

$$\begin{aligned}\omega(v_{p+1},&v_{p+2},\ldots,v_n)\\ &= \epsilon_{\rho_{p+1}\rho_{p+2}\cdots\rho_n}\langle v_{p+1},\theta^{\rho_{p+1}}\rangle\langle v_{p+2},\theta^{\rho_{p+2}}\rangle\cdots\langle v_n,\theta^{\rho_n}\rangle\\ &= \epsilon_{\rho_{p+1}\rho_{p+2}\cdots\rho_n}\theta^{\rho_{p+1}}_{a_{p+1}}\theta^{\rho_{p+2}}_{a_{p+2}}\cdots\theta^{\rho_n}_{a_n}v^{a_{p+1}}_{p+1}v^{a_{p+2}}_{p+2}\cdots v^{a_n}_n\\ &= \omega_{a_{p+1}a_{p+2}\cdots a_n}v^{a_{p+1}}_{p+1}v^{a_{p+2}}_{p+2}\cdots v^{a_n}_n\end{aligned}$$

where

$$\omega_{a_{p+1}a_{p+2}\cdots a_n} = \epsilon_{\rho_{p+1}\rho_{p+2}\cdots\rho_n}\theta^{\rho_{p+1}}_{a_{p+1}}\theta^{\rho_{p+2}}_{a_{p+2}}\cdots\theta^{\rho_n}_{a_n}$$

are the components of ω with respect to a basis for $\mathcal{V}^*$, in which the θ^ρ have components θ^ρ_a, while the v_ρ have components v^a_ρ with respect to the dual basis for $\mathcal{V}$. In this calculation repeated as sum over $1,2,\ldots,n$ and repeated ρs sum over $p+1,p+2,\ldots,n$.

Exercise 20. Verify that this function ω is alternating and multilinear. Show that if $\theta^\rho = k^\rho_\sigma \acute{\theta}^\sigma$, where $\{\acute{\theta}^\sigma\}$ is another basis for forms which annihilate $\mathcal{W}$, and if $\acute{\omega}(v_{p+1}, v_{p+2}, \ldots, v_n) = \det(\langle v_\rho, \acute{\theta}^\sigma \rangle)$, then $\omega = \det(k^\rho_\sigma)\acute{\omega}$. □

Generalising from the nomenclature for a volume function, one calls any multilinear alternating function on $\mathcal{V}$ an *exterior form*; an r-fold exterior form is called an *exterior r-form*, or simply an *r-form*. The number r is called the *degree* of the form. In particular, a covector is also called a *1-form*, and a volume function is an *n-form*. The exterior $(n-p)$-form ω defined, as above, by $\omega(v_{p+1}, v_{p+2}, \ldots, v_n) = \det(\langle v_\rho, \theta^\sigma \rangle)$ is denoted

$$\omega = \theta^{p+1} \wedge \theta^{p+2} \wedge \cdots \wedge \theta^n$$

and called the *exterior product* of the θ^σ. Any exterior form which may be written as an exterior product of 1-forms is called *decomposable*, but not all exterior forms are decomposable.

A decomposable form which is the exterior product of independent constraint forms for an affine subspace is called a *characterising form* for the subspace and for the corresponding vector subspace $\mathcal{W}$. If any one (or more) of the vectors on which a characterising form ω is evaluated lies in the subspace $\mathcal{W}$ which it characterises, then the value is zero. If, for example, w lies in $\mathcal{W}$ and $v_{p+1}, \ldots, v_n$ are arbitrary vectors then $\omega(w, v_{p+2}, \ldots, v_n)$ is a determinant whose first column consists of pairings of w with $\theta^{p+1}, \theta^{p+2}, \ldots, \theta^n$ in turn, and all these pairings yield zeros. Conversely, if $\omega(w, v_{p+2}, \ldots, v_n) = 0$ for some fixed w and all $v_{p+2}, \ldots, v_n$, then w must lie in the subspace $\mathcal{W}$, for otherwise the constraint forms $\theta^{p+1}, \theta^{p+2}, \ldots, \theta^n$ could not be independent.

Exercise 21. Let ω be a decomposable r-form. Show how to recover a basis for the 1-forms of which it is the exterior product. □

Exercise 22. Let π be a permutation of $(1, 2, \ldots, p)$ and let θ^α be 1-forms. Show that $\theta^{\pi(1)} \wedge \theta^{\pi(2)} \wedge \cdots \wedge \theta^{\pi(p)} = \epsilon(\pi)\theta^1 \wedge \theta^2 \wedge \cdots \wedge \theta^p$. □

5. The Correspondence Between Multivectors and Forms

We now display a relation between the characterising multivector and characterising form descriptions of a subspace. Once again let $\mathcal{V}$ be an n-dimensional vector space and let $\mathcal{W}$ be a p-dimensional subspace of it. Let W be a characterising p-vector for $\mathcal{W}$, and ω a characterising $(n-p)$-form. Then there is a basis $\{w_\alpha\}$ for $\mathcal{W}$ such that $W = w_1 \wedge w_2 \wedge \cdots \wedge w_p$; and if $\{\acute{w}_\alpha\}$ is any other basis for $\mathcal{W}$ such that $W = \acute{w}_1 \wedge \acute{w}_2 \wedge \cdots \wedge \acute{w}_p$ then $\acute{w}_\alpha = k^\beta_\alpha w_\beta$ with $\det(k^\beta_\alpha) = 1$.

Now suppose given a volume form Ω on $\mathcal{V}$. Let $\tilde{\omega}$ be the function on $\mathcal{V}^{n-p}$ defined by filling in the first p arguments of the given volume form with $w_1, w_2, \ldots, w_p$:

$$\tilde{\omega} = \Omega(w_1, w_2, \ldots, w_p, \cdot, \cdot, \ldots, \cdot).$$

Then $\tilde{\omega}$ takes $n-p$ vectors as arguments, and is alternating and multilinear in them, and is therefore an $(n-p)$-form on $\mathcal{V}$. Moreover, if $\tilde{\omega}' = \Omega(\acute{w}_1, \acute{w}_2, \ldots, \acute{w}_p, \cdot, \cdot, \ldots, \cdot)$ is the $(n-p)$-form constructed in the same way from the basis $\{\acute{w}_\alpha\}$ then $\tilde{\omega}' = \tilde{\omega}$, as can be seen by filling in the remaining arguments arbitrarily with vectors chosen from $\mathcal{V}$ and recalling that the matrix relating the $\acute{w}_\alpha$ to the w_α has determinant 1. Hence $\tilde{\omega}$ is completely determined by the characterising p-vector W, irrespective of the choice of basis for which $W = w_1 \wedge w_2 \wedge \cdots \wedge w_p$. We shall show now that $\tilde{\omega}$ is a multiple of the characterising $(n-p)$-form ω.

To this end complete $\{w_\alpha\}$ arbitrarily to a basis $\{w_a\}$ for $\mathcal{V}$ by specifying $n-p$ further vectors $w_{p+1}, w_{p+2}, \ldots, w_n$. Let $\{\theta^a\}$ denote the dual basis for $\mathcal{V}^*$. Then $\langle w_\alpha, \theta^\rho \rangle = 0$ for each $\alpha = 1, 2, \ldots, p$ and each $\rho = p+1, p+2, \ldots, n$. Therefore the θ^ρ are constraint forms for the subspace $\mathcal{W}$, and $\theta^{p+1} \wedge \theta^{p+2} \wedge \cdots \wedge \theta^n$ is a characterising form for it. By the result of Exercise 20, $\theta^{p+1} \wedge \theta^{p+2} \wedge \cdots \wedge \theta^n = c\omega$ for some (non-zero) number c. Moreover, since $\{\theta^a\}$ is a basis, $\Omega = c'\theta^1 \wedge \theta^2 \wedge \cdots \wedge \theta^n$ for some (non-zero) number c'. Now let $v_{p+1}, v_{p+2}, \ldots, v_n$ be any $n-p$ vectors in $\mathcal{V}$ and evaluate $\tilde{\omega}$ on them:

$$\begin{aligned}\tilde{\omega}(v_{p+1}, v_{p+2}, \ldots, v_n) &= \Omega(w_1, w_2, \ldots, w_p, v_{p+1}, v_{p+2}, \ldots, v_n) \\ &= c'(\theta^1 \wedge \theta^2 \wedge \cdots \wedge \theta^n)(w_1, w_2, \ldots, v_n).\end{aligned}$$

Apart from the factor c', this expression is a determinant, which we divide into $p \times p$, $p \times (n-p)$, $(n-p) \times p$ and $(n-p) \times (n-p)$ blocks:

$$\frac{1}{c'}\tilde{\omega}(v_{p+1}, v_{p+2}, \ldots, v_n) = \begin{vmatrix} (\langle w_\alpha, \theta^\beta \rangle) & (\langle v_\rho, \theta^\beta \rangle) \\ (\langle w_\alpha, \theta^\sigma \rangle) & (\langle v_\rho, \theta^\sigma \rangle) \end{vmatrix} = \begin{vmatrix} I_p & (\langle v_\rho, \theta^\beta \rangle) \\ 0 & (\langle v_\rho, \theta^\sigma \rangle) \end{vmatrix}$$

since $\{w_a\}$ and $\{\theta^a\}$ are dual bases. Thus

$$\begin{aligned}\tilde{\omega}(v_{p+1}, v_{p+2}, \ldots, v_n) &= c' \det(\langle v_\rho, \theta^\sigma \rangle) \\ &= c'(\theta^{p+1} \wedge \theta^{p+2} \wedge \cdots \wedge \theta^n)(v_{p+1}, v_{p+2}, \ldots, v_n) \\ &= cc'\omega(v_{p+1}, v_{p+2}, \ldots, v_n).\end{aligned}$$

Thus $\tilde{\omega}$ is indeed a non-zero multiple of the characterising $(n-p)$-form ω, and is therefore itself a characterising form for $\mathcal{W}$.

We have shown that every characterising multivector determines a characterising form, by the construction

$$\tilde{\omega} = \Omega(w_1, w_2, \ldots, w_p, \cdot, \cdot, \ldots, \cdot),$$

the w_α being vectors which can be determined from the characterising form by the construction explained in the last section. The characterising $(n-p)$-form constructed in this way is called the *dual* of the characterising p-vector W with respect to the given volume form Ω. Choice of a different volume form will yield a dual which differs from this one by a scalar factor. It follows from these constructions that if a volume form is given on an affine space then it may be used to establish a 1 : 1 correspondence between decomposable p-vectors and decomposable $(n-p)$-forms.

Exercise 23. Let $W^{a_1 a_2 \ldots a_p}$ be the components of a given decomposable p-vector W relative to a chosen basis $\{e_a\}$ and let Ω be any volume form such that $\Omega(e_1, e_2, \ldots, e_n) = 1$. Show that

$$\omega_{a_{p+1} a_{p+2} \ldots a_n} = \epsilon_{a_1 a_2 \ldots a_p a_{p+1} \ldots a_n} W^{a_1 a_2 \ldots a_p}$$

are the components of the dual ω of W relative to Ω. □

Exercise 24. Let ω be a characterising $(n-p)$-form for a p-plane. Show that

$$W^{a_1 a_2 \ldots a_p} = \epsilon^{a_1 a_2 \ldots a_p a_{p+1} \ldots a_n} \omega_{a_{p+1} a_{p+2} \ldots a_n}$$

are components of a characterising p-vector W for the p-plane, and that ω is a multiple of the dual of W. □

6. Sums and Intersections of Subspaces

We now explain, by means of some examples, how the sums and intersections of subspaces may be characterised. Let $\mathcal{A}$ be an affine space of dimension n, modelled on a vector space $\mathcal{U}$, and let $\mathcal{B}$ and $\mathcal{C}$ be affine subspaces of $\mathcal{A}$, modelled on vector subspaces $\mathcal{V}$ and $\mathcal{W}$ respectively. We shall assume that $\mathcal{B}$ and $\mathcal{C}$ have in common at least one point x_0, and so consider them to have been constructed by the attachment of $\mathcal{V}$ and $\mathcal{W}$ to $\mathcal{A}$ at that point. The *intersection* $\mathcal{B} \cap \mathcal{C}$ of $\mathcal{B}$ and $\mathcal{C}$ means the largest affine subspace of $\mathcal{A}$ which lies in both; it comprises their common points, and is constructed by attaching $\mathcal{V} \cap \mathcal{W}$ at x_0. The *sum* $\mathcal{B} + \mathcal{C}$ of $\mathcal{B}$ and $\mathcal{C}$ means the smallest affine subspace of $\mathcal{A}$ which contains them both; it comprises points which can be reached from x_0 by a displacement in $\mathcal{B}$ followed by a displacement parallel to $\mathcal{C}$ or *vice versa*, and is constructed by attaching $\mathcal{V} + \mathcal{W}$ at x_0. The formula

$$\dim(\mathcal{B} + \mathcal{C}) + \dim(\mathcal{B} \cap \mathcal{C}) = \dim \mathcal{B} + \dim \mathcal{C}$$

follows from the corresponding one for vector subspaces (Chapter 1, Section 3).

(1) Suppose that $\mathcal{B}$ and $\mathcal{C}$ are distinct hyperplanes. Let η and ς be constraint forms, hence characterising forms, for $\mathcal{B}$ and $\mathcal{C}$ respectively. Points in $\mathcal{B} \cap \mathcal{C}$ must satisfy both $\langle x - x_0, \eta \rangle = 0$ and $\langle x - x_0, \varsigma \rangle = 0$; and therefore $\eta \wedge \varsigma$ is a characterising form for $\mathcal{B} \cap \mathcal{C}$, which must be an $(n-2)$-plane. (In this case $\mathcal{B} + \mathcal{C}$ is the whole space $\mathcal{A}$, as follows from the dimension formula).

(2) Suppose that $\mathcal{B}$ is a p-plane and $\mathcal{C}$ a q-plane, and that $\mathcal{B} + \mathcal{C}$ is the whole of $\mathcal{A}$, which entails $p + q > n$. Then $\dim(\mathcal{B} \cap \mathcal{C}) = p + q - n$. We shall characterise $\mathcal{B} \cap \mathcal{C}$. Any characterising form for $\mathcal{B}$ is a decomposable $(n-p)$-form, say $\omega = \eta^{p+1} \wedge \eta^{p+2} \wedge \cdots \wedge \eta^n$, and for $\mathcal{C}$, an $(n-q)$-form, say $\chi = \varsigma^{q+1} \wedge \varsigma^{q+2} \wedge \cdots \wedge \varsigma^n$. A point of $\mathcal{B} \cap \mathcal{C}$ must satisfy $\langle x - x_0, \eta^\rho \rangle = 0$, where $\rho = p+1, p+2, \ldots, n$, and $\langle x - x_0, \varsigma^\sigma \rangle = 0$, where $\sigma = q+1, q+2, \ldots, n$. If these constraints are linearly independent then there are $2n - (p+q)$ of them, so that they are satisfied on a subspace of dimension $n - \bigl(2n - (p+q)\bigr) = (p+q) - n$, which is exactly the dimension of $\mathcal{B} \cap \mathcal{C}$. Hence a characterising form for $\mathcal{B} \cap \mathcal{C}$ is

$$\eta^{p+1} \wedge \eta^{p+2} \wedge \cdots \wedge \eta^n \wedge \varsigma^{q+1} \wedge \cdots \wedge \varsigma^n,$$

formed by taking the exterior product of all the constraint forms for the two subspaces. This product is written $\omega \wedge \chi$ and called the *exterior product* of these two exterior forms.

(3) Suppose that $\mathcal{B}$ is a p-plane and $\mathcal{C}$ a line through x_0, and that one wishes to find their sum. If the line lies in the p-plane then the sum is the p-plane itself, but if it does not, then the sum is a $(p+1)$-plane, and the intersection is just the point x_0. We show how to find the sum in this case. Let ω be a characterising $(n-p)$-form for the p-plane and let v be a (non-zero) vector tangent to the line.

Since v does not lie in the p-plane, $\omega(v, v_{p+2}, \ldots, v_n)$ does not vanish for every choice of $v_{p+2}, \ldots, v_n$. It does vanish, however, if one or more of $v_{p+2}, \ldots, v_n$ is a linear combination of v and a vector which lies in the p-plane. Moreover, v being kept fixed, $\omega(v, v_{p+2}, \ldots, v_n)$ is alternating and multilinear in $v_{p+2}, \ldots, v_n$. Thus, v being kept fixed, $\omega(v, \cdot, \ldots, \cdot)$ is an exterior $(n-p-1)$-form which vanishes precisely when one or more of its arguments lies in $\mathcal{B} + \mathcal{C}$; it is therefore a characterising form for $\mathcal{B} + \mathcal{C}$. This $(n - p - 1)$-form is denoted $v \lrcorner \omega$ ($\lrcorner$ is read "hook"); once more, explicitly,

$$(v \lrcorner \omega)(v_{p+2}, \ldots, v_n) = \omega(v, v_{p+2}, \ldots, v_n).$$

The form $v \lrcorner \omega$ is called the *interior product* of v and ω (if the line lies in the p-plane, then $v \lrcorner \omega$, so defined, is zero).

These examples do not exhaust the possibilities for the intersections and sums of subspaces; but we have done enough to indicate how intersections of subspaces may be characterised by exterior products, and the sum of a subspace and a line by an interior product.

7. Volume in a Subspace

The idea of volume introduced in Section 1 may be extended to the case of an affine space which is a subspace of a larger space. Suppose that $\mathcal{B}$ is a p-dimensional affine subspace of an affine space $\mathcal{A}$ constructed by attaching the vector subspace $\mathcal{W}$ of $\mathcal{V}$ (the space on which $\mathcal{A}$ is modelled) to $\mathcal{A}$ at some chosen point x_0. Then a volume form $\Omega_{\mathcal{W}}$ may be chosen on $\mathcal{W}$ and used to compute the volumes of parallepipeds in $\mathcal{B}$. Since $\mathcal{B}$ is p-dimensional, $\Omega_{\mathcal{W}}$ must be a p-form, but it will differ from the p-forms, with $p < n$, introduced so far, because it will need to be defined only on vectors in $\mathcal{W}$ (and be non-zero on every basis) whereas a p-form which arises as a characterising form is defined on all vectors in the ambient space $\mathcal{V}$, and vanishes on those which lie in the $(n - p)$-dimensional subspace which it characterises.

It is a straightforward matter to define a p-form on $\mathcal{W}$, given a p-form ω on $\mathcal{V}$, with the help of the inclusion map $i: \mathcal{B} \hookrightarrow \mathcal{A}$ and the corresponding inclusion map $\mathcal{W} \hookrightarrow \mathcal{V}$ which (in view of the developments in Chapter 2) we denote i_*. Given any p-form ω on $\mathcal{V}$ one can define a p-form on $\mathcal{W}$, denoted $i^*\omega$ and called the *restriction* of ω to $\mathcal{W}$, by

$$i^*\omega(w_1, w_2, \ldots, w_p) = \omega(i_* w_1, i_* w_2, \ldots, i_* w_p).$$

Note the role of the inclusion map in distinguishing two rather different objects: $i^*\omega$, whose arguments may come only from $\mathcal{W}$, and ω, whose arguments may be any elements of $\mathcal{V}$.

Now $i^*\omega$ will serve as a volume form on $\mathcal{W}$ only if it does not vanish on a basis of $\mathcal{W}$. This is easy to ensure, as follows. Let $\mathcal{U}$ be any subspace of $\mathcal{V}$ complementary to $\mathcal{W}$, so that $\mathcal{V} = \mathcal{W} \oplus \mathcal{U}$, and let ω be a characterising form for $\mathcal{U}$. Since $\dim \mathcal{W} = p$, $\dim \mathcal{U} = n - p$, so that ω is a p-form. We show that $i^*\omega$ is a volume form for $\mathcal{W}$. Let $\{e_\alpha\}$, $\alpha = 1, 2, \ldots, p$, be a basis for $\mathcal{W}$ and $\{e_\rho\}$, $\rho = p+1, p+2, \ldots, n$, a basis for $\mathcal{U}$, so that $\{e_a\}$, $a = 1, 2, \ldots, n$, is a basis for $\mathcal{V}$. Let $\{\theta^a\}$ be the dual basis. Then $\theta^1, \theta^2, \ldots, \theta^p$ are constraint forms for $\mathcal{U}$, and so $\omega = c\theta^1 \wedge \theta^2 \wedge \cdots \wedge \theta^p$, where

c is some non-zero number. Therefore

$$i^*\omega(e_1, e_2, \ldots, e_p) = c \det(\langle e_\alpha, \theta^\beta \rangle) = c,$$

and so $i^*\omega$ is a volume form. A different choice of complementary subspace will yield a different ω, but $i^*\omega$ can change only by a scalar factor.

Thus the characterising form for any subspace will serve, by restriction, as volume form for a complementary subspace. Suppose that ω and χ are characterising forms for complementary subspaces $\mathcal{U}$ and $\mathcal{W}$ respectively, so that the restriction of ω to $\mathcal{W}$ is a volume form on that subspace, as is the restriction of χ to $\mathcal{U}$. Then $\omega \wedge \chi$ is an n-form on $\mathcal{V}$, which is non-zero (it is a characterising form for the zero subspace!); thus $\omega \wedge \chi$ is a volume form on $\mathcal{V}$. If $(x_0; v_1, v_2, \ldots, v_n)$ is a parallelepiped whose first p edges belong to $\mathcal{W}$ and whose last $n - p$ edges belong to $\mathcal{U}$ then its volume as measured by $\omega \wedge \chi$ is just $\omega(v_1, v_2, \ldots, v_p)\chi(v_{p+1}, v_{p+2}, \ldots, v_n)$. One may regard $(x_0; v_1, v_2, \ldots, v_p)$ as a parallelepiped in $\mathcal{B}$ and $(x_0; v_{p+1}, v_{p+2}, \ldots, v_n)$ as a parallelepiped in $\mathcal{C}$. The forms ω and χ define (by restriction) volume functions in $\mathcal{B}$ and $\mathcal{C}$, and then $\omega(v_1, v_2, \ldots, v_p)$ and $\chi(v_{p+1}, v_{p+2}, \ldots, v_n)$ are the volumes of these parallelepiped faces in $\mathcal{B}$ and $\mathcal{C}$ respectively. The exterior product generalises, in this sense, the familiar formula "base area$\times$height" for the volume of a 3-dimensional box.

An important instance of these ideas arises when $\mathcal{W}$ is of dimension $n - 1$, so that the complementary subspace $\mathcal{U}$ may be any 1-dimensional subspace not lying in $\mathcal{W}$. In this case a basis for $\mathcal{U}$ consists of a single vector u. The dual of u with respect to a volume form Ω on $\mathcal{V}$, which is a characterising form for $\mathcal{W}$, is simply the interior product $u \lrcorner \Omega$. In this case the restriction of $u \lrcorner \Omega$ is a volume form for $\mathcal{W}$.

Finally, we consider the orientation of a hyperplane. An orientation for $\mathcal{W}$ is a set of volume forms on $\mathcal{W}$ which differ from one another by positive factors. Thus the vector u determines one orientation of $\mathcal{W}$, that corresponding to $u \lrcorner \Omega$, and $-u$ determines the opposite orientation. Thus if an orientation has been chosen for $\mathcal{V}$, and $\mathcal{W}$ is a subspace of $\mathcal{V}$ of codimension 1, then any vector u not tangent to $\mathcal{W}$ determines an orientation for $\mathcal{W}$, called the orientation of $\mathcal{W}$ *induced* by u and the orientation of $\mathcal{V}$: if Ω determines the orientation of $\mathcal{V}$, then the restriction of $u \lrcorner \Omega$ to $\mathcal{W}$ determines the induced orientation of $\mathcal{W}$. Likewise, if $\mathcal{A}$ is an affine space modelled on $\mathcal{V}$ and $\mathcal{B}$ a hyperplane in $\mathcal{A}$ modelled on $\mathcal{W}$, if Ω is a volume function on $\mathcal{A}$, and if u is any vector given at a point of $\mathcal{B}$ but not tangent to it, then $u \lrcorner \Omega$ determines an orientation of $\mathcal{B}$, called the orientation induced by u and by the orientation of $\mathcal{A}$.

Notice that the establishment of an induced orientation, as here set out, has nothing to do with whether u is orthogonal to $\mathcal{W}$: only the sense of u is relevant, and the concept of orthogonality is not required.

8. Exterior Algebra

The definition of volume and the characterisation of subspaces are only two of the many applications of multivectors and exterior forms in geometry. We therefore supply, in the rest of this chapter, the apparatus for these applications, in an exposition of the algebraic properties of multivectors and exterior forms and of exterior and interior products. Here we give prior place to algebra, not geometry. One of the principal ideas of this section is that forms or multivectors of a given degree may be considered to constitute a vector space.

We begin with a recapitulation of the ideas introduced so far. An alternating multilinear function on a vector space $\mathcal{V}$ is called an exterior p-form on $\mathcal{V}$. An alternating multilinear function on $\mathcal{V}^*$ is called a p-vector on $\mathcal{V}$. The p-form $(v_1, v_2, \ldots, v_p) \mapsto \det(\langle v_\alpha, \theta^\beta \rangle)$, where $\theta^1, \theta^2, \ldots, \theta^p$ are given 1-forms, is denoted $\theta^1 \wedge \theta^2 \wedge \cdots \wedge \theta^p$ and called a decomposable p-form; it is the exterior product of the θ^α. A p-vector $(\eta^1, \eta^2, \ldots, \eta^p) \mapsto \det(\langle w_\alpha, \eta^\beta \rangle)$, where $w_1, w_2, \ldots, w_p$ are given vectors, is denoted $w_1 \wedge w_2 \wedge \cdots \wedge w_p$ and called a decomposable p-vector; it is the exterior product of the w_α. If ω is any p-form, and v a vector, then $v \lrcorner \omega$ is the $(p-1)$-form defined by $v \lrcorner \omega(v_1, v_2, \ldots, v_{p-1}) = \omega(v, v_1, \ldots, v_{p-1})$. If ω is any decomposable p-form and χ is any decomposable q-form, say $\omega = \eta^1 \wedge \eta^2 \wedge \cdots \eta^p$ and $\chi = \varsigma^1 \wedge \varsigma^2 \wedge \cdots \wedge \varsigma^q$, and if $\{\eta^1, \ldots, \eta^p, \varsigma^1, \ldots, \varsigma^q\}$ are linearly independent, then $\omega \wedge \chi$ is the $(p+q)$-form $\eta^1 \wedge \eta^2 \wedge \cdots \wedge \eta^p \wedge \varsigma^1 \wedge \varsigma^2 \wedge \cdots \wedge \varsigma^q$.

The symmetry between $\mathcal{V}$ and $\mathcal{V}^*$ expressed by $(\mathcal{V}^*)^* = \mathcal{V}$ entails that a p-vector on $\mathcal{V}$ may also be regarded as a p-form on $\mathcal{V}^*$, and that a p-form on $\mathcal{V}$ may also be regarded as a p-vector on $\mathcal{V}^*$, but to reduce confusion we shall write "p-form" to mean "p-form on $\mathcal{V}$" only, and "p-vector" to mean "p-vector on $\mathcal{V}$" only. By far the greater part of our treatment will refer only to forms; it should be clear how, by interchanging the roles of $\mathcal{V}$ and $\mathcal{V}^*$, it could be extended to multivectors.

Vector spaces of multilinear maps. Let ω_1 and ω_2 be p-forms, and let c_1 and c_2 be numbers. A p-form $c_1\omega_1 + c_2\omega_2$ is defined by

$$(c_1\omega_1 + c_2\omega_2)(v_1, v_2, \ldots, v_p) = c_1\omega_1(v_1, v_2, \ldots, v_p) + c_2\omega_2(v_1, v_2, \ldots, v_p)$$

for all $v_1, v_2, \ldots, v_p \in \mathcal{V}$.

Exercise 25. Check that $c_1\omega_1 + c_2\omega_2$, so defined, is alternating and multilinear. □

Exercise 26. Let W_1 and W_2 be p-vectors, and let c_1 and c_2 be numbers. Devise a definition of $c_1W_1 + c_2W_2$ on the model of the definition for forms, and confirm that the object thus defined is indeed a p-vector. □

Exercise 27. Check that these definitions make the p-forms on $\mathcal{V}$, and the p-vectors on $\mathcal{V}$, into vector spaces. □

The vector space of p-vectors on $\mathcal{V}$ will be denoted $\bigwedge^p \mathcal{V}$, and the vector space of p-forms on $\mathcal{V}$ will be denoted $\bigwedge^p \mathcal{V}^*$. It can be shown that $\bigwedge^p \mathcal{V}^*$ is naturally isomorphic to the dual of $\bigwedge^p \mathcal{V}$, but we shall not prove this here. Note that $\bigwedge^1 \mathcal{V}$ is just $\mathcal{V}$ itself, and $\bigwedge^1 \mathcal{V}^*$ is just $\mathcal{V}^*$.

The advantage of forming a vector space of (say) p-forms is that one may apply the methods and results of linear algebra to it. The disadvantage is that not all of its elements have the simple geometrical interpretation which we have described in earlier sections. For example, there is no reason to suppose that a linear combination of two (decomposable) p-forms which characterise distinct $(n-p)$-dimensional subspaces characterises any subspace at all. This observation (which raises the question of how one determines which forms are decomposable) is considered again in Section 12.

The multilinearity is more fundamental than the alternating property in the construction of these vector spaces. In fact, if T_1 and T_2 are any two p-fold multilinear maps $\mathcal{V}^p \to \mathbf{R}$, not necessarily alternating, and c_1 and c_2 are numbers, then $c_1T_1 + c_2T_2$, defined by

$$(c_1T_1 + c_2T_2)(v_1, v_2, \ldots, v_p) = c_1T_1(v_1, v_2, \ldots, v_p) + c_2T_2(v_1, v_2, \ldots, v_p)$$

is also a p-fold multilinear map, and the set of all p-fold multilinear maps is made into a vector space by this definition.

Multilinear maps may also be multiplied together. If S is a p-fold multilinear map $\mathcal{V}^p \to \mathbf{R}$ and T is a q-fold multilinear map $\mathcal{V}^q \to \mathbf{R}$ then their *tensor product* $S \otimes T$ is the $(p+q)$-fold multilinear map $\mathcal{V}^{p+q} \to \mathbf{R}$ defined by

$$S \otimes T(v_1, v_2, \ldots, v_{p+q}) = S(v_1, v_2, \ldots, v_p)T(v_{p+1}, v_{p+2}, \ldots, v_{p+q}).$$

Exercise 28. Check that $S \otimes T$ is multilinear. □

For the purposes of exterior algebra it is necessary to be able to pick out the alternating part of an arbitrary multilinear map. For example, one may form the tensor product of two alternating maps, but the result, though multilinear, will not be alternating. The extraction of the alternating part of a multilinear map is achieved by an operation which is a generalisation of the construction of a determinant. Let T be a p-fold multilinear map $\mathcal{V}^p \to \mathbf{R}$. Define the *alternating part* of T, $\operatorname{alt} T$, by

$$\operatorname{alt} T(v_1, v_2, \ldots, v_p) = \frac{1}{p!}\sum_{\pi} \epsilon(\pi) T(v_{\pi(1)}, v_{\pi(2)}, \ldots, v_{\pi(p)}),$$

the sum being taken over all permutations π of $(1, 2, \ldots, p)$.

Exercise 29. Show that $\operatorname{alt} T$ is alternating, that $\operatorname{alt}(\operatorname{alt} T) = \operatorname{alt} T$, and that if ω is a p-form then $\operatorname{alt}\omega = \omega$. □

Now let $\{e_a\}$ be a basis for $\mathcal{V}$. The *components* of T relative to this basis are the numbers

$$T_{a_1 a_2 \ldots a_p} = T(e_{a_1}, e_{a_2}, \ldots, e_{a_p}).$$

There is a special notation for the components of $\operatorname{alt} T$: instead of $(\operatorname{alt} T)_{a_1 a_2 \ldots a_p}$ one writes

$$T_{[a_1 a_2 \ldots a_p]}.$$

Exercise 30. Show that if T_{ab} are the components of a 2-fold multilinear (bilinear) map T then

$$T_{[ab]} = \tfrac{1}{2}(T_{ab} - T_{ba});$$

while if T is 3-fold multilinear (trilinear)

$$T_{[abc]} = \tfrac{1}{6}(T_{abc} + T_{bca} + T_{cab} - T_{acb} - T_{bac} - T_{cba}).$$ □

The exterior product. A general definition of the exterior product may be expressed in terms of the operation alt. Let ω be any p-form and χ any q-form. Their exterior product $\omega \wedge \chi$ is defined by

$$\omega \wedge \chi = \frac{(p+q)!}{p!\,q!}\operatorname{alt}(\omega \otimes \chi).$$

The awkward numerical factor, if suppressed here, will pop up elsewhere. In terms of its action on vectors, one may read off $\omega \wedge \chi$ from the definitions of $\otimes$ and alt:

$$(\omega \wedge \chi)(v_1, v_2, \ldots, v_{p+q}) = \frac{1}{p!\,q!}\sum_{\pi} \epsilon(\pi)\omega\big(v_{\pi(1)}, v_{\pi(2)}, \ldots, v_{\pi(p)}\big)\chi\big(v_{\pi(p+1)}, v_{\pi(p+2)}, \ldots, v_{\pi(p+q)}\big)$$

where the sum is taken over all permutations π of $(1, 2, \ldots, p+q)$.

Exercise 31. Show that (according to this formula) if η^1 and η^2 are two 1-forms then

$$(\eta^1 \wedge \eta^2)(v_1, v_2) = \langle v_1, \eta^1\rangle\langle v_2, \eta^2\rangle - \langle v_1, \eta^2\rangle\langle v_2, \eta^1\rangle = \det(\langle v_\alpha, \eta^\beta\rangle)$$

(where $\alpha, \beta = 1, 2$). Show further that if η is a 1-form and ω a 2-form then

$$(\eta \wedge \omega)(u, v, w) = \langle u, \eta\rangle\omega(v, w) + \langle v, \eta\rangle\omega(w, u) + \langle w, \eta\rangle\omega(u, v).$$

Deduce that if η^1, η^2, η^3 are any three 1-forms then

$$(\eta^1 \wedge (\eta^2 \wedge \eta^3))(v_1, v_2, v_3) = \det(\langle v_\alpha, \eta^\beta\rangle)$$

(where now $\alpha, \beta = 1, 2, 3$). □

We now discuss some of the algebraic properties of the exterior product. It is clear from the definition that it is distributive:

$$(c_1\omega_1 + c_2\omega_2) \wedge \chi = c_1\omega_1 \wedge \chi + c_2\omega_2 \wedge \chi$$
$$\omega \wedge (c_1\chi_1 + c_2\chi_2) = c_1\omega \wedge \chi_1 + c_2\omega \wedge \chi_2$$

for any numbers c_1, c_2. The exterior product is also associative:

$$\omega \wedge (\chi \wedge \psi) = (\omega \wedge \chi) \wedge \psi$$

for any three forms ω, χ, ψ. This is not so easy to see, though Exercise 31 gives a clue as to what happens when the forms are all 1-forms: it turns out that in this case $\big((\eta^1 \wedge \eta^2) \wedge \eta^3\big)(v_1, v_2, v_3) = \det(\langle v_\alpha, \eta^\beta\rangle)$ also. The basic idea of the proof of associativity is to show that $\omega \wedge (\chi \wedge \psi)$ and $(\omega \wedge \chi) \wedge \psi$ are both equal to $\operatorname{alt}(\omega \otimes \chi \otimes \psi)$ (apart from numerical factors, the same in each case). It depends on the fact, obvious from the definition, that the tensor product of arbitrary multilinear maps $\mathcal{V}^r \to \mathbf{R}$, $r = 1, 2, \ldots$ is associative.

First, let S be any p-fold multilinear function and T any q-fold multilinear function on $\mathcal{V}$, neither necessarily alternating. We show that if $\operatorname{alt} S = 0$ then $\operatorname{alt}(S \otimes T) = 0$. Partition the symmetric group on $p+q$ elements into equivalence classes by the rule that $\pi' \sim \pi$ if $\pi'(1, 2, \ldots, p+q)$ has the same last q entries as $\pi(1, 2, \ldots, p+q)$, in the same order. Then in each equivalence class all permutations of the first p entries occur, and so since $\operatorname{alt} S = 0$ the contribution of each equivalence class to the sum in the evaluation of $\operatorname{alt}(S \otimes T)$ is zero. In components, this amounts to

$$S_{[a_1 a_2 \ldots a_p} T_{a_{p+1} a_{p+2} \ldots a_{p+q}]} = 0 \quad \text{if} \quad S_{[a_1 a_2 \ldots a_p]} = 0.$$

Now write $\operatorname{alt}(\omega \otimes \chi) - \omega \otimes \chi$ in place of S in this argument, and ψ in place of T. Then $\operatorname{alt}\bigl(\operatorname{alt}(\omega \otimes \chi) - \omega \otimes \chi\bigr) = \operatorname{alt}(\omega \otimes \chi) - \operatorname{alt}(\omega \otimes \chi) = 0$, so that

$$0 = \operatorname{alt}\bigl((\operatorname{alt}(\omega \otimes \chi) - \omega \otimes \chi) \otimes \psi\bigr) = \operatorname{alt}\bigl(\operatorname{alt}(\omega \otimes \chi) \otimes \psi\bigr) - \operatorname{alt}(\omega \otimes \chi \otimes \psi).$$

(The associativity of the tensor product has been assumed here.) But this says (putting in the numerical factor)

$$(\omega \wedge \chi) \wedge \psi = \frac{(p+q+r)!}{p!\,q!\,r!} \operatorname{alt}(\omega \otimes \chi \otimes \psi)$$

(where ω is a p-form, χ a q-form, ψ an r-form). Now apply the corresponding argument to the formation of $\omega \wedge (\chi \wedge \psi)$.

Now that associativity has been established it is no longer necessary to include the brackets in expressions such as $\omega \wedge (\chi \wedge \psi)$; and this is equally true of products involving more than three terms. In particular for any 1-forms $\eta^1, \eta^2, \ldots, \eta^r$ the expression $\eta^1 \wedge \eta^2 \wedge \cdots \wedge \eta^r$ is unambiguous, and in fact

$$\begin{aligned}
(\eta^1 \wedge \eta^2 \wedge \cdots \wedge \eta^r)&(v_1, v_2, \ldots, v_r) \\
&= \frac{r!}{1!\,1!\cdots 1!} \operatorname{alt}(\eta^1 \otimes \eta^2 \otimes \cdots \otimes \eta^r)(v_1, v_2, \ldots, v_r) \\
&= \sum_\pi \epsilon(\pi) \langle v_{\pi(1)}, \eta^1 \rangle \langle v_{\pi(2)}, \eta^2 \rangle \cdots \langle v_{\pi(r)}, \eta^r \rangle \\
&= \det\bigl(\langle v_\alpha, \eta^\beta \rangle\bigr)
\end{aligned}$$

where $\alpha, \beta = 1, 2, \ldots, r$ and the summation is over all permutations π of $(1, 2, \ldots, r)$. In this way we recover the formula for the exterior product of 1-forms used extensively in previous sections of this chapter.

Though exterior multiplication shares with ordinary multiplication of numbers the properties of distributivity and associativity, it is not commutative. In fact if $\omega \in \bigwedge^p \mathcal{V}^*$ and $\chi \in \bigwedge^q \mathcal{V}^*$ then

$$\chi \wedge \omega = (-1)^{pq} \omega \wedge \chi.$$

This may be seen as follows: if π is the permutation

$$(1, 2, \ldots, p+1, p+2, \ldots, p+q) \mapsto (p+1, p+2, \ldots, p+q, 1, 2, \ldots, p)$$

then

$$\begin{aligned}
(\chi \wedge \omega)&(v_1, v_2, \ldots, v_p, v_{p+1}, v_{p+2}, \ldots, v_{p+q}) \\
&= \epsilon(\pi)(\chi \wedge \omega)(v_{p+1}, v_{p+2}, \ldots, v_{p+q}, v_1, v_2, \ldots, v_p) \\
&= \epsilon(\pi)(\omega \wedge \chi)(v_1, v_2, \ldots, v_p, v_{p+1}, v_{p+2}, \ldots, v_{p+q});
\end{aligned}$$

and $\epsilon(\pi) = (-1)^{pq}$ since π involves transposing each of $v_{p+1}, v_{p+2}, \ldots, v_{p+q}$ with each of $v_p, v_{p-1}, \ldots, v_1$ in turn. Thus for any two 1-forms η, ς

$$\varsigma \wedge \eta = -\eta \wedge \varsigma;$$

and in particular the exterior product of a 1-form with itself is 0.

Exercise 32. Show that, more generally, if ω is a p-form and p is odd then $\omega \wedge \omega = 0$. □

With these algebraic properties of the exterior product to hand it is easy to calculate exterior products of forms directly, without having to appeal to the definitions in terms of alternating multilinear maps. As an example we consider the direct calculation of the exterior product of three 1-forms on a 3-dimensional vector space. Let $\{\theta^a\}$ be a basis for 1-forms on the vector space and let

$$\omega = (\eta_1^1\theta^1 + \eta_2^1\theta^2 + \eta_3^1\theta^3) \wedge (\eta_1^2\theta^1 + \eta_2^2\theta^2 + \eta_3^2\theta^3) \wedge (\eta_1^3\theta^1 + \eta_2^3\theta^2 + \eta_3^3\theta^3)$$

where the η_b^a are numbers. Carrying out the multiplication of the first two factors one obtains

$$\begin{aligned} &\eta_1^1\eta_1^2\theta^1 \wedge \theta^1 + \eta_1^1\eta_2^2\theta^1 \wedge \theta^2 + \eta_1^1\eta_3^2\theta^1 \wedge \theta^3 \\ &+ \eta_2^1\eta_1^2\theta^2 \wedge \theta^1 + \eta_2^1\eta_2^2\theta^2 \wedge \theta^2 + \eta_2^1\eta_3^2\theta^2 \wedge \theta^3 \\ &+ \eta_3^1\eta_1^2\theta^3 \wedge \theta^1 + \eta_3^1\eta_2^2\theta^3 \wedge \theta^2 + \eta_3^1\eta_3^2\theta^3 \wedge \theta^3. \end{aligned}$$

Deleting terms with repeated factors and rearranging the others so that all the basis forms appear in order one obtains from this

$$(\eta_1^1\eta_2^1 - \eta_2^1\eta_1^2)\theta^1 \wedge \theta^2 + (\eta_1^1\eta_3^2 - \eta_3^1\eta_1^2)\theta^1 \wedge \theta^3 + (\eta_2^1\eta_3^2 - \eta_3^1\eta_2^2)\theta^2 \wedge \theta^3.$$

Multiplying this by the last factor and omitting terms containing a repeated factor one obtains

$$\begin{aligned} &(\eta_1^1\eta_2^2 - \eta_2^1\eta_1^2)\eta_3^3(\theta^1 \wedge \theta^2) \wedge \theta^3 + (\eta_1^1\eta_3^2 - \eta_3^1\eta_1^2)\eta_2^3(\theta^1 \wedge \theta^3) \wedge \theta^2 \\ &+(\eta_2^1\eta_3^2 - \eta_3^1\eta_3^2)\eta_1^3(\theta^2 \wedge \theta^3) \wedge \theta^1. \end{aligned}$$

Again rearranging factors so that the basis forms appear in order one is left with

$$(\eta_1^1\eta_2^2\eta_3^3 - \eta_2^1\eta_1^2\eta_3^3 + \eta_3^1\eta_1^2\eta_2^3 - \eta_1^1\eta_3^2\eta_2^3 + \eta_2^1\eta_3^2\eta_1^3 - \eta_3^1\eta_2^2\eta_1^3)\theta^1 \wedge \theta^2 \wedge \theta^3$$

(a change of sign occurs for the middle term since $\theta^1 \wedge \theta^3 \wedge \theta^2 = -\theta^1 \wedge \theta^2 \wedge \theta^3$). The result is in fact $\det(\eta_b^a)\theta^1 \wedge \theta^2 \wedge \theta^3$.

Exercise 33. Let $\{\theta^a\}$ be a basis for 1-forms on a 4-dimensional vector space, and let $\eta = \theta^1 + 2\theta^2 - \theta^3$ and $\omega = \theta^1 \wedge \theta^3 + \theta^2 \wedge \theta^4$. Compute $\eta \wedge \omega$ and $\omega \wedge \omega$. □

Exercise 34. Show that if $\eta^1, \eta^2, \ldots, \eta^p$ are linearly dependent 1-forms then $\eta^1 \wedge \eta^2 \wedge \cdots \wedge \eta^p = 0$. □

Exercise 33 provides an example of a form ω such that $\omega \wedge \omega \neq 0$; it is, necessarily, a form of even degree.

9. Bases and Dimensions

Several of our earlier calculations and results should suggest how one may construct a basis for p-forms. In fact one may construct a basis for $\bigwedge^p \mathcal{V}^*$ out of a basis $\{\theta^a\}$ for $\mathcal{V}^*$, as follows. For each collection of distinct integers $a_1, a_2, \ldots, a_p$ (with $1 \leq a_i \leq n = \dim \mathcal{V}$) the p-form $\theta^{a_1} \wedge \theta^{a_2} \wedge \cdots \wedge \theta^{a_p}$ is non-zero: for if $\{e_a\}$ is the basis for $\mathcal{V}$ dual to $\{\theta^a\}$ then

$$(\theta^{a_1} \wedge \theta^{a_2} \wedge \cdots \wedge \theta^{a_p})(e_{a_1}, e_{a_2}, \ldots, e_{a_p}) = 1$$

(no summation) by the determinant formula. The p-forms $\theta^{b_1} \wedge \theta^{b_2} \wedge \cdots \wedge \theta^{b_p}$ and $\theta^{a_1} \wedge \theta^{a_2} \wedge \cdots \wedge \theta^{a_p}$ differ only in sign, if at all, when $(b_1, b_2, \ldots, b_p)$ is merely a permutation of $(a_1, a_2, \ldots, a_p)$. Consider, therefore, the p-forms

$$\{\, \theta^{a_1} \wedge \theta^{a_2} \wedge \cdots \wedge \theta^{a_p} \mid 1 \leq a_1 < a_2 < \cdots < a_p \leq n \,\}.$$

Each of these forms is certainly non-zero, and no coincidences among them can arise from reordering. We shall show that they form a basis for $\bigwedge^p \mathcal{V}^*$. To do so we must show that they are linearly independent, and that every p-form may be expressed as a linear combination of them.

Note first of all that, by the determinant formula, if $1 \leq b_1 < b_2 < \cdots < b_p \leq n$ then

$$(\theta^{a_1} \wedge \theta^{a_2} \wedge \cdots \wedge \theta^{a_p})(e_{b_1}, e_{b_2}, \ldots, e_{b_p}) = 0$$

unless $a_1 = b_1$, $a_2 = b_2, \ldots, b_p = a_p$. Suppose that

$$\sum k_{a_1 a_2 \ldots a_p} \theta^{a_1} \wedge \theta^{a_2} \wedge \cdots \wedge \theta^{a_p} = 0,$$

the sum being taken over all p-tuples of integers $(a_1, a_2, \ldots, a_p)$ with $1 \leq a_1 < a_2 < \cdots < a_p \leq n$, the ks being certain numerical coefficients. The expression on the left is a p-form, and asserting that it is zero is equivalent to asserting that the result of evaluating it on any p vectors is zero. But

$$\sum k_{a_1 a_2 \ldots a_p} (\theta^{a_1} \wedge \theta^{a_2} \wedge \cdots \wedge \theta^{a_p})(e_{b_1}, e_{b_2}, \ldots, e_{b_p}) = k_{b_1 b_2 \ldots b_p}$$

if $1 \leq b_1 < b_2 < \cdots < b_p \leq n$. Thus the p-form can be zero only if all its coefficients are zero; and so the forms $\{\theta^{a_1} \wedge \theta^{a_2} \wedge \cdots \wedge \theta^{a_p}\}$ are linearly independent.

If ω is any p-form then (since it is multilinear) the value of ω on any collection of arguments is known if its value is known whenever its arguments are basis vectors. But (since ω is alternating) its value on any collection of p basis vectors is known if it is known when the basis vectors are distinct and arranged in increasing order of their suffices. It follows that

$$\omega = \sum \omega(e_{a_1}, e_{a_2}, \ldots, e_{a_p}) \theta^{a_1} \wedge \theta^{a_2} \wedge \cdots \wedge \theta^{a_p},$$

the sum being again taken over all $1 \leq a_1 < a_2 < \cdots < a_p \leq n$.

We have shown that $\{\, \theta^{a_1} \wedge \theta^{a_2} \wedge \cdots \wedge \theta^{a_p} \mid 1 \leq a_1 < a_2 < \cdots < a_p \leq n \,\}$ is a linearly independent set of p-forms that spans $\bigwedge^p \mathcal{V}^*$, and so is a basis for that space. The dimension of $\bigwedge^p \mathcal{V}^*$ is thus the number of ways of choosing integers $a_1, a_2, \ldots, a_p$ to satisfy $1 \leq a_1 < a_2 < \cdots < a_p \leq n$, which is $n!/p!\,(n-p)!$ (or the binomial coefficient $\binom{n}{p}$ or ${}_nC_p$), provided $p \leq n$. If $p > n$ there are no non-zero p-forms. The space $\bigwedge^n \mathcal{V}^*$ is 1-dimensional with basis the single n-form $\theta^1 \wedge \theta^2 \wedge \cdots \wedge \theta^n$ which confirms the observations of Section 2. At the other end of the scale, the space $\bigwedge^1 \mathcal{V}^*$ has dimension n according to this result, as is required by its identification with $\mathcal{V}^*$. It is frequently convenient to regard $\mathbf{R}$ itself as constituting the space of 0-forms: with the usual interpretation of $0!$ as 1, the formula $n!/0!\,(n-0)!$ gives the correct dimension; moreover the exterior product rules continue to apply if exterior multiplication by a 0-form is taken to be scalar multiplication.

It is sometimes advantageous to lump all forms, of degrees 0 to n inclusive, together, to form one big space, which we denote $\bigwedge \mathcal{V}^*$. This space is defined as the direct sum of the spaces $\bigwedge^p \mathcal{V}^*$ from $p = 0$ to n: it is thus a vector space of dimension $\sum_{p=0}^{n} \binom{n}{p} = 2^n$. Its elements may be thought of as formal sums of p-forms of varying degrees (the use of the word "formal" is intended to indicate that such a sum may not be considered as an alternating multilinear map if terms of different degrees are involved). It is, moreover, equipped with an associative, bilinear (but non-commutative) product, the exterior product; and is therefore an example of an algebra. It is called the *exterior algebra* of $\mathcal{V}$.

Finally, a computational point. The summation convention has been in abeyance during this discussion of bases of spaces of p-forms, and it would be convenient to restore its use. But then certain adjustments must be made to the coefficients of forms to take account of the fact that without the condition $a_1 < a_2 < \cdots < a_p$ the p-forms $\{\theta^{a_1} \wedge \theta^{a_2} \wedge \cdots \wedge \theta^{a_p}\}$ are no longer independent, though they do span $\bigwedge^p \mathcal{V}^*$. The following two exercises deal with this matter.

Exercise 35. Let ω be a p-form, $\{e_a\}$ a basis for $\mathcal{V}$, $\{\theta^a\}$ the dual basis for $\mathcal{V}^*$, and let $\omega(e_{a_1}, e_{a_2}, \ldots, e_{a_p}) = \omega_{a_1 a_2 \ldots a_p} = \omega_{[a_1 a_2 \ldots a_p]}$ be the components of ω relative to $\{e_a\}$. Show that

$$\omega = \frac{1}{p!}\omega_{a_1 a_2 \ldots a_p}\theta^{a_1} \wedge \theta^{a_2} \wedge \cdots \wedge \theta^{a_p}$$

(summation over repeated indices intended). □

Exercise 36. Show that if $\omega_{ab}\theta^a \wedge \theta^b = 0$ for linearly independent θ^a, then $\omega_{ba} = \omega_{ab}$, but it is not necessarily the case that $\omega_{ab} = 0$. □

(Alternative treatments of the numerical factors in wedge products are described in Note 3.)

10. The Interior Product

The interior product may also be generalised to arbitrary exterior forms. Let ω be any p-form and v a vector. The *interior product* of v and ω is the $(p-1)$-form $v \,\lrcorner\, \omega$ (read "v hook ω") defined by

$$(v \,\lrcorner\, \omega)(v_1, v_2, \ldots, v_{p-1}) = \omega(v, v_1, v_2, \ldots, v_{p-1})$$

for all choices of the vectors $v_1, v_2, \ldots, v_{p-1}$. (The interior product is also frequently denoted $i_v\omega$.) It has the properties

(1) if $p = 1$ then $v \,\lrcorner\, \omega = \langle v, \omega \rangle$
(2) $v \,\lrcorner\, (c_1\omega_1 + c_2\omega_2) = c_1(v \,\lrcorner\, \omega_1) + c_2(v \,\lrcorner\, \omega_2)$
(3) if ω is a p-form, and χ a form of any degree, then

$$v \,\lrcorner\, (\omega \wedge \chi) = (v \,\lrcorner\, \omega) \wedge \chi + (-1)^p \omega \wedge (v \,\lrcorner\, \chi).$$

The first two of these follow immediately from the definition. To prove the third we observe that as a result of the second it is enough to know that it holds when ω and χ are exterior products of 1-forms, that is, when they are decomposable. Now if $\omega = \eta^1 \wedge \eta^2 \wedge \cdots \wedge \eta^p$ say then for any vectors $v_2, v_3, \ldots, v_p$, $(v \,\lrcorner\, \omega)(v_2, v_3, \ldots, v_p)$

is a determinant whose first column has the entries $\langle v,\eta^1\rangle$, $\langle v,\eta^2\rangle,\dots,\langle v,\eta^p\rangle$. Expanding by the first column, alternating the signs in the usual way, we obtain

$$\begin{aligned}(v \lrcorner \omega)&(v_2,v_3,\dots,v_p)\\ &= (\eta^1\wedge\eta^2\wedge\cdots\wedge\eta^p)(v,v_2,\dots,v_p)\\ &= \sum_{i=1}^{p}(-1)^{i-1}\langle v,\eta^i\rangle(\eta^1\wedge\cdots\widehat{\eta^i}\cdots\wedge\eta^p)(v_2,v_3,\dots,v_p)\end{aligned}$$

where, in the sum, the caret mark $\widehat{\ }$ is used to indicate that the 1-form below is to be omitted from the exterior product. Thus

$$v \lrcorner(\eta^1\wedge\eta^2\wedge\cdots\wedge\eta^p) = \sum_{i=1}^{p}(-1)^{i-1}\langle v,\eta^i\rangle(\eta^1\wedge\cdots\widehat{\eta^i}\cdots\wedge\eta^p).$$

Now suppose that $\chi = \varsigma^1\wedge\varsigma^2\wedge\cdots\wedge\varsigma^q$. Then

$$\begin{aligned}v \lrcorner(\omega\wedge\chi) &= \left(\sum_{i=1}^{p}(-1)^{i-1}\langle v,\eta^i\rangle\eta^1\wedge\cdots\widehat{\eta^i}\cdots\wedge\eta^p\right)\wedge\chi\\ &\qquad + \omega\wedge\left(\sum_{j=1}^{q}(-1)^{p+j-1}\langle v,\varsigma^j\rangle\varsigma^1\wedge\cdots\widehat{\varsigma^j}\cdots\wedge\varsigma^p\right)\\ &= (v \lrcorner\omega)\wedge\chi + (-1)^p\omega\wedge(v \lrcorner\chi)\end{aligned}$$

as asserted.

Exercise 37. Show that, if $\dim\mathcal{V} = 4$, if $\omega = \theta^1\wedge\theta^2 + \theta^3\wedge\theta^4$, and if $v = e_1$ (where $\{e_a\}$ and $\{\theta^a\}$ are dual bases) then $v \lrcorner\omega = \theta^2$, while if $v = e_2 + e_3$ then $v \lrcorner\omega = -\theta^1 + \theta^4$. □

Exercise 38. Show that if $v = v^a e_a$ with respect to a basis $\{e_a\}$ for $\mathcal{V}$ (and $\{\theta^a\}$ is the dual basis for $\mathcal{V}^*$) then

$$v \lrcorner\omega = \frac{1}{(p-1)!}v^a\omega_{aa_1a_2\dots a_{p-1}}\theta^{a_1}\wedge\theta^{a_2}\wedge\cdots\wedge\theta^{a_{p-1}}$$

when ω is expressed as in Exercise 35. □

Exercise 39. Show that for any $\omega\in\bigwedge^p\mathcal{V}^*$, $\theta^a\wedge(e_a \lrcorner\omega) = p\omega$ (summation intended). (This formula is analogous to Euler's formula for derivatives of a homogeneous polynomial). □

Note the following useful property of the interior product: for fixed $\omega\in\bigwedge^p\mathcal{V}^*$ one may regard the rule $v\mapsto v \lrcorner\omega$ as defining a map $\mathcal{V}\to\bigwedge^{p-1}\mathcal{V}^*$; this is a linear map, as follows immediately from the multilinearity of ω.

11. Induced Maps of Forms

Linear maps of vectors and covectors may readily be extended to multivectors and exterior forms. The induced maps of forms are of much the greater importance, and we confine our exposition to them. We have already given two examples: the induced map of a volume form in Section 3 above, and the restriction of a decomposable form by the inclusion map in Section 7.

Recall that a linear map $\lambda\colon \mathcal{V} \to \mathcal{W}$ of vector spaces induces a contragredient linear map λ^* of the dual spaces, by $\langle v, \lambda^*(\eta)\rangle = \langle \lambda(v), \eta\rangle$. This construction may be extended to p-forms, to give a linear map of forms, also denoted λ^* and also contragredient, defined as follows. For $\chi \in \bigwedge^p \mathcal{W}^*$ set

$$(\lambda^*\chi)(v_1, v_2, \ldots, v_p) = \chi(\lambda(v_1), \lambda(v_2), \ldots, \lambda(v_p))$$

for all $v_1, v_2, \ldots, v_p \in \mathcal{V}$. Evidently $\lambda^*\chi$, so defined, is alternating multilinear and is therefore an element of $\bigwedge^p \mathcal{V}^*$. Moreover, $\lambda^*\colon \bigwedge^p \mathcal{W}^* \to \bigwedge^p \mathcal{V}^*$ is a linear map. It is called a map of forms *induced* by λ.

Note that if $\chi \in \bigwedge^p \mathcal{W}^*$ and $p > \dim \mathcal{V}$ then $\lambda^*\chi$ is necessarily zero.

The induced map of a composite of two linear maps is the composite of the induced maps in the opposite order, as is required by contragredience: if $\kappa\colon \mathcal{U} \to \mathcal{V}$ and $\lambda\colon \mathcal{V} \to \mathcal{W}$ are linear, then $(\lambda\circ\kappa)^*\colon \bigwedge^p \mathcal{W}^* \to \bigwedge^p \mathcal{U}^*$ is given by $(\lambda\circ\kappa)^* = \kappa^*\circ\lambda^*$.

One important property of induced maps of forms is that they preserve the exterior product, in the sense that $\lambda^*(\chi \wedge \psi) = (\lambda^*\chi) \wedge (\lambda^*\psi)$. We now prove this result. Suppose that $\chi \in \bigwedge^p \mathcal{W}^*$ and $\psi \in \bigwedge^q \mathcal{W}^*$. Then

$$\begin{aligned}
\lambda^*(\chi \wedge \psi)(v_1, v_2, \ldots, v_{p+q}) &= (\chi \wedge \psi)(\lambda(v_1), \lambda(v_2), \ldots, \lambda(v_{p+q})) \\
&= \frac{1}{p!\,q!} \sum_\pi \epsilon(\pi) \big(\lambda(v_{\pi(1)}), \lambda(v_{\pi(2)}), \ldots, \lambda(v_{\pi(p)})\big) \\
&\qquad \times \psi\big(\lambda(v_{\pi(p+1)}), \lambda(v_{\pi(p+2)}), \ldots, \lambda(v_{\pi(p+q)})\big) \\
&= \frac{1}{p!\,q!} \sum_\pi \epsilon(\pi) (\lambda^*\chi)\big(v_{\pi(1)}, v_{\pi(2)}, \ldots, v_{\pi(p)}\big)(\lambda^*\psi)\big(v_{\pi(p+1)}, v_{\pi(p+2)}, \ldots, v_{\pi(p+q)}\big) \\
&= (\lambda^*\chi) \wedge (\lambda^*\psi)(v_1, v_2, \ldots, v_{p+q}).
\end{aligned}$$

The sum is taken over all permutations π of $(1, 2, \ldots, p+q)$.

These ideas may be extended to any affine map of affine spaces by taking its linear part for λ.

12. Decomposable Forms

The forms which we introduced in Sections 1 to 7 were all decomposable, that is, exterior products of 1-forms. In particular, a characterising form of a subspace is the exterior product of 1-forms which vanish on the subspace. In general, elements of $\bigwedge^p \mathcal{V}^*$ are not decomposable: each is a linear combination of decomposable p-forms, since the basis we constructed consists of decomposable forms, but this is the most that can be said. We now explain how the decomposable forms may be singled out in a convenient way.

The problem is that it is not immediately apparent from the expression for a form in terms of a basis (for example) whether or not the form may be expressed as an exterior product of 1-forms. Consider a 4-dimensional vector space $\mathcal{V}$, and compare (to take a simple example) the 2-forms $\theta^1 \wedge \theta^2 + \theta^2 \wedge \theta^4$ and $\theta^1 \wedge \theta^2 + \theta^3 \wedge \theta^4$. (Here the θ^a are supposed to constitute a basis for $\mathcal{V}^*$.) It takes only a moment's thought to realise that the first of these 2-forms is decomposable (it may be written $(\theta^1 - \theta^4) \wedge \theta^2$). But what of the second?

Suppose that $\omega = \theta^1 \wedge \theta^2 + \theta^3 \wedge \theta^4$ were decomposable, so that one could find two linearly independent 1-forms η^1, η^2 such that $\omega = \eta^1 \wedge \eta^2$. Then ω would be the characterising 2-form of a 2-dimensional subspace, spanned by any pair of linearly independent vectors v_1, v_2 such that $\langle v_a, \eta^b \rangle = 0$, $a, b = 1, 2$. Then $v_1 \lrcorner \omega = v_2 \lrcorner \omega = 0$. In fact, the linear map $\mathcal{V} \to \bigwedge^1 \mathcal{V}^*$ by $v \mapsto v \lrcorner \omega$ would have kernel of dimension (at least) 2. On the other hand, considering the same map with ω expressed in its original form, one sees that with $\{e_a\}$ the basis of $\mathcal{V}$ dual to $\{\theta^a\}$ the forms $\{e_a \lrcorner \omega\}$ are linearly independent: they are $\{\theta^2, -\theta^1, \theta^4, -\theta^3\}$. It follows that the map $v \mapsto v \lrcorner \omega$ is actually an isomorphism. The hypothesis that ω is decomposable is therefore untenable.

A vector $v \in \mathcal{V}$ is called *characteristic* for the exterior form ω if $v \lrcorner \omega = 0$. The set of characteristic vectors of a given form, being the kernel of a linear map, is a subspace of $\mathcal{V}$. This subspace is called the *characteristic subspace* of ω and will be denoted $\operatorname{char}\omega$. If ω is a decomposable p-form, which is a characterising form for an $(n-p)$-dimensional subspace $\mathcal{W}$ of $\mathcal{V}$, then $\operatorname{char}\omega = \mathcal{W}$. For suppose that $\omega = \eta^1 \wedge \eta^2 \wedge \cdots \wedge \eta^p$, where $w \in \mathcal{W}$ if and only if $\langle w, \eta^1 \rangle = \langle w, \eta^2 \rangle = \cdots \langle w, \eta^p \rangle = 0$. Then for any $v \in \mathcal{V}$

$$v \lrcorner \omega = \langle v, \eta^1 \rangle \eta^2 \wedge \cdots \wedge \eta^p - \langle v, \eta^2 \rangle \eta^1 \wedge \eta^3 \wedge \cdots\cdots \wedge \eta^p + \cdots + (-1)^{p-1} \langle v, \eta^p \rangle \eta^1 \wedge \eta^2 \wedge \cdots \wedge \eta^{p-1};$$

the $(p-1)$ forms occurring on the right hand side are linearly independent, since the η^α are, and so $v \in \operatorname{char}\omega$ if and only if $v \in \mathcal{W}$.

Thus a decomposable p-form on an n-dimensional space has characteristic subspace of dimension $n-p$. Other p-forms may have characteristic subspace of smaller dimension—indeed, we have given an example whose characteristic subspace has dimension 0.

Exercise 40. Show that if $\omega \in \bigwedge^p \mathcal{V}^*$ and the dimension of $\operatorname{char}\omega$ is greater than $n-p$ then $\omega = 0$. □

We show now that the decomposable forms are precisely those non-zero forms whose characteristic subspaces have maximal dimension. We show, in fact, that if $\omega \in \bigwedge^p \mathcal{V}^*$ and $\dim \operatorname{char}\omega = n-p$ then ω may be written as an exterior product of p 1-forms. Let $\{e_a\}$ be a basis for $\mathcal{V}$ such that $\{e_{p+1}, e_{p+2}, \ldots, e_n\}$ is a basis for $\operatorname{char}\omega$; then $\omega(e_1, e_2, \ldots, e_p) \neq 0$ (or e_1 would also be a characteristic vector) and so we may assume, without loss of generality, that $\omega(e_1, e_2, \ldots, e_p) = 1$. Let $\{\theta^a\}$ be the dual basis for $\mathcal{V}^*$. Then $\omega = \theta^1 \wedge \theta^2 \wedge \cdots \wedge \theta^p$, since ω may certainly be expressed as a linear combination of terms $\theta^{a_1} \wedge \theta^{a_2} \wedge \cdots \wedge \theta^{a_p}$ with $1 \leq a_1 < a_2 < \cdots < a_p \leq n$, but the occurrence of any $a_k > p$ is prevented by the fact that the vectors $e_{p+1}, e_{p+2}, \ldots, e_n$ are characteristic, while $(\theta^1 \wedge \theta^2 \wedge \cdots \wedge \theta^p)(e_1, e_2, \ldots, e_p) = 1$.

Exercise 41. Show that two decomposable forms have the same characteristic subspace if and only if one is a scalar multiple of the other. □

Exercise 42. Show that if Ω is a fixed non-zero n-form on an n-dimensional vector space then the map $v \mapsto v \lrcorner \Omega$ is an isomorphism of vector spaces of dimension n. Deduce that every $(n-1)$-form is decomposable; and in particular, if $n = 3$ then every p-form is decomposable, $p = 1, 2, 3$; while if $n = 4$ the only non-decomposable p-forms occur when

$p = 2$, and any such may be expressed in the form $\theta^1 \wedge \theta^2 + \theta^3 \wedge \theta^4$ with $\{\theta^a\}$ a suitable basis for 1-forms. □

The *rank* of a p-form ω is the codimension of its characteristic subspace, namely $n - \dim \operatorname{char}\omega$. The *annihilator* of the characteristic subspace, $\operatorname{ann}\omega$, is the subspace of $\mathcal{V}^*$ consisting of those η such that $\langle v, \eta \rangle = 0$ for all $v \in \operatorname{char}\omega$.

Exercise 43. Show that the dimension of $\operatorname{ann}\omega$ is equal to the rank of ω. Show that $\operatorname{ann}\omega$ is the subspace of $\mathcal{V}^*$ spanned by all elements of the form $v_1 \lrcorner (v_2 \lrcorner (\ldots \lrcorner (v_{p-1} \lrcorner \omega) \ldots))$ for any $v_1, v_2, \ldots, v_{p-1} \in \mathcal{V}$. Show that ω may be expressed as a linear combination of p-fold exterior products of elements of a basis for $\operatorname{ann}\omega$, which is the smallest subspace of $\mathcal{V}^*$ with this property: thus the rank of ω is the smallest number of linearly independent 1-forms required to express it. □

Exercise 44. Show that a non-zero $(n-2)$-form is of rank either $n - 2$ or n. □

Exercise 45. Show that if $\omega = \frac{1}{2}\omega_{ab}\theta^a \wedge \theta^b$ $(\omega_{ba} = -\omega_{ab})$ is a 2-form such that $\omega_{12} \neq 0$, and if $\chi = \omega - (\omega_{12})^{-1}(\omega_{1a}\theta^a \wedge \omega_{2b}\theta^b)$, then $e_1 \lrcorner \chi = e_2 \lrcorner \chi = 0$, where $\{e_a\}$ and $\{\theta^a\}$ are dual bases. Deduce that there is a basis $\{\phi^a\}$ of $\mathcal{V}^*$ such that

$$\omega = \phi^1 \wedge \phi^2 + \phi^3 \wedge \phi^4 + \cdots + \phi^{2r-1} \wedge \phi^{2r}$$

where $2r$ is the rank of ω (and $\phi^1 = (\omega_{12})^{-1}(\omega_{1a}\theta^a)$ and $\phi^2 = \omega_{2b}\theta^b$, for example). Show that $\omega \wedge \omega \wedge \cdots \wedge \omega = r!\phi^1 \wedge \phi^2 \wedge \cdots \wedge \phi^{2r}$ (there being r factors on the left hand side); deduce that a 2-form ω has rank $2r$ if and only if $\omega \wedge \omega \wedge \cdots \wedge \omega \neq 0$ when there are r factors, $= 0$ for $r + 1$ or more factors. Show that the map $\mathcal{V} \to \mathcal{V}^*$ by $v \mapsto v \lrcorner \omega$ has rank (as a linear map) the rank of ω (as a 2-form). Show that this map can never be an isomorphism if n is odd; if $n = 2k$ is even, then the map is an isomorphism if and only if $\omega \wedge \omega \wedge \cdots \wedge \omega$ (k factors) is a volume. □

Exercise 46. Show that if $\mathcal{W}$ is the characteristic subspace of a p-form ω and θ is a 1-form such that $\theta \wedge \omega = 0$ then θ is a constraint form for $\mathcal{W}$. Show that the converse is not true by considering the 2-form $\omega = \theta^1 \wedge \theta^2 + \theta^3 \wedge \theta^4$ (where $\{\theta^a\}$ is a basis for 1-forms on a 4-dimensional vector space), whose characteristic subspace consists of just the zero vector: show that $\theta \wedge \omega$ is never zero for any non-zero 1-form θ. Show that if ω is decomposable, on the other hand, then every constraint form θ for its characteristic subspace satisfies $\theta \wedge \omega = 0$. □

Exercise 47. Show that ω is a characterising form for a subspace $\mathcal{W}$ if and only if $\theta \wedge \omega = 0$ for every constraint 1-form θ for $\mathcal{W}$. □

Exercise 48. Let χ be a p-form on $\mathcal{V}$ which is zero when restricted to a subspace $\mathcal{W}$ of $\mathcal{V}$, and let ω be a characterising form for $\mathcal{W}$: show that $\chi \wedge \omega = 0$. Show that, conversely, if ω is a characterising form for $\mathcal{W}$ and χ a form such that $\chi \wedge \omega = 0$ then χ restricted to $\mathcal{W}$ is zero. □

13. An Extension Principle for Constructing Linear Maps of Forms

The most approachable p-forms (in concept) are the decomposable ones; as we have mentioned, we have built the linear spaces $\bigwedge^p \mathcal{V}^*$, whose elements are linear combinations of decomposable p-forms, mainly in order to take advantage of the convenience of linearity (compare the case of tangent spaces, where it is very useful to be able to add tangent vectors, though there is no natural way of combining curves which results in the addition of vectors tangent to them). Accordingly, one is often faced with constructions which appear natural in terms of decomposable p-forms, which one wishes to extend to the whole space $\bigwedge^p \mathcal{V}^*$ in a linear way. One might try to tackle this head on, but that would involve a complicated check of consistency because a non-decomposable p-form may be written as a linear combination of decomposable p-forms in many different ways. We now present a useful technical lemma which states, roughly speaking, that a map of decomposable p-forms into a vector space $\mathcal{W}$ may be extended to a linear map of $\bigwedge^p \mathcal{V}^*$ into $\mathcal{W}$.

Each decomposable p-form is the exterior product of p linear forms, that is, p elements of $\mathcal{V}^*$. The construction of decomposable p-forms is therefore represented by a map $\delta: \mathcal{V}^{*p} \to \bigwedge^p \mathcal{V}^*$, where $\delta(\eta^1, \eta^2, \ldots, \eta^p) = \eta^1 \wedge \eta^2 \wedge \cdots \wedge \eta^p$. It follows from the properties of the exterior product that δ is multilinear, which means linear in each variable separately, and alternating. Suppose, for the purposes of illustration, that one is given a linear map $\lambda: \bigwedge^p \mathcal{V}^* \to \mathcal{W}$: then $\lambda \circ \delta$ in a way represents the restriction of λ to decomposable p-forms, though it is in fact a map $\mathcal{V}^{*p} \to \mathcal{W}$ and as such is again multilinear and alternating. The result we shall prove is essentially the converse of this: if μ is any alternating multilinear map $\mathcal{V}^{*p} \to \mathcal{W}$ then there is a unique linear map $\lambda: \bigwedge^p \mathcal{V}^* \to \mathcal{W}$ such that $\lambda \circ \delta = \mu$. Thus it is enough to check, for a map defined on decomposable p-forms, that it is alternating and multilinear, to know that it extends to a linear map of the whole space $\bigwedge^p \mathcal{V}^*$.

The proof depends on the observation that the set of all alternating multilinear maps $\mathcal{V}^{*p} \to \mathcal{W}$ and the set of all linear maps $\bigwedge^p \mathcal{V}^* \to \mathcal{W}$ are both vector spaces, that these vector spaces have the same dimension, and that composition with δ is a linear map from the second space to the first. We denote by $A^p(\mathcal{V}^*, \mathcal{W})$ the set of all alternating multilinear maps $\mathcal{V}^{*p} \to \mathcal{W}$ and by $L(\bigwedge^p \mathcal{V}^*, \mathcal{W})$ the set of all linear maps $\bigwedge^p \mathcal{V}^* \to \mathcal{W}$.

Exercise 49. Show that taking linear combinations of images imposes on each of these sets the structure of a vector space. □

Exercise 50. Show that the dimension of the space of linear maps $L(\bigwedge^p \mathcal{V}^*, \mathcal{W})$ is given by $\dim \bigwedge^p \mathcal{V}^* \times \dim \mathcal{W} = \binom{n}{p} \dim \mathcal{W}$. □

Exercise 51. Show that $\dim A^p(\mathcal{V}^*, \mathbf{R}) = \binom{n}{p}$. □

From the last exercise it follows that $\dim A^p(\mathcal{V}^*, \mathcal{W}) = \binom{n}{p} \dim \mathcal{W}$, for if $\{e_\alpha\}$ is a basis for $\mathcal{W}$, where $\alpha = 1, 2, \ldots, \dim \mathcal{W}$, then each $\mu \in A^p(\mathcal{V}^*, \mathcal{W})$ determines uniquely $\dim \mathcal{W}$ elements of $A^p(\mathcal{V}^*, \mathbf{R})$, its components with respect to $\{e_\alpha\}$, and conversely. Thus $A^p(\mathcal{V}^*, \mathcal{W})$ has the same dimension as $L(\bigwedge^p \mathcal{V}^*, \mathcal{W})$.

Now the map which associates with each element λ of $L(\bigwedge^p \mathcal{V}^*, \mathcal{W})$ the element $\lambda \circ \delta$ of $A^p(\mathcal{V}^*, \mathcal{W})$ is evidently a linear one. Moreover, its kernel is just the zero element of $L(\bigwedge^p \mathcal{V}^*, \mathcal{W})$, for if $\lambda \circ \delta = 0$ then, for any basis $\{\theta^a\}$ of $\mathcal{V}^*$ and for any $1 \leq a_1 < a_2 < \cdots < a_p \leq n$, $\lambda(\theta^{a_1} \wedge \theta^{a_2} \wedge \cdots \wedge \theta^{a_p}) = 0$, and so $\lambda = 0$ since these p-forms constitute a basis for $\bigwedge^p \mathcal{V}^*$. It follows that $\lambda \mapsto \lambda \circ \delta$ is a bijective map, and so given any $\mu \in A^p(\mathcal{V}^*, \mathcal{W})$ there is a unique $\lambda \in L(\bigwedge^p \mathcal{V}^*, \mathcal{W})$ such that $\mu = \lambda \circ \delta$.

As an example of the application of this result, we consider once again the linear map of forms induced by a linear map of vector spaces. Let $\kappa: \mathcal{U} \to \mathcal{V}$ be a linear map. The construction is based on the adjoint map $\kappa^*: \mathcal{V}^* \to \mathcal{U}^*$. For any linear forms $\eta^1, \eta^2, \ldots, \eta^p \in \mathcal{V}$ set

$$\hat{\kappa}^*(\eta^1, \eta^2, \ldots, \eta^p) = \kappa^*(\eta^1) \wedge \kappa^*(\eta^2) \wedge \cdots \wedge \kappa^*(\eta^p).$$

Then by linearity of κ^*, $\hat{\kappa}^*$ is multilinear; it is evidently also alternating, so $\hat{\kappa}^* \in A^p(\mathcal{V}^*, \bigwedge^p \mathcal{U}^*)$. There is thus a unique linear map $\bigwedge^p \mathcal{V}^* \to \bigwedge^p \mathcal{U}^*$, also denoted κ^*, such that

$$\kappa^*(\eta^1 \wedge \eta^2 \wedge \cdots \wedge \eta^p) = \kappa^*(\eta^1) \wedge \kappa^*(\eta^2) \wedge \cdots \wedge \kappa^*(\eta^p).$$

Note that, for any $u_1, u_2, \ldots, u_p \in \mathcal{U}$,

$$\kappa^*(\eta^1 \wedge \eta^2 \wedge \cdots \wedge \eta^p(u_1, u_2, \ldots, u_p) \\ = (\eta^1 \wedge \eta^2 \wedge \cdots \wedge \eta^p)(\kappa(u_1), \kappa(u_2), \ldots, \kappa(u_p)),$$

and so the same holds true for any p-form ω on $\mathcal{V}$:

$$(\kappa^*\omega)(u_1, u_2, \ldots, u_p) = \omega(\kappa(u_1), \kappa(u_2), \ldots, \kappa(u_p)).$$

So the definition given in Section 11 is recovered. From this new point of view the property of preserving exterior products plays the key role.

We shall have occasion to use this extension result again in Chapter 7.

Summary of Chapter 4

A map $\mathcal{V}^p \to \mathbf{R}$ is said to be multilinear if it is linear in each argument. A multilinear map is called alternating if interchange of any two arguments changes the sign. An alternating multilinear map $\mathcal{V}^p \to \mathbf{R}$ is called an (exterior) p-form on $\mathcal{V}$. An alternating multilinear map $\mathcal{V}^{*p} \to \mathbf{R}$ is called a p-vector on $\mathcal{V}$. The p-forms comprise a vector space $\bigwedge^p \mathcal{V}^*$ of dimension $\binom{n}{p}$ (where $n = \dim \mathcal{V}$); likewise the p-vectors comprise a vector space, of the same dimension; $\bigwedge^0 \mathcal{V}^* = \mathbf{R}$, $\bigwedge^1 \mathcal{V}^* = \mathcal{V}^*$, and $\bigwedge^p \mathcal{V}^*$ consists of just the zero vector for $p > n$.

The p-vector $(\eta^1, \eta^2, \ldots, \eta^p) \mapsto \det(\langle w_\alpha, \eta^\beta \rangle)$ is denoted $w_1 \wedge w_2 \wedge \cdots \wedge w_p$, and the p-form $(w_1, w_2, \ldots, w_p) \mapsto \det(\langle w_\alpha, \eta^\beta \rangle)$ is denoted $\eta^1 \wedge \eta^2 \wedge \cdots \wedge \eta^p$. The p-vector $w_1 \wedge w_2 \wedge \cdots \wedge w_p$ is a characterising p-vector for the p-dimensional subspace spanned by its constituent vectors; the p-form $\eta^1 \wedge \eta^2 \wedge \cdots \wedge \eta^p$ is a characterising form for the $(n-p)$-dimensional subspace for which its constituent covectors are constraint forms.

The exterior product $\omega \wedge \chi$ of a p-form ω and a q-form χ is a $(p+q)$-form defined by

$$(\omega \wedge \chi)(v_1, v_2, \ldots, v_{p+q}) \\ = \frac{1}{p!\,q!} \sum_\pi \epsilon(\pi)\omega(v_{\pi(1)}, v_{\pi(2)}, \ldots, v_{\pi(p)})\chi(v_{\pi(p+1)}, v_{\pi(p+2)}, \ldots, v_{\pi(p+q)}),$$

the sum being over all permutations π of $(1, 2, \ldots, p+q)$, the sign $\epsilon(\pi)$ of a permutation π being $+1$ if π may be represented as a product of an even number of transpositions, -1 otherwise. The exterior product is distributive and associative but not commutative: $\chi \wedge \omega = (-1)^{pq}\omega \wedge \chi$.

The inner product of a vector v and a 1-form ω is the $(p-1)$-form $v \lrcorner \omega$ such that $(v \lrcorner \omega)(v_1, v_2, \ldots, v_{p-1}) = \omega(v, v_1, v_2, \ldots, v_{p-1})$. The set of vectors v such that $v \lrcorner \omega = 0$ is called the characteristic subspace of ω, $\operatorname{char} \omega$; if $\omega \neq 0$ its dimension is at most $n - p$, and ω is decomposable when it is equal to $n - p$. Not every p-form is decomposable, but every p-form may be expressed as the sum of decomposable

p-forms. In fact if $\{\theta^a\}$ is a basis for $\mathcal{V}^*$ then $\{\theta^{a_1} \wedge \theta^{a_2} \wedge \cdots \wedge \theta^{a_p} \mid 1 < a_1 < a_2 < \cdots < a_p \leq n\}$ is a basis for $\bigwedge^p \mathcal{V}^*$ each element of which is decomposable.

A non-zero n-form Ω on a vector space of dimension n defines a volume function on any affine space modelled on it: the volume of a parallelepiped with sides $v_1, v_2, \ldots, v_n$ is $\Omega(v_1, v_2, \ldots, v_n)$. This vanishes if the parallelepiped is degenerate because its sides are linearly dependent; otherwise, its sign determines the orientation of the sides of the parallelepiped, in the order given. The facts that Ω is multilinear and alternating correspond to properties of volume, and this prescription generalises the determinant rule for the volume of a parallelepiped in Euclidean space; in fact Ω is a basis-independent version of the determinant, and if $\{e_a\}$ is a basis for $\mathcal{V}$ such that $\Omega(e_1, e_2, \ldots, e_n) = 1$ and if $v_a = k_a^b e_b$ then $\Omega(v_1, v_2, \ldots, v_n) = \det(k_a^b)$. The exterior product of a decomposable p-form and a decomposable q-form, where $p + q = n$, reproduces the principle "volume=base area×height". With respect to a given volume form a $1:1$ correspondence may be established between p-vectors and $(n-p)$-forms.

If there is given any alternating p-fold multilinear map from the dual of one vector space to another one then there is a unique linear map from the space of p-forms on the first vector space to the second which agrees with the given map on decomposable forms.

Notes to Chapter 4

1. The symmetric group. The group of permutations π of $(1, 2, \ldots, n)$ with composition of permutations as the group multiplication is called the *symmetric group* on n objects. It has $n!$ elements. A pair of numbers (i, j) is called an inversion for the permutation π if $i < j$ and $\pi(i) > \pi(j)$, the total number of inversions for π is denoted $\#\pi$, and the *sign* of π is $\epsilon(\pi) = (-1)^{\#\pi}$. A permutation of sign $+1$ is called *even* and a permutation of sign -1 is called *odd*. The map $\pi \mapsto \epsilon(\pi)$ is a homomorphism from the symmetric group to the multiplicative group with two elements $\{+1, -1\}$.

A permutation which interchanges two numbers without other change is called a *transposition*. Every transposition is odd, and every permutation may be expressed as a product of transpositions—an even permutation as the product of an even number of transpositions, an odd permutation as the product of an odd number.

See MacLane and Birkhoff [1967], pp 91–96, for proofs and further developments.

2. Determinants. Let A be an $n \times n$ square matrix, and let A_c^b denote the element of A in the bth row and cth column. The determinant of A, denoted $\det A$ or $\det(A_c^b)$, is the number

$$\det A = \sum_\pi \epsilon(\pi) A_1^{\pi(1)} A_2^{\pi(2)} \cdots A_n^{\pi(n)}$$

the sum being over all permutations π of $(1, 2, \ldots, n)$. The determinant has the following properties:

(1) if two columns of A are interchanged, $\det A$ is multiplied by -1

(2) if a column of A is multiplied by a number k, $\det A$ is multiplied by k

(3) if a multiple of one column of A is added to another column, $\det A$ is unaltered

(4) $\det A^T = \det A$, where A^T is the *transpose* of A, obtained by interchanging its rows and columns

(5) $\det A \neq 0$ is a necessary and sufficient condition for the existence of an inverse matrix A^{-1} such that $AA^{-1} = A^{-1}A = I_n$

(6) $\det AB = \det A \det B$.

See MacLane and Birkhoff Chapter 9, pp 294*ff*, for proofs and further developments.

3. Two conventions for exterior algebra. As almost any excursion into the literature will show, there are two different conventions for numerical coefficients in exterior algebra. An author's choice of convention may be identified from his or her definition of either the interior or the exterior product. All authors seem to agree that the alternating part of a p-fold multilinear form T should be defined by the formula

$$\operatorname{alt} T(v_1, v_2, \ldots, v_p) = \frac{1}{p!} \sum_{\pi} \epsilon(\pi) T(v_{\pi(1)}, v_{\pi(2)}, \ldots, v_{\pi(p)}).$$

Any other numerical factor would lead to failure of the formula $\operatorname{alt}\operatorname{alt} T = \operatorname{alt} T$. The two conventions may then be established by setting either $e = 0$ or $e = 1$ in the formula

$$\omega \wedge \chi = \left(\frac{(p+q)!}{p!\,q!}\right)^e \operatorname{alt}(\omega \otimes \chi)$$

for the exterior product of a p-form and a q-form. It is then necessary to define

$$(v \lrcorner \omega)(v_1, v_2, \ldots, v_{p-1}) = p^{1-e} \omega(v, v_1, v_2, \ldots, v_{p-1}).$$

Further differences arise in exterior calculus (see Chapter 5). In this book we have adopted the convention $e = 1$.

4. The isomorphism between $\bigwedge^p \mathcal{V}^*$ and $(\bigwedge^p \mathcal{V})^*$ asserted in Section 8 is proved in Sternberg [1964] Chapter 1, for example.

5. CALCULUS OF FORMS

We move now from the algebraic properties of forms to their differential properties. The first step is similar to steps we have taken before: from affine lines, and affine maps generally, to smooth curves and smooth maps, for example. In place of the forms on vector spaces of Chapter 4 we now consider fields of forms, which, as maps of tangent spaces, share the multilinearity properties of forms, but may vary from point to point of the affine space. Next we exhibit the exterior derivative, which is a generalisation from the operators curl and div of vector calculus. We go on to discuss the relationships between the exterior, covariant and Lie derivatives. Finally, we prove a generalisation of the result of vector calculus that if the curl of a vector field is zero on a suitable domain then the vector field is a gradient.

1. Fields of Forms

The tangent space $T_x\mathcal{A}$ at any point x of an n-dimensional affine space $\mathcal{A}$ is a vector space. One may therefore construct the space of p-forms $\bigwedge^p(T_x^*\mathcal{A})$, for each integer p between 0 and n. The elements of $\bigwedge^p(T_x^*\mathcal{A})$ are alternating multilinear maps $(T_x\mathcal{A})^p \to \mathbf{R}$, for $p > 1$; $\bigwedge^0(T_x^*\mathcal{A}) = \mathbf{R}$, while $\bigwedge^1(T_x^*\mathcal{A}) = T_x^*\mathcal{A}$. For each p, $\bigwedge^p(T_x^*\mathcal{A})$ is a vector space of dimension $\binom{n}{p}$, and has a basis constructed by taking exterior products of basis elements for $T_x^*\mathcal{A}$. As basis for $T_x^*\mathcal{A}$ it is often convenient to choose the coordinate differentials $\{dx^a\}$ where (x^a) are coordinate functions for some system of coordinates around x.

Exercise 1. Show that $\{\, dx^{a_1} \wedge dx^{a_2} \wedge \cdots \wedge dx^{a_p} \mid 1 \leq a_1 < a_2 < \cdots < a_p \leq n \,\}$ is a basis for $\bigwedge^p(T_x^*\mathcal{A})$. □

A *field of p-forms* ω on $\mathcal{A}$ is a choice of an element ω_x of $\bigwedge^p(T_x^*\mathcal{A})$ for each point $x \in \mathcal{A}$. One tests a field of p-forms for smoothness by reducing the question to another which one knows already how to deal with, as follows. If $V_1, V_2, \ldots, V_p$ are vector fields and ω a field of p-forms then $\omega(V_1, V_2, \ldots, V_p)$ is a function on $\mathcal{A}$ whose value at x is $\omega_x(V_{1x}, V_{2x}, \ldots, V_{px})$. The field of p-forms ω is said to be *smooth* if this function is smooth for every choice of smooth vector field arguments. A smooth p-form field on $\mathcal{A}$ is usually called, for brevity, a *p-form* on $\mathcal{A}$.

Exercise 2. Show that a 0-form is a smooth function and that a 1-form is a smooth covector field. □

In the case of a 1-form θ, the value of θ on a vector field V may be denoted with angle brackets, thus: $\langle V, \theta \rangle$, as well as $\theta(V)$.

Two p-forms may be added, and a p-form may be multiplied by a smooth function, to give in each case another p-form; these operations are carried out point by point. Moreover, one may define the exterior product of a p-form and a q-form, again on a point by point basis; the result is a $(p+q)$-form. Again, given a vector field and a p-form one may construct a $(p-1)$-form by taking the interior product point by point. Such operations are often referred to as *pointwise* operations.

As we noted in Chapter 3, Section 10, associated with any smooth function f there is a 1-form, whose value at a point is the covector df at that point: we shall denote this 1-form df also.

Exercise 3. Show that for any vector field V, $\langle V, df\rangle = df(V) = Vf$. □

Sometimes it will be necessary to deal with objects which behave like forms, but which are not defined on the whole of $\mathcal{A}$, only on some open subset of it. As an example consider dx^a, where x^a is a coordinate function: if the coordinates in question are global (affine, for example) then dx^a is a 1-form; but in general for curvilinear coordinates this will not be the case. We call such objects *local forms*. Thus if the x^a are coordinate functions of some (not necessarily global) coordinate system we may build up p-forms (local if the coordinates are not global) by taking exterior products of 1-forms chosen from the dx^a, multiplying the results by smooth functions (or smooth local functions, with domain containing the coordinate patch), and taking sums. The operations concerned work as well for local forms as for globally defined ones, but one has to bear in mind the possibility that the domain may turn out to be significant.

The construction of forms from coordinate 1-forms indicated above generates all (local) forms. As we have already pointed out, the coordinate covectors may be used to define a basis for $\bigwedge^p(T_x^*\mathcal{A})$ at each point x; and so each p-form whose domain includes the coordinate patch of the coordinates (x^a) may be expressed uniquely (on the patch) as a linear combination of the (local) p-forms $dx^{a_1}\wedge dx^{a_2}\wedge\cdots\wedge dx^{a_p}$, $1\leq a_1 < a_2 < \cdots < a_p \leq n$, with coefficients which are local functions.

Let ω be a p-form. The *components* of ω relative to the given coordinate system are defined by

$$\omega_{a_1 a_2 \ldots a_p} = \omega(\partial_{a_1}, \partial_{a_2}, \ldots, \partial_{a_p}).$$

They are smooth functions on the coordinate patch and satisfy $\omega_{a_1 a_2 \ldots a_p} = \omega_{[a_1 a_2 \ldots a_p]}$ (the bracket notation for indices is explained in Chapter 4, Section 8). The p-form ω may be written out in terms of its components in two different ways, as has already been indicated in Chapter 4 for the case of forms on a vector space. If the summation convention is suspended, as one finds in many books, then ω may be written as a linear combination of basis p-forms, each occurring once:

$$\omega = \sum \omega_{a_1 a_2 \ldots a_p} dx^{a_1}\wedge dx^{a_2}\wedge\cdots\wedge dx^{a_p}$$

the sum being over all $(a_1, a_2, \ldots, a_p)$ with $1\leq a_1 < a_2 < \cdots < a_p \leq n$. Restoring the summation convention, and allowing each basis p-form to recur $p!$ times with its indices in all possible orders, one obtains

$$\omega = \frac{1}{p!}\omega_{a_1 a_2 \ldots a_p} dx^{a_1}\wedge dx^{a_2}\wedge\cdots\wedge dx^{a_p}.$$

Where this coordinate patch overlaps another one, with coordinates $(\acute{x}^a)$, the change of components is given by

$$\omega_{a_1 a_2 \ldots a_p} = \frac{\partial \acute{x}^{b_1}}{\partial x^{a_1}}\frac{\partial \acute{x}^{b_2}}{\partial x^{a_2}}\cdots\frac{\partial \acute{x}^{b_p}}{\partial x^{a_p}}\acute{\omega}_{b_1 b_2 \ldots b_p},$$

the Jacobians and the functions all being evaluated pointwise.

The definition of a p-form in tensor calculus begins with this last formula. One specifies a set of smooth functions—the components—in one or more coordinate patches, in such a way that the change of components is given by this formula in any overlapping patches; one may then determine the p-form from its components by multiplying by exterior products of coordinate differentials and adding, as explained above.

Exercise 4. Show that a p-form (in the sense of Chapter 4) on a vector space $\mathcal{V}$ defines a p-form (in the present sense) on an affine space $\mathcal{A}$ of which $\mathcal{V}$ is the underlying vector space, whose coefficients with respect to affine coordinates are constants. □

Exercise 5. Show that if, with respect to arbitrary coordinates,

$$\alpha = x^1 dx^2 \wedge dx^3 - x^2 dx^1 \wedge dx^3 + x^3 dx^1 \wedge dx^2$$

and

$$\beta = x^1 dx^1 + x^2 dx^2 + x^3 dx^3$$

then

$$\alpha \wedge \beta = \left((x^1)^2 + (x^2)^2 + (x^3)^2\right) dx^1 \wedge dx^2 \wedge dx^3.$$ □

The algebra of forms on an affine space is like the algebra of forms on a vector space as described in Chapter 4, with the smooth functions replacing the real numbers in the role of scalars. In fact a p-form ω on $\mathcal{A}$ may be considered as an alternating multilinear map of the $\mathcal{F}(\mathcal{A})$-module $\mathcal{X}(\mathcal{A})^p$ to $\mathcal{F}(\mathcal{A})$, where now multilinearity is over $\mathcal{F}(\mathcal{A})$ rather than $\mathbf{R}$: for any vector fields V_i and V_i' and functions f and f'

$$\begin{aligned}\omega(V_1, V_2, \ldots, fV_i + f'V_i', \ldots, V_p)& \\ = f\omega(V_1, V_2, \ldots, V_i, \ldots, V_p) &+ f'\omega(V_1, V_2, \ldots, V_i', \ldots, V_p),\end{aligned}$$

and similarly for the other arguments. Conversely, any alternating, and in this sense multilinear, map of the $\mathcal{F}(\mathcal{A})$-module $\mathcal{X}(\mathcal{A})$ to $\mathcal{F}(\mathcal{A})$ is a form. We denote by $\bigwedge^p \mathcal{A}^*$ the space of p-forms on $\mathcal{A}$.

We described in Section 11 of Chapter 4 the construction from a linear map of vector spaces of an induced linear map of forms over those vector spaces, which acts contragrediently to the initial linear map. Any smooth map $\phi\colon \mathcal{A} \to \mathcal{B}$ of affine spaces induces a linear map $\phi_{*x}\colon T_x\mathcal{A} \to T_{\phi(x)}\mathcal{B}$, and this may be used to induce a further map of forms, which again acts contragrediently. This construction works as follows. If ω is a p-form on $\mathcal{B}$ define, for each $x \in \mathcal{A}$, an element $(\phi^*\omega)_x$ of $\bigwedge^p(T_x^*\mathcal{A})$ by

$$(\phi^*\omega)_x(v_1, v_2, \ldots, v_p) = \omega_{\phi(x)}(\phi_{*x}v_1, \phi_{*x}v_2, \ldots, \phi_{*x}v_p)$$

where $v_1, v_2, \ldots, v_p \in T_x\mathcal{A}$. Then $\phi^*\omega$, the p-form field whose value at x is $(\phi^*\omega)_x$, is smooth; it is often called the *pull-back* of ω by ϕ.

Exercise 6. Show that $\phi^*(\omega_1 + \omega_2) = \phi^*\omega_1 + \phi^*\omega_2$; that $\phi^*(f\omega) = (f \circ \phi)\phi^*\omega$; that $\phi^*(\omega \wedge \chi) = (\phi^*\omega) \wedge (\phi^*\chi)$; and that $\phi^*(df) = d(f \circ \phi)$, where ω_1, ω_2, ω and χ are forms on $\mathcal{B}$, ω_1 and ω_2 having the same degree, and $f \in \mathcal{F}(\mathcal{B})$. □

If (x^a) are the coordinate functions of a coordinate system on $\mathcal{A}$ and (y^α) those of a coordinate system on $\mathcal{B}$ then

$$\begin{aligned}
&\phi^*\left(\frac{1}{p!}\omega_{\alpha_1\alpha_2\ldots\alpha_p}dy^{\alpha_1}\wedge dy^{\alpha_2}\wedge\cdots\wedge dy^{\alpha_p}\right)\\
&\quad=\frac{1}{p!}(\omega_{\alpha_1\alpha_2\ldots\alpha_p}\circ\phi)\phi^*dy^{\alpha_1}\wedge\phi^*dy^{\alpha_2}\wedge\cdots\wedge\phi^*dy^{\alpha_p}\\
&\quad=\frac{1}{p!}(\omega_{\alpha_1\alpha_2\ldots\alpha_p}\circ\phi)d\phi^{\alpha_1}\wedge d\phi^{\alpha_2}\wedge\cdots\wedge d\phi^{\alpha_p}\\
&\quad=\frac{1}{p!}(\omega_{\alpha_1\alpha_2\ldots\alpha_p}\circ\phi)\frac{\partial\phi^{\alpha_1}}{\partial x^{a_1}}\frac{\partial\phi^{\alpha_2}}{\partial x^{a_2}}\cdots\frac{\partial\phi^{\alpha_p}}{\partial x^{a_p}}dx^{a_1}\wedge dx^{a_2}\wedge\cdots\wedge dx^{a_p},
\end{aligned}$$

where the ϕ^α are the functions which represent ϕ with respect to the two coordinate systems. The calculation of $\phi^*\omega$ in coordinates is therefore very straightforward: one substitutes ϕ^α for y^α wherever it appears in ω, including those places where it appears as dy^α; in the latter case one evaluates $d\phi^\alpha$, regarding ϕ^α as a function, expressing the answer in terms of the x^a; finally one carries out any necessary algebra. Thus, for example, if

$$\omega = dy^1\wedge dy^2 \qquad\text{and}\qquad \phi(x^1,x^2)=(x^1\cos x^2, x^1\sin x^2)$$

then

$$\begin{aligned}
\phi^*\omega &= d(x^1\cos x^2)\wedge d(x^1\sin x^2)\\
&=(\cos x^2dx^1 - x^1\sin x^2dx^2)\wedge(\sin x^2dx^1 + x^1\cos x^2dx^2)\\
&=x^1(\cos^2 x^2+\sin^2 x^2)dx^1\wedge dx^2 = x^1dx^1\wedge dx^2.
\end{aligned}$$

Exercise 7. Show that if $\omega = dy^1\wedge dy^2\wedge dy^3$ and

$$\phi(x^1,x^2,x^3)=(x^1\sin x^2\cos x^3, x^1\sin x^2\sin x^3, x^1\cos x^2)$$

then

$$\phi^*\omega = (x^1)^2\sin x^2dx^1\wedge dx^2\wedge dx^3.$$ □

Exercise 8. Show that if ϕ is a smooth map of $\mathcal{A}$ to itself and $\Omega = f dx^1\wedge dx^2\wedge\cdots\wedge dx^n$ is an n-form on $\mathcal{A}$ ($n=\dim\mathcal{A}$) then

$$\phi^*\omega = (f\circ\phi)(\det\phi_*)dx^1\wedge dx^2\wedge\cdots\wedge dx^n$$

where the function $\det\phi_*$ on $\mathcal{A}$ has as its coordinate representation with respect to any coordinate system for $\mathcal{A}$ the determinant of $(\partial_b\phi^a)$, which is the Jacobian matrix of the coordinate representation of ϕ. □

2. The Exterior Derivative

We have already observed that given a function (or 0-form) f on an affine space we may define a 1-form df, its differential. We also call df the *exterior derivative* of f; its expression in coordinates is $df = (\partial_a f)dx^a$, and so the operation of forming the exterior derivative of a function is closely related to the operation of taking a gradient in vector calculus. Our intention now is to show how the exterior derivative may be extended so as to apply to a form of any degree. The exterior derivative of a p-form will be a $(p+1)$-form; in the case of 1- and 2-forms in a 3-dimensional space the resulting operations will have very close affinities with curl and div of vector calculus.

Consider, first, a p-form ω whose expression in terms of some affine coordinates (x^a) takes the simple form $\omega = f dx^{a_1} \wedge dx^{a_2} \wedge \cdots \wedge dx^{a_p}$ for some function f. A straightforward way of extending the exterior derivative of functions so as to apply to such ω suggests itself: construct the form

$$df \wedge dx^{a_1} \wedge dx^{a_2} \wedge \cdots \wedge dx^{a_p} = (\partial_a f) dx^a \wedge dx^{a_1} \wedge dx^{a_2} \wedge \cdots \wedge dx^{a_p}.$$

This is a $(p+1)$-form. Since every p-form is a sum of p-forms of this type it is easy to extend the construction to an arbitrary p-form, to obtain from

$$\omega = \frac{1}{p!}\omega_{a_1 a_2 \ldots a_p} dx^{a_1} \wedge dx^{a_2} \wedge \cdots \wedge dx^{a_p}$$

the $(p+1)$-form

$$\frac{1}{p!} d\omega_{a_1 a_2 \ldots a_p} \wedge dx^{a_1} \wedge dx^{a_2} \wedge \cdots \wedge dx^{a_p}.$$

So far, the construction may appear to depend on a particular choice of affine coordinates. Suppose, however, that $\acute{x}^a = k^a_b x^b + c^a$ are new affine coordinates. Then

$$\acute{f} d\acute{x}^{a_1} \wedge d\acute{x}^{a_2} \wedge \cdots \wedge d\acute{x}^{a_p} = (k^{a_1}_{b_1} k^{a_2}_{b_2} \cdots k^{a_p}_{b_p} \acute{f}) dx^{b_1} \wedge dx^{b_2} \wedge \cdots \wedge dx^{b_p}$$

and

$$\begin{aligned} d(k^{a_1}_{b_1} k^{a_2}_{b_2} \cdots k^{a_p}_{b_p} \acute{f}) \wedge dx^{b_1} \wedge dx^{b_2} \wedge \cdots \wedge dx^{b_p} \\ = (k^{a_1}_{b_1} k^{a_2}_{b_2} \cdots k^{a_p}_{b_p}) d\acute{f} \wedge dx^{b_1} \wedge dx^{b_2} \wedge \cdots \wedge dx^{b_p} \\ = d\acute{f} \wedge d\acute{x}^{a_1} \wedge d\acute{x}^{a_2} \wedge \cdots \wedge d\acute{x}^{a_p}. \end{aligned}$$

Thus carrying out the prescribed construction in any affine coordinate system gives the same answer. We may therefore define the *exterior derivative operator*, d, as follows: for any p-form ω, whose expression in affine coordinates (x^a) is

$$\omega = \frac{1}{p!}\omega_{a_1 a_2 \ldots a_p} dx^{a_1} \wedge dx^{a_2} \wedge \cdots \wedge dx^{a_p},$$

the exterior derivative $d\omega$ is the $(p+1)$-form defined by

$$d\omega = \frac{1}{p!} d\omega_{a_1 a_2 \ldots a_p} \wedge dx^{a_1} \wedge dx^{a_2} \wedge \cdots \wedge dx^{a_p}.$$

Strictly speaking one should distinguish notationally between the exterior derivative operators for forms of different degrees (by writing, say, d_p for the operator on p-forms), thus making it clear for example that the d on the left hand side of the definition (since it operates on a p-form) is a different operator from those on the right hand side (which all operate on functions, the case which is assumed already known). However this distinction is rarely if ever enforced, and indeed the various operators are so similar that the distinction is hardly necessary.

Exercise 9. Show that, in dimension 2, if $\omega = p_1 dx^1 + p_2 dx^2$ then
$$d\omega = \left(\frac{\partial p_2}{\partial x^1} - \frac{\partial p_1}{\partial x^2}\right) dx^1 \wedge dx^2;$$
in dimension 3, if $\omega = p_1 dx^1 + p_2 dx^2 + p_3 dx^3$ then
$$d\omega = \left(\frac{\partial p_3}{\partial x^2} - \frac{\partial p_2}{\partial x^3}\right) dx^2 \wedge dx^3 - \left(\frac{\partial p_1}{\partial x^3} - \frac{\partial p_3}{\partial x^1}\right) dx^1 \wedge dx^3 + \left(\frac{\partial p_2}{\partial x^1} - \frac{\partial p_1}{\partial x^2}\right) dx^1 \wedge dx^2,$$
while if
$$\omega = p_1 dx^2 \wedge dx^3 + p_2 dx^3 \wedge dx^1 + p_3 dx^1 \wedge dx^2$$
then
$$d\omega = \left(\frac{\partial p_1}{\partial x^1} + \frac{\partial p_2}{\partial x^2} + \frac{\partial p_3}{\partial x^3}\right) dx^1 \wedge dx^2 \wedge dx^3.$$
□

Exercise 10. Let
$$\theta = x^2x^3x^4dx^1 + x^1x^3x^4dx^2 + x^1x^2x^4dx^3 + x^1x^2x^3dx^4$$
$$\eta = x^3x^4dx^1 \wedge dx^2 + x^1x^2dx^3 \wedge dx^4$$
$$\varsigma = x^1dx^2 \wedge dx^3 \wedge dx^4 + x^2dx^1 \wedge dx^3 \wedge dx^4 + x^3dx^1 \wedge dx^2 \wedge dx^4 + x^4dx^1 \wedge dx^2 \wedge dx^3$$
$$\omega = x^1dx^1 + x^2dx^2 + x^3dx^3 + x^4dx^4.$$
Show that $d\theta = 0$; $d\eta = \varsigma$; $d\varsigma = 0$; $d\omega = 0$; $d(\eta \wedge \eta) = 0$; $d(\theta \wedge \omega \wedge \eta) = 0$; and that $d(\omega \wedge \eta) = ((x^1)^2 - (x^2)^2 + (x^3)^2 - (x^4)^2)dx^1 \wedge dx^2 \wedge dx^3 \wedge dx^4$. □

Exercise 11. Show that the exterior derivative of any n-form on an n-dimensional affine space is zero. □

The results of Exercise 9 reveal the similarities between d and the operations curl and div (the similarity between d operating on functions and grad has already been remarked on several times). Thus up to a point vector calculus is subsumed in exterior calculus. However, some caution is necessary, because here the operands are not vector fields, nor are the results of carrying out the operations. To recover the operations of vector calculus in their entirety one needs to use the metric structure of Euclidean space.

We have so far dealt with the expressions for exterior derivatives of forms only in terms of affine coordinates. Conveniently, and remarkably, the same expressions apply in any coordinate system (this fact lends emphasis to the cautionary comments in the previous paragraph). For consider the p-form $\omega = \acute{f} d\acute{x}^{a_1} \wedge d\acute{x}^{a_2} \wedge \cdots \wedge d\acute{x}^{a_p}$ as before, except that the coordinates $(\acute{x}^a)$ are no longer assumed affine (and may indeed be defined only locally). In terms of some affine coordinates (x^a) we have
$$\omega = \left(\frac{\partial \acute{x}^{a_1}}{\partial x^{b_1}} \frac{\partial \acute{x}^{a_2}}{\partial x^{b_2}} \cdots \frac{\partial \acute{x}^{a_p}}{\partial x^{b_p}} \acute{f}\right) dx^{b_1} \wedge dx^{b_2} \wedge \cdots \wedge dx^{b_p}.$$

In computing $d\omega$ we must now (in contrast to the case of a coordinate transformation between two sets of affine coordinates) take into account the partial derivatives of the terms $\partial \acute{x}^a / \partial x^b$. The derivative of the first such term in the expression for ω contributes a term
$$\left(\frac{\partial^2 \acute{x}^{a_1}}{\partial x^b \partial x^{b_1}} \frac{\partial \acute{x}^{a_2}}{\partial x^{b_2}} \cdots \frac{\partial \acute{x}^{a_p}}{\partial x^{b_p}} \acute{f}\right) dx^b \wedge dx^{b_1} \wedge dx^{b_2} \wedge \cdots \wedge dx^{b_p}$$
to $d\omega$; this term is actually zero, since the second partial derivative is symmetric in b and b_1 and so
$$\frac{\partial^2 \acute{x}^{a_1}}{\partial x^b \partial x^{b_1}} dx^b \wedge dx^{b_1} = 0.$$

Similarly, all the other contributions to $d\omega$ arising in this way vanish, and all that remains is

$$\left(\frac{\partial \acute{x}^{a_1}}{\partial x^{b_1}} \frac{\partial \acute{x}^{a_2}}{\partial x^{b_2}} \cdots \frac{\partial \acute{x}^{a_p}}{\partial x^{b_p}}\right) d\acute{f} \wedge dx^{b_1} \wedge dx^{b_2} \wedge \cdots \wedge dx^{b_p}$$
$$= d\acute{f} \wedge d\acute{x}^{a_1} \wedge d\acute{x}^{a_2} \wedge \cdots \wedge d\acute{x}^{a_p}$$

as before. Thus the formula for the exterior derivative given previously for affine coordinates applies in fact for any coordinates.

This calculation reveals what is perhaps the key factor behind the simplicity and utility of exterior calculus, namely the way in which the alternating character of forms eliminates second partial derivatives from consideration. The reader will notice several occurrences of the same effect below.

3. Properties of the Exterior Derivative

We now display the most important properties of exterior differentiation.

From the definition it is clear that exterior differentiation is **R**-linear:

$$d(k_1\omega_1 + k_2\omega_2) = k_1 d\omega_1 + k_2\omega_2 \qquad k_1, k_2 \in \mathbf{R}.$$

It is not however $\mathcal{F}(\mathcal{A})$-linear: in fact

$$d(f\omega) = f d\omega + df \wedge \omega \qquad f \in \mathcal{F}(\mathcal{A}).$$

This follows from the rule for evaluating the differential of a product of functions, $d(fg) = f dg + (df)g$: for if $\omega = g dx^{a_1} \wedge dx^{a_2} \wedge \cdots \wedge dx^{a_p}$ then

$$d(f\omega) = d(fg) \wedge dx^{a_1} \wedge dx^{a_2} \wedge \cdots \wedge dx^{a_p} = f d\omega + df \wedge \omega.$$

With respect to the exterior product the exterior derivative obeys a rule something like Leibniz's rule, except for some differences in matters of sign: if ω is a p-form, and χ another form, whose degree is unimportant, then

$$d(\omega \wedge \chi) = d\omega \wedge \chi + (-1)^p \omega \wedge d\chi.$$

This again is a consequence of the Leibniz property of the differential, but now the properties of the exterior product also come into play. If $\omega = f dx^{a_1} \wedge dx^{a_2} \wedge \cdots \wedge dx^{a_p}$ and $\chi = g dx^{b_1} \wedge dx^{b_2} \wedge \cdots \wedge dx^{b_q}$ then

$$\omega \wedge \chi = fg dx^{a_1} \wedge dx^{a_2} \wedge \cdots \wedge dx^{a_p} \wedge dx^{b_1} \wedge dx^{b_2} \wedge \cdots \wedge dx^{b_q}$$

and so

$$\begin{aligned} &d(\omega \wedge \chi) \\ &\quad = ((df)g + f dg) \wedge dx^{a_1} \wedge dx^{a_2} \wedge \cdots \wedge dx^{a_p} \wedge dx^{b_1} \wedge dx^{b_2} \wedge \cdots \wedge dx^{b_q} \\ &\quad = (df \wedge dx^{a_1} \wedge dx^{a_2} \wedge \cdots \wedge dx^{a_p}) \wedge (g dx^{b_1} \wedge dx^{b_2} \wedge \cdots \wedge dx^{b_q}) \\ &\qquad + (-1)^p (f dx^{a_1} \wedge dx^{a_2} \wedge \cdots \wedge dx^{a_p}) \wedge (dg \wedge dx^{b_1} \wedge dx^{b_2} \wedge \cdots \wedge dx^{b_q}) \\ &\quad = d\omega \wedge \chi + (-1)^p \omega \wedge d\chi \end{aligned}$$

since p interchanges are required to move dg into position. The full result follows, again, by linearity. The exterior derivative is said to be an *anti-derivation* of the algebra of forms.

Finally, the exterior derivative enjoys an important property which generalises the familiar facts that curl(grad f) $= \mathbf{0}$ and div(curl $\mathbf{X}$) $= 0$: for any form ω, $d(d\omega) = 0$, or in short,

$$d^2 = 0.$$

(This means, if d_p represents the exterior derivative of p-forms, that $d_{p+1} \circ d_p = 0$.) This is again a consequence of the symmetry of second partial derivatives: for if $\omega = f dx^{a_1} \wedge dx^{a_2} \wedge \cdots \wedge dx^{a_p}$, so that

$$d\omega = df \wedge dx^{a_1} \wedge dx^{a_2} \wedge \cdots \wedge dx^{a_p} = \frac{\partial f}{\partial x^a} dx^a \wedge dx^{a_1} \wedge dx^{a_2} \wedge \cdots \wedge dx^{a_p},$$

then

$$\begin{aligned} d(d\omega) &= d\left(\frac{\partial f}{\partial x^b}\right) \wedge dx^b \wedge dx^{a_1} \wedge dx^{a_2} \wedge \cdots \wedge dx^{a_p} \\ &= \frac{\partial^2 f}{\partial x^a \partial x^b} dx^a \wedge dx^b \wedge dx^{a_1} \wedge dx^{a_2} \wedge \cdots \wedge dx^{a_p} = 0. \end{aligned}$$

A coordinate independent expression for d. The exterior derivative has been introduced in a coordinate-dependent form. For aesthetic reasons, and for many theoretical purposes, it is desirable to have a definition which is independent of coordinates. This we now explain.

Such a definition uses vector fields in a "catalytic" role, and uses the facts that if $\omega \in \bigwedge^p \mathcal{A}$ and $V_1, V_2, \ldots, V_p$ are smooth vector fields, then $\omega(V_1, V_2, \ldots, V_p)$ is a smooth function, that ω is an alternating $\mathcal{F}(\mathcal{A})$-multilinear map from $\mathcal{X}(\mathcal{A})^p$ to $\mathcal{F}(\mathcal{A})$; and that any such map defines a p-form. The shape that the required formula might take is suggested by the following exercises.

Exercise 12. Show that the components of $d\omega$ are given by the following equivalent expressions:

$$d\omega\left(\partial_{a_1}, \partial_{a_2}, \ldots, \partial_{a_{p+1}}\right) = \sum_{r=1}^{p+1} (-1)^{r+1} \partial_{a_r}\left(\omega(\partial_{a_1}, \partial_{a_2}, \ldots \widehat{\partial_{a_r}} \ldots, \partial_{a_{p+1}})\right)$$

$$(d\omega)_{a_1 a_2 \ldots a_{p+1}} = (p+1) \partial_{[a_1} \omega_{a_2 \ldots a_{p+1}]}.$$

(The caret indicates a term to be omitted.) □

Exercise 13. In the case of a 1-form ω

$$d\omega(\partial_a, \partial_b) = \partial_a(\omega(\partial_b)) - \partial_b(\omega(\partial_a)).$$

Show that direct transliteration of the right hand side of this expression, when the coordinate vector fields are replaced by arbitrary vector fields V, W, namely $V(\omega(W)) - W(\omega(V))$, fails to satisfy the correct rule for the effect of multiplying a vector field by a function, and does not represent a 2-form. Show that χ, given by

$$\chi(V, W) = V(\omega(W)) - W(\omega(V)) - \omega([V, W]),$$

does satisfy the rules for a 2-form however. Conclude that since the bracket of coordinate vector fields vanishes, χ agrees with $d\omega$ on coordinate vector fields, and that $\chi = d\omega$. □

The required formula for the exterior derivative of a p-form ω is

$$d\omega(V_1, V_2, \ldots, V_{p+1}) = \sum_{r=1}^{p+1} (-1)^{r+1} V_r\big(\omega(V_1, V_2, \ldots \widehat{V_r} \ldots, V_{p+1})\big) + \sum_{1 \le r \le s \le p+1} (-1)^{r+s} \omega\big([V_r, V_s], V_1, \ldots \widehat{V_r} \ldots \widehat{V_s} \ldots, V_{p+1}\big).$$

This agrees with the first formula in Exercise 12 when the arguments are coordinate vector fields; it agrees with the formula obtained in Exercise 13 when $p = 1$, and reduces to the definition of the exterior derivative of a function when $p = 0$. To complete the proof of the formula (that is, the confirmation that the expression on the right hand side does indeed give the $(p + 1)$-form $d\omega$ as defined previously) it remains to show that the right hand side is $\mathcal{F}(\mathcal{A})$-multilinear and alternating and so corresponds to a $(p + 1)$-form. This is more complicated, but no more difficult in principle, than the particular case already tackled in Exercise 13. We therefore leave it as a further exercise; the reader may find it helpful to consider the case $p = 2$ first.

Exercise 14. Show that

$$\begin{aligned} d\omega(U, V, W) =& U(\omega(V, W)) + V(\omega(W, U)) + W(\omega(U, V)) \\ & - \omega([U, V], W) - \omega([V, W], U) - \omega([W, U], V) \end{aligned}$$

for any 2-form ω, by showing that the right hand side defines a 3-form, and then evaluating this 3-form with coordinate vector fields for arguments. Then complete the proof of the general formula for $d\omega$ when ω is a p-form. □

Exercise 15. Use the coordinate-free definition of d to show that $d^2 = 0$, first for $p = 0$, 1, and 2, and then in general. □

The exterior derivative and smooth maps. Let $\phi: \mathcal{A} \to \mathcal{B}$ be a smooth map of affine spaces and let $\omega \in \bigwedge^p \mathcal{B}$. We explained in Section 1 how to define and calculate the pull-back $\phi^*\omega \in \bigwedge^p \mathcal{A}$. We now show that

$$\phi^*(d\omega) = d(\phi^*\omega);$$

in other words, that the pull-back operation commutes with exterior differentiation. We may for simplicity assume, as before, that $\omega = f dy^{\alpha_1} \wedge dy^{\alpha_2} \wedge \cdots \wedge dy^{\alpha_p}$; the general case follows by linearity. Then

$$\phi^*\omega = \left(\frac{\partial \phi^{\alpha_1}}{\partial x^{a_1}} \frac{\partial \phi^{\alpha_2}}{\partial x^{a_2}} \cdots \frac{\partial \phi^{\alpha_p}}{\partial x^{a_p}} f \circ \phi \right) dx^{a_1} \wedge dx^{a_2} \wedge \cdots \wedge dx^{a_p}$$

and so

$$\begin{aligned} d(\phi^*\omega) &= d\left(\frac{\partial \phi^{\alpha_1}}{\partial x^{a_1}} \frac{\partial \phi^{\alpha_2}}{\partial x^{a_2}} \cdots \frac{\partial \phi^{\alpha_p}}{\partial x^{a_p}} f \circ \phi \right) \wedge dx^{a_1} \wedge dx^{a_2} \wedge \cdots \wedge dx^{a_p} \\ &= \left(\sum_{r=1}^{p} \frac{\partial \phi^{\alpha_1}}{\partial x^{a_1}} \cdots \frac{\partial^2 \phi^{\alpha_r}}{\partial x^a \partial x^{a_r}} \cdots \frac{\partial \phi^{\alpha_p}}{\partial x^{a_p}} \right) (f \circ \phi) dx^a \wedge dx^{a_1} \wedge \cdots \wedge dx^{a_p} \\ &\qquad + \left(\frac{\partial \phi^{\alpha_1}}{\partial x^{a_1}} \frac{\partial \phi^{\alpha_2}}{\partial x^{a_2}} \cdots \frac{\partial \phi^{\alpha_p}}{\partial x^{a_p}} \right) \phi^*(df) \wedge dx^{a_1} \wedge dx^{a_2} \wedge \cdots \wedge dx^{a_p} \\ &= \phi^*(df \wedge dy^{\alpha_1} \wedge dy^{\alpha_2} \wedge \cdots \wedge dy^{\alpha_p}) = \phi^*(d\omega), \end{aligned}$$

the terms under the summation sign contributing nothing because of the symmetry of second partial derivatives.

4. Lie Derivatives of Forms

The definition of the Lie derivative of a covector field (in present terminology, a 1-form) which was given in Chapter 3, Section 6 is easily adapted to apply to forms of other degrees. The definition given in Chapter 3 was applied to a form defined along an integral curve of a vector field, and then (in Section 10) by an obvious extension to a form defined all over the affine space (or at least on some open subset of it). In the present discussion we shall concentrate from the start on the latter situation; it is easy to see how to recover the former, if necessary.

Let ω be a p-form on an affine space and V a vector field which generates a one-parameter group ϕ_t. The *Lie derivative* of the p-form ω with respect to V, $\mathcal{L}_V\omega$, is the p-form given by

$$\mathcal{L}_V\omega = \frac{d}{dt}\left(\phi_t{}^*\omega\right)_{t=0}.$$

The value of $\mathcal{L}_V\omega$ at the point x is thus

$$(\mathcal{L}_V\omega)_x = \frac{d}{dt}\left(\phi_t{}^*(\omega_{\phi_t(x)})\right)_{t=0} = \lim_{t\to 0}\frac{1}{t}\left(\phi_t{}^*(\omega_{\phi_t(x)}) - \omega_x\right).$$

This rather more complicated formula serves to define $\mathcal{L}_V\omega$ in the case that V generates, not a one-parameter group, but only a flow.

The Lie derivative measures the rate of change of a form under the action of the flow of a vector field: so, for example, if ω is a p-form invariant under the flow ($\phi_t{}^*(\omega_{\phi_t(x)}) = \omega_x$ for all t) then $\mathcal{L}_V\omega = 0$. The converse is true, as will be shown below.

We now list the main properties of the Lie derivative of forms, most of which are consequences of the definition or known properties of the induced map as given in Exercise 6 and Section 3.

From the definition and from the linearity of induced maps it is clear that the Lie derivative is $\mathbf{R}$-linear in ω:

$$\mathcal{L}_V(k_1\omega_1 + k_2\omega_2) = k_1\mathcal{L}_V\omega_1 + k_2\mathcal{L}_V\omega_2 \qquad k_1, k_2 \in \mathbf{R}.$$

It is not $\mathcal{F}(\mathcal{A})$-linear; rather,

$$\mathcal{L}_V(f\omega) = f\mathcal{L}_V\omega + (Vf)\omega \qquad f \in \mathcal{F}(\mathcal{A}).$$

This follows from the fact that $\phi_t{}^*(f\omega) = (f\circ\phi_t)\phi_t{}^*\omega$. Unlike the exterior derivative, the Lie derivative is a derivation of the algebra of forms:

$$\mathcal{L}_V(\omega\wedge\chi) = (\mathcal{L}_V\omega)\wedge\chi + \omega\wedge(\mathcal{L}_V\chi).$$

This is a consequence of the fact that $\phi_t{}^*(\omega\wedge\chi) = (\phi_t{}^*\omega)\wedge(\phi_t{}^*\chi)$. From the commutativity of pull-back and exterior derivative, $d(\phi_t{}^*\omega) = \phi_t{}^*(d\omega)$, it follows that Lie and exterior derivatives commute:

$$d(\mathcal{L}_V\omega) = \mathcal{L}_V(d\omega).$$

This has already been pointed out, in the case where ω is a 0-form, or function, in Section 7 of Chapter 3, where we stated that $\mathcal{L}_V(df) = d(Vf)$; Lie derivative and directional derivative coincide for functions.

With these rules at one's disposal, the calculation of Lie derivatives in coordinates becomes quite straightforward. A p-form is a sum of terms like $f dx^{a_1} \wedge dx^{a_2} \wedge \cdots \wedge dx^{a_p}$; by linearity of the Lie derivative, it is permissible to deal with each term separately and sum the results, as before. Now

$$\begin{aligned}&\mathcal{L}_V(f dx^{a_1} \wedge dx^{a_2} \wedge \cdots \wedge dx^{a_p})\\&\qquad = f\mathcal{L}_V(dx^{a_1} \wedge dx^{a_2} \wedge \cdots \wedge dx^{a_p}) + (Vf)dx^{a_1} \wedge dx^{a_2} \wedge \cdots \wedge dx^{a_p}.\end{aligned}$$

The second term on the right hand side may be computed as it stands; to simplify the first, the derivation property is used:

$$\begin{aligned}&\mathcal{L}_V(dx^{a_1} \wedge dx^{a_2} \wedge \cdots \wedge dx^{a_p})\\&\quad = \mathcal{L}_V(dx^{a_1}) \wedge dx^{a_2} \wedge \cdots \wedge dx^{a_p} + dx^{a_1} \wedge \mathcal{L}_V(dx^{a_2}) \wedge \cdots \wedge dx^{a_p} + \cdots\\&\qquad\qquad + dx^{a_1} \wedge dx^{a_2} \wedge \cdots \wedge \mathcal{L}_V(dx^{a_p}).\end{aligned}$$

Now the commutativity of $\mathcal{L}_V$ and d comes into play:

$$\mathcal{L}_V(dx^a) = d(Vx^a) = dV^a$$

where $V = V^a\partial_a$. Thus

$$\begin{aligned}&\mathcal{L}_V(f dx^{a_1} \wedge dx^{a_2} \wedge \cdots \wedge dx^{a_p})\\&\quad = (Vf)dx^{a_1} \wedge dx^{a_2} \wedge \cdots \wedge dx^{a_p} + f\sum_{i=1}^{p} dx^{a_1} \wedge \cdots \wedge dV^{a_i} \wedge \cdots \wedge dx^{a_p}.\end{aligned}$$

Exercise 16. Show that if $V = (x^1)^2\partial_1 + (x^2)^2\partial_2 + (x^3)^2\partial_3$ and $\omega = x^2x^3dx^1 + x^1x^3dx^2 + x^1x^2dx^3$ then

$$\mathcal{L}_V\omega = x^2x^3(2x^1 + x^2 + x^3)dx^1 + x^1x^3(x^1 + 2x^2 + x^3)dx^2 + x^1x^2(x^1 + x^2 + 2x^3)dx^3$$

while if $V = x^2\partial_1 - x^1\partial_2$ and $\omega = ((x^1)^2 + (x^2)^2)dx^1 \wedge dx^2$ then

$$\mathcal{L}_V\omega = 0. \qquad \square$$

Exercise 17. Show that if $\Omega = \rho dx^1 \wedge dx^2 \wedge \cdots \wedge dx^n$ is an n-form on an n-dimensional space and $V = V^a\partial_a$ then

$$\mathcal{L}_V\Omega = (\rho\partial_a V^a + V\rho)dx^1 \wedge dx^2 \wedge \cdots \wedge dx^n. \qquad \square$$

Exercise 18. Show that if $\omega = \frac{1}{p!}\omega_{a_1a_2\ldots a_p}dx^{a_1} \wedge dx^{a_2} \wedge \cdots \wedge dx^{a_p}$ then

$$\mathcal{L}_V\omega = \frac{1}{p!}\partial_{[a_1}(V^a\omega_{|a|a_2\cdots a_p]})dx^{a_1} \wedge dx^{a_2} \wedge \cdots \wedge dx^{a_p},$$

where the bars around the suffix a indicate that it is to be omitted from the skew symmetrisation. $\square$

We next show how $\mathcal{L}_V\omega$ depends on V. This is not so straightforward to derive; to obtain the required results we first generalise the formula for the Lie derivative of a 1-form α given in Chapter 3, Exercise 47, which may be written

$$(\mathcal{L}_V\alpha)(W) = V(\alpha(W)) - \alpha([V,W]).$$

This throws V into more tractable positions, from which such a property as the $\mathbf{R}$-linearity of $\mathcal{L}_V\alpha$ in V becomes obvious. This result follows from the fact that ϕ_{t*} is the adjoint of $\phi_t{}^*$, and so an expression involving $\phi_t{}^*$ acting on a 1-form may be converted into one involving ϕ_{t*} acting on a vector argument introduced for the purpose. The same strategy works in the general case, and produces a somewhat similar result. We show that for a p-form ω and vector fields $W_1, W_2, \ldots, W_p$

$$(\mathcal{L}_V\omega)(W_1, W_2, \ldots, W_p) = V\big(\omega(W_1, W_2, \ldots, W_p)\big) - \sum_{r=1}^{p} \omega(W_1, \ldots, [V, W_r], \ldots, W_p).$$

By definition,

$$\big(\phi_t{}^*\omega_{\phi_t(x)}\big)(w_1, w_2, \ldots, w_p) = \omega_{\phi_t(x)}(\phi_{t*}w_1, \phi_{t*}w_2, \ldots, \phi_{t*}w_p)$$

where $w_1, w_2, \ldots, w_p$ are the values of $W_1, W_2, \ldots, W_p$ at a point x. But $\phi_{t*}w_r$ is approximately the value of the vector field $W_r - t\mathcal{L}_V W_r$ at $\phi_t(x)$, from which the result follows. An alternative method of obtaining the same result is to adapt the method used to obtain the cordinate-free definition of d in Section 3, as in the following exercise.

Exercise 19. Show that if

$$\Theta(W_1, W_2, \ldots, W_p) = V\big(\omega(W_1, W_2, \ldots, W_p)\big) - \sum_{r=1}^{p} \omega(W_1, \ldots, [V, W_r], \ldots, W_p)$$

then Θ is $\mathcal{F}(\mathcal{A})$-multilinear and alternating and is therefore a p-form. By evaluating this expression when $W_1, W_2, \ldots, W_p$ are coordinate vector fields, show that $\Theta = \mathcal{L}_V\omega$. □

It follows immediately from this formula that $\mathcal{L}_V\omega$ is $\mathbf{R}$-linear in V:

$$\mathcal{L}_{(k_1V_1 + k_2V_2)}\omega = k_1\mathcal{L}_{V_1}\omega + k_2\mathcal{L}_{V_2}\omega.$$

Other properties may be deduced from the same formula.

It is convenient at this point to make use of the *interior product* of a form by a vector field. As with all the other algebraic operations involving forms, this involves nothing more than applying the corresponding vector space concept from Chapter 4 pointwise. Thus, if ω is a p-form and V a vector field, then $V \lrcorner \omega$ is the $(p-1)$-form defined by $(V \lrcorner \omega)_x = V_x \lrcorner \omega_x$. For any vector fields $W_1, W_2, \ldots, W_{p-1}$

$$(V \lrcorner \omega)(W_1, W_2, \ldots, W_{p-1}) = \omega(V, W_1, W_2, \ldots, W_{p-1}).$$

Exercise 20. Show that, for any smooth function f,

$$\mathcal{L}_{fV}\omega = f\mathcal{L}_V\omega + df \wedge (V \lrcorner \omega)$$

(use the formula for $\mathcal{L}_V\omega$ given in Exercise 19 and above). □

Exercise 21. Show that

$$\mathcal{L}_V(\mathcal{L}_W\omega) - \mathcal{L}_W(\mathcal{L}_V\omega) = \mathcal{L}_{[V,W]}\omega$$

and that

$$\mathcal{L}_V(W \lrcorner \omega) = (\mathcal{L}_V W) \lrcorner \omega + W \lrcorner \mathcal{L}_V\omega.$$

□

The significance of vanishing Lie derivative. We now consider the consequences of the condition $\mathcal{L}_V\omega = 0$. We pointed out above that if ω is invariant under the flow generated by V (in other words, if ω is Lie transported by this flow) then $\mathcal{L}_V\omega = 0$; we seek now to prove the converse. In order to do so we first derive a result concerning the behaviour of the Lie derivative under smooth invertible transformations.

Let Φ be a smooth invertible map with smooth inverse. We wish to derive an alternative expression for $\mathcal{L}_V(\Phi^*\omega)$. Now if V generates a one-parameter group ϕ_t, then

$$\phi_t{}^*(\Phi^*\omega) = (\Phi \circ \phi_t)^*\omega = \Phi^*(\Phi \circ \phi_t \circ \Phi^{-1})^*\omega,$$

from which the required formula will follow by differentiation with respect to t. The reason for choosing this rearrangement of terms is that $\Phi \circ \phi_t \circ \Phi^{-1}$ is also a one-parameter group: its generator is the vector field V^Φ defined (Chapter 3, Exercise 28) by

$$(V^\Phi)_x = \Phi_{*x}\left(V_{\Phi^{-1}(x)}\right).$$

Thus

$$\mathcal{L}_V(\Phi^*\omega) = \Phi^*(\mathcal{L}_{V^\Phi}\omega).$$

In particular, taking for Φ an element of the one-parameter group generated by V, and using the fact that V is invariant by such a transformation, one obtains

$$\mathcal{L}_V(\phi_s{}^*\omega) = \phi_s{}^*(\mathcal{L}_V\omega).$$

If $\mathcal{L}_V\omega = 0$ it then follows that $\mathcal{L}_V(\phi_s{}^*\omega) = 0$ for all s and therefore that $\phi_t{}^*\omega$ is independent of t. Thus in this case $\phi_t{}^*\omega = \phi_0{}^*\omega = \omega$, and so ω is left invariant by the one-parameter group generated by V. (If V should generate only a flow the result remains true though the argument must be modified.)

5. Volume Forms and the Divergence of a Vector Field

A volume form, in the terminology of Chapter 4, is simply an n-form on an n-dimensional affine space arising from an n-form on its underlying vector space. Such a form defines a volume element on the affine space (to use the appropriate phrase from multiple integration) which is invariant under translations.

Exercise 22. Show that an n-form Ω on an n-dimensional affine space is invariant under all translations if and only if its (single) component with respect to affine coordinates is constant. □

In the present context it is appropriate to generalise this concept and to call any nowhere vanishing n-form on an n-dimensional affine space a *volume form*. A volume form determines an orientation of the affine space. Relative to positively oriented affine coordinates it may be written $\rho dx^1 \wedge dx^2 \wedge \cdots \wedge dx^n$ where ρ is a positive function. If one thinks of $dx^1 \wedge dx^2 \wedge \cdots \wedge dx^n$ as determining a volume in the usual geometric sense then it is natural to interpret ρ as (for example) a density function.

In Exercise 17 the following formula for the Lie derivative of an n-form was obtained: if $\Omega = \rho dx^1 \wedge dx^2 \wedge \cdots \wedge dx^n$ and $V = V^a\partial_a$ then

$$\mathcal{L}_V\Omega = \left(\rho\frac{\partial V^a}{\partial x^a} + V\rho\right) dx^1 \wedge dx^2 \wedge \cdots \wedge dx^n.$$

If Ω is a volume form, so that ρ never vanishes, this may be written

$$\mathcal{L}_V\Omega = \left(\frac{\partial V^a}{\partial x^a} + \frac{1}{\rho}V\rho\right)\Omega = \frac{1}{\rho}\frac{\partial}{\partial x^a}(\rho V^a)\Omega.$$

In the case $\rho = 1$ the coefficient on the right hand side is just the usual expression for the divergence of the vector field V. More generally, if Ω is invariant under the one-parameter group generated by V, then

$$\frac{\partial}{\partial x^a}(\rho V^a) = 0,$$

which is the continuity equation of fluid dynamics. So much may be said under the assumption that the coordinates are affine, but the formula holds for any coordinates. In fact there are present here the basic ideas for a coordinate independent definition of the divergence of a vector field.

Given a volume form Ω on an affine space $\mathcal{A}$ of dimension n any n-form on $\mathcal{A}$ may be expressed uniquely as a multiple of Ω with a coefficient which will in general be a smooth function on $\mathcal{A}$. In other words Ω will serve as a basis for the module of n-forms on $\mathcal{A}$ over $\mathcal{F}(\mathcal{A})$. In particular the n-form $\mathcal{L}_V\Omega$, for any vector field V, may be so expressed: the coefficient in this case is called the *divergence* of V with respect to Ω, written $\operatorname{div}_\Omega V$. Thus

$$\mathcal{L}_V\Omega = (\operatorname{div}_\Omega V)\Omega.$$

By its very definition, $\operatorname{div}_\Omega V$ describes how the volume form Ω is changed under the action of the flow of V; in particular, $\operatorname{div}_\Omega V = 0$ is the necessary and sufficient condition for the vector field to be volume-preserving.

Exercise 23. Prove that if Ω is a volume form and V is a vector field whose flow is volume-preserving, so that $\operatorname{div}_\Omega V = 0$, and if W is a vector field whose flow consists of symmetries of V, that is, transformations which leave V invariant, so that $\mathcal{L}_W V = 0$, then $\operatorname{div}_\Omega W$ is constant along the integral curves of V, that is, $V(\operatorname{div}_\Omega W) = 0$. □

We now have two ways of constructing quantities which generalise the divergence of vector calculus: this, and the exterior derivative of an $(n-1)$-form, as exemplified in Exercise 9. In fact a natural way of expressing an $(n-1)$-form ω is as follows:

$$\begin{aligned}\omega &= p_1 dx^2 \wedge dx^3 \wedge \cdots \wedge dx^n - p_2 dx^1 \wedge dx^3 \wedge \cdots \wedge dx^n + \cdots \\ &\qquad\qquad + (-1)^{n-1} p_n dx^1 \wedge dx^2 \wedge \cdots \wedge dx^{n-1} \\ &= \sum_{r=1}^{n} (-1)^{r-1} p_r dx^1 \wedge dx^2 \wedge \cdots \widehat{dx^r} \cdots \wedge dx^n.\end{aligned}$$

Then in the computation of $d\omega$ the only derivative of p_r which contributes is the one with respect to x_r, and in fact

$$d\omega = \left(\sum_{r=1}^{n} \frac{\partial p_r}{\partial x^r}\right) dx^1 \wedge dx^2 \wedge \cdots \wedge dx^n.$$

The two constructions of the divergence are related as follows. The particular expression for ω used above is just what would be obtained as a result of taking the interior product of the n-form $dx^1 \wedge dx^2 \wedge \cdots \wedge dx^n$ with a suitable vector field. In fact, given a volume form Ω, any vector field V determines an $(n-1)$-form $V \lrcorner \Omega$, and it is easy to see, either directly or by considerations of dimension (compare Chapter 4, Exercise 42), that every $(n-1)$-form may be expressed uniquely in this way. In the case in which the n-form is just $dx^1 \wedge dx^2 \wedge \cdots \wedge dx^n$ (and the coordinates are taken to be affine) the $(n-1)$-form corresponding to the vector field $V = V^a \partial_a$ is

$$\sum_{r=1}^{n} (-1)^{r-1} V^r dx^1 \wedge dx^2 \wedge \cdots \widehat{dx^r} \cdots \wedge dx^n.$$

As has already been shown, the exterior derivative of this $(n-1)$-form is indeed the divergence of V times $dx^1 \wedge dx^2 \wedge \cdots \wedge dx^n$. Thus, with $\Omega_0 = dx^1 \wedge dx^2 \wedge \cdots \wedge dx^n$, we may write

$$\mathcal{L}_V \Omega_0 = d(V \lrcorner \Omega_0).$$

In fact this formula holds with any volume form in place of Ω_0. This result may be obtained directly, or alternatively by the following argument. Let $\Omega = \rho\Omega_0$, where ρ is a non-vanishing smooth function. Then

$$\mathcal{L}_V \Omega = \mathcal{L}_V(\rho\Omega_0) = (V\rho)\Omega_0 + \rho\mathcal{L}_V \Omega_0$$

while

$$d(V \lrcorner \Omega) = d(V \lrcorner \rho\Omega_0) = d\rho \wedge (V \lrcorner \Omega_0) + \rho d(V \lrcorner \Omega_0).$$

The expression $d\rho \wedge (V \lrcorner \Omega_0)$ may now be simplified by the following trick. Since Ω_0 is an n-form $d\rho \wedge \Omega_0$, being an $(n+1)$-form on an n-dimensional space, is zero. Thus

$$0 = V \lrcorner (d\rho \wedge \Omega_0) = (V\rho)\Omega_0 - d\rho \wedge (V \lrcorner \Omega_0).$$

On combining these various expressions we obtain

$$\mathcal{L}_V \Omega = d(V \lrcorner \Omega) = (\operatorname{div}_\Omega V)\Omega.$$

Exercise 24. By using similar arguments show that for any volume form Ω and any non-vanishing function f

$$\operatorname{div}_{f\Omega} V = \frac{1}{f} \operatorname{div}_\Omega fV. \qquad \square$$

6. A Formula Relating Lie and Exterior Derivatives

The formula $\mathcal{L}_V\Omega = d(V \lrcorner \Omega)$ obtained above is the particular case (for n-forms) of a simple and important general formula relating Lie and exterior derivatives and the interior product. In order to derive it we consider the expression $\mathcal{L}_V\omega - d(V \lrcorner \omega)$, where ω is a p-form; we use the coordinate independent expressions for Lie and exterior derivatives to evaluate this expression. For any vector fields $W_1, W_2, \ldots, W_p$ we have

$$\begin{aligned}
&\mathcal{L}_V\omega(W_1,W_2,\ldots,W_p) - d(V \lrcorner \omega)(W_1,W_2,\ldots,W_p) \\
&\quad = V\left(\omega(W_1,W_2,\ldots,W_p)\right) - \sum_{r=1}^{p} \omega(W_1,\ldots,[V,W_r],\ldots,W_p) \\
&\qquad - \sum_{r=1}^{p}(-1)^{r+1}W_r\left(\omega(V,W_1,\ldots\widehat{W_r}\ldots,W_p)\right) \\
&\qquad - \sum_{1\leq r\leq s\leq p}(-1)^{r+s}\omega(V,[W_r,W_s],\ldots\widehat{W_r}\ldots\widehat{W_s}\ldots,W_p) \\
&\quad = V\left(\omega(W_1,W_2,\ldots,W_p)\right) + \sum_{r=1}^{p}(-1)^{r+2}W_r\left(\omega(V,W_1,\ldots\widehat{W_r}\ldots,W_p)\right) \\
&\qquad + \sum_{r=1}^{p}(-1)^{r+2}\omega([V,W_r],W_1,\ldots\widehat{W_r}\ldots,W_p) \\
&\qquad + \sum_{1\leq r\leq s\leq p}(-1)^{r+s+2}\omega([W_r,W_s],V,\ldots\widehat{W_r}\ldots\widehat{W_s}\ldots,W_p).
\end{aligned}$$

Careful inspection of this final expression reveals that it is just an exterior derivative: it is in fact $d\omega(V,W_1,\ldots,W_p)$. Thus on elimination of the "catalytic" vector fields $W_1,W_2,\ldots,W_p$ and rearranging one obtains

$$\mathcal{L}_V\omega = d(V \lrcorner \omega) + V \lrcorner d\omega.$$

The final term is missing when ω is an n-form on an n-dimensional space. Note that in that case this formula slightly generalises the one obtained in the previous section, because ω is not restricted to be nowhere vanishing.

Exercise 25. Show that on evaluating the expression $d(V \lrcorner \omega) + V \lrcorner d\omega$ with coordinate vector fields for arguments one recovers the expression for the components of $\mathcal{L}_V\omega$ given in Exercise 18. □

Exercise 26. Repeat the calculations of Exercise 16 using this expression for the Lie derivative. □

7. Exterior Derivative and Covariant Derivative

Among the properties of the covariant derivative operator there are two, namely

$$\begin{aligned}
\nabla_{U+V}W &= \nabla_U W + \nabla_V W \\
\nabla_{fV} &= f\nabla_V W,
\end{aligned}$$

which show that, for a fixed vector field W and 1-form θ, the map $V \mapsto \langle \nabla_V W, \theta \rangle$ is a 1-form. We shall rewrite some of the theory of the covariant derivative in a way which takes advantage of this observation.

Let $\{U_a\}$ be a basis of vector fields and $\{\theta^a\}$ the dual basis of 1-forms. Then for each U_a the map $V \mapsto \langle \nabla_V U_b, \theta^a \rangle$ is, as we have remarked, a 1-form, which we shall denote ω_b^a. The 1-forms ω_b^a are called the *connection forms* associated with the basis of vector fields U_a. For any vector field V,

$$\nabla_V U_b = \langle V, \omega_b^a \rangle U_a$$

and therefore

$$\nabla_V W = \big(V(W^a) + \langle V, \omega_b^a \rangle W^b\big) U_a \qquad \text{where } W = W^a U_a.$$

Also (Chapter 3, Section 11)

$$\langle U_c, \omega_b^a \rangle U_a = \nabla_{U_c} U_b = \gamma_{bc}^a U_a$$

and therefore

$$\omega_b^a = \gamma_{bc}^a \theta^c.$$

Exercise 27. Show that if $U_a = U_a^b \partial_b$ with respect to a basis of affine coordinate vector fields then $\omega_b^a = (U^{-1})_c^a dU_b^c$ or in matrix notation $\omega = U^{-1} dU$. □

Now the components W^b of the vector field W with respect to the basis $\{U_a\}$ are given by $W^b = \langle W, \theta^b \rangle$. Thus

$$\nabla_V W = \big(V\langle W, \theta^a \rangle + \langle V, \omega_b^a \rangle \langle W, \theta^b \rangle\big) U_a.$$

We may express the first order commutation relation $\nabla_V W - \nabla_W V = [V, W]$ as follows:

$$V\langle W, \theta^a \rangle - W\langle V, \theta^a \rangle + \langle V, \omega_b^a \rangle \langle W, \theta^b \rangle - \langle W, \omega_b^a \rangle \langle V, \theta^b \rangle = \langle [V, W], \theta^a \rangle$$

which on rearrangement gives

$$d\theta^a(V, W) + (\omega_b^a \wedge \theta^b)(V, W) = 0.$$

It follows that

$$d\theta^a + \omega_b^a \wedge \theta^b = 0.$$

Exercise 28. The same result may derived in another way. Show from the relation $\nabla_V W - \nabla_W V - [V, W] = 0$ that for any 1-form θ, $d\theta(V, W) = \langle W, \nabla_V \theta \rangle - \langle V, \nabla_W \theta \rangle$. Show that $\langle V, \omega_a^b \rangle = -\langle U_a, \nabla_V \theta^b \rangle$ and deduce that $d\theta^a = -\omega_b^a \wedge \theta^b$. □

The second order commutation relation $\nabla_V \nabla_W - \nabla_W \nabla_V - \nabla_{[V,W]} = 0$ for an affine space may be expressed as follows. For each basis vector field U_b

$$\begin{aligned}
\nabla_V(\nabla_W U_b) &= \nabla_V\big(\langle W, \omega_b^a \rangle U_a\big) \\
&= V\big(\langle W, \omega_b^a \rangle\big) U_a + \langle W, \omega_b^a \rangle \nabla_V U_a \\
&= \big(V\big(\langle W, \omega_b^a \rangle\big) + \langle V, \omega_c^a \rangle \langle W, \omega_b^c \rangle\big) U_a
\end{aligned}$$

which may be written

$$\langle \nabla_V(\nabla_W U_b), \theta^a \rangle = V\big(\langle W, \omega_b^a \rangle\big) + \langle V, \omega_c^a \rangle \langle W, \omega_b^c \rangle.$$

Subtraction of the similar term with V and W interchanged and of the term $\langle \nabla_{[V,W]} U_b, \theta^a \rangle$ yields

$$V(\langle W, \omega^a_b \rangle) + \langle V, \omega^a_c \rangle \langle W, \omega^c_b \rangle - W(\langle V, \omega^a_b \rangle) - \langle W, \omega^a_c \rangle \langle V, \omega^c_b \rangle - \langle [V,W], \omega^a_b \rangle = 0.$$

This is to say

$$d\omega^a_b(V,W) + (\omega^a_c \wedge \omega^c_b)(V,W) = 0.$$

Since this holds for every pair of vector field arguments

$$d\omega^a_b + \omega^a_c \wedge \omega^c_b = 0.$$

The equations $d\theta^a + \omega^a_b \wedge \theta^b = 0$ and $d\omega^a_b + \omega^a_c \wedge \omega^c_b = 0$ are called the first and second *structure equations* for the connection with respect to the local vector field basis.

8. Closed and Exact Forms

A p-form given on an affine space is said to be *closed* if its exterior derivative is zero, and to be *exact* if it is itself the exterior derivative of a $(p-1)$-form. An exact form is necessarily closed, since if $\omega = d\chi$ then $d\omega = d(d\chi) = 0$. We shall show that in an affine space an everywhere-defined closed form is necessarily exact.

Exercise 29. Show that with respect to any coordinates the condition for a 1-form $\alpha = \alpha_a dx^a$ to be closed is that $\partial_a \alpha_b = \partial_b \alpha_a$. Show that for it to be exact there must be some function f such that $\alpha_a = \partial_a f$. □

The general result includes, in a sense, the following results from vector calculus: that a vector field is a gradient if and only if its curl is zero, and that a vector field is a curl if and only if its divergence is zero. However, we deal here only with forms, whereas the classical results make implicit use of the metric of Euclidean space to identify forms with vector fields. The classical results are developed in Chapter 7.

In vector calculus, if $\operatorname{curl}\mathbf{X} = \mathbf{0}$, then a potential function ϕ for which $\mathbf{X} = \operatorname{grad}\phi$ is constructed by setting

$$\phi(P) = \int_{P_0}^{P} \mathbf{X} \cdot d\mathbf{l}$$

where P_0 is a conveniently chosen point and the line integral is taken along any smooth curve from P_0 to P. It follows from Stokes's theorem that the function so obtained is independent of the choice of path of integration. This method is not directly applicable to exterior forms of degree greater than 1, but suitably reformulated it yields a construction which can be generalised to treat such forms.

Expressing a closed 1-form as an exterior derivative. Consider, therefore, on an affine space $\mathcal{A}$ a closed 1-form α. We shall construct a function f such that $\alpha = df$.

Choose any point x_0 of $\mathcal{A}$. This point may be joined to any other point x by an affine line segment $\ell_x : t \mapsto x_0 + t(x - x_0)$. In affine coordinates (x^a) with x_0 as origin, one may write $\alpha = \alpha_a dx^a$, and the line ℓ_x is $t \mapsto (tx^a)$. We define a function f by the formula

$$f^x(x^a) = \int_0^1 \alpha_b(tx^c)x^b dt.$$

In the integration the x^a, which are the coordinates of the endpoint of the segment, must be understood to be constants: only t is a variable of integration. In effect we are computing the line integral of α along the line segment ℓ_x. The function f is smooth, if α is, and its exterior derivative may be computed directly:

$$df = \frac{\partial f^x}{\partial x^a}dx^a = \left\{\int_0^1 \left(\alpha_a(tx^c) + \frac{\partial \alpha_b}{\partial x^a}(tx^c)tx^b\right) dt\right\} dx^a,$$

the second term being rearranged by change of dummy indices to yield dx^a as common factor. However, since α is closed, $\partial_b \alpha_a = \partial_a \alpha_b$ (Exercise 29), and so the second term may be rewritten again to yield

$$df = \left\{\int_0^1 \left(\alpha_a(tx^c) + \frac{\partial \alpha_a}{\partial x^b}(tx^c)tx^b\right) dt\right\} dx^a.$$

But this integrand is the derivative with respect to t of the function $t \mapsto t\alpha_a(tx^c)$; thus

$$df = \left\{\int_0^1 \frac{d}{dt}(t\alpha_a(tx^c))dt\right\} dx^a = \alpha_a(x^c)dx^a = \alpha.$$

This establishes the construction for a 1-form.

Exercise 30. Show that the 1-form $x^2dx^1 + x^1dx^2$ is closed ; construct by the above method a function f for which $df = x^2dx^1 + x^1dx^2$. □

Exercise 31. Show that the 1-form $-x^2dx^1 + x^1dx^2$ is not closed; verify that the integral defined above is zero for this form. □

Exercise 32. Show that the 1-form

$$\alpha = \frac{-x^2dx^1 + x^1dx^2}{(x^1)^2 + (x^2)^2}$$

on $\mathbf{R}^2$ is smooth except at the origin, and closed wherever defined. Verify that, although $\int_0^1 \alpha_a(tx^c)x^a dt$ is not defined, $\lim_{\epsilon\to 0} \int_\epsilon^1 \alpha_a(tx^c)x^a dt = 0$. Show that, for $x^1 > 0$, $\alpha = df$ where $f = \arctan(x^2/x^1)$, and show how to extend this function to one which is smooth everywhere except for $x^2 = 0$, $x^1 \leq 0$. □

There is in fact no smooth function f on $\mathbf{R}^2$, or even on $\mathbf{R}^2$ with the origin removed, such that df is the 1-form α of Exercise 32, even though α is closed. Even so, one frequently writes $d\vartheta$ for α, where ϑ is the angle of plane polar coordinates! The moral of Exercise 32 is that for locally defined forms closure does not necessarily imply exactness.

Expressing a closed form of any degree as an exterior derivative. The integral defined above is meaningful, and yields a smooth function f, whenever α is an everywhere smooth 1-form, whether or not it is closed. Moreover, there is a relation between df, α and $d\alpha$ which suggests a way of generalising the construction to forms of higher degree. To discover this relation, compute df in affine coordinates, as before, but no longer assuming that α is closed:

$$df = \left\{ \int_0^1 \left(\alpha_a(tx^c) + \partial_a\alpha_b(tx^c)tx^b\right)dt \right\} dx^a,$$

and add and subtract a $\partial_b\alpha_a$ term inside the round brackets:

$$\begin{aligned} df &= \left\{ \int_0^1 \left(\alpha_a(tx^c) + \partial_b\alpha_a(tx^c)tx^b\right)dt \right\} dx^a \\ &\quad - \left\{ \int_0^1 \left(\partial_b\alpha_a(tx^c) - \partial_a\alpha_b(tx^c)\right)tx^b dt \right\} dx^a \\ &= \alpha - \left\{ \int_0^1 (d\alpha)_{ab}(tx^c)tx^a dt \right\} dx^b. \end{aligned}$$

The final term on the right is constructed from $d\alpha$ in much the same way as f is constructed from α. On the route to generalisation, we define linear maps $h_1: \bigwedge^1 \mathcal{A} \to \bigwedge^0 \mathcal{A}$ and $h_2: \bigwedge^2 \mathcal{A} \to \bigwedge^1 \mathcal{A}$ by

$$h_1(\alpha) = \int_0^1 \alpha_a(tx^c)x^a dt \qquad \text{where } \alpha = \alpha_a dx^a$$

$$h_2(\beta) = \left\{ \int_0^1 \beta_{ab}(tx^c)tx^a dt \right\} dx^b$$

where $\beta = \frac{1}{2}\beta_{ab}dx^a \wedge dx^b$. Then

$$d(h_1(\alpha)) = \alpha - h_2(d\alpha),$$

so that $d \circ h_1 + h_2 \circ d$ is the identity map on 1-forms.

The key step in the generalisation is to construct a linear map $h_p: \bigwedge^p \mathcal{A} \to \bigwedge^{p-1} \mathcal{A}$ for each $p = 1, 2, \ldots, n$, similar to h_1 and h_2, such that $d \circ h_p + h_{p+1} \circ d$ is the identity map on p-forms. If this composite map is applied to a closed form ω it will yield $d(h_p(\omega)) = \omega$, showing that ω is exact, as required.

A clue to the construction of the maps h_p is gained from an analysis of h_1 and h_2. Each of these maps is effected by carrying out the following steps: first, contract the coefficients of the given form with x^a; then change the argument to tx^c and multiply by a suitable power of t; then integrate with respect to t. Now the process of contracting the given form with x^a is equivalent to taking its interior product with the vector field $\Delta = x^a\partial_a$. This vector field generates the one-parameter group of dilations $\delta_t: x \mapsto x_0 + e^t(x - x_0)$ or, in affine coordinates based on x_0 (as we have been using), $(x^a) \mapsto (e^t x^a)$. The integral curve of Δ through x is (almost) the line segment ℓ_x used in the construction, though differently parametrised. However, the origin of affine coordinates x_0 does not lie on the integral curve of Δ through any other point x, but is itself a (degenerate) integral curve. On the other hand, for each point x, the limit of $\delta_t(x)$ as $t \to -\infty$ is x_0. If we change the variable in the integrals defining h_1 and h_2 to e^t we obtain

$$h_1(\alpha) = \int_{-\infty}^0 \alpha_a(e^t x^c)e^t x^a dt$$

$$h_2(\beta) = \left\{ \int_{-\infty}^{0} \beta_{ab}(e^t x^c) e^{2t} x^a dt \right\} dx^b.$$

Now

$$\delta_t{}^* \alpha = \alpha_a(e^t x^c) e^t dx^a$$
$$\delta_t{}^* \beta = \tfrac{1}{2}\beta_{ab}(e^t x^c) e^{2t} dx^a \wedge dx^b.$$

On taking the interior products of these forms with Δ we recover the integrands of the above integrals. We may therefore express h_1 and h_2 as follows:

$$h_1(\alpha) = \int_{-\infty}^{0} (\Delta \lrcorner \delta_t{}^* \alpha) dt$$
$$h_2(\beta) = \int_{-\infty}^{0} (\Delta \lrcorner \delta_t{}^* \beta) dt.$$

Integration here means integration with respect to the parameter t, the coordinates x^a being regarded for this purpose as constants. Therefore any operation which affects only the coordinates may be interchanged with the integration. Using this fact, and also the fact that the exterior derivative commutes with the pull-back, we write the formula $\alpha = d(h_1(\alpha)) + h_2(d\alpha)$ in the following way:

$$\alpha = d\left(\Delta \lrcorner \int_{-\infty}^{0} (\delta_t{}^* \alpha) dt\right) + \Delta \lrcorner \int_{-\infty}^{0} (\delta_t{}^* d\alpha) dt.$$

Compare this with the Lie derivative formula

$$\mathcal{L}_\Delta \alpha = d(\Delta \lrcorner \alpha) + \Delta \lrcorner d\alpha.$$

It is apparent that, suitably formulated, the integration process is simply the inverse of the Lie derivative along the generator of dilations.

A similar construction works for a form of arbitrary degree. If ω is a p-form then

$$\delta_t{}^* \omega = \frac{1}{p!} \omega_{a_1 a_2 \ldots a_p}(e^t x^c) e^{pt} dx^{a_1} \wedge dx^{a_2} \wedge \cdots \wedge dx^{a_p},$$

and so $\lim_{t \to -\infty} \delta_t{}^* \omega = 0$. Moreover, from the definition of the Lie derivative,

$$\mathcal{L}_\Delta(\delta_t{}^* \omega) = \frac{d}{ds}(\delta_{s+t}{}^* \omega)_{s=0} = \frac{d}{dt}(\delta_t{}^* \omega)$$

and so on the one hand

$$\int_{-\infty}^{0} \frac{d}{dt}(\delta_t{}^* \omega) dt = \delta_0{}^* \omega = \omega,$$

while on the other hand

$$\int_{-\infty}^{0} \frac{d}{dt}(\delta_t{}^* \omega) dt = \int_{-\infty}^{0} \mathcal{L}_\Delta(\delta_t{}^* \omega) dt = \int_{-\infty}^{0} \big(d(\Delta \lrcorner \delta_t{}^* \omega) + \Delta \lrcorner d(\delta_t{}^* \omega)\big) dt$$
$$= d(\Delta \lrcorner \int_{-\infty}^{0} (\delta_t{}^* \omega) dt) + \Delta \lrcorner \int_{-\infty}^{0} (\delta_t{}^*(d\omega)) dt,$$

again using the fact that integration here is with respect to t to interchange the order of operations. Therefore

$$\omega = d\left(\Delta \lrcorner \int_{-\infty}^{0} (\delta_t{}^*\omega)dt\right) + \Delta \lrcorner \int_{-\infty}^{0} \left(\delta_t{}^*(d\omega)\right)dt.$$

Dropping any notational reference to degree, as is customary also for d, we denote the map $\omega \mapsto \Delta \lrcorner \int_{-\infty}^{0}(\delta_t{}^*\omega)dt$ by h, and infer that for any p, $d \circ h + h \circ d$ is the identity on p-forms. Therefore if $d\omega = 0$ then $\omega = d(h(\omega))$ and every closed form on an affine space is exact.

Note that for a given closed p-form ω there are many $(p-1)$-forms χ such that $d\chi = \omega$: the addition of a closed $(p-1)$-form to any such form χ will produce another with the same property. This operation is sometimes called a "gauge transformation".

Exercise 33. Let A_1, A_2 and A_3 be given functions, smooth everywhere on a 3-dimensional affine space. Show that the necessary and sufficient condition for there to exist functions f_1, f_2 and f_3 such that (in affine coordinates)

$$A_1 = \frac{\partial f_2}{\partial x^3} - \frac{\partial f_3}{\partial x^2} \qquad A_2 = \frac{\partial f_3}{\partial x^1} - \frac{\partial f_1}{\partial x^3} \qquad A_3 = \frac{\partial f_1}{\partial x^2} - \frac{\partial f_2}{\partial x^1}$$

is that

$$\frac{\partial A_1}{\partial x^1} + \frac{\partial A_2}{\partial x^2} + \frac{\partial A_3}{\partial x^3} = 0.$$

□

Exercise 34. An open subset of an affine space is said to be star-shaped with respect to a point x_0 in it if for each point x which lies in the subset the line segment joining x_0 to x also lies in it. Show that any local form, defined on a star-shaped set, which is closed is necessarily exact. □

Exercise 35. Show that the Lie derivative with respect to the dilation field is an isomorphism of forms: that is to say, if $\mathcal{L}_\Delta \omega = 0$ then $\omega = 0$; and if ω is a form specified everywhere then there is a form χ of the same degree such that $\omega = \mathcal{L}_\Delta \chi$. Show by example that this is not true if the form ω is undefined at some point (consider the form α of Exercise 32). □

Summary of Chapter 5

A p-form on an affine space is a choice of element of $\bigwedge^p(T_x^*\mathcal{A})$, the vector space of alternating $\mathbf{R}$-multilinear forms of degree p on $T_x\mathcal{A}$, for each point $x \in \mathcal{A}$. If ω is a p-form and $V_1, V_2, \ldots, V_p$ are smooth vector fields then $\omega(V_1, V_2, \ldots, V_p)$ is a function on $\mathcal{A}$, and ω is smooth for all choices of arguments. Alternatively a p-form is an $\mathcal{F}(\mathcal{A})$-multilinear alternating map $\mathcal{X}(\mathcal{A})^p \to \mathcal{F}(\mathcal{A})$. The operations of exterior algebra (including the exterior and interior products) are carried out on forms pointwise. A p-form ω may be expressed in terms of coordinates in the following way: $\omega = \frac{1}{p!}\omega_{a_1 a_2 \ldots a_p} dx^{a_1} \wedge dx^{a_2} \wedge \cdots \wedge dx^{a_p}$, where the functions $\omega_{a_1 a_2 \ldots a_p}$ satisfy $\omega_{a_1 a_2 \ldots a_p} = \omega_{[a_1 a_2 \ldots a_p]}$. A smooth map ϕ induces a map of forms by $\omega \mapsto \phi^*\omega$ where $(\phi^*\omega)_x(v_1, v_2, \ldots, v_p) = \omega_{\phi(x)}(\phi_* v_1, \phi_* v_2, \ldots, \phi_* v_p)$; this map is contragredient to ϕ, is $\mathbf{R}$-linear, and satisfies $\phi^*(\omega_1 \wedge \omega_2) = (\phi^*\omega_1) \wedge (\phi^*\omega_2)$.

The exterior derivative of a p-form ω is the $(p+1)$-form $d\omega$ given by $d\omega = df \wedge dx^{a_1} \wedge dx^{a_2} \wedge \cdots \wedge dx^{a_p}$ if $\omega = f dx^{a_1} \wedge dx^{a_2} \wedge \cdots \wedge dx^{a_p}$, and extended to arbitrary forms by linearity. A coordinate-free definition is

$$d\omega(V_1, V_2, \ldots, V_{p+1}) = \sum_{r=1}^{p+1} (-1)^{r+1} V_r\big(\omega(V_1, \ldots \widehat{V_r} \ldots, V_{p+1})\big)$$
$$+ \sum_{1 \le r \le s \le p+1} (-1)^{r+s} \omega\big([V_r, V_s], V_1, \ldots \widehat{V_r} \ldots \widehat{V_s} \ldots, V_{p+1}\big).$$

The exterior derivative generalises the operations of vector calculus. It has the following important properties:

$$d(k_1\omega_1 + k_2\omega_2) = k_1 d\omega_1 + k_2\omega_2 \qquad d(f\omega) = f d\omega + df \wedge \omega$$
$$d(\omega \wedge \chi) = d\omega \wedge \chi + (-1)^p \omega \wedge d\chi \qquad \text{if } \omega \text{ is a } p\text{-form}$$
$$d^2 = 0 \qquad \phi^*(d\omega) = d(\phi^*\omega).$$

The Lie derivative of a p-form ω by a vector field V is the p-form $\mathcal{L}_V\omega$ given by $d/dt(\phi_t{}^*\omega)_{t=0}$ where V generates the flow ϕ. Alternatively,

$$(\mathcal{L}_V\omega)(W_1, W_2, \ldots, W_p) = V\big(\omega(W_1, W_2, \ldots, W_p)\big) - \sum_{r=1}^{p} \omega(W_1, \ldots, [V, W_r], \ldots, W_p).$$

The Lie derivative measures the rate of change of ω under the action of the flow of V. It has the following important properties: it is $\mathbf{R}$-linear in both ω and V; $\mathcal{L}_V$ commutes with d;

$$\mathcal{L}_V(f\omega) = f\mathcal{L}_V\omega + (Vf)\omega \qquad \mathcal{L}_V(\omega_1 \wedge \omega_2) = (\mathcal{L}_V\omega_1) \wedge \omega_2 + \omega_1 \wedge (\mathcal{L}_V\omega_2)$$
$$\mathcal{L}_{fV}\omega = f\mathcal{L}_V\omega + df \wedge (V \lrcorner \omega) \qquad \mathcal{L}_V(\mathcal{L}_W\omega) - \mathcal{L}_W(\mathcal{L}_V\omega) = \mathcal{L}_{[V,W]}\omega$$
$$\mathcal{L}_V(W \lrcorner \omega) = (\mathcal{L}_V W) \lrcorner \omega = W \lrcorner \mathcal{L}_V\omega.$$

The Lie and exterior derivatives are related *via* the interior product by

$$\mathcal{L}_V\omega = d(V \lrcorner \omega) + V \lrcorner d\omega.$$

The connection 1-forms ω_a^b associated with a basis of vector fields $\{U_a\}$ are defined by $\langle V, \omega_a^b\rangle = \langle \nabla_V U_a, \theta^b\rangle$ where $\{\theta^a\}$ is the basis of 1-forms dual to $\{U_a\}$. The connection forms satisfy $d\theta^a + \omega_b^a \wedge \theta^b = 0$, $d\omega_b^a + \omega_c^a \wedge \omega_b^c = 0$, the structure equations for the vector field basis.

A form ω is closed if $d\omega = 0$ and is exact if $\omega = d\chi$ for some χ of degree one less. Every exact form is necessarily closed; for forms globally defined on an affine space the converse is true, as may be shown by constructing a family of linear operators h such that $d \circ h + h \circ d$ is the identity on p-forms for each p. However, a local form may be closed without being exact.

6. FROBENIUS'S THEOREM

If $\phi: \mathcal{B} \rightarrow \mathcal{A}$ is a smooth map of affine spaces then, for any $y \in \mathcal{B}$, the set of vectors $\{ \phi_* w \mid w \in T_y \mathcal{B} \}$ is a linear subspace of $T_{\phi(y)} \mathcal{A}$. It would be natural to think of this vector subspace as consisting of those vectors in $T_{\phi(y)} \mathcal{A}$ which are tangent to the image $\phi(\mathcal{B})$ of $\mathcal{B}$ under ϕ. In general this idea presents difficulties, which will be explained in later chapters; but one case of particular interest, in which the notion is a sensible one, arises when ϕ_{*y} is an injective map for all $y \in \mathcal{B}$, so that the space $\{ \phi_* w \mid w \in T_y \mathcal{B} \}$ has the same dimension as $\mathcal{B}$ for all y. In this case we call the image $\phi(\mathcal{B})$ a *submanifold* of $\mathcal{A}$ (this terminology anticipates developments in Chapter 10 and is used somewhat informally in the present chapter). Since it has an m-dimensional tangent space at each point (where $m = \dim \mathcal{B}$) the submanifold $\phi(\mathcal{B})$ is regarded as an m-dimensional object. Our assumption of injectivity entails that $m \leq n = \dim \mathcal{A}$.

A curve (other than one which degenerates to a point) defines a submanifold of dimension 1, the injectivity of the tangent map corresponding in this case to the assumption that the tangent vector to the curve never vanishes. We regard $\mathbf{R}$, for this purpose, as a 1-dimensional affine space. In the present context the image of the curve will play a more important role than the curve (as map) itself. A congruence of curves on an affine space $\mathcal{A}$ defines a vector field on $\mathcal{A}$; if the parametrisation of the curves is disregarded, then one obtains a collection of 1-dimensional submanifolds of $\mathcal{A}$, exactly one through each point, and associated with it there is a field, not of vectors, but of 1-dimensional subspaces of the tangent spaces to $\mathcal{A}$. Again, degenerate curves are not allowed.

An obvious, and as it turns out, important, generalisation of this idea is to consider collections of submanifolds of $\mathcal{A}$, exactly one through each point of $\mathcal{A}$, and all of the same dimension m, but with this common dimension not necessarily being 1. Such a collection of submanifolds defines on $\mathcal{A}$ a field of m-dimensional subspaces of the tangent spaces to $\mathcal{A}$, which we call a distribution of dimension m. The m-dimensional subspace of $T_x \mathcal{A}$ determined by the distribution is just the subspace consisting of vectors tangent to the submanifold through x. So in this way, starting with a suitable collection of submanifolds one may construct a distribution. But one may imagine a distribution to have been defined initially without reference to any submanifolds; the question then arises, is there even so a collection of submanifolds whose spaces of tangent vectors coincide with the given distribution? When $m = 1$ there will be such a collection of submanifolds, as follows from (though it is not quite equivalent to) the theorem on the existence of integral curves of a vector field. But in the more general case the answer to the question is: not necessarily. A certain condition must be satisfied by the distribution to ensure the existence of submanifolds with the required property—integral submanifolds we shall call them, in a natural extension of the terminology for vector fields and curves.

In this chapter we describe the geometry of distributions, state the condition for the existence of integral submanifolds, prove its sufficiency, a result known as Frobenius's Theorem, and make some applications.

1. Distributions and Integral Submanifolds

Distributions. An m-dimensional *distribution* $\mathcal{D}$ on an affine space $\mathcal{A}$ is an assignment, to each point $x \in \mathcal{A}$, of an m-dimensional subspace $\mathcal{D}_x$ of $T_x\mathcal{A}$. Naturally, we shall be concerned only with distributions in which $\mathcal{D}_x$ varies smoothly, in some appropriate sense, from place to place. Before making this precise, however, we consider the various ways in which a distribution may be specified, making use of geometric objects already defined.

Since we are concerned with subspaces of vector spaces, the methods of Chapter 4 suggest themselves. Thus for each x we may specify $\mathcal{D}_x$ simply by giving a basis for it (consisting of elements of $T_x\mathcal{A}$), or by giving a suitable m-vector at x. Alternatively we may use a dual approach, and specify $\mathcal{D}_x$ by giving $n - m$ linearly independent constraint 1-forms at x for it, or by giving a characterising $(n - m)$-form at x for it. Each of these methods is useful in an appropriate context. For the present we concentrate on the specification of a distribution using forms.

Given a distribution $\mathcal{D}$, we call any local 1-form θ which vanishes on $\mathcal{D}$ (in the sense that at each point x, θ_x vanishes when restricted to $\mathcal{D}_x$) a *constraint 1-form* for $\mathcal{D}$. An m-dimensional distribution on an n-dimensional affine space is *smooth* if one can find $n - m$ smooth local 1-forms θ^ρ (where $\rho = m+1, m+2, \ldots, n$) such that, for each x, the θ^ρ_x constitute a basis for the constraint 1-forms for $\mathcal{D}$ at x. Given $n - m$ smooth constraint 1-forms θ^ρ as described in this definition, one can express any other smooth constraint 1-form θ uniquely in the form $\theta = f_\rho \theta^\rho$ with smooth local functions f_ρ for coefficients. Because of this we shall call such a set of constraint 1-forms $\{\theta^\rho\}$ a *basis* for the constraint 1-forms for $\mathcal{D}$. Bases of constraint 1-forms for distributions are not uniquely determined: if $\{\theta^\rho\}$ is one basis and if $(n - m)^2$ smooth local functions A^ρ_σ are given, such that for each x the matrix $(A^\rho_\sigma(x))$ is non-singular, then $\{A^\rho_\sigma \theta^\sigma\}$ is another basis for the constraint 1-forms for the same distribution, and any two bases are related in this way on their common domain.

Almost all of the indeterminacy inherent in the use of 1-forms to specify a distribution may be avoided by using instead a characterising $(n - m)$-form. A smooth local $(n-m)$-form ω is called a *characterising form* for a smooth distribution $\mathcal{D}$ if ω_x is characterising for $\mathcal{D}_x$ for all x. Any basis $\{\theta^\rho\}$ for the constraint 1-forms for $\mathcal{D}$ defines a characterising $(n-m)$-form $\omega = \theta^{m+1} \wedge \theta^{m+2} \wedge \cdots \wedge \theta^n$. Conversely, any characterising form must be decomposable, in the sense that it may be expressed as an exterior product of 1-forms, and these are then constraint 1-forms for the distribution.

Exercise 1. Consider the 1-form $\theta = -x^2 dx^1 + x^1 dx^2 + dx^3$ in a 3-dimensional affine space. Show that θ is a constraint 1-form for a 2-dimensional distribution $\mathcal{D}$, and that the vectors $\partial_1 + x^2\partial_3$, $\partial_2 - x^1\partial_3$ constitute at each point x a basis for $\mathcal{D}_x$. □

Exercise 2. Consider the two 1-forms $\theta^4 = -x^2dx^1 + x^1dx^2 + dx^3$, $\theta^5 = -x^4dx^3 + x^3dx^4 + dx^5$ in a 5-dimensional affine space. Show that they constitute a constraint basis for a distribution $\mathcal{D}$, and find a characterising form for it. Find a vector basis for $\mathcal{D}_x$ in terms of coordinate vectors. □

Exercise 3. Find a basis for the constraint 1-forms for the distribution (on a 4-dimensional affine space with affine coordinates (x^1, x^2, x^3, x^4)) which has the (decomposable) 2-form $dx^1 \wedge dx^2 + x^3dx^1 \wedge dx^3 - x^2x^3dx^2 \wedge dx^3$ as a characterising form. □

Exercise 4. The 1-forms $dx^1 + x^2dx^3$, $dx^1 + x^3dx^2$ do not, as constraint 1-forms, define a distribution: why not? □

Exercise 5. Suppose that ω is an $(n-m)$-form on an n-dimensional affine space, which is nowhere vanishing and is decomposable (in the sense explained above). Show that if, for each point x, $\mathcal{D}_x$ is the characteristic subspace of ω_x then the $\mathcal{D}_x$ constitute a smooth m-dimensional distribution. □

Exercise 6. Show that two $(n-m)$-forms, both decomposable, determine one and the same m-dimensional distribution if and only if one is a multiple of the other by a nowhere vanishing smooth function. □

Constraint forms and characterising forms for the same m-dimensional distribution are related as follows: if ω is a characterising $(n-m)$-form, and therefore decomposable, then θ is a constraint 1-form if and only if $\theta \wedge \omega = 0$ (Exercise 46 of Chapter 4); again, ω is a characterising form if and only if $\theta \wedge \omega = 0$ for every constraint 1-form θ (Exercise 47 of Chapter 4). This can be taken somewhat further. We shall say that the distribution $\mathcal{D}$ is *isotropic* for a p-form χ if, for every point x, χ_x is zero when restricted to $\mathcal{D}_x$ (that is, $\chi_x(v_1, v_2, \ldots, v_p) = 0$ when all of the p arguments $v_1, v_2, \ldots, v_p$ lie in $\mathcal{D}_x$). Thus in particular $\mathcal{D}$ is isotropic for all its constraint 1-forms and for any linear combinations of p-fold exterior products of them, including its characterising $(n-m)$-forms; but more generally than this, $\mathcal{D}$ is isotropic (for example) for $\alpha \wedge \theta$ where θ is a constraint 1-form and α is any form whatsoever. The case of a characterising form is rather special: it gives zero when just one of its arguments is taken from $\mathcal{D}$. The forms for which $\mathcal{D}$ is isotropic have a significant role to play in the argument. They may be specified as follows: $\mathcal{D}$ is isotropic for χ if and only if $\chi \wedge \omega = 0$, where ω is a characterising form for $\mathcal{D}$ (Exercise 48 of Chapter 4).

Exercise 7. Let $\{\theta^\rho\}$ be a basis for the constraint 1-forms for a distribution $\mathcal{D}$, and let χ be a p-form for which $\mathcal{D}$ is isotropic. Show that there are $(p-1)$-forms λ_ρ such that $\chi = \lambda_\rho \wedge \theta^\rho$. □

A set of forms (of differing degrees) is called an *ideal* if it has the property that for every form χ it contains, it also contains $\lambda \wedge \chi$ for every form λ (this includes the possibility of λ being a 0-form, that is, a function). Every set of forms, even a finite set, is contained in some ideal, though possibly the only ideal containing it is the whole algebra of forms. The smallest ideal containing a given finite set of forms is said to be *generated* by it.

Exercise 8. Let $\{\theta^\rho\}$ be a basis for the constraint 1-forms for a distribution $\mathcal{D}$: show that the ideal generated by this finite set of forms consists of all the forms for which $\mathcal{D}$ is isotropic. □

Integral submanifolds of a distribution. We have introduced above the idea of a submanifold, and pointed out its relation to the idea of a curve. Just as it is desirable to allow the domain of a curve to be an open interval of $\mathbf{R}$ and not necessarily the whole of it, so it is desirable to allow the domain of a map defining a submanifold to be an open subset of an affine space and not necessarily the whole of it. Allowing for this, we define a *submanifold of an affine space* $\mathcal{A}$ as follows. Let $\mathcal{S}$ be a subset of $\mathcal{A}$. We call $\mathcal{S}$ a submanifold of $\mathcal{A}$ if there is another affine space $\mathcal{B}$, an open subset $\mathcal{O}$ of $\mathcal{B}$, and a smooth map $\phi: \mathcal{O} \to \mathcal{A}$ such that $\mathcal{S}$ is the image of $\mathcal{O}$ under ϕ and that for every $y \in \mathcal{O}$ the linear map $\phi_*: T_y\mathcal{B} \to T_{\phi(y)}\mathcal{A}$ is injective.

The map ϕ in the definition is not unique. In particular, if $\Psi: \mathcal{O} \to \mathcal{P}$ is a smooth map of the open set $\mathcal{O}$ to an open set $\mathcal{P}$ in $\mathcal{B}$ which is bijective and has a smooth inverse, then $\phi: \mathcal{O} \to \mathcal{A}$ and $\psi = \phi \circ \Psi^{-1}: \mathcal{P} \to \mathcal{A}$ determine the same submanifold. We call any map ϕ with the properties of the definition a *parametrisation* of the submanifold $\mathcal{S}$, and any map ψ as just described a *reparametrisation* of ϕ.

If $\phi: \mathcal{O} \to \mathcal{A}$ is a parametrisation of a submanifold $\mathcal{S}$ of $\mathcal{A}$ then, for any $y \in \mathcal{O}$, ϕ_* maps $T_y\mathcal{B}$ linearly and injectively into $T_{\phi(y)}\mathcal{A}$, and so $\phi_*(T_y\mathcal{B})$ is a subspace of $T_x\mathcal{A}$ which is isomorphic to $T_y\mathcal{B}$. Moreover, if ψ is a reparametrisation of ϕ, then ϕ and ψ determine, in this manner, the same subspace of $T_x\mathcal{A}$ at each point x of $\mathcal{S}$. We call this subspace of $T_x\mathcal{A}$ the *tangent space to the submanifold* $\mathcal{S}$ at the point $x \in \mathcal{S}$. All of the tangent spaces to $\mathcal{S}$ have the same dimension, namely $m = \dim \mathcal{B}$, so we say that $\mathcal{S}$ has dimension m. The tangent space to $\mathcal{S}$ at x will be denoted $T_x\mathcal{S}$.

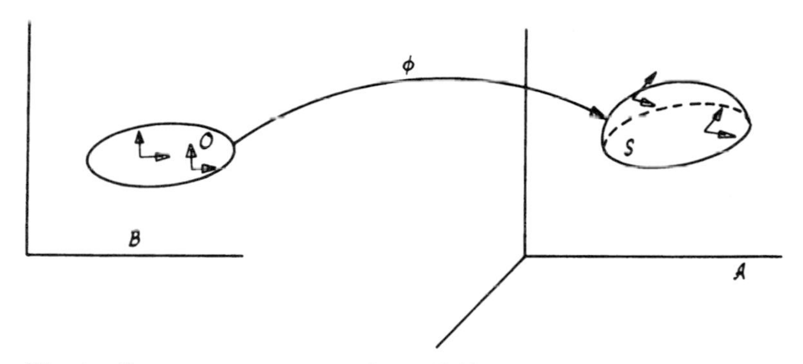

Fig. 1 Tangent spaces to a submanifold.

Exercise 9. Show that the map given by $(y^1, y^2) \mapsto (\cos y^1 \sin y^2, \sin y^1 \sin y^2, \cos y^2)$ for $0 < y^2 < \pi$ is a parametrisation of a submanifold, but that it is not so on any larger domain. □

Exercise 10. Show that if $\mathbf{R}$ is regarded as a 1-dimensional affine space then any curve with non-vanishing tangent vector is a parametrisation of a 1-dimensional submanifold, and that its tangent space at any point is the 1-dimensional space spanned by the tangent vector to the curve at that point. □

Exercise 11. Show that any affine p-plane in an affine space is a p-dimensional submanifold of it. □

Exercise 12. Show that the tangent space at x to a submanifold $\mathcal{S}$ of $\mathcal{A}$, through x, consists of the tangent vectors at x to all curves in $\mathcal{A}$ which lie in $\mathcal{S}$. □

Exercise 13. Let $\phi\colon \mathcal{O} \to \mathcal{A}$ be a parametrisation of a submanifold $\mathcal{S}$ of $\mathcal{A}$ (where $\mathcal{O}$ is an open subset of $\mathcal{B}$) and let $y \in \mathcal{O}$. Define an affine map $\Phi\colon \mathcal{B} \to \mathcal{A}$ in terms of the linear map $\phi_*\colon T_y\mathcal{B} \to T_{\phi(y)}\mathcal{A}$ as follows. For each $z \in \mathcal{B}$ regard $z - y$ as an element of $T_y\mathcal{B}$ and set $\Phi(z) = \phi(y) + \phi_*(z - y)$, where $\phi_*(z - y) \in T_{\phi(y)}\mathcal{A}$ is to be identified with an element of the vector space on which $\mathcal{A}$ is modelled. Show that Φ attaches $\mathcal{B}$ at $\phi(y)$ as an m-dimensional affine subspace of $\mathcal{A}$, and that this affine m-plane touches $\mathcal{S}$ at $\phi(y)$ in the sense that its tangent space there coincides with $T_{\phi(y)}\mathcal{S}$. □

We call this m-plane the *tangent plane* to $\mathcal{S}$ at $\phi(y)$.

Suppose now that there is given, on an n-dimensional affine space $\mathcal{A}$, an m-dimensional smooth distribution $\mathcal{D}$. An m-dimensional submanifold $\mathcal{S}$ of $\mathcal{A}$ is called an *integral submanifold* of $\mathcal{D}$ if at each point $x \in \mathcal{S}$, $T_x\mathcal{S} = \mathcal{D}_x$. Thus $\mathcal{S}$ is an integral submanifold of $\mathcal{D}$ if at each point x through which it passes its tangent space is just the subspace $\mathcal{D}_x$ of $T_x\mathcal{A}$ already given there by $\mathcal{D}$.

It is not necessarily the case that a distribution admits integral submanifolds: our aim is to give a necessary and sufficient condition for it to do so. A distribution $\mathcal{D}$ such that through every point of $\mathcal{A}$ there is an integral submanifold of $\mathcal{D}$ is said to be *integrable*.

A simple example of an integrable distribution is the one defined on a 3-dimensional affine space by the single constraint 1-form given in affine coordinates by dx^3. Its integral submanifolds may be parametrised by $(y^1, y^2) \mapsto (y^1, y^2, c)$, where c is a constant, different integral submanifolds being obtained for different values of c. The integral submanifolds are simply the 2-planes parallel to the x^1x^2-plane. There is one and only one integral submanifold through each point of the 3-dimensional affine space: since the x^3-axis cuts each integral submanifold just once it is convenient to distinguish the integral submanifolds from each other (in this example) by using the points on that axis, and in fact the constant c in the parametrisation given above is just the x^3 coordinate of the point in which the integral submanifold intersects the x^3-axis.

On the other hand, the distribution (on the same space and with the same coordinates) defined by the constraint 1-form $-x^2dx^1 + x^1dx^2 + dx^3$ (Exercise 1) is not integrable. We may indicate why by assuming that it has an integral submanifold through some point and deriving a contradiction. This is particularly easy to do when the point in question is the coordinate origin. As in the previous example, the constraint 1-form at that point is just dx^3, and so the tangent plane to the integral submanifold through the coordinate origin (supposing one to exist) would have to be the x^1x^2-plane. Accordingly it would be possible to use for parameters on the integral submanifold the first two coordinates of the points on it—at least for points close enough to the origin. In other words it would be possible to regard the integral submanifold as the graph of some function f on $\mathbf{R}^2$, and having therefore the parametrisation $\phi\colon (y^1, y^2) \mapsto (y^1, y^2, f(y^1, y^2))$. The function f is required to satisfy $f(0,0) = 0$, $\partial_1 f(0,0) = 0$, $\partial_2 f(0,0) = 0$. Now

$$\phi_*\left(\frac{\partial}{\partial y^1}\right) = \frac{\partial}{\partial x^1} + \frac{\partial f}{\partial y^1}\frac{\partial}{\partial x^3}$$

$$\phi_*\left(\frac{\partial}{\partial y^2}\right) = \frac{\partial}{\partial x^2} + \frac{\partial f}{\partial y^2}\frac{\partial}{\partial x^3}$$

and these vectors belong to the distribution at $\phi(y)$ (as they must if ϕ is to be a parametrisation of an integral submanifold) if and only if

$$-y^2 + \partial_1 f = 0, \qquad y^1 + \partial_2 f = 0.$$

Cross-differentiating makes it clear that there is no smooth function f satisfying both these equations simultaneously. The distribution cannot be integrable.

Before turning to the question of the necessary and sufficient conditions for integrability, we point out that when they do exist integral submanifolds of a distribution share one property with integral curves of a vector field: there is one, and only one, integral submanifold passing through each point of the space. If one has found a parametrisation ϕ for the integral submanifold through a point x_1, and x_2 is some other point on it, then a parametrisation for the integral submanifold through x_2 can differ from ϕ only by being a reparametrisation of it. In the case of integral curves of a vector field one is able to go further and say that such a reparametrisation can be no more than a change of origin, but this depends on the fact that one is dealing with a vector field rather than a 1-dimensional distribution, and there is no analogue of that particular aspect of the congruence property of integral curves which applies to integral submanifolds of a distribution.

2. Necessary Conditions for Integrability

The task now is to find conditions on a distribution necessary and sufficient for the spaces that make it up to fit together to form the tangent spaces to a collection of submanifolds, in other words, for it to be integrable.

It is easy to derive a necessary condition, as follows. Suppose that $\mathcal{D}$ is an integrable distribution, and that $\phi\colon \mathcal{O} \to \mathcal{A}$ (where $\mathcal{O}$ is an open subset of an affine space $\mathcal{B}$) is a parametrisation of one of its integral submanifolds, so that for any $y \in \mathcal{O}$, $\phi_*(T_y\mathcal{B}) = \mathcal{D}_{\phi(y)}$. Then if θ is any constraint 1-form for $\mathcal{D}$, and w any element of $T_y\mathcal{B}$, we have $\langle \phi_* w, \theta_{\phi(y)} \rangle = 0$; from which it follows that $(\phi^*\theta)_y = 0$. This holds for all y, so that $\phi^*\theta = 0$. Since the exterior derivative commutes with the pull-back, it follows that $\phi^* d\theta = 0$. Then for any $y \in \mathcal{O}$ and any $w_1, w_2 \in T_y\mathcal{B}$, $d\theta_{\phi(y)}(\phi_* w_1, \phi_* w_2) = 0$, which means that the restriction of $d\theta_{\phi(y)}$ to $\mathcal{D}_{\phi(y)}$ is zero. Since by assumption there is an integral submanifold through every point of $\mathcal{A}$, this property holds at every point of $\mathcal{A}$. Thus if $\mathcal{D}$ is integrable, and θ is any one of its constraint 1-forms, then $\mathcal{D}$ is isotropic for $d\theta$. Using the results of Section 1 this conclusion may be equivalently expressed in several different ways: if $\mathcal{D}$ is integrable then

(1) if ω is a characterising form for $\mathcal{D}$ and θ a constraint 1-form then

$$d\theta \wedge \omega = 0$$

(2) if $\{\theta^\rho\}$ is a basis for the constraint 1-forms for $\mathcal{D}$ then

$$d\theta^\rho \wedge \theta^{m+1} \wedge \theta^{m+2} \wedge \cdots \wedge \theta^n = 0$$

(3) if $\{\theta^\rho\}$ is a basis for the constraint 1-forms for $\mathcal{D}$ then there are 1-forms λ^ρ_σ such that

$$d\theta^\rho = \lambda^\rho_\sigma \wedge \theta^\sigma.$$

This gives, in various forms, a necessary condition for the integrability of $\mathcal{D}$: it is known as the *Frobenius integrability condition.*

Exercise 14. Show that the distribution defined by the 1-form $-x^2dx^1 + x^1dx^2 + dx^3$ (Exercise 1) fails to meet the Frobenius integrability condition in any of its forms. □

Exercise 15. Show that the Frobenius integrability condition for the distribution defined on a 3-dimensional affine space with affine coordinates (x^1, x^2, x^3) by the constraint 1-form $P_1dx^1 + P_2dx^2 + P_3dx^3$ (where the coefficient functions P_a do not all vanish simultaneously) amounts to the condition

$$P_1(\partial_3P_2 - \partial_2P_3) + P_2(\partial_1P_3 - \partial_3P_1) + P_3(\partial_2P_1 - \partial_1P_2) = 0.$$ □

(In more classical language, with the P_a identified as components of a vector field $\mathbf{P}$, this would be written $\mathbf{P} \cdot \operatorname{curl} \mathbf{P} = 0$, and regarded as a necessary condition for the family of 2-planes orthogonal to $\mathbf{P}$ to be integrable.)

Exercise 16. Show that the three conditions given above are indeed equivalent to each other and to the condition that $\mathcal{D}$ be isotropic for $d\theta$. □

Exercise 17. Show that the Frobenius integrability condition is automatically satisfied for any 1-dimensional distribution. □

Exercise 18. The 1-forms λ^ρ_σ in condition (3) above are not uniquely determined. Verify that this is so by showing that if $\mu^\rho_\sigma = \lambda^\rho_\sigma + L^\rho_{\sigma\tau}\theta^\tau$, where the functions $L^\rho_{\sigma\tau}$ satisfy $L^\rho_{\tau\sigma} = L^\rho_{\sigma\tau}$, then $d\theta^\rho = \mu^\rho_\sigma \wedge \theta^\sigma$ also. Show that conversely, if $d\theta^\rho = \lambda^\rho_\sigma \wedge \theta^\sigma = \mu^\rho_\sigma \wedge \theta^\sigma$, then λ^ρ_σ and μ^ρ_σ must be related in the way just described. □

Exercise 19. Show that if $d\theta^\rho = \lambda^\rho_\sigma \wedge \theta^\sigma$, and $\hat{\theta}^\rho = A^\rho_\sigma\theta^\sigma$, where the functions A^ρ_σ are the elements of a non-singular matrix (so that $\{\hat{\theta}^\rho\}$ is another basis for constraint 1-forms) then

$$d\hat{\theta}^\rho = \hat{\lambda}^\rho_\sigma \wedge \hat{\theta}^\sigma \quad \text{where} \quad \hat{\lambda}^\rho_\sigma = dA^\rho_\tau(A^{-1})^\tau_\sigma + A^\rho_\pi\lambda^\pi_\tau(A^{-1})^\tau_\sigma$$

(up to multiples of $\hat{\theta}^\rho$ as in Exercise 18), $(A^{-1})^\rho_\sigma$ being the elements of the matrix inverse to (A^ρ_σ). □

Exercise 20. Show that if $\mathcal{D}$ is integrable and χ is any form for which it is isotropic then $\mathcal{D}$ is also isotropic for $d\chi$. Deduce that if $\mathcal{D}$ is integrable then the ideal of forms generated by any basis for its constraint 1-forms contains the exterior derivative of every form in it. □

3. Sufficient Conditions for Integrability

As it turns out, the Frobenius integrability condition (in any of its equivalent forms) is sufficient, as well as necessary, for the distribution to be integrable. This result is known as Frobenius's theorem. We now embark on the proof of sufficiency.

It should be stated at the outset that several steps in the proof work only locally: that is to say, they involve assumptions or known results which may hold only in a neighbourhood of a point and not all over the ambient space. The result is therefore also local: it guarantees the existence of an integral submanifold through every point, but only in a neighbourhood of the point.

We deal with an m-dimensional distribution $\mathcal{D}$ on an n-dimensional affine space $\mathcal{A}$. It will be convenient to use affine coordinates throughout, and to employ indices

α, β in the range $1, 2, \ldots, m$

ρ, σ in the range $m + 1, m + 2, \ldots, n$

a, b in the range $1, 2, \ldots, n$.

The distribution $\mathcal{D}$ is assumed to satisfy the Frobenius integrability condition: its most convenient expression for our purposes is the one given under (3) in the last section. The aim is to show that through every point of $\mathcal{A}$ there passes an integral submanifold of $\mathcal{D}$. The strategy of the proof is to construct an integral submanifold of $\mathcal{D}$ through an arbitrary point x_0 of $\mathcal{A}$, starting off rather in the manner proposed for the 1-form $\theta = -x^2 dx^1 + x^1 dx^2 + dx^3$ towards the end of Section 1 (though in that case the construction turned out to be unsuccessful for reasons now clear). That method depended rather heavily on a particular property of θ, namely that it contains a coordinate 1-form with coefficient 1, to which θ reduces at the origin. The first step in the general construction is to take advantage of the freedom of choice of a basis for constraint 1-forms (underlined by Exercise 19) to pick a basis with an analogous property.

Let $\{\hat{\theta}^\rho\}$ be any basis of constraint 1-forms for $\mathcal{D}$, with $\hat{\theta}^\rho = \hat{\theta}^\rho_a dx^a$. The fact that the $\hat{\theta}^\rho$ are linearly independent implies that at each point $x \in \mathcal{A}$ the matrix of coefficients $(\hat{\theta}^\rho_a(x))$ has rank $n-m$, that is, has $n-m$ linearly independent columns. After renumbering the coordinates if necessary, it may be arranged that at x_0 the last $n-m$ columns of this matrix are linearly independent; the same will remain true in some neighbourhood of x_0. Then the $(n-m) \times (n-m)$ matrix $(\hat{\theta}^\rho_\sigma)$ will be non-singular on this neighbourhood. Let A^ρ_σ be functions such that the matrix (A^ρ_σ) is inverse to $(\hat{\theta}^\rho_\sigma)$: then the 1-forms $\theta^\rho = A^\rho_\sigma \hat{\theta}^\sigma$, which also constitute a basis for constraint forms, have the expression

$$\theta^\rho = \theta^\rho_\alpha dx^\alpha + dx^\rho$$

for certain functions θ^ρ_α. Furthermore, it may be arranged, by an affine transformation of coordinates, that x_0 is at the origin of coordinates, and that each constraint 1-form θ^ρ actually reduces to dx^ρ there.

Exercise 21. Show that, supposing the origin already to have been fixed, the affine coordinate transformation $\hat{x}^\alpha = x^\alpha$, $\hat{x}^\rho = x^\rho + \theta^\rho_\alpha(x_0)x^\alpha$ has the required effect. □

After these adjustments to constraint 1-forms and coordinates have been made we are left with a basis of constraint 1-forms $\{\theta^\rho\}$ such that $\theta^\rho = \theta^\rho_\alpha dx^\alpha + dx^\rho$ with $\theta^\rho_\alpha(x_0) = 0$ and with x_0 as origin of coordinates. It is required to find an integral submanifold $\mathcal{S}$ of $\mathcal{D}$ through x_0. As a consequence of our choice of $\{\theta^\rho\}$ the tangent plane to $\mathcal{S}$ at x_0 must consist of the coordinate m-plane spanned by the x^α, and given by $x^\rho = 0$. In constructing the integral submanifold $\mathcal{S}$ we shall use the x^α as parameters: that is to say, we shall give a parametrisation of $\mathcal{S}$ in the coordinate form

$$(y^1, y^2, \ldots, y^m) \mapsto (y^1, y^2, \ldots, y^m, \xi^{m+1}(y^\alpha), \xi^{m+2}(y^\alpha), \ldots, \xi^n(y^\alpha))$$

for certain functions ξ^ρ. Represented in this way, the integral submanifold may be thought of as a graph: it is the graph of the map $\xi\colon \mathbf{R}^m \to \mathbf{R}^{n-m}$ whose components are the functions ξ^ρ.

We denote by $\mathcal{B}$ the coordinate m-plane $x^\rho = 0$, considered as an affine space; we denote by π the projection map $\mathcal{A} \to \mathcal{B}$ which maps each point of $\mathcal{A}$ to the point with the same first m coordinates in $\mathcal{B}$; we denote by $\mathcal{N}$ the neighbourhood of x_0 in

$\mathcal{A}$ on which the constraint 1-forms θ^ρ are defined. Observe that at any point $x \in \mathcal{A}$, a vector $v = v^\alpha \partial_\alpha + v^\rho \partial_\rho$ belongs to $\mathcal{D}_x$ if and only if

$$v^\rho = -\theta^\rho_\alpha(x) v^\alpha,$$

and that $v^\alpha \partial_\alpha = \pi_* v \in T_{\pi(x)}\mathcal{B}$. Conversely, given any vector $v^\alpha \partial_\alpha \in T_{\pi(x)}\mathcal{B}$ there is a unique vector $v \in \mathcal{D}_x$ such that $\pi_* v = v^\alpha \partial_\alpha$, namely

$$v = v^\alpha \partial_\alpha - \theta^\rho_\alpha(x) v^\alpha \partial_\rho.$$

We call this vector the *lift* of $v^\alpha \in T_{\pi(x)}\mathcal{B}$ to x. Given any point $y \in \pi(\mathcal{N}) \subset \mathcal{B}$, one may construct the lifts of a vector at y to all the points x in $\mathcal{N}$ such that $\pi(x) = y$. The "lift" of a vector from the coordinate origin of $\mathcal{B}$ to the coordinate origin, x_0, of $\mathcal{A}$ coincides with the vector itself.

This lifting construction is the basis for our construction of an integral submanifold. First, we show how it may be extended to curves in $\mathcal{B}$. The aim, given a curve γ in $\mathcal{B}$, is to find a curve Γ in $\mathcal{A}$ which projects onto γ ($\pi \circ \Gamma = \gamma$) and which is tangent to $\mathcal{D}$ ($\dot\Gamma(t) \in \mathcal{D}_{\Gamma(t)}$). But then the tangent vectors to Γ must be lifts of tangent vectors to γ. Thus the coordinate functions for Γ must satisfy

$$\Gamma^\alpha(t) = \gamma^\alpha(t)$$
$$\dot\Gamma^\rho(t) = -\theta^\rho_\alpha(\Gamma^a(t))\dot\gamma^\alpha(t).$$

The latter equations constitute a system of $n - m$ first order ordinary differential equations for the functions Γ^ρ, and therefore admit a unique solution with specified initial conditions. Thus given a curve γ in $\pi(\mathcal{N})$, a number t_0 in the domain of γ, and a point x in $\mathcal{N}$ such that $\pi(x) = \gamma(t_0)$, there is a unique curve Γ in $\mathcal{N}$ such that $\pi \circ \Gamma = \gamma$, $\dot\Gamma(t) \in \mathcal{D}_{\Gamma(t)}$, and $\Gamma(t_0) = x$. We call Γ the *lift* of γ through x. One further useful property of the lift of a curve stems from the fact that the defining system of differential equations is linear in $\dot\gamma^\alpha$. It follows from this that if $\tilde\gamma = \gamma \circ h$ is a reparametrisation of γ, its lift $\tilde\Gamma$ is obtained by applying the same reparametrisation to the lift of γ: $\tilde\Gamma = \Gamma \circ h$.

We now use this lifting construction for curves to construct a parametrisation ϕ of a submanifold of $\mathcal{A}$, as follows. Let (y^α) be the coordinates of a point y in $\pi(\mathcal{N}) \subset \mathcal{B}$ and let γ be the curve given by $\gamma^\alpha(t) = ty^\alpha$, that is, the radial line joining the origin to the given point. The idea is to take for $\phi(y^\alpha) \in \mathcal{A}$ that point on the lift of γ through x_0 which projects onto y, that is, the point $\Gamma(1)$. To investigate the validity of this process, we must look more closely at the properties of Γ in this context.

We shall denote by $t \mapsto \Gamma(t, y^\beta)$ the lift through x_0 of the radial line $t \mapsto (ty^\alpha)$ in $\mathcal{B}$, to make clear its dependence on (y^α). Its component functions satisfy the equations

$$\Gamma^\alpha(t, y^\beta) = ty^\alpha$$
$$\frac{\partial\Gamma^\rho}{\partial t}(t, y^\beta) = -\theta^\rho_\alpha(ty^\beta, \Gamma^\sigma)y^\alpha$$
$$\Gamma^a(0, y^\beta) = 0.$$

The differential equations are to be regarded as ordinary differential equations, as before, in which the y^α are regarded as parameters. The existence of a solution

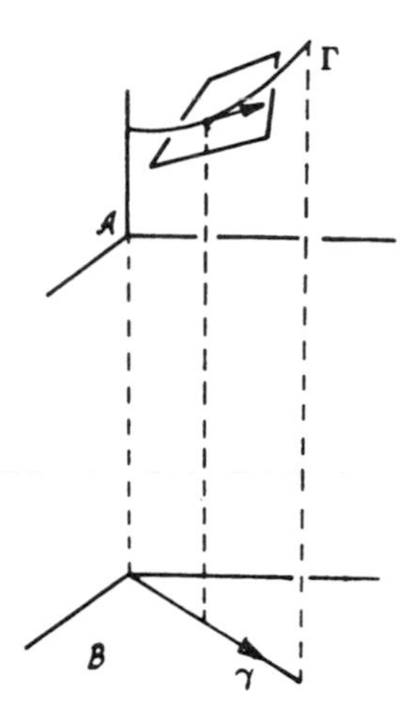

Fig. 2 Lifting a radial line.

is guaranteed, for each (y^α), for t in some open interval containing 0. Moreover, the right hand sides of these differential equations are assumed to depend smoothly on the y^α; the solutions of a system of differential equations depending smoothly on parameters will themselves depend smoothly on those parameters, so that the Γ^a may be regarded as smooth functions of the y^α. Furthermore, because of the reparametrisation property mentioned above,

$$\Gamma(t, ky^\beta) = \Gamma(kt, y^\beta) \qquad \text{for any } k \in \mathbf{R}.$$

As suggested above, we wish to define ϕ by $\phi(y^\alpha) = \Gamma(1, y^\alpha)$; but there remains one technicality to be dealt with before we can do so, which is concerned with the domain of ϕ. The problem arises that for arbitrary (y^α) there is no guarantee that $\Gamma(t, y^\alpha)$ is defined for $t = 1$: we know only that it is defined for $-\varepsilon < t < \varepsilon$ for some positive ε. But here the reparametrisation property for lifted curves comes to the rescue, for although $\Gamma(1, y^\alpha)$ may not be defined, it is true that (for example) $\Gamma(1, \frac{1}{2}\varepsilon y^\alpha)$ is defined: in fact, $\Gamma(1, \frac{1}{2}\varepsilon y^\alpha) = \Gamma(\frac{1}{2}\varepsilon, y^\alpha)$. Thus there are points y in each direction from the coordinate origin in $\mathcal{B}$ for which $\Gamma(1, y^\alpha)$ is defined, and in fact there is an open neighbourhood $\mathcal{O}$ of the origin in $\mathcal{B}$ such that $\Gamma(1, y^\alpha)$ is defined when $y \in \mathcal{O}$.

We may accordingly define a map $\phi: \mathcal{O} \to \mathcal{A}$ by $\phi(y^\alpha) = \Gamma(1, y^\alpha)$. The map ϕ is smooth; it is given in coordinates by

$$(y^\alpha) \mapsto (y^\alpha, \Gamma^\rho(1, y^\alpha))$$

and is therefore the graph of a map $\xi: \mathbf{R}^m \to \mathbf{R}^{n-m}$, where $\xi^\rho(y^\alpha) = \Gamma(1, y^\alpha)$. It follows that ϕ_* is necessarily injective at each point, so that ϕ is a parametrisation of a submanifold $\mathcal{S}$ of $\mathcal{A}$. The submanifold $\mathcal{S}$ certainly passes through the coordinate origin x_0. At each point x on it, its tangent space $T_x\mathcal{S}$ has at least a 1-dimensional subspace in common with $\mathcal{D}_x$, namely that spanned by the lift of the radial vector at $\pi(x)$; while its tangent space at x_0 actually coincides with the space determined there by $\mathcal{D}$. It remains to show that S is actually an integral submanifold of $\mathcal{D}$.

For this purpose we shall take a point $x \in S$ and a vector $v \in T_x S$ and show that v is annihilated by the constraint 1-forms θ^ρ_x. To do so we consider how these 1-forms vary along the lifts of radial curves which were used to define S. It is therefore desirable to introduce a vector field tangent to these curves. Now the dilation field $\Delta = y^\alpha \partial_\alpha$ on $\mathcal{B}$ has for its integral curves the radial lines in $\mathcal{B}$, albeit reparametrised exponentially (apart from the origin, which is itself a degenerate integral curve). Let $\hat{\Delta}$ be the lift of Δ to $\mathcal{N} \subset \mathcal{A}$: the integral curve $\hat{\delta}_x$ of $\hat{\Delta}$ through a point $x = \phi(y^\alpha) \in S$ is given by $\hat{\delta}_x(t) = \phi(e^t y^\alpha)$. This curve again degenerates to a point, namely x_0, when $y^\alpha = 0$; otherwise, it does not pass through x_0, but $\hat{\delta}_x(t) \to x_0$ as $t \to -\infty$. At each point x of S other than x_0, $\hat{\Delta}_x$ spans the known common 1-dimensional subspace of $T_x S$ and $\mathcal{D}_x$.

For any $v \in T_x S$, we define a vector field V along $\hat{\delta}_x$ by Lie transporting v by $\hat{\Delta}$: thus $V(0) = v$, and $\mathcal{L}_{\hat{\Delta}} V = 0$. The curve $\hat{\delta}_x$ lies in S, $\hat{\Delta}$ is tangent to S, and so is v: it follows that $V(t)$ is tangent to S at $\hat{\delta}_x(t)$ for all t. Thus $\lim_{t\to-\infty} V(t)$ is tangent to S at x_0. Since the tangent space to S and the space determined by $\mathcal{D}$ coincide at x_0, it follows that $\lim_{t\to-\infty}\langle V, \theta^\rho\rangle_{\hat{\delta}_x(t)} = 0$. We consider next how $\langle V, \theta^\rho\rangle$ varies along $\hat{\delta}_x$, using the Frobenius integrability condition $d\theta^\rho = \lambda^\rho_\sigma \wedge \theta^\sigma$:

$$\begin{aligned}
\frac{d}{dt}\langle V, \theta^\rho\rangle &= \hat{\Delta}(\langle V, \theta^\rho\rangle) \\
&= \langle V, \mathcal{L}_{\hat{\Delta}}\theta^\rho\rangle && \text{since } \mathcal{L}_{\hat{\Delta}} V = 0 \\
&= \langle V, d(\hat{\Delta} \lrcorner \theta^\rho) + \hat{\Delta} \lrcorner d\theta^\rho\rangle \\
&= \langle V, \hat{\Delta} \lrcorner (\lambda^\rho_\sigma \wedge \theta^\sigma)\rangle && \text{since } \langle \hat{\Delta}, \theta^\rho\rangle = 0 \\
&= \langle \hat{\Delta}, \lambda^\rho_\sigma\rangle \langle V, \theta^\sigma\rangle.
\end{aligned}$$

The functions $\langle V, \theta^\rho\rangle$ therefore satisfy a set of linear ordinary differential equations. Furthermore, $\hat{\Delta}$ enters linearly on the right hand sides of the equations, and therefore a reparametrisation $s = e^t$ of $\hat{\delta}_x$ will not change the form of the equations. The reparametrised curve is just $s \mapsto \phi(sy^\alpha) = \Gamma(s, y^\alpha)$, with tangent vector $\dot{\Gamma}$: thus

$$\frac{d}{ds}\langle V, \theta^\rho\rangle = \langle \dot{\Gamma}, \lambda^\rho_\sigma\rangle \langle V, \theta^\sigma\rangle.$$

Now $s \to 0$ as $t \to -\infty$, and therefore $\langle V, \theta^\rho\rangle = 0$ at $s = 0$. But the uniqueness of solutions of systems of ordinary differential equations implies, for linear equations, that a solution which vanishes anywhere vanishes everywhere. Thus $\langle V, \theta^\rho\rangle = 0$ all along the curve, and in particular $\langle v, \theta^\rho\rangle = 0$.

We have shown that any vector at x tangent to S lies in $\mathcal{D}_x$, and so $T_x S$ coincides with $\mathcal{D}_x$. Thus S is indeed an integral submanifold of $\mathcal{D}$, and the proof is complete.

Exercise 22. Show that the single 1-form θ on a 3-dimensional affine space given in affine coordinates by $\theta = x^3 dx^1 + x^3 dx^2 - dx^3$ satisfies the Frobenius integrability condition. Use the construction given in the proof in the text to show that the integral submanifold through the point with coordinates $(0,0,c)$ is given by $\phi(y^1, y^2) = (y^1, y^2, c\exp(y^1 + y^2))$. □

Exercise 23. In the above proof of Frobenius's theorem the integrability condition is not used in the construction of the submanifold $\mathcal{S}$, only in showing that $\mathcal{S}$ is an integral submanifold. Carry out the construction of the submanifold through the origin of coordinates for the distribution defined by the 1-form $-x^2dx^1 + x^1dx^2 + dx^3$ of Exercise 1, and investigate why it fails to be an integral submanifold. □

A vector field version of the integrability condition. So far we have worked entirely with the specification of a distribution using forms. Dually, one may specify a distribution using vector fields, and the Frobenius integrability conditions may also be stated conveniently in terms of vector fields.

A vector field V on $\mathcal{A}$ is said to *belong to a distribution* $\mathcal{D}$ if for every $x \in \mathcal{A}$, $V_x \in \mathcal{D}_x$. The distribution being smooth, one can find (at least locally) a set of m vector fields $V_1, V_2, \ldots, V_m$ which serve as a *basis for the distribution*, in the sense that the vectors $V_{\alpha x}$ form a basis for $\mathcal{D}_x$ at each point x. This follows from the fact that any basis for the constraint 1-forms $\{\theta^\rho\}$ may be extended to a basis of 1-forms on $\mathcal{A}$; the first m members of the dual basis of vector fields will serve the purpose. One says also that such vector fields *span the distribution*. Any vector field belonging to the distribution may be uniquely expressed as a linear combination (with variable coefficients) of basis vector fields.

Exercise 24. Find a vector field basis for each of the distributions of Exercises 2 and 3. □

Exercise 25. Show that the following is an alternative definition of the smoothness of a distribution: $\mathcal{D}$ is smooth if it has everywhere a basis of smooth local vector fields. □

Now let V, W be vector fields belonging to a distribution $\mathcal{D}$ and let θ be any constraint 1-form for it. Then $\langle V, \theta\rangle = \langle W, \theta\rangle = 0$ because θ is a constraint form for $\mathcal{D}$: it follows that

$$d\theta(V, W) = -\langle [V, W], \theta\rangle.$$

Thus if $\mathcal{D}$ is integrable, so that it is isotropic for $d\theta$, then $\langle [V, W], \theta\rangle = 0$; since this holds for any constraint 1-form θ, it follows that $[V, W]$ belongs to $\mathcal{D}$. So if $\mathcal{D}$ is integrable, the bracket of any pair of vector fields belonging to $\mathcal{D}$ also belongs to $\mathcal{D}$. Conversely, if this condition is satisfied, it follows that $\mathcal{D}$ is isotropic for every one of its constraint 1-forms, and so $\mathcal{D}$ is integrable. The Frobenius integrability condition may therefore be stated in the following way: a distribution $\mathcal{D}$ is integrable if and only if the bracket of every pair of vector fields belonging to $\mathcal{D}$ also belongs to it.

When $\mathcal{D}$ is integrable the vector fields belonging to it are tangent to its integral submanifolds.

Exercise 26. Show that this vector field version of the Frobenius integrability condition is equivalent to the following (more operational) one: $\mathcal{D}$ is integrable if and only if, given a basis $\{V_\alpha\}$ for it, there are functions $f_{\alpha\beta}^\gamma$ such that $[V_\alpha, V_\beta] = f_{\alpha\beta}^\gamma V_\gamma$. Investigate how these functions are affected by a change of basis for $\mathcal{D}$. □

Exercise 27. Confirm the integrability or otherwise of the distributions in Exercises 1, 2, 3 and 22 by using the vector field criterion of Exercise 26. □

Exercise 28. Derive the necessity of the bracket condition for integrability in another way, as follows. Suppose that $\phi\colon \mathcal{O} \to \mathcal{A}$ defines an integral submanifold of $\mathcal{D}$. Let V, W be vector fields belonging to $\mathcal{D}$: deduce from the injectivity of ϕ_* that there are vector fields V', W' on $\mathcal{O}$ to which V, W are ϕ-related (Chapter 3, Section 9). Conclude that $[V, W]$ must also belong to $\mathcal{D}$, at least on the image of ϕ. □

The two versions of Frobenius's theorem which have been presented here link up with different parts of the book. The vector field version is related to the ideas of Chapter 3, as the last exercise shows. Again, it will be recalled that it was shown in that chapter that the bracket of two vector fields may be interpreted as the (second-order) tangent vector to a curve built out of the flows generated by the vector fields. This result is closely related to the present discussion: for if two vector fields are tangent to a submanifold, then their integral curves through points of the submanifold lie in it, and so their bracket at any point of it is also tangent to it. Frobenius's theorem shows in effect that a converse to this assertion holds true. The form version of Frobenius's theorem, on the other hand, is related to material on connections to be found in Chapter 11 and beyond.

4. Special Coordinate Systems

In this section we shall show how to construct special curvilinear coordinate systems adapted to distributions.

We consider first of all a 1-dimensional distribution, and show that curvilinear coordinates may be introduced such that the first coordinate vector field ∂_1 spans the distribution.

There will be locally a non-vanishing vector field V which spans the distribution. Local coordinates are constructed as follows. A point x_0 is chosen, and through it a hyperplane $\mathcal{B}$ such that V_{x_0} is not tangent to $\mathcal{B}$. It will remain true that V is not tangent to $\mathcal{B}$ in some neighbourhood of x_0: we say that $\mathcal{B}$ is *transverse* to V. Affine coordinates $(\acute{x}^a)$ are now chosen in such a way that x_0 is the origin, $\mathcal{B}$ is the coordinate hyperplane $\acute{x}^1 = 0$, and V_{x_0} coincides with $\acute{\partial}_1$. A map Φ of some open subset of $\mathbf{R}^n$ to $\mathbf{R}^n$ is defined by setting $\left(\Phi^a(\xi^1, \xi^2, \ldots, \xi^n)\right)$ equal to the affine coordinates of the point $\phi_{\xi^1}(\acute{x})$ where ϕ is the flow of V, and $\acute{x}$ is the point with affine coordinates $(0, \xi^2, \ldots, \xi^n)$ and is therefore a point of $\mathcal{B}$. The Jacobian matrix of Φ at the origin is easily seen to be the identity, and Φ therefore defines a coordinate transformation to local curvilinear coordinates, say (x^a). In terms of these coordinates V is the generator of the flow $(t, x^1, x^2, \ldots, x^n) \mapsto (x^1 + t, x^2, \ldots, x^n)$: thus $V = \partial_1$ everywhere on the coordinate patch.

Note that as a result of this construction we may assert that, given a point x_0, there is an open set $\mathcal{O}$ containing x_0 and a hyperplane $\mathcal{B}$ through x_0 such that each integral submanifold of the 1-dimensional distribution spanned by V intersects $\mathcal{B}$ in $\mathcal{O}$ once and once only. The special coordinates we have constructed may be described as follows: for any $x \in \mathcal{O}$ let $\acute{x}$ be the point of $\mathcal{O} \cap \mathcal{B}$ on whose orbit under the flow of V the point x lies; then $(x^2, \ldots, x^n)$ are the affine coordinates of $\acute{x}$ and x^1 is the parameter distance from $\acute{x}$ to x.

Exercise 29. Show that if $\{\theta^2, \theta^3, \ldots, \theta^n\}$ are constraint 1-forms for a 1-dimensional distribution then it is possible to find functions A^ρ_σ forming the entries of a non-singular matrix, and coordinates (x^a), such that the 1-forms $\hat{\theta}^\rho = A^\rho_\sigma \theta^\sigma$ are given by $\hat{\theta}^\rho = dx^\rho$ $(\rho = 2, 3, \ldots, n)$. □

The straightening-out lemma. A 1-dimensional distribution is not quite the same thing as a vector field, since at each point it determines, not a single vector, but a 1-dimensional subspace of the tangent space. However, the result given above applies equally well to a single vector field provided that the vector field is nowhere zero. In other words, given a vector field V and a point x such that $V_x \neq 0$ there is a neighbourhood of x (on which V remains non-zero) and a coordinate system for this neighbourhood with respect to which $V = \partial_1$.

It follows that two vector fields, on affine spaces of the same dimension, are locally equivalent, near points at which they are non-zero, in the following sense: if $\mathcal{A}$ and $\mathcal{B}$ are affine spaces of the same dimension, if V and W are vector fields on $\mathcal{A}$ and $\mathcal{B}$ respectively, and if x and y are points at which V and W are non-zero respectively, then there are neighbourhoods $\mathcal{O}$ and $\mathcal{P}$ of x and y and a smooth, bijective map $\psi: \mathcal{O} \to \mathcal{P}$ with smooth inverse such that $\psi_* V_x = W_{\psi(x)}$ at each point x of $\mathcal{O}$. In fact V and W are each locally equivalent to the generator of translations parallel to the x^1-axis in an affine coordinate system. This result is therefore often called the "straightening out lemma".

In passing, we point out that the distinctive local features of a vector field are therefore to be found in its behaviour near its zeros, in other words the points at which it vanishes; these are the fixed points of its flow. At a zero x of a vector field V with flow ϕ, for each t the induced map ϕ_{t*} is a linear transformation of the tangent space $T_x\mathcal{A}$ to itself: in fact ϕ_{t*} is a one-parameter group of linear transformations. It is thus the exponential of the linear transformation given by $d/dt(\phi_{t*})_{t=0}$. Study of this map will give information about the behaviour of V near x. Note that, in terms of any local coordinate system, $d/dt(\phi_{t*})_{t=0}$ is represented by the matrix $(\partial_b V^a)(x)$, where $V = V^a\partial_a$.

Exercise 30. Use the straightening out lemma for the vector field V to show that $\mathcal{L}_V W = [V, W]$ at points at which $V \neq 0$, and use the remark immediately preceding this exercise to complete a new proof of this important result. □

Coordinates adapted to an integrable distribution. We next extend the straightening out lemma to obtain a special coordinate system in a given region adapted to a number of linearly independent vector fields. By virtue of their linear independence, none of the vector fields can vanish anywhere in the region. We seek a curvilinear coordinate system in which these vector fields are coordinate vector fields. Now any two coordinate vector fields commute (have zero bracket); we must therefore assume the same holds for the given vector fields. So suppose that $V_1, V_2, \ldots, V_m$ are linearly independent vector fields on an open subset of an affine space $\mathcal{A}$, such that $[V_\alpha, V_\beta] = 0$ for all $\alpha, \beta = 1, 2, \ldots, m$. It follows that the flows generated by any pair of these vector fields commute (Section 12 of Chapter 3): if ϕ_α is the flow generated by V_α then $\phi_{\alpha,s} \circ \phi_{\beta,t} = \phi_{\beta,t} \circ \phi_{\alpha,s}$ for all $s, t \in \mathbf{R}$ for which both transformations are defined. It also follows that the distribution generated by the V_α is integrable. It is easy to identify its integral submanifolds in terms of the flows ϕ_α: the integral submanifold through a point x consists of all points $\phi_{1,t^1} \circ \phi_{2,t^2} \circ \cdots \circ \phi_{m,t^m}(x)$ for $(t^1, t^2, \ldots, t^m) \in \mathbf{R}^m$. For the map $\mathbf{R}^m \to \mathcal{A}$ by

$$(t^1, t^2, \ldots, t^m) \mapsto \phi_{1,t^1} \circ \phi_{2,t^2} \circ \cdots \circ \phi_{m,t^m}(x)$$

certainly defines a submanifold; and if

$$\acute{x} = \phi_{1,t^1} \circ \phi_{2,t^2} \circ \cdots \circ \phi_{m,t^m}(x)$$

lies in it then so does

$$\phi_{\alpha,s}(\acute{x}) = \phi_{1,t^1} \circ \phi_{2,t^2} \circ \cdots \circ \phi_{\alpha,s+t^\alpha} \circ \cdots \circ \phi_{m,t^m}(x)$$

for any $\alpha = 1, 2, \ldots, m$, from the commutativity of the flows, and so the vector field V_α is tangent to the submanifold for each α. To construct the required coordinate system one chooses a point x_0 and an $(n-m)$-plane $\mathcal{B}$ through x_0 transverse to the distribution in the sense that the subspace of $T_{x_0}\mathcal{A}$ defined by $\mathcal{B}$ is complementary to that defined by the distribution. This complementarity will persist at points of $\mathcal{B}$ in some neighbourhood of x_0. Choose affine coordinates $(\acute{x}^a)$ such that x_0 is the origin, $\mathcal{B}$ is the coordinate $(n-m)$-plane $\acute{x}^1 = \acute{x}^2 + \cdots = \acute{x}^m = 0$, and $\acute{\partial}_1, \acute{\partial}_2, \ldots, \acute{\partial}_m$ coincide with the given vector fields at x_0. Define a map Φ of some open subset of $\mathbf{R}^n$ to $\mathbf{R}^n$ by setting $(\Phi^a(\xi^1, \xi^2, \ldots, \xi^n))$ equal to the affine coordinates of the point $\phi_{1,\xi^1} \circ \phi_{2,\xi^2} \circ \cdots \circ \phi_{m,\xi^m}(\acute{x})$ where $\acute{x}$ is the point with affine coordinates $(0, \ldots, 0, \xi^{m+1}, \ldots, \xi^n)$, which is therefore a point of $\mathcal{B}$. Then the Jacobian matrix of Φ at the origin is the identity, so that Φ defines a coordinate transformation to local curvilinear coordinates (x^a). In terms of these coordinates $V_\alpha = \partial_\alpha$ eveywhere on the coordinate patch, and the integral submanifolds of the distribution spanned by the V_α are given by $x^{m+1} = \text{constant}$, $x^{m+2} = \text{constant}, \ldots, x^n = \text{constant}$.

Exercise 31. Let $\hat{W}_1, \hat{W}_2, \ldots, \hat{W}_m$ be linearly independent, not necessarily commuting, vector fields which satisfy the Frobenius integrability conditions. Show that local functions A_α^β can be found, forming the elements of a nowhere singular matrix, so that (possibly after renumbering) the vector fields $W_\alpha = A_\alpha^\beta \hat{W}_\beta$, which span the same distribution, take the form $W_\alpha = \partial_\alpha + \psi_\alpha^\rho \partial_\rho$. Show that by virtue of the integrability conditions the W_α commute pairwise. Hence give a new proof of Frobenius's theorem, using vector fields. □

Exercise 32. Let $\{\theta^\rho\}$, $\rho = m+1, m+2, \ldots, n$, be a system of 1-forms satisfying the Frobenius integrability conditions. Show that there are functions A_σ^ρ, such that on some neighbourhood (A_σ^ρ) is a non-singular matrix, and coordinates (x^a), such that the 1-forms $A_\sigma^\rho \theta^\sigma$, which generate the same system, are given by $A_\sigma^\rho \theta^\sigma = dx^\rho$. Deduce that the integrable submanifolds of any m-dimensional integrable distribution may be expressed in the form $x^{m+1} = \text{constant}$, $x^{m+2} = \text{constant}, \ldots, x^n = \text{constant}$. □

Coordinates adapted to a smooth map. Let $\psi: \mathcal{A} \to \mathcal{B}$ be a smooth map, not necessarily satisfying the submanifold property: thus ψ_{*x} is not necessarily injective for $x \in \mathcal{A}$. Instead, let the dimension of the kernel of this linear map be the same at all points x, or at least at all points x in some open subset of $\mathcal{A}$. We show how local coordinates may be constructed on $\mathcal{A}$ and $\mathcal{B}$ in terms of which ψ takes a particularly simple form.

We set $\mathcal{D}_x = \{ v \in T_x\mathcal{A} \mid \psi_{*x} v = 0 \}$. The condition on constancy of dimension ensures that this defines a distribution. The smoothness of ψ ensures that it is smooth. Furthermore, it is integrable, as the following simple argument shows: any local vector field belonging to $\mathcal{D}$ is ψ-related to the zero vector field on $\mathcal{B}$, and therefore the bracket of two such vector fields is also ψ-related to the zero vector field on $\mathcal{B}$, which implies that the bracket belongs to $\mathcal{D}$.

Exercise 33. Confirm this result by a coordinate argument. □

It follows that there are coordinates (x^a) on $\mathcal{A}$ such that the coordinate fields ∂_ρ, $\rho = m+1, m+2, \ldots, n$, form a basis for $\mathcal{D}$, whose dimension is therefore $n-m$. But then

$$0 = \psi_*\left(\frac{\partial}{\partial x^\rho}\right) = \frac{\partial \psi^i}{\partial x^\rho}\frac{\partial}{\partial y^i},$$

where (y^i) are coordinates on $\mathcal{B}$, $i = 1, 2, \ldots, p = \dim \mathcal{B}$. Thus $\partial\psi^i/\partial x^\rho = 0$, and so the components of ψ are functions only of (x^α), $\alpha = 1, 2, \ldots, m$. In terms of these coordinates we may therefore express ψ in the form $\phi \circ \Pi$ where Π is the projection $\mathbf{R}^n \to \mathbf{R}^m$ onto the first m factors and ϕ is an m-dimensional submanifold map.

We now turn our attention to the image of ψ in $\mathcal{B}$. We have shown that it is a submanifold of dimension m. We introduce new coordinates on $\mathcal{B}$ as follows. Choose a point x_0 in the domain of ψ and make it the origin of the adapted coordinates in $\mathcal{A}$. Choose a subspace $\mathcal{W}$ of $T_{\psi(x_0)}\mathcal{B}$ complementary to the tangent space to the image of ψ, with basis $\{w_r\}$, $r = m+1, m+2, \ldots, p$. Define a map $\Psi\colon \mathbf{R}^p \to \mathcal{B}$ by setting $\Psi(z^i) = \psi(z^\alpha) + z^r w_r$. Thus the first m entries in (z^i) are used to determine a point on the image of ψ and the remaining ones are the components of a vector which translates that point off the image. The Jacobian matrix of Ψ at the origin is easily seen to be non-singular because of the fact that $\{\psi_*(\partial_\alpha), w_r\}$ is a basis for $T_{\psi(x_0)}\mathcal{B}$. Thus Ψ determines a local coordinate system about $\psi(x_0)$ for $\mathcal{B}$. With respect to the new coordinates the map ψ is given by

$$\psi^1(x^a) = x^1 \quad \psi^2(x^a) = x^2 \quad \ldots \quad \psi^m(x^a) = x^m$$
$$\psi^{m+1}(x^a) = \psi^{m+2}(x^a) = \cdots = \psi^n(x^a) = 0.$$

These are the required coordinates.

5. Applications: Partial Differential Equations

In this section and the two following we continue the development of the ideas of this chapter by applying them in three specific contexts: the theory of partial differential equations, Darboux's theorem, and Hamilton-Jacobi theory.

Integrability conditions for systems of first order partial differential equations. As a first application of Frobenius's theorem to the theory of partial differential equations we consider systems of first order partial differential equations of the form

$$\frac{\partial \xi^\rho}{\partial x^\alpha} = \Theta^\rho_\alpha(x^\beta, \xi^\sigma)$$

for the unknown functions $\xi^\rho(x^\alpha)$. The functions Θ^ρ_α are given functions of n variables, there are m independent variables x^α, and $n-m$ dependent variables ξ^ρ. As before, we shall use indices α, β with the range $1, 2, \ldots, m$ and ρ, σ with the range $m+1, m+2, \ldots, n$.

There are $m(n-m)$ equations in all, and so the number of equations exceeds the number of unknowns except when $m = 1$. In the latter case the equations reduce to a system of ordinary differential equations, and are always soluble as a consequence of the existence theorem for solutions of such a system. Otherwise, the equations form what is known as an overdetermined system, and will not in general be soluble unless some integrability conditions on the Θ^ρ_α are satisfied. A necessary condition is easily found, by differentiating the equations and using the symmetry of second partial derivatives: there results the condition

$$\frac{\partial\Theta^\rho_\alpha}{\partial x^\beta} + \frac{\partial\Theta^\rho_\alpha}{\partial x^\sigma}\Theta^\sigma_\beta = \frac{\partial\Theta^\rho_\beta}{\partial x^\alpha} + \frac{\partial\Theta^\rho_\beta}{\partial x^\sigma}\Theta^\sigma_\alpha.$$

It is a consequence of Frobenius's theorem that these are also sufficient conditions for the system of equations to be soluble. The connection with Frobenius's theorem is achieved by consideration of the 1-forms (on an n-dimensional affine space $\mathcal{A}$ with affine coordinates (x^a), $a = 1, 2, \ldots, n$)

$$\theta^\rho = \Theta^\rho_\alpha(x^\beta, x^\sigma)dx^\alpha - dx^\rho.$$

A solution (ξ^ρ) of the system of partial differential equations may be regarded as defining a submanifold of $\mathcal{A}$, in the form of its graph $(x^\alpha) \mapsto (x^\alpha, \xi^\rho(x^\alpha))$. The pull-back of θ^ρ by this map is just

$$\left(\Theta^\rho_\alpha(x^\beta, \xi^\sigma) - \frac{\partial\xi^\rho}{\partial x^\alpha}\right) dx^\alpha,$$

and so (ξ^ρ) is a solution of the system of partial differential equations if and only if the submanifold is an integral submanifold of the distribution defined by the 1-forms θ^ρ as constraint forms. These 1-forms have (apart from the sign of dx^ρ) just the same structure as those used in the proof of sufficiency of the Frobenius integrability condition in Section 3. The integrability condition for these 1-forms is derived as follows:

$$\begin{aligned}
d\theta^\rho &= -\frac{\partial\Theta^\rho_\alpha}{\partial x^\beta}dx^\alpha \wedge dx^\beta - \frac{\partial\Theta^\rho_\alpha}{\partial x^\sigma}dx^\alpha \wedge dx^\sigma \\
&= -\frac{\partial\Theta^\rho_\alpha}{\partial x^\beta}dx^\alpha \wedge dx^\beta - \frac{\partial\Theta^\rho_\alpha}{\partial x^\sigma}dx^\alpha \wedge \left(\Theta^\sigma_\beta dx^\beta - \theta^\sigma\right) \\
&= \frac{\partial\Theta^\rho_\alpha}{\partial x^\sigma}dx^\alpha \wedge \theta^\sigma - \left(\frac{\partial\Theta^\rho_\alpha}{\partial x^\beta} + \frac{\partial\Theta^\rho_\alpha}{\partial x^\sigma}\Theta^\sigma_\beta\right) dx^\alpha \wedge dx^\beta
\end{aligned}$$

and therefore $d\theta^\rho = \lambda^\rho_\sigma \wedge \theta^\sigma$ (with $\lambda^\rho_\sigma = (\partial\Theta^\rho_\alpha/\partial x^\sigma)dx^\alpha$) if

$$\left(\frac{\partial\Theta^\rho_\alpha}{\partial x^\beta} + \frac{\partial\Theta^\rho_\alpha}{\partial x^\sigma}\Theta^\sigma_\beta\right) dx^\alpha \wedge dx^\beta = 0.$$

But this is precisely the same as the condition obtained by cross-differentiation. That condition is therefore necessary and sufficient for the solubility of the system of partial differential equations.

Exercise 34. Show that if vector fields $W_\alpha = \partial_\alpha + \Psi^\rho_\alpha\partial_\rho$ commute pairwise then the functions Ψ^ρ_α satisfy the condition that ensures the solubility of the system of partial differential equations $\partial\xi^\rho/\partial x^\alpha = \Psi^\rho_\alpha(x^\beta, \xi^\sigma)$. □

Characteristics of first order partial differential equations. One is very often confronted with the problem of finding submanifolds to which a given vector field is tangent. Thus baldly stated, the problem leads to a partial differential equation. Let V be a vector field given on an affine space $\mathcal{A}$, and let $f =$ constant be the equation of a submanifold (of codimension 1) to which it is tangent. Then $Vf = 0$ at each point. This equation has the coordinate presentation $V^a \partial_a f = 0$ and may thus be considered to be a partial differential equation for the function f.

Very often, the partial differential equation is the starting point, and the geometrical problem is more-or-less disguised. In the theory of such partial differential equations the integral curves of V are called the *characteristics* of the partial differential equation. A standard method for solving the partial differential equation begins with the construction of the characteristics, which is to say, the solution of the system of ordinary differential equations $d\gamma^a/dt = V^a$ for the integral curves γ of V.

Importance is often attached to some functions, known or unknown, which are constant along the integral curves of a given vector field. In the physical context, these functions would represent *conserved quantities*, or *constants of the motion*, which are independent of the time (if the parameter t is so interpreted). Any conserved quantity f must satisfy the equation $Vf = 0$, which is the equation from which we started. Sometimes it is easier to solve the equations of characteristics first, sometimes the partial differential equation.

From the geometrical point of view, it is easy to see why such partial differential equations have "so many" solutions. For example, let $\mathcal{A}$ be a 3-dimensional affine space, V a given vector field in $\mathcal{A}$, and σ any curve transverse to V. Now transport σ along V: as parameter time t elapses, σ (a wisp of smoke) moves along the integral curves of V (the wind). This generates a 2-dimensional submanifold to which V is tangent, made up of a one-parameter family of integral curves of V, each of which intersects σ. If this 2-submanifold has the equation $f =$ constant then f satisfies the equation $Vf = 0$. All this will be true, whatever the initial choice of the curve σ.

In an n-dimensional space, σ has to be $(n-2)$-dimensional, and again V must not be tangent to it. Transport of σ along V generates a submanifold of dimension $n-1$ whose equation $f =$ constant again determines f such that $Vf = 0$. Care has to be taken about smoothness, but the present argument is anyhow only intended to be heuristic.

The specification of σ appears in the physical context as the specification of initial data for the physical problem. It might happen, say in the 3-dimensional case, that the initial data were such that σ was an integral curve of V—that V was everywhere tangent to it. Then σ would no longer determine a unique one-parameter family of integral curves, and any one-parameter family including it would yield a solution of the equation $Vf = 0$ consistent with the initial data. In this case the problem to be solved is called a *characteristic initial value problem*. Evidently intermediate cases are possible, in which V is tangent to σ at some points but not at others. Some more general characteristic initial value problems are of great physical importance.

6. Application: Darboux's Theorem

In this section we describe how coordinates may be chosen so that a closed 2-form takes a particularly simple form.

The characteristic subspace of a 2-form ω on a vector space (Section 12 of Chapter 4) is the space of vectors v satisfying $v \lrcorner \omega = 0$. (This use of the word "characteristic" is not to be confused with its use at the end of the immediately preceding section.) A 2-form on an affine space determines at each point a characteristic subspace of the tangent space at that point. We shall suppose that the dimension of this subspace does not vary from point to point, for the given 2-form ω. We do not, however, assume that ω is decomposable.

We show first that when ω is closed the distribution defined by its characteristic subspaces satisfies the Frobenius integrability conditions. Thus we must show that if V and W are vector fields such that $V \lrcorner \omega = W \lrcorner \omega = 0$ then $[V,W] \lrcorner \omega = 0$ also. Observe that since ω is closed

$$\mathcal{L}_V \omega = d(V \lrcorner \omega) + V \lrcorner d\omega = 0.$$

Using a result from Exercise 21 of Chapter 5 we obtain

$$\mathcal{L}_V(W \lrcorner \omega) = 0 = [V,W] \lrcorner \omega + W \lrcorner \mathcal{L}_V \omega = [V,W] \lrcorner \omega$$

as required. Coordinates $(y^1, y^2, \ldots, y^p)$ may therefore be introduced so that the coordinate vectors $\partial_1, \partial_2, \ldots, \partial_p$ span the distribution of characteristic vectors (where p is the dimension of the characteristic subspaces).

Exercise 35. Show that, with respect to these coordinates, ω depends only on the remaining coordinates: that is to say, if $(x^1, x^2, \ldots, x^m)$ complete the set of coordinates (so that $m + p = n$, the dimension of the space) then $\omega = \omega_{\alpha\beta} dx^\alpha \wedge dx^\beta$, where $\alpha, \beta = 1, 2, \ldots, m$, the $\omega_{\alpha\beta}$ being functions of (x^γ). Deduce that m must be even, from the fact that there are no non-zero characteristic vectors of ω in the space spanned by the $\partial/\partial x^\alpha$. □

We may as well suppose, then, for the rest of the argument, that we are dealing with a 2-form, on a space $\mathcal{A}$ of even dimension $m = 2k$, which has no non-zero characteristic vectors. At each point $x \in \mathcal{A}$ one may define a linear map $T_x\mathcal{A} \to T_x^*\mathcal{A}$, by means of the 2-form ω, by $v \mapsto v \lrcorner \omega$. Since (as we now assume) ω has no non-zero characteristic vectors, this map is injective; and therefore, since $T_x\mathcal{A}$ and $T_x^*\mathcal{A}$ have the same dimension, it is an isomorphism. Thus, given any 1-form θ on $\mathcal{A}$, there is a vector field V such that $V \lrcorner \omega = \theta$.

Exercise 36. Show that if $\theta = df$ is exact and $V \lrcorner \omega = df$ then $\mathcal{L}_V \omega = 0$ and $Vf = 0$. □

We now begin the construction of the required coordinate system. Choose some function f such that df is nowhere zero (or at least such that $df \neq 0$ at some point x; the argument then provides suitable coordinates in a neighbourhood of x, which is the most that can be expected anyway), and let V be the vector field determined in this way by df. Then V is nowhere zero, and so coordinates (y^α) may be found, about any point, such that $V = \partial/\partial y^1$. With respect to these coordinates, $\partial f/\partial y^1 = 0$, and since $\mathcal{L}_V \omega = 0$, both f and the coefficients of ω are independent of y^1. Let W be the vector field defined by $W \lrcorner \omega = -dy^1$. Then $Wy^1 = 0$,

$$Wf = \langle W, df \rangle = \langle W, \partial_1 \lrcorner \omega \rangle = -\langle \partial_1, W \lrcorner \omega \rangle = \langle \partial_1, dy^1 \rangle = 1,$$

and

$$[\partial_1, W] \lrcorner \omega = \mathcal{L}_{\partial_1}(W \lrcorner \omega) - W \lrcorner \mathcal{L}_{\partial_1}\omega = -\mathcal{L}_{\partial_1}(dy^1) = 0,$$

so that $[\partial_1, W] = 0$ since ω has no non-zero characteristic vectors. Thus W contains no term in ∂_1, and its coefficients are also independent of y^1. It is therefore possible to make a further change of coordinates, without affecting y^1, such that $W = \partial/\partial y^2$. Moreover, by choosing the coordinate hypersurface $y^2 = 0$ to be a level surface of f, it can be ensured that the coordinate expression for f is just y^2.

Consider now the 2-form $\hat{\omega} = \omega - dy^1 \wedge dy^2$. It is closed. The vector fields $\partial/\partial y^1$ and $\partial/\partial y^2$ are characteristic for $\hat{\omega}$, and every characterstic vector field is a linear combination of these two: for if $V \lrcorner \hat{\omega} = 0$ then $V \lrcorner \omega$ is a linear combination of dy^1 and dy^2. Moreover, $\mathcal{L}_{\partial_1}\hat{\omega} = \mathcal{L}_{\partial_2}\hat{\omega} = 0$. Thus $\hat{\omega}$ depends only on $y^3, y^4, \ldots, y^{2k}$ and has no non-zero characteristic vector fields among the vector fields spanned by $\partial_3, \partial_4, \ldots, \partial_{2k}$. We may therefore repeat the argument to find coordinates such that $\omega - dy^1 \wedge dy^2 - dy^3 \wedge dy^4$ depends only on $y^5, y^6, \ldots, y^{2k}$ and has no characteristic vector fields, other than zero, among the vector fields spanned by $\partial_5, \partial_6, \ldots, \partial_{2k}$. Continuing in this way, we may at each stage make coordinate transformations which do not affect the coordinates already fixed so as to eliminate two more coordinates from consideration. Eventually all the coordinates will be used up, and so coordinates will have been found such that

$$\omega = dy^1 \wedge dy^2 + dy^3 \wedge dy^4 + \cdots + dy^{2k-1} \wedge dy^{2k}.$$

This result is known as Darboux's theorem: stated in full, it says that if ω is a closed 2-form on an affine space, such that the codimension of the space of characteristic vectors of ω, necessarily even, does not vary from point to point, and takes the value $2k$, then locally coordinates may be found such that ω takes the form given above. The number $2k$, the codimension of the space of characteristic vectors, is the rank of ω, introduced (in the vector space context) in Chapter 4, Section 12: see in particular Exercise 45 there, which is a parallel to the present result. Note that a 2-form which has no non-zero characteristic vectors must have rank equal to the dimension of the space on which it resides; and this dimension must therefore be even. So given a closed 2-form ω of maximal rank on a $2n$-dimensional space, coordinates may be found such that

$$\omega = dy^1 \wedge dy^2 + dy^3 \wedge dy^4 + \cdots + dy^{2k-1} \wedge dy^{2n}.$$

It is more convenient to separate the even and odd coordinates: if one uses $p_1, p_2, \ldots, p_n$ for the odd coordinates, and $q^1, q^2, \ldots, q^n$ for the even ones, then

$$\omega = dp_a \wedge dq^a$$

(the positions of the indices are chosen to make the use of the summation convention possible; there are also other, more compelling, reasons for the choice which will become apparent in a later chapter). Notice that in terms of these special coordinates it is simple to give a 1-form θ whose exterior derivative is the closed 2-form ω: if $\theta = p_a dq^a$ then $d\theta = dp_a \wedge dq^a = \omega$.

Exercise 37. Using the argument in the text, show that a closed 2-form of rank 2 on a 2-dimensional space may be expressed in the form $dp \wedge dq$ (that is, give the final stage in the proof of Darboux's theorem). □

Exercise 38. Show that given any function h, the integral curves of the vector field V_h defined by $V_h \lrcorner \omega = -dh$, where ω has maximal rank, satisfy Hamilton's equations

$$\dot{q}^a = \frac{\partial h}{\partial p_a} \qquad \dot{p}_a = -\frac{\partial h}{\partial q^a},$$

when expressed in terms of coordinates (q^a, p_a) such that $\omega = dp_a \wedge dq^a$. □

Exercise 39. If θ is a 1-form such that $d\theta$ has rank $2n$, on a space of dimension greater than $2n$, then coordinates may be found such that $\theta = p_a dq^a + df$ for some function f. Show that df is dependent on, or linearly independent of, $dp_1, dp_2, \dots, dp_n, dq^1, dq^2, \dots, dq^n$ according as $\theta \wedge d\theta \wedge \cdots \wedge d\theta$ (with n factors $d\theta$) is or is not zero. Show that in the latter case f may be chosen as one of the coordinates q^α, p_α with $\alpha > n$; in the former case the coordinates (q^a, p_a) may be chosen such that $\theta = p_a dq^a$. □

7. Application: Hamilton-Jacobi Theory

In this section we draw together the considerations of the previous two in the study of a particular kind of partial differential equation, well known in classical mechanics, known generically as the Hamilton-Jacobi equation. We begin, however, with a special case of the result of the last section.

Suppose there is given, on a $2n$-dimensional affine space $\mathcal{A}$, a 1-form θ whose exterior derivative $\omega = d\theta = dp_a \wedge dq^a$ is already in the Darboux form with respect to affine coordinates (q^a, p_a). Consider $\mathcal{A}$ as an affine product $\mathcal{Q} \times \mathcal{P}$ where (q^a) are coordinates on $\mathcal{Q}$ and (p_a) on $\mathcal{P}$, both spaces being n-dimensional. We shall be concerned with smooth maps $\phi: \mathcal{Q} \to \mathcal{P}$. As in the discussion of Frobenius's theorem, such a map defines a submanifold of $\mathcal{A} = \mathcal{Q} \times \mathcal{P}$, namely its graph, parametrised by the map $\hat{\phi}: \mathcal{Q} \to \mathcal{A}$ given by

$$\hat{\phi}(q) = (q, \phi(q)).$$

One way of constructing such a map is to take a function f on $\mathcal{Q}$ and set $\phi_a = \partial f / \partial q^a$. In this case, $\hat{\phi}^* \theta = df$, and therefore $\hat{\phi}^* \omega = 0$. Conversely, if $\hat{\phi}^* \omega = 0$ then $\hat{\phi}^* \theta = df$ for some function f on $\mathcal{Q}$ and ϕ is constructed from f in the manner described.

Suppose further that there is given a smooth function h on $\mathcal{A}$. The coordinate expression of this function with respect to (q^a, p_a) may be used to define a first order partial differential equation, in general non-linear, by

$$h\left(q^a, \frac{\partial f}{\partial q^a}\right) = 0.$$

This equation (an equation for the unknown function f) is the *Hamilton-Jacobi equation* for h. Note that the function f on $\mathcal{Q}$ will be a solution of this differential equation if the graph map $\hat{\phi}$ generated by f maps $\mathcal{Q}$ into the level surface $h = 0$. Conversely, any graph map $\hat{\phi}$ which maps $\mathcal{Q}$ into the level surface $h = 0$ of h and satisfies $\hat{\phi}^* \omega = 0$ will generate a solution to the differential equation.

We shall assume that $dh \neq 0$ where $h = 0$. The vector field V_h defined by $V_h \lrcorner \omega = -dh$ is called the *characteristic vector field* of the partial differential equation defined by h. Since $V_h h = 0$ this vector field is tangent to the level surfaces of

h, and its flow maps each level surface into itself. Furthermore, it maps the graph of any solution of the differential equation into itself. For by Exercise 38

$$V_h = \frac{\partial h}{\partial p_a}\frac{\partial}{\partial q^a} - \frac{\partial h}{\partial q^a}\frac{\partial}{\partial p_a},$$

and on the solution graph $p_a = \phi_a(q^b)$ where $\partial\phi_a/\partial q^b = \partial\phi_b/\partial q^a$. Thus

$$\begin{aligned} V_h(p_a - \phi_a) &= -\left(\frac{\partial h}{\partial q^a} + \frac{\partial h}{\partial p_b}\frac{\partial \phi_a}{\partial q^b}\right) = -\left(\frac{\partial h}{\partial q^a} + \frac{\partial h}{\partial p_b}\frac{\partial \phi_b}{\partial q^a}\right) \\ &= -\frac{\partial}{\partial q^a}\left(h(q^b, \phi_b)\right) = 0 \end{aligned}$$

since $h(q^b, \phi_b)$ is constant.

Conversely, the characteristic vector fields may be used to generate full solutions of the partial differential equation from partial solutions, as follows. Take an $(n-1)$-dimensional submanifold $\mathcal{S}$ in $\mathcal{A}$ which lies in the level surface $h = 0$ of h, on which ω vanishes, which projects down onto an $(n-1)$-dimensional submanifold of $\mathcal{A}$ under projection onto the first factor, and which is transverse to the vector field V_h. Define an n-dimensional submanifold $\hat{\phi}\colon \mathcal{S} \times I \to \mathcal{A}$, where $I \subset \mathbf{R}$ is an open interval, by $\hat{\phi}(x,t) = \psi_t(x)$ where ψ is the flow generated by V_h. Then from the fact that $V_h h = 0$ it follows that $\hat{\phi}$ also lies in the level surface $h = 0$ of h; and from the fact that $\mathcal{L}_{V_h}\omega = 0$ it follows that $\hat{\phi}^*\omega = 0$. It will be true that $\hat{\phi}$ is actually a graph for small enough values of t, provided that V_h is nowhere tangent to the $\mathcal{P}$ factor on $\mathcal{S}$; but it may not be possible to extend $\hat{\phi}$ to a graph all over $\mathcal{Q}$ even though $\mathcal{S}$ and V_h are perfectly well behaved: this corresponds to the occurrence of singularities in the solution of the partial differential equation.

In practice, solution of the ordinary differential equations to find the integral curves of the characteristic vector field may be no easier than solution of the partial differential equation itself. In Hamiltonian mechanics, in fact, the process may be reversed: the Hamilton-Jacobi equation may be used as a means of solving Hamilton's equations, which are the differential equations for the integral curves of the characteristic vector field. In fact, knowing a so-called complete solution of the Hamilton-Jacobi equation is equivalent to knowing a coordinate system on $\mathcal{A}$ in which the characteristic vector field is straightened out. The method involves exploiting the fact that if f is a solution of the Hamilton-Jacobi equation $h(q^a, \partial f/\partial q^a) = 0$ then V_h is tangent to the submanifold $p_a = \partial f/\partial q^a$. By finding sufficiently many such submanifolds one is able to tie V_h down completely. First, though, the notion of a Hamilton-Jacobi equation must be generalised slightly. It is clearly not desirable to have to restrict attention only to the level surface $h = 0$; nor is it necessary, for if f is a solution of the equation $h(q^a, \partial f/\partial q^a) = c$ for any constant c, then V_h is tangent also to the submanifold $p_a = \partial f/\partial q^a$ (the argument given earlier still applies). Every point of $\mathcal{A}$ lies on a level surface of h, and V_h is tangent to the level surfaces. We shall therefore deal with all partial differential equations of the form $h(q^a, \partial f/\partial q^a) = c$, calling them collectively the Hamilton-Jacobi equations. We shall now, however, have to make the restriction that dh is never zero.

A *complete solution* of the Hamilton-Jacobi equations is a collection of smooth submanifolds, of dimension n, one through each point of $\mathcal{A}$, non-intersecting, such that each submanifold is the graph of a map $\phi: \mathcal{Q} \to \mathcal{P}$, lying in a level surface of h, and satisfying the "integrability condition" that ω vanishes when restricted to it. Then each such submanifold is generated by a function f on $\mathcal{Q}$ which is a solution of the Hamilton-Jacobi equation $h(q^a, \partial f/\partial q^a) = c$ for the appropriate constant c. Suppose now that new coordinates are introduced into $\mathcal{A}$, say (q^a, k_a), such that the coordinate n-submanifolds k_a = constant are the submanifolds of a complete solution of the Hamilton-Jacobi equations (the q^a being, as before, coordinates on $\mathcal{Q}$). For each fixed (k^a) there is a function on $\mathcal{Q}$ which is the solution of the Hamilton-Jacobi equation giving the corresponding submanifold. There is thus a function on $\mathcal{A}$ whose coordinate expression with respect to the new coordinates, say F, has the property that for each fixed (k_a) the function $(q^a) \mapsto F(q^a, k_a)$ is the solution of the Hamilton-Jacobi equation giving the corresponding submanifold. This function is called the *generating function* of the complete solution.

Since V_h is tangent to the submanifold corresponding to a solution of a Hamilton-Jacobi equation the coordinates k_a will be constant along any integral curve of V_h. We may go further. The 2-form ω will not take the Darboux form when expressed in terms of (q^a) and (k_a). It is possible, however, to make a further change of coordinates, this time leaving the k_a unchanged, so that ω does take the Darboux form with respect to the new coordinates; and this new coordinate system is the one we want. It is defined by

$$\acute{q}^a = \frac{\partial F}{\partial k_a} \qquad \acute{p}_a = k_a.$$

This does not define $\acute{q}^a$ and $\acute{p}_a$ explicitly in terms of q^a and p_a, since we do not (and cannot) have an explicit expression for k_a in terms of the q^a and p_a. However, the definition of the k_a defines them implicitly in terms of the q^a and p_a: the submanifold k_a = constant is given, in terms of q^a and p_a, by $p_a = (\partial F/\partial q^a)(q^b, k_b)$, and these are the required relations. Now

$$\begin{aligned}
\omega = dp_a \wedge dq^a &= \left(\frac{\partial^2 F}{\partial q^a \partial q^b} dq^b + \frac{\partial^2 F}{\partial q^a \partial k_b} dk_b\right) \wedge dq^a \\
&= \frac{\partial^2 F}{\partial q^a \partial k_b} d\acute{p}_b \wedge dq^a = d\acute{p}_b \wedge \left(d(\frac{\partial F}{\partial k_b}) - \frac{\partial^2 F}{\partial k_a \partial k_b} d\acute{p}_a\right) \\
&= d\acute{p}_b \wedge d\acute{q}^b
\end{aligned}$$

as required. Finally, note that one may choose h as one of the coordinates $\acute{p}_a$ (so long as $dh \neq 0$): suppose we set $h = \acute{p}_1$, then in terms of $(\acute{q}^a, \acute{p}_a)$,

$$V_h \lrcorner \omega = V_h \lrcorner (d\acute{p}_a \wedge d\acute{q}^a) = -dh = -d\acute{p}_1,$$

from which it follows that $V_h = \partial/\partial \acute{q}^1$.

Summary of Chapter 6

A smooth m-dimensional distribution $\mathcal{D}$ on an n-dimensional affine space $\mathcal{A}$ is a choice of subspace $\mathcal{D}_x$ of $T_x\mathcal{A}$ at each $x \in \mathcal{A}$, of dimension m, which varies smoothly from point to point of $\mathcal{A}$. A distribution may be specified in a number of ways: by giving $n-m$ independent 1-forms $\{\theta^\rho\}$ which are constraint forms for it; by giving a nowhere zero decomposable $(n-m)$-form ω which is a characterising form for it; or by giving m independent vector fields $\{V_\alpha\}$ which span it. Smoothness of $\mathcal{D}$ corresponds to smoothness of the geometric object used to specify it.

A submanifold $\mathcal{S}$ of $\mathcal{A}$ is the image of a smooth map ϕ of some open subset $\mathcal{O}$ of an affine space $\mathcal{B}$ into $\mathcal{A}$, for which $\phi_*: T_y\mathcal{B} \to T_{\phi(y)}\mathcal{A}$ is injective for all $y \in \mathcal{O}$. Such a ϕ is a parametrisation of $\mathcal{S}$; $\phi_*(T_y\mathcal{B}) = T_{\phi(y)}\mathcal{S}$ is the tangent space to $\mathcal{S}$ at $\phi(y)$; the dimension of $\mathcal{S}$ is the dimension of each of the tangent spaces, namely $\dim \mathcal{B}$.

A submanifold $\mathcal{S}$ of dimension m is an integral submanifold of a distribution $\mathcal{D}$ of the same dimension if for each $x \in \mathcal{S}$, $T_x\mathcal{S} = \mathcal{D}_x$. A given distribution need not have integral submanifolds. A necessary and sufficient condition for the existence of integral submanifolds, one through each point of $\mathcal{A}$, is the Frobenius integrability condition, which may be equivalently stated in several different ways: for a basis for constraint 1-forms, $d\theta^\rho = \lambda^\rho_\sigma \wedge \theta^\sigma$ for some 1-forms λ^ρ_σ; for a characterising form, $d\theta \wedge \omega = 0$ for any constraint 1-form; for a vector field basis, $[V_\alpha, V_\beta] = f^\gamma_{\alpha\beta} V_\gamma$ for some functions $f^\gamma_{\alpha\beta}$. When the integrability condition is satisfied, an integral submanifold may be constructed through a given point x in the form of a graph, defined over the affine m-plane through x defined by the distribution at that point. The construction is based on a method of lifting vectors tangent to the m-plane into vectors in $\mathcal{A}$ tangent to the distribution. The theorem which establishes the sufficiency of the integrability condition is Frobenius's theorem.

Given a vector field V there is, in a neighbourhood of any point at which it is non-zero, a coordinate system (x^a) such that $V = \partial_1$. Further, given m vector fields $V_1, V_2, \ldots, V_m$ which are linearly independent and commute pairwise there is locally a coordinate system in which $V_\alpha = \partial_\alpha$. Since it may be shown that for a distribution satisfying the Frobenius integrability condition one may always find a basis for the distribution consisting of pairwise commuting local vector fields this gives another proof of Frobenius's theorem: the integral submanifolds are given by $x^\rho = $ constant. Furthermore, if constraint 1-forms θ^ρ satisfy the Frobenius integrability conditions, then there are functions A^ρ_σ such that the matrix (A^ρ_σ) is everywhere non-singular, and coordinates (x^a), such that $A^\rho_\sigma \theta^\sigma = dx^\rho$ (so that, in particular, the 1-forms $A^\rho_\sigma \theta^\sigma$ are exact).

Coordinates (y^a) may be found so that a given closed 2-form ω of constant rank $2r$ takes the form $\omega = dy^1 \wedge dy^2 + dy^3 \wedge dy^4 + \cdots + dy^{2r-1} \wedge dy^{2r}$. This result is Darboux's theorem. The 2-form $dp_a \wedge dq^a$ on an even dimensional space $\mathcal{A} = \mathcal{Q} \times \mathcal{P}$, which is closed and has rank $2n$, plays a key role in Hamiltonian mechanics and in the solution of the Hamilton-Jacobi equation for a function h, which is the partial differential equation $h(q^a, \partial f/\partial q^a) = 0$. The vector field V_h determined by $V_h \lrcorner \omega = -dh$ defines Hamilton's equations of mechanics, and is also the characteristic vector field of the Hamilton-Jacobi equation.

7. METRICS ON AFFINE SPACES

The ordinary scalar product of vectors which one encounters in elementary mechanics and geometry may be generalised to affine spaces of dimension other than 3. In elementary Euclidean geometry one is concerned mostly with the use of the scalar product to measure lengths of, and angles between, displacements; in mechanics one is also concerned with magnitudes of, and angles between, velocity vectors. In either case the scalar product comes from an operation in a vector space $\mathcal{V}$, which may be transferred to an affine space $\mathcal{A}$ modelled on it. The realisation of $\mathcal{V}$ as tangent space to $\mathcal{A}$ at each point generalises to manifolds, as will be explained in Chapters 9 and, especially, 10, but the realisation as space of displacements does not. We shall therefore give preference to the tangent space realisation in this chapter.

The structure on $\mathcal{A}$ determined in this way by a scalar product on $\mathcal{V}$ is usually called a metric. It is unfortunate, but now irremediable, that the word is used in a different sense in topology. In the case of Euclidean space the two meanings are closely related; however, as well as the generalisation to arbitrary dimension, we shall also consider the generalisation of the concept of a scalar product in a different direction, which includes the space-time of special relativity, and in this case the relation between the two meanings of the word metric is not close.

When an affine space is equipped with a metric it becomes possible to establish a 1 : 1 correspondence between vectors and covectors, which we have been at pains to keep separate until now. As a result, in Euclidean 3-dimensional space the various operations of the exterior derivative may be made to assume the familiar forms of vector calculus in their entirety.

We begin the chapter by discussing the algebraic properties of scalar products.

1. Scalar Products on Vector Spaces

The ordinary scalar product $\mathbf{a} \cdot \mathbf{b}$ of vectors in Euclidean 3-dimensional space may be defined in either of two ways: trigonometrically, as the perpendicular projection of $\mathbf{a}$ on $\mathbf{b}$, multiplied by the magnitude of $\mathbf{b}$; or algebraically, as $a_1b_1 + a_2b_2 + a_3b_3$, where $\mathbf{a} = (a_1, a_2, a_3)$, $\mathbf{b} = (b_1, b_2, b_3)$. It is convenient to start with the algebraic definition, to identify its main properties as a basis for generalisation, and to derive the trigonometrical constructions afterwards.

The ordinary scalar product of two vectors $\mathbf{a}$, $\mathbf{b}$ is a real number, and has these properties:

(1) bilinearity:

$$(k_1\mathbf{a}_1 + k_2\mathbf{a}_2) \cdot \mathbf{b} = k_1(\mathbf{a}_1 \cdot \mathbf{b}) + k_2(\mathbf{a}_2 \cdot \mathbf{b})$$
$$\mathbf{a} \cdot (k_1\mathbf{b}_1 + k_2\mathbf{b}_2) = k_1(\mathbf{a} \cdot \mathbf{b}_1) + k_2(\mathbf{a} \cdot \mathbf{b}_2)$$

(2) symmetry:

$$\mathbf{b} \cdot \mathbf{a} = \mathbf{a} \cdot \mathbf{b}$$

(3) non-degeneracy:

$$\text{if } \mathbf{a} \cdot \mathbf{b} = 0 \text{ for all } \mathbf{b} \text{ then } \mathbf{a} = \mathbf{0}$$

(4) positive-definiteness:

$$\mathbf{a} \cdot \mathbf{a} \geq 0, \text{ and if } \mathbf{a} \cdot \mathbf{a} = 0 \text{ then } \mathbf{a} = \mathbf{0}.$$

There is nothing characteristically 3-dimensional about these properties, and so they may be used to generalise the notion of a scalar product to affine spaces of other dimensions. Note that the "vectors" appearing in the definition are displacement vectors or tangent vectors, that is, elements of the underlying vector space, which we may take in this case to be $\mathbf{R}^3$.

In applications to special relativity both the physical interpretation of the scalar product, and the mathematical formulation which reflects it, make it appropriate to give up the requirement of positive-definiteness. There are occasions when even the requirement of non-degeneracy has to be given up. One is led, therefore, to consider a construction like the scalar product but satisfying the conditions of symmetry and bilinearity only.

Bilinear and quadratic forms. We therefore define a *symmetric bilinear form* on a vector space $\mathcal{V}$ as a map $g: \mathcal{V} \times \mathcal{V} \to \mathbf{R}$ such that

$$\begin{aligned} g(k_1 v_1 + k_2 v_2, w) &= k_1 g(v_1, w) + k_2 g(v_2, w) \\ g(v, k_1 w_1 + k_2 w_2) &= k_1 g(v, w_1) + k_2 g(v, w_2) \end{aligned} \tag{1}$$

$$g(w, v) = g(v, w) \tag{2}$$

for all $v, w, v_1, v_2, w_1, w_2 \in \mathcal{V}$ and $k_1, k_2 \in \mathbf{R}$. Thus g is multilinear in just the same way as the forms considered in Chapter 4 are, but differs from them in being symmetric instead of alternating.

The *components* of g with respect to a basis $\{e_a\}$ for $\mathcal{V}$ are the numbers

$$g_{ab} = g(e_a, e_b).$$

Note that the symmetry condition implies that $g_{ba} = g_{ab}$, while from the bilinearity it follows that the g_{ab} determine g: if $v = v^a e_a$ and $w = w^a e_a$ then

$$g(v, w) = g_{ab} v^a w^b.$$

In dealing with bilinear forms it is often convenient to employ matrix notation: this last formula may be written

$$g(v, w) = v^T G w$$

where on the right G denotes the square matrix, necessarily symmetric, with entries g_{ab}, v and w denote the column vectors with entries v^a and w^a respectively, and the superscript T denotes the transpose. Thus $G^T = G$.

Exercise 1. Show that if $\acute{e}_a = h_a^b e_a$ are the elements of a new basis then the components of g with respect to the $\acute{e}_a$ are given by $\acute{g}_{ab} = h_a^c h_b^d g_{cd}$, or $\acute{G} = H^T G H$. □

The function $v \mapsto g(v,v)$ is called a *quadratic form* on $\mathcal{V}$. If the quadratic form is given, then the bilinear form may be recovered with the help of the identity

$$g(v,w) = \tfrac{1}{2}\big(g(v+w, v+w) - g(v,v) - g(w,w)\big).$$

Thus the specification of a symmetric bilinear form and of a quadratic form are entirely equivalent, and the theory of these objects is often developed in the language of quadratic forms. It is known from the theory of quadratic forms that if g is given then there is a basis for $\mathcal{V}$, which we call a *standard basis*, with respect to which

$$g(v,w) = v^1w^1 + v^2w^2 + \cdots + v^rw^r - v^{r+1}w^{r+1} - \cdots - v^{r+s}w^{r+s}.$$

The corresponding matrix G has on the main diagonal first r ones, then s minus ones, then $n-(r+s)$ zeros, where $n = \dim \mathcal{V}$, and it has zeros everywhere else. The choice of standard basis in which g takes this form is by no means unique; however, Sylvester's theorem of inertia asserts that the numbers of diagonal ones, minus ones and zeros are independent of the choice of standard basis.

A symmetric bilinear form g on $\mathcal{V}$ is called *non-degenerate* if

$$g(v,w) = 0 \text{ for all } w \text{ implies that } v = 0. \tag{3}$$

If g is non-degenerate then there are no zeros on the diagonal in its expression with respect to a standard basis, $r+s=n$, and g is characterised by r, the number of ones (the dimension n of $\mathcal{V}$ having been fixed).

Exercise 2. Show that g is a non-degenerate symmetric bilinear form if and only if G is non-singular; in such a case, the matrix H relating two bases (as defined in Exercise 1) satisfies $(\det H)^2 = \det \acute{G}/\det G$. □

We shall call a non-degenerate symmetric bilinear form a *scalar product*; sometimes a symmetric bilinear form is called a scalar product even if it is degenerate, but in this book we maintain the distinction. A scalar product g is given with respect to a standard basis by

$$g(v,w) = v^1w^1 + v^2w^2 + \cdots + v^rw^r - v^{r+1}w^{r+1} - \cdots - v^nw^n.$$

It is said to have *signature* $(r, n-r)$, or simply signature r.

2. Euclidean and Pseudo-Euclidean Spaces

Euclidean space. The standard scalar product on $\mathbf{R}^n$, given by

$$g(v,w) = v^1w^1 + v^2w^2 + \cdots + v^nw^n,$$

has signature n. Any scalar product of signature n is called *Euclidean*, and a vector space with Euclidean scalar product is called a *Euclidean (vector) space*. A scalar product is called *positive-definite* if

$$g(v,v) \geq 0 \text{ for all } v, \text{ and } g(v,v) = 0 \text{ only if } v = 0. \tag{4}$$

It is clear that g is positive-definite if and only if it is Euclidean.

Exercise 3. Let g be a Euclidean scalar product. Show that for any vectors v, w, and any real number t,

$$g(tv + w, tv + w) = t^2 g(v,v) + 2tg(v,w) + g(w,w).$$

Deduce from the positive-definiteness of g that the discriminant of the right hand side, considered as a quadratic in t, cannot be positive for $v \neq 0$, and deduce the Schwartz inequality

$$|g(v,w)| \leq \sqrt{g(v,v)}\sqrt{g(w,w)},$$

with equality if and only if v and w are linearly dependent. Show that if v and w are both non-zero then $|g(v,w)|/\sqrt{g(v,v)}\sqrt{g(w,w)}$ is the cosine of exactly one angle ϑ such that $0 \leq \vartheta \leq \pi$ (this angle is then defined to be the angle between v and w). Show further that

$$\sqrt{g(v+w, v+w)} \leq \sqrt{g(v,v)} + \sqrt{g(w,w)}$$

(the "triangle inequality"). □

Orthonormality. Generalising from the Euclidean case, one says, for any scalar product g, that vectors v and w are *orthogonal* if $g(v,w) = 0$, and a vector v is a *unit vector* if $|g(v,v)| = 1$. Notice that v is called a unit vector whether $g(v,v) = 1$ or $g(v,v) = -1$. A basis is called *orthogonal* if the vectors in it are mutually orthogonal, and *orthonormal* if they are also unit vectors. Thus a basis in which g takes the standard form (appropriate to its signature) is orthonormal, and conversely.

Exercise 4. Infer from Exercises 1 and 2 that the matrix H of a change of orthonormal basis for a Euclidean space must be orthogonal, which is to say that $H^T H = I_n$, and deduce that $\det H = +1$. □

Exercise 5. Show that for any symmetric bilinear form g, if a vector v is orthogonal to vectors $w_1, w_2, \ldots, w_m$, then it is orthogonal to every vector in the subspace spanned by $w_1, w_2, \ldots, w_m$. □

Vectors $v_1, v_2, \ldots, v_m$ form an *orthogonal set* if they are mutually orthogonal and an *orthonormal set* if they are, in addition, unit vectors.

Exercise 6. Show that vectors of an orthogonal set are necessarily linearly independent. □

Exercise 7. Let $\mathcal{V}$ be an n-dimensional vector space with Euclidean scalar product g and let $\mathcal{W}$ be a p-dimensional subspace of $\mathcal{V}$. Let $\mathcal{W}^\perp$ denote the set of vectors orthogonal to every vector in $\mathcal{W}$. Show that $\mathcal{W}^\perp$ is a subspace of $\mathcal{V}$ of dimension $n - p$. Show that $(\mathcal{W}^\perp)^\perp = \mathcal{W}$. Show that $\mathcal{V}$ is a direct sum $\mathcal{V} = \mathcal{W} \oplus \mathcal{W}^\perp$. □

The subspace $\mathcal{W}^\perp$ is called the *orthogonal complement* to $\mathcal{W}$ in $\mathcal{V}$.

Lorentzian scalar products. A scalar product which is non-degenerate, but not necessarily positive-definite, is said to be *pseudo-Euclidean*. The case of greatest interest is the scalar product in the Minkowski space of special relativity theory, which is generally rearranged, still with signature $(1,3)$, to

$$g(v,w) = -v^1w^1 - v^2w^2 - v^3w^3 + v^4w^n4,$$

although in a standard basis it should be written $v^1w^1 - v^2w^2 - v^3w^3 - v^4w^4$. It is sometimes reversed to the signature $(3,1)$ form $v^1w^1 + v^2w^2 + v^3w^3 - v^4w^4$, and in older books is found in the form $v^0w^0 - v^1w^1 - v^2w^2 - v^3w^3$. In this book we shall adhere to the first form displayed above.

It is in most respects as easy to discuss the scalar product of signature $(1, n-1)$ or $(n-1, 1)$ on an n-dimensional vector space as the 4-dimensional example. For any

n, a scalar product with this signature is said to be *hyperbolic normal* or *Lorentzian*. There is a conventional choice of sign to be made when dealing with Lorentzian scalar products: we shall choose always the signature $(1, n-1)$.

In a space with pseudo-Euclidean, but not Euclidean, scalar product one can find a non-zero vector v for which $g(v, v)$ has any chosen real value, positive, negative or zero. With our choice of signature a non-zero vector v in a space with Lorentzian scalar product is called

$$\left\{\begin{array}{l} \textit{timelike} \\ \textit{null} \text{ or } \textit{lightlike} \\ \textit{spacelike} \end{array}\right\} \quad \text{if} \quad \left\{\begin{array}{l} g(v,v) > 0 \\ g(v,v) = 0 \\ g(v,v) < 0 \end{array}\right\}.$$

These names arise from the physical interpretation in the 4-dimensional (Lorentzian or Minkowskian) case of special relativity theory: a timelike vector is a possible 4-momentum vector for a massive particle; a lightlike vector is a possible 4-momentum vector for a massless particle such as a photon; a spacelike vector will lie in the instantaneous rest space of any timelike vector to which it is orthogonal. Although this physical interpretation cannot be maintained in a space with Lorentzian scalar product if the dimension of the space is greater than 4, nevertheless the image is very convenient.

The vectors orthogonal, with respect to a Lorentzian scalar product, to a given non-zero vector form a subspace of codimension 1, called its *orthogonal subspace*. The orthogonal subspace is called

$$\left\{\begin{array}{l} \textit{spacelike} \\ \textit{null} \\ \textit{timelike} \end{array}\right\} \quad \text{if the given vector is} \quad \left\{\begin{array}{l} \text{timelike} \\ \text{null} \\ \text{spacelike} \end{array}\right\}.$$

The Lorentzian scalar product induces a symmetric bilinear form on a subspace, by restriction: if a subspace of codimension 1 is spacelike, then the induced bilinear form, with the sign reversed, is a Euclidean scalar product; if the subspace is null then the bilinear form is degenerate, while if the subspace is timelike and of dimension greater than 1 then the induced bilinear form is again a Lorentzian scalar product.

Exercise 8. Show that in an n-dimensional vector space with Lorentzian scalar product the timelike vectors are separated into two disjoint sets by the set of null vectors (the null cone), while in a vector space with pseudo-Euclidean scalar product of signature $(p, n-p)$ where $1 < p < n-1$ there is no such separation for vectors v with $g(v, v) > 0$. □

The disjunction in the Lorentzian case corresponds to the distinction between past and future.

Exercise 9. Let $\mathcal{V}$ be an n-dimensional vector space with Lorentzian scalar product g. Show that if v and w are timelike vectors both pointing to the future or both pointing to the past then $g(v, w) > 0$, whereas if one points to the future and one to the past then $g(v, w) < 0$. Show that in no case can two timelike vectors be orthogonal; show that a non-zero vector orthogonal to a timelike vector must be spacelike. □

Exercise 10. Show that all vectors in a spacelike subspace of codimension 1 of a vector space with Lorentzian scalar product are spacelike. Show that if a vector is null then it lies in its own orthogonal subspace, while all vectors in that subspace which are linearly independent of it are spacelike. Show that provided its dimension is greater than 1 a timelike subspace contains vectors of all three types. □

Exercise 11. Let $\mathcal{V}$ be an n-dimensional vector space with Lorentzian scalar product and $\mathcal{W}$ a 2-dimensional subspace of $\mathcal{V}$. Show that if $\mathcal{W}$ contains two linearly independent null vectors then it has an orthogonal basis consisting of a timelike and a spacelike vector; if $\mathcal{W}$ contains a non-zero null vector and no other null vector linearly independent of that one, then it has an orthogonal basis consisting of a null and a spacelike vector; while if $\mathcal{W}$ contains no non-zero null vectors then it has an orthogonal basis consisting of two spacelike vectors. □

3. Scalar Products and Dual Spaces

We have been careful, in earlier chapters, to draw a sharp distinction between vector spaces and their duals. The need for this distinction is clear, for example, in the case of a tangent space and its dual, a cotangent space: elements of the two spaces play quite different geometric roles. However, specification of a bilinear form on a vector space allows one to define a linear map from the vector space to its dual, and if the bilinear form is non-degenerate, whatever its signature, this linear map is a bijection and may be used to identify the two spaces in a manner which does not depend on a particular choice of basis.

Suppose, first of all, that g is a symmetric bilinear form on a vector space $\mathcal{V}$. For any fixed $v \in \mathcal{V}$ the map $\mathcal{V} \to \mathbf{R}$ by $w \mapsto g(v,w)$ is linear, because g is bilinear, and therefore defines an element of $\mathcal{V}^*$. This element of $\mathcal{V}^*$ will be denoted $g(v)$, so that

$$\langle w, g(v)\rangle = g(v,w)$$

for all $w \in \mathcal{V}$. Here g is used in two different senses: on the left, with one argument, to denote a linear map from $\mathcal{V}$ to $\mathcal{V}^*$; on the right, with two arguments, to denote a bilinear form on $\mathcal{V}$. No confusion need arise from this.

If $\{e_a\}$ is a basis for $\mathcal{V}$, and $\{\theta^a\}$ the dual basis for $\mathcal{V}^*$, then $g(e_a) = g_{ab}\theta^b$ where g_{ab} are the components of the bilinear form g, and so if $v = v^a e_a$ then $g(v) = g_{ab}v^a\theta^b$; which is to say that the components of $g(v)$ are $g_{ab}v^a$. It is usual, when a bilinear form g has been fixed once for all, to write v_b for $g_{ab}v^a$. The position of the index is important: except in special cases $v_a \neq v^a$.

Because of the relation between the components, this process of constructing an element of $\mathcal{V}^*$ from an element of $\mathcal{V}$ with the help of g is called *lowering the index*. In matrix notation, the map from components of elements of $\mathcal{V}$, expressed as column vectors, to components of elements of $\mathcal{V}^*$, expressed as row vectors, is given by $v \mapsto v^T G = (Gv)^T$.

If the bilinear form g is non-degenerate an inverse map $g^{-1}\colon \mathcal{V}^* \to \mathcal{V}$ may be defined such that, for any $\alpha \in \mathcal{V}^*$ and any $v \in \mathcal{V}$,

$$\langle v, \alpha\rangle = g\big(g^{-1}(\alpha), v\big).$$

In matrix notation, with respect to the same bases as before, if now α is the row vector of components of an element of $\mathcal{V}^*$ then the corresponding element of $\mathcal{V}$ has components $G^{-1}\alpha^T = (\alpha G^{-1})^T$, where G^{-1} is the matrix inverse to the (non-singular) matrix G. As is customary, we denote by g^{ab} the entries in G^{-1}. The components α^a of $g^{-1}(\alpha)$ are given by $\alpha^a = \alpha_b g^{ba}$. The g^{ab} and g_{ab} are related by

$$g^{ac}g_{cb} = g_{bc}g^{ca} = \delta^a_b.$$

The matrix G^{-1} is symmetric; that is, $g^{ba} = g^{ab}$ (see Exercise 16).

The map $g^{-1}: \mathcal{V}^* \to \mathcal{V}$ is called *raising the index*.

Exercise 12. Show that in $\mathbf{R}^3$, with standard basis and standard Euclidean scalar product, the map $v \mapsto g(v)$ is given by

$$(v^1, v^2, v^3)^T \mapsto (v^1, v^2, v^3),$$

but that if the Lorentzian scalar product $g(v,w) = -v^1w^1 - v^2w^2 + v^3w^3$ is used then the map $v \mapsto g(v)$ is given by

$$(v^1, v^2, v^3)^T \mapsto (-v^1, -v^2, v^3). \qquad \square$$

Exercise 13. Show that if $\{e_a\}$ is a basis for $\mathcal{V}$ with scalar product g and $\{e^a\}$ the set of elements of $\mathcal{V}^*$ given by $e^a = g(e_a)$, then $\{e^a\}$ is a basis for $\mathcal{V}^*$. Show that the matrix of the map $v \mapsto g(v)$ with respect to these bases is the identity matrix. Show that in contrast the matrix of the same map with respect to $\{e_a\}$ and $\{\theta^a\}$, the dual basis for $\mathcal{V}^*$, is (g_{ab}). Show that if g is Euclidean then $\{e_a\}$ and $\{e^a\}$ are dual if and only if $\{e_a\}$ is orthonormal. $\square$

Exercise 14. A bilinear form B, which is not symmetric, on a vector space $\mathcal{V}$ determines two linear maps $\mathcal{V} \to \mathcal{V}^*$, since for fixed $v \in \mathcal{V}$ the two linear forms $w \mapsto B(v,w)$ and $w \mapsto B(w,v)$ may be distinct. Find the components of the image of v, and the matrix representation of the map with respect to dual bases, in the two cases. $\square$

Exercise 15. Confirm that the linear map $v \mapsto g(v)$ is bijective if and only if g is non-degenerate. $\square$

Exercise 16. If g is a scalar product then a bilinear form g^* may be defined on $\mathcal{V}^*$ by $g^*(\alpha,\beta) = g(g^{-1}(\alpha), g^{-1}(\beta))$. Show that g^* is symmetric and non-degenerate and of the same signature as g. Show that if $\{e_a\}$ and $\{\theta^a\}$ are dual bases then $g^*(\theta^a, \theta^b) = g^{ab}$ as defined above. Conclude that $g^{ba} = g^{ab}$. Show that $G^* = G^{-1}$. $\square$

Exercise 17. Let T be any p-multilinear form on $\mathcal{V}^*$. Show that a multilinear form $g(T)$ may be defined on $\mathcal{V}$ by

$$g(T)(v_1, v_2, \ldots, v_p) = T(g(v_1), g(v_2), \ldots, g(v_p)).$$

Show in particular that if T is a p-vector then $g(T)$ is an exterior p-form. Show similarly that if g is non-degenerate and if S is any p-multilinear form on $\mathcal{V}$ then $g^{-1}(S)$ defined by

$$g^{-1}(S)(\alpha^1, \alpha^2, \ldots, \alpha^p) = S(g^{-1}(\alpha^1), g^{-1}(\alpha^2), \ldots, g^{-1}(\alpha^p))$$

is a p-multilinear form on $\mathcal{V}^*$ and that if S is an exterior p-form then $g^{-1}(S)$ is a p-vector. $\square$

4. The Star Operator

The ordinary space of classical mechanics is an affine space modelled on $\mathbf{R}^3$ with a Euclidean scalar product. In this context one encounters such formulae as, for example,

$\mathbf{a} \cdot (\mathbf{b} \times \mathbf{c})$ is the volume of the parallepiped with edges $\mathbf{a}$, $\mathbf{b}$ and $\mathbf{c}$

$(\mathbf{a} \times \mathbf{b}) \cdot (\mathbf{c} \times \mathbf{d}) = (\mathbf{a} \cdot \mathbf{c})(\mathbf{b} \cdot \mathbf{d}) - (\mathbf{a} \cdot \mathbf{d})(\mathbf{b} \cdot \mathbf{c})$.

We shall show how to establish corresponding formulae in any vector space or affine space with a scalar product.

The construction of the vector $\mathbf{a} \times \mathbf{b}$ from the vectors $\mathbf{a}$ and $\mathbf{b}$ may be understood, in terms of the results of this chapter and of Chapter 4, as follows:

(1) construct 1-forms $g(\mathbf{a})$ and $g(\mathbf{b})$
(2) take their exterior product, obtaining the 2-form $g(\mathbf{a}) \wedge g(\mathbf{b})$
(3) define the vector $\mathbf{a} \times \mathbf{b}$ by the rule

$$(\mathbf{a} \times \mathbf{b}) \lrcorner \Omega = g(\mathbf{a}) \wedge g(\mathbf{b})$$

where Ω is the volume form which assigns unit volume to a parallepiped with orthonormal edges and the usual orientation. Thus the vector $\mathbf{a} \times \mathbf{b}$ is a characterising vector for a line which has the 2-form $g(\mathbf{a}) \wedge g(\mathbf{b})$ as characterising form.

Exercise 18. Confirm that the effect of carrying out these operations is indeed to produce the vector product. □

The fact that the result of this sequence of operations is a vector depends crucially on the dimension of the space being 3: in no other case will the final operation produce a vector. It is not to be expected, therefore, that the vector product generalises as such to spaces of other dimension.

The construction of a characterising p-vector from a characterising $(n-p)$-form, as in step (3) above, becomes possible when a scalar product has been chosen, because (as we shall show) a scalar product fixes a volume form, up to a sign. Moreover, the availability of a scalar product makes possible new constructions which cannot be achieved with a volume form alone, such as the formation of a scalar product of two p-forms, for any p. This formation generalises the formula for the scalar product of two vector products, displayed above.

Volume forms related to a scalar product. Again let $\mathcal{V}$ be an n-dimensional vector space with Euclidean scalar product g. The matrix relating any two orthogonal bases of $\mathcal{V}$ is orthogonal, and therefore has determinant ± 1 (Exercise 4). Now let Ω be a volume form, which is to say, a non-zero n-form, on $\mathcal{V}$. It follows that the value of Ω can at most change sign under a change of orthonormal basis: if $\{e_a\}$ and $\{\acute{e}_a\}$ are orthonormal bases then

$$\Omega(\acute{e}_1, \acute{e}_2, \ldots, \acute{e}_n) = \pm\Omega(e_1, e_2, \ldots, e_n).$$

Given a particular orthonormal basis $\{e_a\}$ there is just one volume form Ω such that $\Omega(e_1, e_2, \ldots, e_n) = 1$; then Ω and $-\Omega$ take the values ± 1 on every orthonormal basis, and are the only two volume forms to do so. There are therefore exactly two volume forms which take the values ± 1 on every orthonormal basis. These are the volume forms determined by the scalar product, in which the volume of a unit hypercube has absolute value 1. Choosing between them amounts to deciding which orthonormal bases are positively oriented. We assume now that this has been done.

Recall one motive for the definition of the exterior product: to generalise the elementary geometrical idea that

$$\text{volume} = \text{area of base} \times \text{height}.$$

In a Euclidean space of any dimension one may extend this by establishing a formula which is in effect a generalisation of

$$\text{area of base} = \text{volume} \div \text{height}.$$

The construction is a combination of the linear map g, extended to forms and multivectors, and the dual map between forms and multivectors, introduced in

Section 5 of Chapter 4. However, it may be introduced in a more overtly geometrical way, which we now describe.

First, we show how to specialise to Euclidean spaces a construction of a volume form on a vector space which we described in Chapter 4, Section 7. In this construction one takes the exterior product of characterising forms for a pair of complementary subspaces. Consider a p-dimensional subspace $\mathcal{W}$ of the n-dimensional space $\mathcal{V}$ (which is supposed equipped with a Euclidean scalar product), and its orthogonal complement $\mathcal{W}^\perp$, a subspace of dimension $n-p$. The scalar product on $\mathcal{V}$ induces Euclidean scalar products on $\mathcal{W}$ and $\mathcal{W}^\perp$, by restriction. Any characterising p-form for $\mathcal{W}^\perp$ defines a volume p-form on $\mathcal{W}$, by restriction, and so there are two characterising p-forms for $\mathcal{W}^\perp$ whose restrictions to $\mathcal{W}$ coincide with the volume p-forms defined on that space by its Euclidean scalar product: we denote them $\pm\Omega_{\mathcal{W}}$. Likewise, there are two characterising $(n-p)$-forms for $\mathcal{W}$, say $\pm\Omega_{\mathcal{W}^\perp}$, whose restrictions to $\mathcal{W}^\perp$ coincide with the volume $(n-p)$-forms defined there by its Euclidean scalar product. The exterior products of these p- and $(n-p)$-forms,

$$(\pm\Omega_{\mathcal{W}}) \wedge (\pm\Omega_{\mathcal{W}^\perp}) = \pm(\Omega_{\mathcal{W}} \wedge \Omega_{\mathcal{W}^\perp}) = \pm\Omega$$

say, are volume forms on $\mathcal{V}$: in fact they are just the two volume forms determined by the Euclidean scalar product on $\mathcal{V}$, as may easily be seen by evaluating Ω or $-\Omega$ on any orthonormal basis for $\mathcal{V}$, p of whose members lie in $\mathcal{W}$ and $n-p$ in $\mathcal{W}^\perp$.

Secondly, we point out that this construction may be reversed, in a sense, again by making use of the Euclidean scalar product. The forms $\Omega_{\mathcal{W}}$ and $\Omega_{\mathcal{W}^\perp}$ are decomposable. Suppose now that there is given a decomposable p-form ω on $\mathcal{V}$. Let $\mathcal{W}$ be the orthogonal complement of the characteristic subspace of ω: it is a p-dimensional subspace of $\mathcal{V}$, and the restriction of ω to $\mathcal{W}$ is a volume p-form on $\mathcal{W}$. We shall suppose for the present that the restriction of ω to $\mathcal{W}$ actually coincides with one of the volume p-forms defined by its scalar product (also obtained by restriction, as before). We seek to construct an $(n-p)$-form on $\mathcal{V}$, which we shall denote $*\omega$, such that

$$\omega \wedge *\omega = \Omega,$$

where Ω is the volume n-form on $\mathcal{V}$ defined by the Euclidean scalar product and the orientation, supposed already chosen. The form $*\omega$ may be determined up to sign by taking it to be a characterising $(n-p)$-form for $\mathcal{W}$, whose restriction to $\mathcal{W}^\perp$ (the characteristic subspace of ω) is one of the volume $(n-p)$-forms defined there by the scalar product.To complete its definition we have merely to choose its sign so that $\omega \wedge *\omega = \Omega$ rather than $-\Omega$. Note that $*\omega$, like ω, is decomposable.

To sum up: $\mathcal{V}$ is an oriented vector space with Euclidean scalar product, and ω is a decomposable p-form on $\mathcal{V}$, which coincides with a volume form, determined by restriction of the scalar product, on the orthogonal complement of its characteristic subspace. Then $*\omega$ is a decomposable $(n-p)$-form determined by ω such that the exterior product $\omega \wedge *\omega$ is the volume form on $\mathcal{V}$ determined by the scalar product and the given orientation, and that the characteristic subspaces of the two forms are orthogonal complements of each other. The construction of $*\omega$ from ω, with the interpretation in terms of volume forms, is the analogue of the formula area of base = volume$\div$height from which we started.

This construction determines a map from certain p-forms to $(n-p)$-forms, though so far the forms concerned are of a rather special type. We shall explain shortly how it may be extended to give a linear map $\bigwedge^p \mathcal{V}^* \to \bigwedge^{n-p} \mathcal{V}^*$. The construction, and therefore the resulting map, depends heavily on the use of the scalar product, both through the direct sum decomposition of $\mathcal{V}$ into orthogonal complementary subspaces and through the repeated use of volume forms defined by scalar products.

Exercise 19. Show that if $\{e_a\}$ is a positively oriented orthonormal basis for $\mathcal{V}$ then the volume form determined by the metric is $e^1 \wedge e^2 \wedge \cdots \wedge e^n$, where $e^a = g(e_a)$, as in Exercise 13. Show that if $\{f_a\}$ is an arbitrary positively oriented basis then the volume form is $\sqrt{\det(g_{ab})}\theta^1 \wedge \theta^2 \wedge \cdots \wedge \theta^n$, where $\{\theta^a\}$ is the basis of $\mathcal{V}^*$ dual to $\{f_a\}$ and $g_{ab} = g(f_a, f_b)$. □

Exercise 20. Show that in a 2-dimensional Euclidean space with orthonormal basis $\{e_1, e_2\}$ and volume 2-form $e^1 \wedge e^2$,

$$*e^1 = e^2 \text{ and } *e^2 = -e^1.$$

Show that in a 3-dimensional Euclidean space with orthonormal basis $\{e_1, e_2, e_3\}$ and volume 3-form $e^1 \wedge e^2 \wedge e^3$

$$\begin{array}{lll} *e^1 = e^2 \wedge e^3, & *e^2 = -e^1 \wedge e^3, & *e^3 = e^1 \wedge e^2, \\ *(e^1 \wedge e^2) = e^3, & *(e^1 \wedge e^3) = -e^2, & *(e^2 \wedge e^3) = e^1. \end{array}$$

Show that in the 2-dimensional case $**\omega = -\omega$ for every form ω considered, while in the 3-dimensional case $**\omega = \omega$. □

Exercise 21. Show that choice of the opposite orientation in the definition of $*\omega$ changes its sign. □

Exercise 22. In an n-dimensional Euclidean space $\mathcal{V}$ with Euclidean volume form Ω, let ω be a decomposable p-form whose value on any set of p orthonormal vectors is ± 1 or 0. Show that an orthonormal basis $\{e_a\}$ may be chosen for $\mathcal{V}$ such that $\Omega(e_1, e_2, \ldots, e_n) = 1$, $\omega(e_1, e_2, \ldots, e_p) = 1$ and $\{e_{p+1}, e_{p+2}, \ldots, e_n\}$ is a basis for the characteristic subspace of ω. Show that $\omega = e^1 \wedge e^2 \wedge \cdots \wedge e^p$ and that $*\omega = e^{p+1} \wedge e^{p+2} \wedge \cdots \wedge e^n$, or alternatively $*\omega = e_p \lrcorner (e_{p-1} \lrcorner (\ldots \lrcorner (e_1 \lrcorner \Omega) \ldots))$. □

The star operator in general. The final result of the preceding exercise may suggest how the process of associating $*\omega$ with ω could be extended from decomposable exterior forms to arbitrary ones. What is needed is the extension principle introduced in Chapter 4, Section 13.

Recall the maps $g\colon \mathcal{V} \to \mathcal{V}^*$ and $g^{-1}\colon \mathcal{V}^* \to \mathcal{V}$ determined by the scalar product, which were introduced in Section 3. The map $\mathcal{V}^{*p} \to \bigwedge^{n-p} \mathcal{V}^*$ given by

$$(\eta^1, \eta^2, \ldots, \eta^p) \mapsto g^{-1}(\eta^p) \lrcorner (g^{-1}(\eta^{p-1}) \lrcorner (\ldots \lrcorner (g^{-1}(\eta^1) \lrcorner \Omega) \ldots))$$

is multilinear and alternating, and therefore by the extension principle may be extended to a linear map $\bigwedge^p \mathcal{V}^* \to \bigwedge^{n-p} \mathcal{V}^*$. This map is denoted by $\omega \mapsto *\omega$ for arbitrary forms ω, not only for decomposable ones, and is called the *star operator*. Note that for any decomposable p-form $\eta^1 \wedge \eta^2 \wedge \cdots \wedge \eta^p$

$$*(\eta^1 \wedge \eta^2 \wedge \cdots \wedge \eta^p) = g^{-1}(\eta^p) \lrcorner (g^{-1}(\eta^{p-1}) \lrcorner (\ldots \lrcorner (g^{-1}(\eta^1) \lrcorner \Omega) \ldots))$$

The extended definition subsumes the original one, by Exercise 22.

Exercise 23. Let $\{e_a\}$ be a positively oriented orthonormal basis for $\mathcal{V}$ and let $\{e^a\}$ be the basis for $\mathcal{V}^*$ given by $e^a = g(e_a)$. Show that for each $a_1, a_2, \ldots, a_p$ with $1 \leq a_1 < a_2 < \cdots < a_p \leq n$

$$*(e^{a_1} \wedge e^{a_2} \wedge \cdots \wedge e^{a_p}) = \epsilon e^{b_1} \wedge e^{b_2} \wedge \cdots \wedge e^{b_{n-p}}$$

where $\{b_1, b_2, \ldots, b_{n-p}\}$ is the subset of $\{1, 2, \ldots, n\}$ complementary to $\{a_1, a_2, \ldots, a_p\}$, $1 \leq b_1 < b_2 < \cdots < b_{n-p} \leq n$, and ϵ is the sign of the permutation taking $(1, 2, \ldots, n)$ to $(a_1, a_2, \ldots, a_p, b_1, b_2, \ldots, b_{n-p})$. □

It follows from Exercise 23 that the star operation is a linear isomorphism of the spaces $\bigwedge^p \mathcal{V}^*$ and $\bigwedge^{n-p} \mathcal{V}^*$ (which have the same dimension) since it maps a basis of one to a basis of the other.

If a decomposable p-form ω defines a volume p-form compatible with the scalar product on restriction to the orthogonal complement of its characteristic subspace, then $\omega \wedge *\omega = \Omega$, as follows from the preceding subsection, but in general this is not the case: for example if $\chi = k\omega$ for some $k \in \mathbf{R}$ then by linearity $\chi \wedge *\chi = k^2\Omega$. In general, if ω and χ are any two p-forms, then $\omega \wedge *\chi$ is an n-form and therefore a multiple of Ω. We set

$$\omega \wedge *\chi = g(\omega, \chi)\Omega;$$

then g is a bilinear form on $\bigwedge^p \mathcal{V}^*$. We shall show that it is a Euclidean scalar product, and reduces to g^* on $\bigwedge^1 \mathcal{V}^* = \mathcal{V}^*$.

If $\{e_a\}$ is a positively oriented orthonormal basis for $\mathcal{V}$ then $\{e^a\}$ is a basis for $\mathcal{V}^*$ orthonormal with respect to g^*, and $\{\, e^{a_1} \wedge e^{a_2} \wedge \cdots \wedge e^{a_p} \mid 1 \leq a_1 < a_2 < \cdots < a_p \leq n \,\}$ is a basis for $\bigwedge^p \mathcal{V}^*$. Now if ω is any one of these basis p-forms then $g(\omega, \omega) = 1$, while if ω and χ are distinct basis p-forms then $*\chi$ has at least one of the e_a in common with ω (by Exercise 23), and so $\omega \wedge *\chi = 0$, which means that $g(\omega, \chi) = 0$. Thus the bilinear form g takes the standard Euclidean form with respect to this basis and is therefore a Euclidean scalar product on $\bigwedge^p \mathcal{V}^*$ for each p, with the given basis as an orthonormal basis. In the case $p = 1$ the basis $\{e_a\}$ is thus an orthonormal basis for g, as it is for g^*, and the two therefore coincide. We shall therefore use g henceforth to denote the scalar product on $\mathcal{V}^*$ as well as $\mathcal{V}$ and all the other exterior product spaces.

Exercise 24. Show that in a 3-dimensional vector space $\mathcal{V}$, with Euclidean scalar product and the usual orientation, the map $\mathcal{V} \times \mathcal{V} \to \mathcal{V}$ by $(\mathbf{v}, \mathbf{w}) \mapsto g^{-1}(*(g(\mathbf{v}) \wedge g(\mathbf{w})))$ is the vector product. Show that the vector product of two vectors is an axial vector, that is, changes sign under a change of orientation of $\mathcal{V}$. □

Exercise 25. Show that in a 3-dimensional Euclidean space

$$g(*\omega, *\chi) = g(\omega, \chi)$$

for all p-forms ω, χ, where $p = 1$ or 2, and that for any two decomposable 2-forms $\eta^1 \wedge \eta^2$, $\varsigma^1 \wedge \varsigma^2$,

$$g(\eta^1 \wedge \eta^2, \varsigma^1 \wedge \varsigma^2) = g(\eta^1, \varsigma^1)g(\eta^2, \varsigma^2) - g(\eta^1, \varsigma^2)g(\eta^2, \varsigma^1).$$

Hence show that for any vectors $\mathbf{v}_1, \mathbf{v}_2, \mathbf{w}_1, \mathbf{w}_2$,

$$(\mathbf{v}_1 \times \mathbf{v}_2) \cdot (\mathbf{w}_1 \times \mathbf{w}_2) = (\mathbf{v}_1 \cdot \mathbf{w}_1)(\mathbf{v}_2 \cdot \mathbf{w}_2) - (\mathbf{v}_1 \cdot \mathbf{w}_2)(\mathbf{v}_2 \cdot \mathbf{w}_1).$$ □

Exercise 26. Show that for fixed linear forms $\varsigma^1, \varsigma^2, \ldots, \varsigma^p$ the map $\mathcal{V}^{*p} \to \mathbf{R}$ by $(\eta^1, \eta^2, \ldots, \eta^p) \mapsto \det(g(\eta^\alpha, \varsigma^\beta))$, $\alpha, \beta = 1, 2, \ldots, p$, is multilinear and alternating and therefore extends to a linear form on $\bigwedge^p \mathcal{V}^*$. Deduce that there is a unique symmetric bilinear form on $\bigwedge^p \mathcal{V}^*$ whose value on the decomposable p-forms $\eta^1 \wedge \eta^2 \wedge \cdots \wedge \eta^p$ and

$\varsigma^1 \wedge \varsigma^2 \wedge \cdots \wedge \varsigma^p$ is $\det(g(\eta^\alpha, \varsigma^\beta))$, and by considering its values when η^α and ς^β are chosen from the 1-forms $\{e^a\}$ of an orthogonal basis for $\mathcal{V}^*$ show that this bilinear form is actually g. □

The star operator for indefinite signature. The star operator may also be defined for a scalar product with indefinite signature. In this case intuition is not so reliable a guide to results, so we follow the algebra developed for the Euclidean case in order to arrive at the definitions. The number of minus signs in the signature of the scalar product will play an important role; we denote it s. It is convenient in the course of the construction to choose bases such that the spacelike vectors (those of negative "squared length") take the first s places. The details of the construction are left to the reader, in the following exercises. After the second of these one of the two volume forms determined by the scalar product, and thus an orientation, is chosen for the remainder of the construction.

Exercise 27. Let $\mathcal{V}$ be an n-dimensional vector space with scalar product g. Let $\{e_a\}$ be an orthonormal basis for $\mathcal{V}$ such that $g(e_a, e_a) = -1$ for $1 \leq a \leq s$. Show that $\{e^a\}$, where $e^a = g(e_a)$, is an orthonormal basis for $\mathcal{V}^*$ such that $g(e^a, e^a) = -1$ for $1 \leq a \leq s$, and that $\{\theta^a\} = \{-e^1, -e^2, \ldots, -e^s, e^{s+1}, \ldots, e^n\}$ is the basis dual to $\{e_a\}$. □

Exercise 28. There are two volume forms on $\mathcal{V}$ determined by g. Show that they are given by $\pm\Omega$ where $\Omega = \theta^1 \wedge \theta^2 \wedge \cdots \wedge \theta^n = (-1)^s e^1 \wedge e^2 \wedge \cdots \wedge e^n$, $\{e_a\}$ being a chosen orthonormal basis and $\{\theta^a\}$ the dual basis. Let $\{f_a\}$ be an arbitrary basis for $\mathcal{V}$ with dual basis $\{\iota^a\}$. Show that $\Omega = \pm\sqrt{(-1)^s \det(g_{ab})}\,\iota^1 \wedge \iota^2 \wedge \cdots \wedge \iota^n$ where $g_{ab} = g(f_a, f_b)$. □

Exercise 29. Show that for each p there is a unique symmetric bilinear form g on $\bigwedge^p \mathcal{V}^*$ such that if $\omega = \eta^1 \wedge \eta^2 \wedge \cdots \wedge \eta^p$ and $\chi = \varsigma^1 \wedge \varsigma^2 \wedge \cdots \wedge \varsigma^p$ are decomposable p-forms then $g(\omega, \chi) = \det(g(\eta^\alpha, \varsigma^\beta))$. Show that the standard basis for $\bigwedge^p \mathcal{V}^*$ constructed out of an orthonormal basis for $\mathcal{V}^*$ is orthonormal with respect to g, and infer that g is a scalar product on $\bigwedge^p \mathcal{V}^*$. Show that on $\mathcal{V}^* = \bigwedge^1 \mathcal{V}^*$, $g = g^*$, and that $g(\Omega, \Omega) = (-1)^s$. □

Exercise 30. Let ω be any chosen element of $\bigwedge^p \mathcal{V}^*$ and χ any chosen element of $\bigwedge^{n-p} \mathcal{V}^*$: then $\omega \wedge \chi$ is a multiple of Ω. Writing $\omega \wedge \chi = \gamma_\omega(\chi)\Omega$, show that γ_ω is a linear form on $\bigwedge^{n-p} \mathcal{V}^*$ and from the fact that g is a scalar product infer that there is an element $*\omega$ of $\bigwedge^{n-p} \mathcal{V}^*$ such that $\gamma_\omega(\chi) = g(*\omega, \chi)$. Show that the map $\bigwedge^p \mathcal{V}^* \to \bigwedge^{n-p} \mathcal{V}^*$ by $\omega \mapsto *\omega$ is a linear map. □

Exercise 31. Let $\{e_a\}$ be a positively oriented orthonormal basis for $\mathcal{V}$ such that $g(e_a, e_a) = -1$ for $1 \leq a \leq s$. Show that if $e^a = g(e_a)$ then for $1 \leq a_1 < a_2 < \cdots < a_p \leq n$

$$*(e^{a_1} \wedge e^{a_2} \wedge \cdots \wedge e^{a_p}) = \epsilon e^{b_1} \wedge e^{b_2} \wedge \cdots \wedge e^{b_{n-p}}$$

where $(a_1, a_2, \ldots, a_p, b_1, b_2, \ldots, b_{n-p})$ is chosen to be an even permutation of $(1, 2, \ldots, n)$, and $\epsilon = +1$ if there is an even number of a's not exceeding s and -1 otherwise. □

Exercise 32. Let $\omega \in \bigwedge^p \mathcal{V}^*$. Show that $**\omega = (-1)^{p(n-p)+s}\omega$. □

Exercise 33. Show that $g(*\omega, *\chi) = (-1)^s g(\omega, \chi)$, for any two forms of the same degree. □

Exercise 34. Let $\mathcal{V}$ be a 4-dimensional vector space with Lorentzian scalar product g and positively oriented orthonormal basis $\{e_a\}$ such that $g(e_1, e_1) = g(e_2, e_2) = g(e_3, e_3) = -g(e_4, e_4) = -1$. Show that

$$*(e^2 \wedge e^3) = e^1 \wedge e^4 \qquad *(e^3 \wedge e^1) = e^2 \wedge e^4 \qquad *(e^1 \wedge e^2) = e^3 \wedge e^4$$
$$*(e^1 \wedge e^4) = -e^2 \wedge e^3 \qquad *(e^2 \wedge e^4) = -e^3 \wedge e^1 \qquad *(e^3 \wedge e^4) = -e^1 \wedge e^2$$

while

$$*(\theta^2 \wedge \theta^3) = -\theta^1 \wedge \theta^4 \quad *(\theta^3 \wedge \theta^1) = -\theta^2 \wedge \theta^4 \quad *(\theta^1 \wedge \theta^2) = -\theta^3 \wedge \theta^4$$
$$*(\theta^1 \wedge \theta^4) = \theta^2 \wedge \theta^3 \quad *(\theta^2 \wedge \theta^4) = \theta^3 \wedge \theta^1 \quad *(\theta^3 \wedge \theta^4) = \theta^1 \wedge \theta^2$$

for the dual basis $\{\theta^a\}$. □

Exercise 35. Let $\mathcal{V}$ be a 4-dimensional vector space with Lorentzian scalar product. Find the signature of the induced scalar product on $\bigwedge^2 \mathcal{V}^*$. Let ω be a decomposable 2-form. Find out how the sign of $g(\omega,\omega)$ depends on the nature of the characteristic 2-space of ω, as set out in Exercise 11. □

5. Metrics on Affine Spaces

The length of a curve in ordinary 3-dimensional Euclidean space is given by the integral (between appropriate limits) of the length of its tangent vector. Thus the computation of length of a curve uses the Euclidean scalar product in the underlying vector space *via* its identification with the tangent space at each point. One may use the same method to transfer a scalar product on any vector space to an affine space modelled on it. One obtains in this way a scalar product on each tangent space to the affine space. Such a field of scalar products is called a *metric* on the affine space (whatever its signature). The metric constructed in this way is of a rather special kind, being in a sense constant; however we shall be concerned only with such metrics in this chapter, and call them (where distinction is necessary) *affine metrics*. An affine space with affine metric is called an *affine metric space* and denoted $\mathcal{E}^{r,n-r}$, where $(r, n-r)$ is the signature of the scalar product defining the metric. An affine metric space with positive definite metric is called a *Euclidean space*, whatever its dimension n, and is denoted $\mathcal{E}^n$. We shall write g for the metric as for the scalar product on which it is based.

In an affine coordinate system the components of g with respect to the coordinate vectors, given by

$$g_{ab} = g(\partial_a, \partial_b),$$

are constants; they satisfy the symmetry condition $g_{ba} = g_{ab}$. If the affine coordinates are based on an orthonormal basis for the underlying vector space then (g_{ab}) takes the standard form appropriate to the signature. If, however, the coordinate system is not affine the components of the metric will not be constants, but in general will be functions on the coordinate patch. It is usual to express the metric in the form

$$g = g_{ab} dx^a dx^b :$$

this should be taken to mean that for any vectors v, w,

$$g(v, w) = g_{ab}\langle v, dx^a\rangle\langle w, dx^b\rangle = g_{ab} v^a w^b,$$

where $v = v^a \partial_a$, $w = w^a \partial_a$.

Exercise 36. Show that the components g_{ab}, $\acute{g}_{ab}$ of g with respect to two systems of coordinates (x^a), $(\acute{x}^a)$ are related by

$$g_{ab} = \frac{\partial \acute{x}^c}{\partial x^a}\frac{\partial \acute{x}^d}{\partial x^b}\acute{g}_{cd}.$$ □

Exercise 37. Show that if $\{U_a\}$ is any (local) basis of vector fields, with dual basis of 1-forms $\{\theta^a\}$, then $g = g_{ab}\theta^a\theta^b$ where $g_{ab} = g(U_a, U_b)$: this means that for any vector fields V, W, $g(V,W) = g_{ab}\langle V,\theta^a\rangle\langle W,\theta^b\rangle$. □

The result of Exercise 36 could be written in the form $\acute{g}_{ab}d\acute{x}^a d\acute{x}^b = g_{ab}dx^a dx^b$. In order to find out how to express g in terms of curvilinear coordinates, therefore, it is enough to express the differentials of the orthonormal affine coordinates (x^a) in terms of those of the curvilinear coordinates, which may easily be done from the coordinate transformation equations, and substitute them in the expression

$$(dx^1)^2 + (dx^2)^2 + \cdots + (dx^r)^2 - (dx^{r+1})^2 - \cdots - (dx^n)^2 = g_{ab}dx^a dx^b,$$

employing ordinary algebra to simplify the result (which is valid because of the symmetry of g_{ab}). Thus if (in $\mathcal{E}^2$)

$$x^1 = r\cos\vartheta \qquad x^2 = r\sin\vartheta$$

(polar coordinates) then

$$\begin{aligned}(dx^1)^2 + (dx^2)^2 &= (\cos\vartheta\, dr - r\sin\vartheta\, d\vartheta)^2 + (\sin\vartheta\, dr + r\cos\vartheta\, d\vartheta)^2 \\ &= dr^2 + r^2 d\vartheta^2.\end{aligned}$$

Exercise 38. Show that if

$$x^1 = r\sin\vartheta\cos\varphi \qquad x^2 = r\sin\vartheta\sin\varphi \qquad x^3 = r\cos\vartheta$$

(spherical polar coordinates in $\mathcal{E}^3$) then

$$(dx^1)^2 + (dx^2)^2 + (dx^3)^3 = dr^2 + r^2 d\vartheta^2 + r^2\sin^2\vartheta\, d\varphi^2.$$ □

From such an expression for g one reads off the lengths of, and angles between, the coordinate vector fields of the curvilinear coordinate system. Thus, in polar coordinates in $\mathcal{E}^2$, $g(\partial/\partial r, \partial/\partial r) = 1$, $g(\partial/\partial\vartheta, \partial/\partial\vartheta) = r^2$, $g(\partial/\partial r, \partial/\partial\vartheta) = 0$. Coordinate systems like this in which the coordinate vector fields are orthogonal but not necessarily unit vector fields (and so (g_{ab}) is diagonal but does not necessarily have ± 1 for its diagonal elements) are called *orthogonal coordinates*.

Exercise 39. Show that in spherical polar coordinates the 1-forms $\{dr, r\,d\vartheta, r\sin\vartheta\, d\varphi\}$ constitute an orthonormal basis for $T_x^*\mathcal{E}^3$ at each point x. □

Exercise 40. Generalised spherical polar coordinates for $\mathcal{E}^n$ are defined in terms of orthonormal affine coordinates by

$$\begin{aligned}x^1 &= r\cos\vartheta^1 \\ x^2 &= r\sin\vartheta^1\cos\vartheta^2 \\ x^3 &= r\sin\vartheta^1\sin\vartheta^2\cos\vartheta^3 \\ &\cdots \\ x^{n-1} &= r\sin\vartheta^1\sin\vartheta^2\cdots\sin\vartheta^{n-2}\cos\vartheta^{n-1} \\ x^n &= r\sin\vartheta^1\sin\vartheta^2\cdots\sin\vartheta^{n-2}\sin\vartheta^{n-1}.\end{aligned}$$

Show that generalised spherical polars are orthogonal and that $\partial/\partial r$ is a unit vector field. □

Lengths of curves. The length of a curve σ, between $\sigma(t_1)$ and $\sigma(t_2)$ (where $t_2 \geq t_1$), in a Euclidean space $\mathcal{E}^n$, is defined to be

$$\int_{t_1}^{t_2} \sqrt{g(\dot\sigma(t), \dot\sigma(t))}\, dt.$$

Exercise 41. Show that length is independent of parametrisation, and is unaffected by a change of parameter which reverses the orientation of the curve. □

Thus length is really a function of the path rather than the curve. Provided that $t_2 > t_1$ and that σ is not a constant curve its length is strictly positive. Moreover, if $t_3 > t_2 > t_1$,

$$\int_{t_1}^{t_3} \sqrt{g(\dot\sigma(t),\dot\sigma(t))}\,dt = \int_{t_1}^{t_2} \sqrt{g(\dot\sigma(t),\dot\sigma(t))}\,dt + \int_{t_2}^{t_3} \sqrt{g(\dot\sigma(t),\dot\sigma(t))}\,dt$$

so that length is additive in the expected way. If one defines $s\colon \mathbf{R} \to \mathbf{R}$ by

$$s(t) = \begin{cases} \displaystyle\int_{t_0}^{t} \sqrt{g(\dot\sigma(t),\dot\sigma(t))}\,dt & \text{for } t \geq t_0 \\ \displaystyle -\int_{t}^{t_0} \sqrt{g(\dot\sigma(t),\dot\sigma(t))}\,dt & \text{for } t < t_0 \end{cases}$$

then s is a smooth increasing function, called the *arc length* function of the curve, with initial point $\sigma(t_0)$.

Exercise 42. Show that the arc length functions (of the same curve) with different initial points differ only by a change of origin. □

Reparametrisation of the curve by means of its arc-length function yields a curve with the property that its tangent vector is always a unit vector, since it now satisfies

$$\int_0^s \sqrt{g(\dot\sigma(s),\dot\sigma(s))}\,ds = s \qquad \text{for all } s.$$

This relation between the metric and arc length is the reason for the use of the expression $ds^2 = g_{ab}dx^a dx^b$ for a metric which is often found in tensor calculus books.

The concept of length extends easily to a curve with a finite number of corners, or piecewise-smooth curve: a curve σ is *piecewise-smooth* on a closed interval $[a,b]$ of its domain if it is continuous, and if $[a,b]$ can be subdivided into a finite number of consecutive subintervals $[a,t_1]$, $[t_1,t_2]$, $[t_2,t_3],\ldots,[t_{n-1},t_n]$, $[t_n,b]$, on each of which it is smooth. The length of σ over $[a,b]$ is just the sum of its lengths over the subintervals.

A unit tangent vector u at x_0 determines a line, parametrised by arc length, by $s \mapsto x_0 + su$. Suppose that x_0, x_1 and x_2 are the vertices of a triangle, and that $x_1 = x_0 + s_1u_1$, $x_2 = x_0 + su = x_1 + s_2u_2$, where u, u_1 and u_2 are unit vectors. Then by the triangle inequality (Exercise 3) applied to s_1u_1, s_2u_2 and $su = s_1u_1 + s_2u_2$, $s \leq s_1 + s_2$ with equality only when x_0, x_1 and x_2 are collinear with x_1 between x_0 and x_2.

This familiar result about the shortest distance between two points generalises to cover curves, not just broken line segments. Given a point x_0 in $\mathcal{E}^n$, the function r defined as the Euclidean length of the displacement from x_0 to x, or the length of the line joining x_0 to x, is smooth except at x_0. It is the first coordinate function of generalised spherical polar coordinates based on x_0. If σ is any smooth curve parametrised by arc length joining x_0 to x, with $\sigma(0) = x_0$ and $\sigma(S) = x$, so that

S is the length of the segment of σ from x_0 to x, then by the Schwartz inequality (Exercise 3)

$$\sqrt{g(\dot\sigma(s),\dot\sigma(s))} \geq \left| g\left(\dot\sigma(s), \frac{\partial}{\partial r}\right)\right| \geq g\left(\dot\sigma(s), \frac{\partial}{\partial r}\right)$$

since $g(\partial/\partial r, \partial/\partial r) = 1$. But

$$g\left(\dot\sigma(s), \frac{\partial}{\partial r}\right) = \langle \dot\sigma(s), dr\rangle$$

since generalised spherical polars are orthogonal and $\partial/\partial r$ is unit. Thus

$$S = \int_0^S \sqrt{g(\dot\sigma(s),\dot\sigma(s))}\,ds \geq \int_0^S \langle\dot\sigma(s), dr\rangle ds = \int_0^S \frac{d}{ds}(r\circ\sigma)ds = r(\sigma(S)) = r(x).$$

That is to say, the length S of a smooth curve joining x_0 to x cannot be less than the length $r(x)$ of the straight line joining x_0 to x. Moreover, since equality holds in the Schwartz inequality only if the arguments are linearly dependent, the curve σ can have length $r(x)$ only if it coincides with the straight line. Thus of all smooth curves joining two points it is the straight line which has the shortest length.

We have devoted space to this well-known fact because the corresponding result in Lorentzian geometry, to which we now turn, may be a little unexpected.

Tangent vectors v, and displacements, in $\mathcal{E}^{1,n-1}$, n-dimensional affine space equipped with a metric of Lorentzian signature, are called *timelike*, *null* or *spacelike* according as $g(v,v) > 0$, $= 0$ or < 0. Two points have timelike, null or spacelike separation according as the displacement vector between them is timelike, null or spacelike. At each point one may construct the *null cone*, which consists of all points whose separations from the given one (the vertex) are null. It is a hypersurface, smooth except at the vertex, containing all the lines through the vertex with null tangent vectors. It divides the space into three parts, two of which are the disjoint components of the set of points having timelike separation from the vertex, the other consisting of the points having spacelike separation from the vertex. A choice of component of the timelike-separated points to represent the future of the vertex, made for one point, may be consistently imposed all over the Lorentzian affine space by the rule that parallel displacement vectors, if timelike, are all future or all past pointing. A curve is called timelike, null or spacelike at a point according as its tangent vector there is timelike, null or spacelike.

The definition of arc length in Euclidean space does not carry over directly to a Lorentzian space. The most useful analogous concept is that of proper time along an everywhere timelike curve. Let σ be an everywhere timelike curve: *proper time* along σ from $\sigma(t_0)$ is the parameter τ defined by

$$\tau(t) = \begin{cases} \displaystyle\int_{t_0}^{t} \sqrt{g(\dot\sigma(t),\dot\sigma(t))}\,dt & \text{for } t \geq t_0 \\ \displaystyle -\int_{t}^{t_0} \sqrt{g(\dot\sigma(t),\dot\sigma(t))}\,dt & \text{for } t < t_0. \end{cases}$$

Evidently to attempt a similar definition for an everywhere null curve would be pointless, since the integrals involved would be always zero.

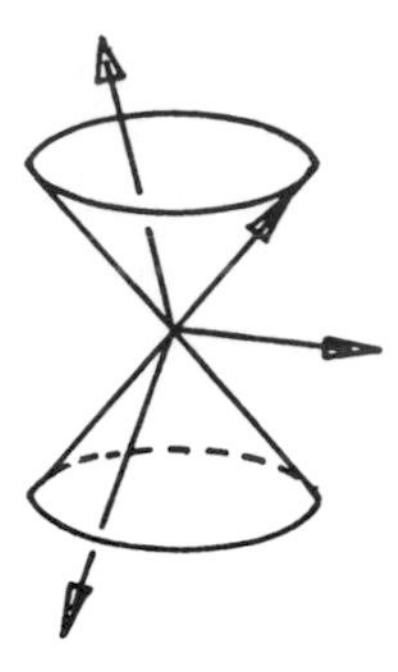

Fig. 1 A null cone.

Exercise 43. Show that in $\mathcal{E}^{1,3}$ with orthonormal affine coordinates, so that $g = -(dx^1)^2 - (dx^2)^2 - (dx^3)^2 + (dx^4)^2$, the helix $\sigma(t) = (a\cos\varpi t, a\sin\varpi t, 0, bt)$ will be timelike if $\varpi < b/a$, null if $\varpi = b/a$ and spacelike if $\varpi > b/a$. □

Exercise 44. Coordinates $(\kappa, \vartheta, \varphi, \varrho)$ are defined with respect to orthonormal affine coordinates (x^1, x^2, x^3, x^4) for $\mathcal{E}^{1,3}$ by

$$x^1 = \varrho\sinh\kappa\cos\vartheta \qquad x^2 = \varrho\sinh\kappa\sin\vartheta\cos\varphi$$
$$x^3 = \varrho\sinh\kappa\sin\vartheta\sin\varphi \qquad x^4 = \varrho\cosh\kappa.$$

Show that they provide, for $\varrho > 0$, coordinates for the future of the origin of the affine coordinates; and that $\varrho(x)$ is the proper time of the point x from the origin along the straight line joining them. Show that

$$-(dx^1)^2 - (dx^2)^2 - (dx^3)^2 + (dx^4)^2$$
$$= -\varrho^2 d\kappa^2 - \varrho^2\sinh^2\kappa d\vartheta^2 - \varrho^2\sinh^2\kappa\sin^2\vartheta d\varphi^2 + d\varrho^2. \quad □$$

If σ is an everywhere timelike curve which is parametrised by proper time, such that $\sigma(0) = x_0$ and $\sigma(T) = x$ (so that T is the proper time from x_0 to x measured along the curve) then by using the coordinates introduced in Exercise 44 and adapting the argument given earlier for Euclidean arc length it is easy to show that $T \leq \varrho(x)$, with equality only if σ is actually the straight line joining x_0 and x. Thus for timelike curves in Lorentzian geometry, the straight line joining two points with timelike separation is the longest (in the sense of proper time) of all timelike curves joining the two points. The reason for this reversal from the Euclidean case is that by introducing more wiggles into a timelike curve one tends to make it more nearly null and therefore to decrease the proper time along it.

Exercise 45. Show that there is a piecewise linear null curve joining two points of timelike separation (two pieces suffice) . Show that one may therefore construct a timelike curve (even a smooth one) of arbitrarily small proper time joining the two points. □

Exercise 46. Show that if x_0, x_1, x_2 are points in $\mathcal{E}^{1,3}$ with x_1 lying to the future of x_0 and x_2 to the future of x_1 (and therefore also to the future of x_0), and if $\tau(x_i, x_j)$ represents the proper time from x_i to x_j along the straight line joining them, then

$$\tau(x_0, x_2) \geq \tau(x_0, x_1) + \tau(x_1, x_2),$$

with equality only if the points are collinear. □

6. Parallelism in Affine Metric Space

As we explained in Chapter 2, the notion of parallelism of tangent vectors at different points in an affine space is independent of any metric structure. However, parallelism does respect any affine metric on the space in the sense that lengths of, and angles between (or more generally, scalar products of) parallel vectors are the same. It follows that if V and W are parallel vector fields along a curve σ then the function $g(V,W)$ (on the domain of σ) is constant:

$$\frac{d}{dt}\big(g(V,W)\big) = 0.$$

More generally, if V and W are any vector fields on σ, not necessarily parallel, then

$$\begin{aligned} g\big(&V(t+h),W(t+h)\big) - g\big(V(t),W(t)\big)\\ &= g\big(V(t+h)_{\|},W(t+h)_{\|}\big) - g\big(V(t),W(t)\big)\\ &= g\big(V(t+h)_{\|} - V(t),W(t+h)_{\|}\big) + g\big(V(t),W(t+h)_{\|} - W(t)\big)\end{aligned}$$

where $V(t+h)_{\|}$ is the vector at $\sigma(t)$ parallel to the vector $V(t+h)$. It follows that

$$\frac{d}{dt}\big(g(V,W)\big) = g\left(\frac{DV}{Dt},W\right) + g\left(V,\frac{DW}{Dt}\right).$$

Exercise 47. Show that if U, V, W are vector fields (not restricted to a curve) then

$$U(g(V,W)) = g(\nabla_U V,W) + g(V,\nabla_U W).$$ □

We defined the connection components Γ^a_{bc} for a (curvilinear) coordinate system by

$$\Gamma^a_{bc}\partial_a = \nabla_{\partial_c}\partial_b.$$

We can derive from the result of Exercise 47 an important relationship between the connection coefficients and the components of the metric with respect to the coordinate vector fields: taking for U, V, W in the formula in the exercise the coordinate fields ∂_a, ∂_b, ∂_c we obtain

$$\begin{aligned}\partial_a g(\partial_b,\partial_c) &= \frac{\partial g_{bc}}{\partial x^a}\\ &= g(\nabla_{\partial_a}\partial_b,\partial_c) + g(\partial_b,\nabla_{\partial_a}\partial_c)\\ &= g(\Gamma^d_{ba}\partial_d,\partial_c) + g(\partial_b,\Gamma^d_{ca}\partial_d)\\ &= g_{cd}\Gamma^d_{ba} + g_{bd}\Gamma^d_{ca}.\end{aligned}$$

Two similar formulae are obtained by cyclically permuting the indices a,b,c:

$$\begin{aligned}\frac{\partial g_{ca}}{\partial x^b} &= g_{ad}\Gamma^d_{cb} + g_{cd}\Gamma^d_{ab}\\ \frac{\partial g_{ab}}{\partial x^c} &= g_{bd}\Gamma^d_{ac} + g_{ad}\Gamma^d_{bc}.\end{aligned}$$

Notice that in view of the symmetry of Γ^a_{bc} in its lower two indices, each term appearing on the right hand side of one of these equations has a matching term in one of the others. By adding the first two and subtracting the last, therefore, they may be solved for $g_{cd}\Gamma^d_{ab}$:

$$g_{cd}\Gamma^d_{ab} = \frac{1}{2}\left(\frac{\partial g_{bc}}{\partial x^a} + \frac{\partial g_{ac}}{\partial x^b} - \frac{\partial g_{ab}}{\partial x^c}\right)$$

and therefore, pre-multiplying by g^{ce}, using $g^{ce}g_{cd} = \delta^e_d$, and relabelling indices, one obtains

$$\Gamma^c_{ab} = \frac{1}{2}g^{cd}\left(\frac{\partial g_{bd}}{\partial x^a} + \frac{\partial g_{ad}}{\partial x^b} - \frac{\partial g_{ab}}{\partial x^d}\right).$$

Thus Γ^c_{ab} is determined by g_{ab} and its derivatives.

Exercise 48. Compute the connection coefficients in polar and spherical polar coordinates. □

Exercise 49. Show that $\Gamma^b_{ab} = \frac{1}{2}\partial_a(\log|\det(g_{cd})|)$. □

Exercise 50. By writing down the two other equations obtained from the one in Exercise 47 by cyclic permutation of U, V, W, show that for any three vector fields

$$2g(\nabla_U V, W) = U(g(V,W)) + V(g(U,W)) - W(g(U,V)) \\ + g([U,V],W) - g([U,W],V) - g(U,[V,W]).$$ □

Exercise 51. Given any (local) basis $\{V_a\}$ of vector fields, not necessarily coordinate fields, so that $[V_a, V_b]$ is not necessarily zero, one may define $g_{ab} = g(V_a, V_b)$ and $\gamma^c_{ab}V_c = \nabla_{V_b}V_a$. Show that

$$\gamma^c_{ab} = \tfrac{1}{2}g^{cd}(V_a g_{bd} + V_b g_{ad} - V_d g_{ab} - f^e_{ad}g_{be} - f^e_{bd}g_{ae} - f^e_{ab}g_{de})$$

where $[V_a, V_b] = f^c_{ab}V_c$. In particular, if $\{V_a\}$ is orthonormal and the space is Euclidean, then

$$\gamma^c_{ab} = -\tfrac{1}{2}\delta^{cd}(f^e_{ad}\delta_{be} + f^e_{bd}\delta_{ae} + f^e_{ab}\delta_{de})$$

where (δ_{ab}) is the unit matrix; and if $\gamma_{abc} = \gamma^d_{ab}\delta_{cd}$ (so that γ_{abc} and γ^c_{ab} have the same values for each a, b, c) and if, likewise, $f_{abc} = f^d_{ab}\delta_{cd}$, then $\gamma_{abc} = -\frac{1}{2}(f_{abc} + f_{acb} + f_{bca})$. □

Exercise 52. Show that the connection forms ω^b_a for an orthonormal basis $\{V_a\}$ of vector fields on a Euclidean space can be thought of as elements of a skew-symmetric matrix of 1-forms, in the sense that $\omega^c_a\delta_{bc} + \omega^c_b\delta_{ac} = 0$. □

7. Vector Calculus

In the presence of an affine metric the operations of the exterior calculus may be applied to vector fields through the medium of raising and lowering indices; the result in $\mathcal{E}^3$ is to give the familiar operations grad, div and curl.

The gradient. The gradient of a function f is the vector field obtained by using the metric to raise the index on df: thus

$$\operatorname{grad} f = g^{-1}(df).$$

Exercise 53. Show that $g(\operatorname{grad} f, V) = Vf$ for every vector field V. Deduce that, in particular, if V is tangent to the level surfaces of f, then $\operatorname{grad} f$ is orthogonal to V; thus $\operatorname{grad} f$ is orthogonal to the level surfaces of f. □

In coordinates,

$$\operatorname{grad} f = g^{ab}\frac{\partial f}{\partial x^a}\frac{\partial}{\partial x^b}.$$

Thus in orthonormal affine coordinates in $\mathcal{E}^n$ the components of grad f are equal to those of df, and consequently the distinction between grad f and df is rarely made in ordinary vector calculus. The distinction becomes necessary, however, even using orthonormal affine coordinates, if the affine metric is non-Euclidean: in $\mathcal{E}^{1,3}$, for example,

$$\operatorname{grad} f = -\frac{\partial f}{\partial x^1}\frac{\partial}{\partial x^1} - \frac{\partial f}{\partial x^2}\frac{\partial}{\partial x^2} - \frac{\partial f}{\partial x^3}\frac{\partial}{\partial x^3} + \frac{\partial f}{\partial x^4}\frac{\partial}{\partial x^4}.$$

In curvilinear coordinates the components of grad f look quite different from those of df because of the effect of the g^{ab}. Thus in polar coordinates in $\mathcal{E}^2$

$$df = \frac{\partial f}{\partial r}dr + \frac{\partial f}{\partial \vartheta}d\vartheta$$

while

$$\operatorname{grad} f = \frac{\partial f}{\partial r}\frac{\partial}{\partial r} + \frac{1}{r^2}\frac{\partial f}{\partial \vartheta}\frac{\partial}{\partial \vartheta}.$$

Exercise 54. Compute the coefficients of grad f in spherical polar coordinates (the metric is given in Exercise 38). □

Exercise 55. Show that with respect to an orthogonal coordinate system $(\operatorname{grad} f)^a = (g_{aa})^{-1}\partial_a f$ (no sum). □

In the case of a pseudo-Euclidean affine metric space the restriction of the tangent space scalar product to the tangent hyperplane of a surface may be positive or negative definite, indefinite but non-degenerate, or even degenerate. In the case of a Lorentzian affine metric space a tangent hyperplane will be called spacelike, timelike or null according as the restriction of the scalar product is negative definite, Lorentzian, or degenerate, as in Section 2. A surface all of whose tangent hyperplanes are of the same type will be said to be of that type: thus a null surface is one whose tangent hyperplanes are all null. The type of a level surface of a function f may be deduced from that of its normal grad f (considered as a vector field over the level surface), and since $g(\operatorname{grad} f, \operatorname{grad} f) = g(df, df)$ the type of the level surface is given by the scalar product of df with itself. Furthermore, $g(\operatorname{grad} f, \operatorname{grad} f) = \langle \operatorname{grad} f, df \rangle = \operatorname{grad} f(f)$, so that in particular a level surface of f is null if and only if grad f is everywhere tangent to it.

The divergence. In Chapter 5 we defined the divergence of a vector field V with respect to a volume n-form Ω by

$$\mathcal{L}_V\Omega = (\operatorname{div}_\Omega V)\Omega,$$

using the fact that the Lie derivative of Ω, being an n-form, must be a multiple of Ω. In the case of an affine metric space the divergence is fixed by the choice of Ω to be a volume form derived from the metric. It does not matter which of the two is chosen, as is clear from the defining equation: thus divergence is unaffected by a change of orientation. With respect to orthonormal affine coordinates $\Omega = \pm dx^1 \wedge dx^2 \wedge \cdots \wedge dx^n$, and from the relation $\mathcal{L}_V\Omega = d(V \lrcorner \Omega)$ one finds that if $V = V^a\partial_a$ then $\operatorname{div} V = \partial V^a/\partial x^a$ (summation intended). Since we now deal with a standard volume (strictly speaking, one or other of the two standard volumes) we drop the practice of subscripting the symbol div to show which volume is in use.

Exercise 56. Using the fact that in curvilinear coordinates (Exercise 28)

$$\Omega = \pm\sqrt{(-1)^s \det(g_{ab})}dx^1 \wedge dx^2 \wedge \cdots \wedge dx^n$$

show that

$$\operatorname{div} V = \frac{1}{\sqrt{(-1)^s \det(g_{bc})}} \frac{\partial}{\partial x^a}\left(\sqrt{(-1)^s \det(g_{bc})}V^a\right). \qquad \square$$

The divergence may also be expressed in terms of the star operator. Given any vector field V, the metric may be used to construct a 1-form $g(V)$ by lowering the index; then $*g(V)$ is an $(n-1)$-form, whose exterior derivative $d{*g}(V)$ is an n-form and therefore a multiple of Ω (we assume now that one particular volume form is chosen).

Exercise 57. Show that $*g(V) = V \lrcorner \Omega$ and thus that $d{*g}(V) = (\operatorname{div} V)\Omega$. $\square$

If we extend the star map to maps $\bigwedge^0 \mathcal{V}^* \to \bigwedge^n \mathcal{V}^*$ and $\bigwedge^n \mathcal{V}^* \to \bigwedge^0 \mathcal{V}^*$ by $*1 = (-1)^s\Omega$ and $*\Omega = 1$ (the occurrence of the factor $(-1)^s$ is required by the fact that $g(\Omega,\Omega) = (-1)^s$: Exercise 29) then

$$\operatorname{div} V = *d{*g}(V).$$

(Note that choosing the opposite orientation changes the sign of the star map, but not that of $\operatorname{div} V$, which confirms an earlier remark).

The divergence of a vector field may also be expressed in terms of its covariant derivative. Given a vector field V, the map of tangent spaces defined by $w \mapsto \nabla_w V$ is a linear one, which we write ∇V. In affine coordinates the matrix of ∇V with respect to the coordinate vectors is just $(\partial_b V^a)$ and so $\operatorname{div} V$ is its trace.

Exercise 58. Show that in general curvilinear coordinates the matrix representation of ∇V has entries $\partial_b V^a + \Gamma^a_{cb}V^c$ and so by using the formula $\Gamma^a_{ba} = \frac{1}{2}\partial_b(\log|\det(g_{cd})|)$ (Exercise 49) confirm that the trace of ∇V is given by the formula for $\operatorname{div} V$ in Exercise 56. $\square$

Exercise 59. Let $\{e_a\}$ be a fixed orthonormal basis for the underlying vector space of an affine metric space and let $\{E_a\}$ be the corresponding orthonormal basis of parallel vector fields on the affine space. Show that for any vector field V

$$\mathcal{L}_V\Omega(E_1, E_2, \ldots, E_n) = \sum_{c=1}^{n} \Omega(E_1, E_2, \ldots, \nabla_{E_c}V, \ldots, E_n)$$

and establish the connection between the Lie derivative and covariant derivative definitions of div. (Exercise 13 of Chapter 4 is relevant.) $\square$

The curl. Unlike grad and div, curl in its usual form can be defined only in $\mathcal{E}^3$.

We gave the definition of the vector product of two vectors in $\mathcal{E}^3$ in terms of the star operator, the raising and lowering of indices, and the exterior product, in Exercise 24. The definition of curl follows the same route. Given a vector field $\mathbf{V}$ in $\mathcal{E}^3$ one constructs the 1-form $g(\mathbf{V})$; then $dg(\mathbf{V})$ is a 2-form; $*dg(\mathbf{V})$ is a 1-form; and $g^{-1}(*dg(\mathbf{V}))$ is a vector field.

Exercise 60. Show that $g^{-1}(*dg(\mathbf{V})) = \operatorname{curl} \mathbf{V}$ by evaluating it in orthonormal affine coordinates. (The rules for the star operation on orthonormal basis vectors in $\mathcal{E}^3$ are given in Exercise 20.) $\square$

Since $\operatorname{curl} \mathbf{V}$ involves the star operation once only, it is an axial vector field: it changes sign when the orientation is reversed.

The special feature of $\mathcal{E}^3$—the fact that $1 = 3 - 2$!—enables one to define $\operatorname{curl}\mathbf{V}$ as a vector field. However, in any dimension, for any vector field V, one may construct the 2-form $dg(V)$, which serves as a generalised curl. This may be expressed in terms of ∇V as follows. We have defined ∇V as a linear map of tangent vectors. By composing this map with g, which lowers indices, we obtain a map of tangent to cotangent vectors, which may in turn be regarded as a bilinear form. This bilinear form is given by

$$(W_1, W_2) \mapsto g(\nabla_{W_1} V, W_2)$$

for any pair of vector fields W_1, W_2. It will not be symmetric, in general, but will have an alternating part given by

$$(W_1, W_2) \mapsto \tfrac{1}{2}\{g(\nabla_{W_1} V, W_2) - g(W_1, \nabla_{W_2} V)\}.$$

Now

$$\begin{aligned} g(\nabla_{W_1} V, W_2) &- g(W_1, \nabla_{W_2} V) \\ &= W_1(g(V, W_2)) - g(V, \nabla_{W_1} W_2) - W_2(g(W_1, V)) + g(\nabla_{W_2} W_1, V) \\ &= W_1(g(V, W_2)) - W_2(g(V, W_1)) - g(V, [W_1, W_2]) \\ &= W_1\langle W_2, g(V)\rangle - W_2\langle W_1, g(V)\rangle - \langle [W_1, W_2], g(V)\rangle \\ &= dg(V)(W_1, W_2). \end{aligned}$$

Thus $dg(V)$ is twice the alternating part of the bilinear form $(W_1, W_2) \mapsto g(\nabla_{W_1} V, W_2)$.

Orthogonal coordinate systems. In vector calculus in $\mathcal{E}^3$, frequent use is made of coordinate systems, such as cylindrical or spherical polars, which are orthogonal without being orthonormal. It is very convenient to have expressions for grad, div and curl with respect to such coordinates; we shall derive these expressions, probably already familiar to the reader, by using the methods developed in this section.

When using orthogonal coordinates (x^a) it is usual to employ an orthonormal basis of vector fields $\{\mathbf{V}_a\}$ derived by normalising the coordinate fields. Thus positive functions h_a are defined by

$$g(\partial_a, \partial_a) = {h_a}^2 \qquad \text{(no summation)}$$

then

$$\mathbf{V}_a = \frac{1}{h_a}\partial_a \qquad \text{(no summation)}$$

and

$$g(\mathbf{V}_a) = h_a dx^a \qquad \text{(no summation).}$$

To avoid having to continually repeat the phrase "no summation", we shall suspend the summation convention for the rest of this section.

Given any function f, we have

$$\begin{aligned} \operatorname{grad} f &= g^{-1}(df) \\ &= g^{-1}\left(\left(\frac{1}{h_1}\frac{\partial f}{\partial x^1}\right)h_1 dx^1 + \left(\frac{1}{h_2}\frac{\partial f}{\partial x^2}\right)h_2 dx^2 + \left(\frac{1}{h_3}\frac{\partial f}{\partial x^3}\right)h_3 dx^3\right) \\ &= \left(\frac{1}{h_1}\frac{\partial f}{\partial x^1}\right)\mathbf{V}_1 + \left(\frac{1}{h_2}\frac{\partial f}{\partial x^2}\right)\mathbf{V}_2 + \left(\frac{1}{h_3}\frac{\partial f}{\partial x^3}\right)\mathbf{V}_3. \end{aligned}$$

The divergence of a vector field $\mathbf{W}$ is calculated as follows. The volume form Ω is given by

$$\Omega = g(\mathbf{V}_1) \wedge g(\mathbf{V}_2) \wedge g(\mathbf{V}_3) = h_1 h_2 h_3 dx^1 \wedge dx^2 \wedge dx^3.$$

Then if $\mathbf{W} = W^a \mathbf{V}_a$,

$$\mathbf{W} \lrcorner \Omega = h_2 h_3 W^1 dx^2 \wedge dx^3 - h_1 h_3 W^2 dx^1 \wedge dx^3 + h_1 h_2 W^3 dx^1 \wedge dx^2$$

and so

$$\begin{aligned} d(\mathbf{W} \lrcorner \Omega) & \\ &= \left(\frac{\partial}{\partial x^1}(h_2 h_3 W^1) + \frac{\partial}{\partial x^2}(h_1 h_3 W^2) + \frac{\partial}{\partial x^3}(h_1 h_2 W^3)\right) dx^1 \wedge dx^2 \wedge dx^3 \\ &= \frac{1}{h_1 h_2 h_3}\left(\frac{\partial}{\partial x^1}(h_2 h_3 W^1) + \frac{\partial}{\partial x^2}(h_1 h_3 W^2) + \frac{\partial}{\partial x^3}(h_1 h_2 W^3)\right) \Omega \end{aligned}$$

and div $\mathbf{W}$ is the coefficient of Ω in this last expression.

Exercise 61. Show that for any vector field $\mathbf{W}$ on $\mathcal{E}^3$, curl $\mathbf{W}$ is given in terms of orthogonal coordinates by

$$\begin{aligned} \operatorname{curl} \mathbf{W} = & \left(\frac{1}{h_2 h_3}\right)\left(\frac{\partial(h_3 W^3)}{\partial x^2} - \frac{\partial(h_2 W^2)}{\partial x^3}\right) \mathbf{V}_1 \\ & + \left(\frac{1}{h_1 h_3}\right)\left(\frac{\partial(h_1 W^1)}{\partial x^3} - \frac{\partial(h_3 W^3)}{\partial x^1}\right) \mathbf{V}_2 \\ & + \left(\frac{1}{h_1 h_2}\right)\left(\frac{\partial(h_2 W^2)}{\partial x^1} - \frac{\partial(h_1 W^1)}{\partial x^2}\right) \mathbf{V}_3. \end{aligned}$$ □

Differential identities. Many differential identities involving the operators grad, div and curl are known in vector calculus: they are often consequences of the facts that $d^2 = 0$ and $** = (-1)^{p(n-p)}$. Clearly

$$\begin{aligned} \operatorname{curl}(\operatorname{grad} f) &= g^{-1} * d g g^{-1}(df) = g^{-1} * d^2 f = 0 \\ \operatorname{div}(\operatorname{curl} \mathbf{V}) &= * d * g g^{-1} * d g(\mathbf{V}) = * d^2 g(\mathbf{V}) = 0. \end{aligned}$$

It is also easy to establish, for example, that

$$\operatorname{div}(f\mathbf{V}) = f \operatorname{div} \mathbf{V} + \operatorname{grad} f \cdot \mathbf{V}$$

as follows:

$$\begin{aligned} \operatorname{div}(f\mathbf{V}) &= * d * g(f\mathbf{V}) = * d * f g(\mathbf{V}) \\ &= f * d * g(\mathbf{V}) + *(df \wedge * g(\mathbf{V})) \\ &= f \operatorname{div} \mathbf{V} + * g(df, g(\mathbf{V}))\Omega \\ &= f \operatorname{div} \mathbf{V} + g(g^{-1}(df), \mathbf{V}) = f \operatorname{div} \mathbf{V} + \operatorname{grad} f \cdot \mathbf{V}. \end{aligned}$$

Exercise 62. Prove the following identities:

$$\begin{aligned} \operatorname{curl}(f\mathbf{V}) &= f \operatorname{curl} \mathbf{V} + \operatorname{grad} f \times \mathbf{V} \\ \operatorname{div}(\mathbf{V} \times \mathbf{W}) &= \mathbf{W} \cdot \operatorname{curl} \mathbf{V} - \mathbf{V} \cdot \operatorname{curl} \mathbf{W}. \end{aligned}$$ □

Summary of Chapter 7

An affine metric on an affine space $\mathcal{A}$ derives from a scalar product on the vector space $\mathcal{V}$ on which it is modelled. A scalar product on a vector space is a bilinear form (that is, a map $\mathcal{V} \times \mathcal{V} \to \mathbf{R}$ linear in each argument) which is symmetric (unchanged in value on interchange of its arguments) and non-degenerate (the matrix representing it with respect to any basis of $\mathcal{V}$ is non-singular). A basis may be found for $\mathcal{V}$ with respect to which the matrix of the scalar product is diagonal and has r ones and $s = n - r$ minus ones along the diagonal ($n = \dim \mathcal{V}$): in this case the scalar product has signature $(r, n - r)$. When $r = n$ the scalar product is Euclidean; when $r = 1$ it is Lorentzian. Any basis for $\mathcal{V}$ in which the scalar product takes its standard form is orthonormal. In the case of a Lorentzian scalar product, vectors of positive "squared length" are timelike, of zero "squared length" null, and of negative "squared length" spacelike.

A scalar product g on $\mathcal{V}$ allows one to identify $\mathcal{V}$ and $\mathcal{V}^*$ by associating with a vector v the covector $w \mapsto g(w, v)$. There is also a volume form Ω, determined up to sign by the condition that the volume of a parallelepiped spanned by an orthonormal basis is ± 1. With these objects one may construct a linear isomorphism of $\bigwedge^p \mathcal{V}^*$ with $\bigwedge^{n-p} \mathcal{V}^*$ for each p, called the star operator, and a scalar product on $\bigwedge^p \mathcal{V}^*$ such that if $\omega, \chi \in \bigwedge^p \mathcal{V}^*$, $\omega \wedge *\chi = g(\omega, \chi)\Omega$.

A scalar product on $\mathcal{V}$ may be transferred to each tangent space of an affine space $\mathcal{A}$ modelled on $\mathcal{V}$, and then defines a metric on $\mathcal{A}$. The length of a curve σ, between $\sigma(t_1)$ and $\sigma(t_2)$, in $\mathcal{E}^n$ (n-dimensional affine space with Euclidean metric) is $\int_{t_1}^{t_2} \sqrt{g(\dot\sigma(t), \dot\sigma(t)}dt$; for a timelike curve in $\mathcal{E}^{1,n-1}$ (affine space with Lorentzian metric) the corresponding quantity is proper time. Straight lines minimise Euclidean length in Euclidean space, but timelike straight lines maximise proper time in a space with Lorentzian metric.

Parallel displacement in an affine space with affine metric preserves scalar products. It follows that for any vector fields U, V, W, $U(g(V,W)) = g(\nabla_U V, W) + g(V, \nabla_U W)$. The connection coefficients are related to the components of the metric, in an arbitrary coordinate system, by $\Gamma^c_{ab} = \frac{1}{2}g^{cd}(\partial_a g_{bd} + \partial_b g_{ad} - \partial_d g_{ab})$.

The operations associated with a metric provide the final link between vector calculus and exterior calculus. The vector product of two vectors $\mathbf{v}$, $\mathbf{w}$ in $\mathcal{E}^3$ is given by $g^{-1}(*(g(\mathbf{v}) \wedge g(\mathbf{w})))$, the gradient of a function f on any affine metric space by $g^{-1}(df)$; the divergence of a vector field V on any affine metric space by $*d*g(V)$; and the curl of a vector field $\mathbf{V}$ on $\mathcal{E}^3$ by $g^{-1}(*dg(\mathbf{V}))$. These coordinate-independent definitions may be used at will to give coordinate expressions for these objects with respect to any coordinate system.

8. ISOMETRIES

In Chapter 1 we introduced the group of affine transformations, which consists of those transformations of an affine space which preserve its affine structure. In Chapter 4 we discussed the idea of volume, and picked out from among all affine transformations the subgroup of those which preserve the volume form, namely those with unimodular linear part. In Chapter 7 we introduced another structure on affine space: a metric. We now examine the transformations which preserve this structure. They are called isometries.

An isometry of an affine space is necessarily an affine transformation. This may be deduced from the precise definition, as we shall show, and need not be imposed as part of it. Isometries form a group. Particular examples of isometry groups which are important and may be familiar are the Euclidean group, the group of isometries of Euclidean space $\mathcal{E}^3$; and the Poincaré group, which is the group of isometries of Minkowski space $\mathcal{E}^{1,3}$. Each of these groups is intimately linked to a group of linear transformations of the underlying vector space, namely the group preserving its scalar product. Such groups are called orthogonal groups (though in the case of $\mathcal{E}^{1,3}$ the appropriate group is more frequently called the Lorentz group).

Any one-parameter group of isometries induces a vector field on the affine metric space on which it acts. This vector field is called an infinitesimal isometry. The infinitesimal isometries of a given affine metric space $\mathcal{A}$ form a finite-dimensional vector space (which is different from the vector space underlying $\mathcal{A}$). Furthermore, the bracket of any two infinitesimal isometries is again an infinitesimal isometry. The space of infinitesimal isometries, equipped with the bracket operation, is an example of what is known as a Lie algebra. It is an infinitesimal, linear counterpart of the isometry group itself, and its study is important for this reason. We shall devote considerable attention to the Lie algebra of infinitesimal isometries here, especially in Sections 2 and 3. In later sections we shall describe some special features of the Euclidean and Poincaré groups, concentrating especially on their linear parts.

1. Isometries Defined

Let $\mathcal{A}$ be an affine metric space. A smooth map $\phi\colon \mathcal{A} \to \mathcal{A}$ is said to be an *isometry* of $\mathcal{A}$ if it preserves the metric g, in the sense that for every pair of vectors u, v tangent to $\mathcal{A}$ at any point of it

$$g(\phi_* u, \phi_* v) = g(u, v).$$

Thus, in the case of a Euclidean space, ϕ is an isometry if ϕ_* preserves the lengths of tangent vectors and the angles between them.

If ϕ is an isometry and σ is any curve then the "length" of any segment of the image curve $\phi \circ \sigma$ is the same as the "length" of the corresponding segment of σ:

$$\int_{t_1}^{t_2} \sqrt{\left|g(\phi_*\dot\sigma(t), \phi_*\dot\sigma(t))\right|}\,dt = \int_{t_1}^{t_2} \sqrt{\left|g(\dot\sigma(t), \dot\sigma(t))\right|}\,dt.$$

(We write "length" so as to include the pseudo-Euclidean case.) In fact this property is equivalent to the condition $g(\phi_* v, \phi_* v) = g(v, v)$ for all tangent vectors v, which in turn is equivalent to the isometry condition. In particular this means that isometries preserve the "lengths" of straight line segments, and therefore of displacements. This implies the equivalent definition: an isometry of an affine metric space is a smooth map which preserves "lengths" or "distances". However, the original definition is the more useful of the two, since it is given in a form which may be readily generalised, while this equivalent one is not.

Any translation of an affine metric space is clearly an isometry. More generally, an affine transformation is an (affine) isometry if and only if its linear part μ satisfies

$$g(\mu(v), \mu(w)) = g(v, w) \qquad \text{for all } v, w.$$

Here g should properly be interpreted as the scalar product on the underlying vector space, of which v and w are arbitrary elements. Note that μ is necessarily non-singular, since if $\mu(v) = 0$ then $g(v, w) = 0$ for all w, whence $v = 0$ since g is non-degenerate. It follows that every affine isometry is invertible. Moreover, since

$$g(\mu^{-1}(v), \mu^{-1}(w)) = g(\mu(\mu^{-1}(v)), \mu(\mu^{-1}(w))) = g(v, w),$$

the inverse transformation is also an isometry.

Exercise 1. Show that the composition of any two affine isometries is also an isometry. □

It follows that the set of affine isometries is a group under composition.

We show next that an isometry ϕ is necessarily an affine transformation. Suppose that, with respect to affine coordinates, $\phi(x^b) = (\phi^a(x^b))$. Then the condition for ϕ to be an isometry is

$$g_{ab}\frac{\partial\phi^a}{\partial x^c}\frac{\partial\phi^b}{\partial x^d} = g_{cd}, \qquad \text{where } g_{ab} = g(\partial_a, \partial_b).$$

Differentiating this equation gives

$$g_{ab}\left(\frac{\partial^2\phi^a}{\partial x^c\partial x^e}\frac{\partial\phi^b}{\partial x^d} + \frac{\partial\phi^a}{\partial x^c}\frac{\partial^2\phi^b}{\partial x^d\partial x^e}\right) = 0.$$

Two further equations may be obtained by cyclic interchange of c, d and e:

$$g_{ab}\left(\frac{\partial^2\phi^a}{\partial x^d\partial x^c}\frac{\partial\phi^b}{\partial x^e} + \frac{\partial\phi^a}{\partial x^d}\frac{\partial^2\phi^b}{\partial x^e\partial x^c}\right) = 0$$

$$g_{ab}\left(\frac{\partial^2\phi^a}{\partial x^e\partial x^d}\frac{\partial\phi^b}{\partial x^c} + \frac{\partial\phi^a}{\partial x^e}\frac{\partial^2\phi^b}{\partial x^c\partial x^d}\right) = 0.$$

On adding these two and subtracting the first, and using the symmetry of g_{ab}, one finds that

$$g_{ab}\frac{\partial^2\phi^a}{\partial x^c\partial x^d}\frac{\partial\phi^b}{\partial x^e} = 0.$$

But (g_{ab}) and $(\partial\phi^b/\partial x^e)$ are non-singular matrices (the latter by the same argument as was used for the linear part of an affine isometry) and therefore

$$\frac{\partial^2\phi^a}{\partial x^c\partial x^d} = 0$$

as required.

The study of the isometries of an affine metric space is therefore reduced to the study of its affine isometry group. Furthermore, the main point of interest about an isometry concerns its linear part μ, which must satisfy the condition

$$g\bigl(\mu(v),\mu(w)\bigr) = g(v,w)$$

for all $v, w \in \mathcal{V}$, the underlying vector space. The set of linear transformations μ satisfying this condition (for a given scalar product g) forms a group, a subgroup of $GL(\mathcal{V})$. The matrix M of μ with respect to an orthonormal basis for $\mathcal{V}$ must satisfy

$$M^T GM = G$$

where G is the diagonal matrix with diagonal entries ± 1 of the appropriate signature. We call the group of matrices satisfying this condition, and by extension the group of linear transformations of $\mathcal{V}$ preserving g, the ***orthogonal group*** of G (or g), and denote it $O(p, n-p)$ where $(p, n-p)$ is the signature of G, or simply $O(n)$ in the positive-definite case. Then the isometry group of $\mathcal{E}^{p,n-p}$ is the semi-direct product of $O(p, n-p)$ and the translation group: it inherits this structure from the group of affine transformations of which it is a subgroup (see Chapter 1, Section 4 for the definition of the semi-direct product in that context).

Exercise 2. Show that if $\{e_a\}$ is an orthonormal basis and μ is orthogonal then $\{\mu(e_a)\}$ is also an orthonormal basis. □

It is clear from the definition (by taking determinants) that any orthogonal linear transformation μ must satisfy

$$(\det \mu)^2 = 1.$$

In fact if Ω is either of the two volume forms consistent with the metric then, since μ maps orthonormal bases to orthonormal bases, for any orthonormal basis $\{e_a\}$ it follows that

$$(\mu^*\Omega)(e_1, e_2, \ldots, e_n) = \Omega\bigl(\mu(e_1), \mu(e_2), \ldots, \mu(e_n)\bigr) = \pm 1.$$

Consequently

$$\mu^*\Omega = (\det \mu)\Omega = \pm\Omega$$

and therefore $\det \mu = 1$ if μ is orientation-preserving, $\det \mu = -1$ if μ is orientation-reversing. In particular, an orientation-preserving isometry also preserves the volume associated with the metric.

The set of orientation-preserving orthogonal matrices is a subgroup of the group $O(p, n-p)$, called the ***special orthogonal group*** and denoted $SO(p, n-p)$. It consists of the elements of $O(p, n-p)$ with determinant 1, and is the intersection of $O(p, n-p)$ with $SL(n, \mathbf{R})$.

Exercise 3. Show that any element of $O(2)$ may be expressed in one of the following forms:

$$\begin{pmatrix} \cos t & -\sin t \\ \sin t & \cos t \end{pmatrix} \quad \text{or} \quad \begin{pmatrix} \cos t & \sin t \\ \sin t & -\cos t \end{pmatrix}$$

the first matrix corresponding to an orientation-preserving, the second to an orientation-reversing transformation. Show that the first matrix leaves no (real) direction fixed, unless it is the identity; but the second leaves the line $x^1 \sin \frac{t}{2} - x^2 \cos \frac{t}{2} = 0$ pointwise fixed (the first matrix represents rotation through t; the second, reflection in the given line). □

Exercise 4. Show that any element of $SO(1,1)$ may be expressed in the form

$$\pm \begin{pmatrix} \cosh t & \sinh t \\ \sinh t & \cosh t \end{pmatrix}.$$

□

Exercise 5. Let $\mathcal{V}$ be a vector space with scalar product g, and let μ be an orthogonal transformation of $\mathcal{V}$. Show that if μ leaves a subspace $\mathcal{W}$ invariant then it also leaves the orthogonal subspace $\mathcal{W}^{\perp}$ invariant. Show that if g induces a scalar product on $\mathcal{W}$ (by restriction) then the restriction of μ to $\mathcal{W}$ is orthogonal with respect to the induced scalar product. □

2. Infinitesimal Isometries

We shall now describe the vector fields induced on an affine metric space by the one-parameter subgroups of its isometry group. They are called *infinitesimal isometries*.

One-parameter groups of isometries and their generators. The set of rotations of $\mathcal{E}^2$ defined in Exercise 3 has been used already as an example of a one-parameter affine group in Section 1 of Chapter 3; its generator was shown to be the vector field $-x^2\partial_1 + x^1\partial_2$.

Exercise 6. Show that the matrices

$$\begin{pmatrix} \cosh t & \sinh t \\ \sinh t & \cosh t \end{pmatrix}$$

of Exercise 4 form a one-parameter group, and that the infinitesimal generator of the corresponding one-parameter group of isometries of $\mathcal{E}^{1,1}$ is $x^2\partial_1 + x^1\partial_2$. □

It follows from the considerations of one-parameter affine groups in Chapter 3, Section 1 that in affine coordinates an infinitesimal isometry must take the form

$$(A^a_b x^b + P^a)\partial_a \qquad (A^a_b,\ P^a \text{ constants})$$

where the matrix $A = (A^a_b)$ is given by $A = d/dt(M_t)(0)$, M_t being the matrix of the linear part μ_t of the one-parameter isometry group. If the coordinates are orthonormal then M_t is a one-parameter group of orthogonal matrices:

$$M_t{}^T G M_t = G.$$

On differentiating this equation with respect to t and setting $t = 0$ one obtains

$$A^T G + GA = 0$$

(since M_0 is the identity).

Exercise 7. Deduce that in the Euclidean case, with respect to orthonormal coordinates, A must be skew-symmetric. □

We shall say, in the general case, that a matrix A satisfying the condition $A^T G + GA = 0$ is *skew-symmetric with respect to the scalar product* defined by G. Thus the matrix A occurring in the expression of an infinitesimal isometry in terms of orthonormal coordinates is skew-symmetric with respect to the scalar product.

Exercise 8. Show that if matrices A, B are skew-symmetric with respect to the same scalar product, so is their commutator $[A, B] = AB - BA$. Show also that if A is skew-symmetric with respect to any scalar product then $\operatorname{tr} A = 0$. □

Conversely, if a matrix A is skew-symmetric with respect to a given scalar product then the one-parameter group of matrices $\exp(tA)$ consists of orthogonal matrices. To see why, consider the matrix function $a(t) = \exp(tA)^T G \exp(tA)$. Its derivative is given by

$$\frac{da}{dt}(t) = \exp(tA)^T \left(A^T G + GA\right) \exp(tA) = 0$$

and so $a(t) = a(0) = G$, and $\exp(tA)$ is orthogonal. It follows that if the matrix A is skew-symmetric (with respect to a given scalar product) then the one-parameter group generated by the vector field $(A^a_b x^b + P^a)\partial_a$ is a one-parameter group of isometries (the coordinates being orthonormal for the given metric). For suppose that τ is the integral curve of the vector field through the origin, so that

$$\frac{d\tau^a}{dt} = A^a_b \tau^b + P^a \qquad \tau^a(0) = 0.$$

Then on any other integral curve σ

$$\frac{d}{dt}(\sigma^a - \tau^a) = (A^a_b \sigma^b + P^a) - (A^a_b \tau^b + P^a) = A^a_b(\sigma^b - \tau^b)$$

and so

$$\sigma^a(t) - \tau^a(t) = \exp(tA)^a_b \left(\sigma^b(0) - \tau^b(0)\right)$$

or

$$\sigma^a(t) = \exp(tA)^a_b \sigma^b(0) + \tau^a(t).$$

This means that the one-parameter (affine) group generated by the vector field has $\exp(tA)$ as its linear part and, since this is always orthogonal, the one-parameter group consists of isometries.

To summarise: the vector field $(A^a_b x^b + P^a)\partial_a$ (in orthonormal coordinates) is an infinitesimal isometry if and only if $A = (A^a_b)$ is skew-symmetric with respect to the scalar product.

3. Killing's Equation and Killing Fields

In deriving these results we have made explicit use of the fact that isometries are affine transformations, and we have employed the special coordinates available in an affine space. There is another approach to the identification of infinitesimal isometries which does not appeal directly to the affine structure of the space and is therefore more suitable for later generalisation. It is based directly on the original definition of an isometry, and leads to an equation for the infinitesimal isometry called Killing's equation. When expressed in terms of coordinates (which need not be affine) Killing's equation becomes a set of partial differential equations, of first order, for the components of the infinitesimal isometry vector field.

Let ϕ_t be a one-parameter group of isometries, and X the vector field which is its generator. Consider the orbit $t \mapsto \phi_t(x)$ of some point x, and let V, W be any two vector fields given along this orbit, at least near x. The required equation is obtained by relating the rate of change of the function $t \mapsto g(V(t), W(t))$ to the Lie derivatives of V and W with respect to X, using the fact that X is an infinitesimal isometry. By the isometry condition for ϕ_{-t}

$$g(V(t), W(t)) = g(\phi_{-t*}V(t), \phi_{-t*}W(t)).$$

On differentiating both sides with respect to t and putting $t = 0$ one obtains

$$X(g(V, W)) = g(\mathcal{L}_X V, W) + g(V, \mathcal{L}_X W).$$

Strictly speaking this argument establishes the result only at the point x; but x, V and W are arbitrary, so the result holds at every point and for every pair of vector fields. Thus if X is an infinitesimal isometry

$$X(g(V, W)) = g([X, V], W) + g(V, [X, W])$$

for every pair of vector fields V, W. This is *Killing's equation*.

Another form of the equation may be obtained by making use of the covariant derivative: using the properties of covariant differentiation in an affine metric space given in Section 6 of Chapter 7 (especially the result of Exercise 47) we have

$$X(g(V, W)) = g(\nabla_X V, W) + g(V, \nabla_X W);$$

but $[X, V] = \nabla_X V - \nabla_V X$, and so Killing's equation may be written

$$g(\nabla_V X, W) + g(V, \nabla_W X) = 0.$$

A solution X of Killing's equation is called a *Killing field*; this term is used interchangeably with the term infinitesimal isometry, as the result of the following exercise justifies.

Exercise 9. By considering vector fields V, W defined along an integral curve of X by Lie transport, show that if X is a Killing field then its flow consists of isometries (use the first stated form of Killing's equation). □

Exercise 10. Show that, with respect to any coordinates, Killing's equation is equivalent to the following set of simultaneous partial differential equations for X^a, the components of X with respect to the coordinate fields:

$$X^a \frac{\partial g_{bc}}{\partial x^a} + g_{ab}\frac{\partial X^a}{\partial x^c} + g_{ac}\frac{\partial X^a}{\partial x^b} = 0.$$ □

Exercise 11. Show that, for $\mathcal{E}^2$ in polar coordinates, the conditions for $X = \xi\partial/\partial r + \eta\partial/\partial\vartheta$ to be an infinitesimal isometry are

$$\frac{\partial \xi}{\partial r} = 0 \qquad \frac{\partial \xi}{\partial \vartheta} + r^2\frac{\partial \eta}{\partial r} = 0 \qquad \xi + r\frac{\partial \eta}{\partial \vartheta} = 0$$

and deduce that

$$X = a\left(\cos\vartheta\frac{\partial}{\partial r} - \frac{1}{r}\sin\vartheta\frac{\partial}{\partial\vartheta}\right) + b\left(\sin\vartheta\frac{\partial}{\partial r} + \frac{1}{r}\cos\vartheta\frac{\partial}{\partial\vartheta}\right) + c\frac{\partial}{\partial\vartheta}$$

for some constants a, b, c. □

Exercise 12. Show that with respect to affine coordinates Killing's equation is equivalent to $g_{ab}\partial_c X^a + g_{ac}\partial_b X^a = 0$. By differentiating and using an argument similar to the one used to show that an isometry is an affine transformation show that $X = (A^a_b x^b + P^a)\partial_a$ where $g_{ab}A^a_c + g_{ac}A^a_b = 0$. □

Exercise 13. Show that the condition for a vector field X to be an infinitesimal isometry of the metric g is that ∇X (Chapter 7, Section 7) be skew-symmetric with respect to the scalar product determined by g. □

Killing's equation and the Lie derivative. In Section 4 of Chapter 5 we defined the Lie derivative of a p-form and derived the equation

$$(\mathcal{L}_V\omega)(W_1, W_2, \dots, W_p)$$
$$= V\left(\omega(W_1, W_2, \dots, W_p)\right) - \sum_{r=1}^{p} \omega(W_1, \dots, [V, W_r], \dots, W_p).$$

The first form of Killing's equation given above is clearly reminiscent of the right hand side of this formula. This suggests that, with the appropriate definition of the Lie derivative of a metric, one should be able to express Killing's equation in the very satisfactory form

$$\mathcal{L}_X g = 0.$$

The appropriate definition is not hard to find, and works in fact in a rather more general context. We digress a little to discuss this now.

In Section 8 of Chapter 4 we defined multilinear functions (on a vector space). A field of p-fold multilinear functions on an affine space $\mathcal{A}$ is an assignment, to each point x, of a p-fold multilinear function on $T_x\mathcal{A}$ (with the same p for all x). If Q is such a field then for each choice of p vector fields $W_1, W_2, \dots, W_p$ there is determined a function $Q(W_1, W_2, \dots, W_p)$ on $\mathcal{A}$, whose value at x is $Q_x(W_{1x}, W_{2x}, \dots, W_{px})$. This construction is an obvious extension of the one defined for p-forms in Chapter 5, Section 1, and reduces to it if Q_x is alternating for each x. The field Q is smooth if the function $Q(W_1, W_2, \dots, W_p)$ is smooth for all smooth vector fields $W_1, W_2, \dots, W_p$. Such a field is usually called a smooth *covariant tensor field of valence p*. An affine metric is a covariant tensor field of valence 2, with some special properties. Other examples of tensor fields (besides metrics and exterior forms) will be introduced later (in Chapter 9, for example).

The definition of the Lie derivative of a form makes no particular use of the alternating property of forms and extends immediately to covariant tensor fields. In fact smooth maps of affine spaces induce maps of covariant tensor fields just as they do of forms. Let Q be a covariant tensor field on an affine space $\mathcal{B}$ and $\phi\colon \mathcal{A} \to \mathcal{B}$ a smooth map. A covariant tensor field ϕ^*Q on $\mathcal{A}$, of the same valence as Q, is defined by

$$(\phi^*Q)_x(w_1, w_2, \dots, w_p) = Q_{\phi(x)}(\phi_* w_1, \phi_* w_2, \dots, \phi_* w_p)$$

where $w_1, w_2, \dots, w_p \in T_x\mathcal{A}$. This new tensor field ϕ^*Q is called the *pull-back* of Q by ϕ. To define the Lie derivative of a tensor field Q with respect to a vector field V one uses the pull-back $\phi_t{}^*Q$ of Q by the map ϕ_t of the flow of V:

$$\mathcal{L}_V Q = \frac{d}{dt}(\phi_t{}^*Q)_{t=0}.$$

(Strictly speaking, if V does not generate a one-parameter group then $\mathcal{L}_V Q$ may have to be defined pointwise, as described for forms in Chapter 5, Section 4.)

Exercise 14. Show that for any vector fields $W_1, W_2, \ldots, W_p$,

$$(\mathcal{L}_V Q)(W_1, W_2, \ldots, W_p) = V(Q(W_1, W_2, \ldots, W_p)) - \sum_{r=1}^{p} Q(W_1, \ldots, [V, W_r], \ldots, W_p).$$

Deduce that a vector field X is an infinitesimal isometry of a metric g if and only if $\mathcal{L}_X g = 0$. □

Exercise 15. Show that for any vector fields U, V and any constants k, l (and any covariant tensor field Q)

$$\mathcal{L}_{kU+lV} Q = k\mathcal{L}_U Q + l\mathcal{L}_V Q$$
$$\mathcal{L}_U(\mathcal{L}_V Q) - \mathcal{L}_V(\mathcal{L}_U Q) = \mathcal{L}_{[U,V]} Q.$$

Deduce that if X and Y are both infinitesimal isometries of a metric g then so are $kX + lY$ (for any constants k and l) and $[X, Y]$. □

The Lie algebra of Killing fields. It follows from Exercise 15 that the set of Killing fields (infinitesimal isometries) of an affine metric space is itself a linear space: if X and Y are Killing fields so is $kX + lY$ for any constants k, l. (The same conclusion may be drawn from the explicit form of a Killing field in affine coordinates.) This linear space is in fact finite-dimensional. A basis for the space may be defined as follows. Choose orthonormal affine coordinates. A Killing field is determined by a matrix A which is skew symmetric with respect to the scalar product of the appropriate signature, and a vector P. For each a, b with $a > b$ there is a skew symmetric matrix A which has $A^a_b = 1$, $A^b_a = \pm 1$, the sign depending on the signature, and all other entries zero. These matrices form a basis for the space of matrices skew symmetric with respect to the scalar product. A basis for Killing fields built out of these matrices is given by $\{ x^b \partial_a \pm x^a \partial_b, \partial_c \}$ where $a, b, c = 1, 2, \ldots, n$ and $a > b$. There are $\frac{1}{2}n(n-1) + n$ elements in this basis, and so the Killing fields form a vector space of dimension $\frac{1}{2}n(n+1)$.

It also follows from Exercise 15 that the space of Killing fields is closed under bracket: if X and Y are Killing fields so is $[X, Y]$. (This can also be seen from the explicit form of a Killing field and from Exercise 8, in view of the relationship between the bracket of affine vector fields and the commutator of matrices given in Chapter 3, Exercise 57.)

A (finite-dimensional) vector space $\mathcal{V}$ equipped with a bracket operation, that is to say, a bilinear map $\mathcal{V} \times \mathcal{V} \to \mathcal{V}$ which is anti-symmetric and satisfies Jacobi's identity, is called a *Lie algebra*. The Killing fields of an affine metric space form a Lie algebra. In this case the properties of the bracket follow from those of the bracket of general vector fields, by restriction. However, Lie algebras arise in other contexts, and not all bracket operations are obtained from the bracket of vector fields in such a transparent way. For example, square matrices of a given size form a Lie algebra with the matrix commutator as bracket, and so do vectors in $\mathcal{E}^3$ with the vector product as bracket (Chapter 3, Exercise 58). The set of all vector fields on an affine space does not form a Lie algebra since it is not finite dimensional over the reals; however, the set of affine vector fields is one. We shall have much more to say about Lie algebras in Chapter 12.

4. Conformal Transformations

As a further application of these ideas we shall briefly discuss the *conformal transformations* of an affine metric space. A conformal transformation is one which preserves the metric up to a scalar factor: the transformation ϕ is conformal if

$$g(\phi_* v, \phi_* w) = \kappa g(v, w)$$

for all $v, w \in T_x \mathcal{A}$ (and all $x \in \mathcal{A}$), where κ is a positive function on $\mathcal{A}$. An isometry is to be regarded as a special case of a conformal transformation. A conformal transformation of Euclidean space preserves the ratios of lengths of vectors (but not, unless it is an isometry, the lengths themselves); and it preserves angles. A conformal transformation of a Lorentzian space maps null vectors to null vectors; in other words, it preserves null cones.

A vector field X is an *infinitesimal conformal transformation*, or *conformal Killing field*, if its flow consists of conformal transformations. In order for X to be a conformal Killing field its flow ϕ must satisfy $\phi_t{}^* g = \kappa_t g$ for some function κ_t, and therefore

$$\mathcal{L}_X g = \rho g$$

where ρ is a function on $\mathcal{A}$, not necessarily positive.

Exercise 16. By adapting the method of Exercise 9, show that if $\mathcal{L}_X g = \rho g$ then the flow of X consists of conformal transformations. □

Exercise 17. Show that if X and Y are conformal Killing fields so is $kX + lY$ for every pair of constants k, l, and so also is $[X, Y]$. □

Exercise 18. Show that in an affine coordinate system the conformal Killing equation $\mathcal{L}_X g = \rho g$ is equivalent to

$$g_{ab}\partial_c X^a + g_{ac}\partial_b X^a = \rho g_{bc}.$$ □

A simple example of a conformal Killing field which is not a Killing field is the dilation field $\Delta = x^a \partial_a$. For $\partial_c \Delta^a = \delta^a_c$ and so

$$g_{ab}\partial_c \Delta^a + g_{ac}\partial_b \Delta^a = 2g_{bc}$$

and Δ is therefore a conformal Killing field with ρ the constant function 2. A conformal transformation for which the factor ρ is a constant is called a *homothety*: the dilation field is an infinitesimal homothety.

Exercise 19. By adapting the argument of Exercise 12, show that any infinitesimal homothety of an affine metric space is (apart from an additive Killing field) a constant multiple of the dilation field. □

Exercise 20. Confirm that for any constants α_a the vector field X given in affine coordinates by

$$X = (\alpha_b x^b x^a - \tfrac{1}{2} g_{bc} x^b x^c g^{ad} \alpha_d)\partial_a$$

is a conformal Killing field for the metric g, with $\rho = \alpha_a x^a$. □

It may be shown, by an extension of the method of Exercise 12, that the vector fields of Exercises 19 and 20, together with all Killing fields, give all conformal Killing fields in an affine metric space of dimension greater than 2. It follows that when the dimension of the space is greater than 2 the set of conformal Killing fields is a Lie algebra. However, in dimension 2 the space of conformal Killing fields is not finite-dimensional, though it is a linear space closed under bracket. A conformal Killing field, unlike a Killing field, need not be affine.

5. The Rotation Group

We shall devote the rest of this chapter to a closer investigation of the two affine isometry groups of greatest interest, namely the Euclidean group (isometries of the Euclidean space $\mathcal{E}^3$) and (in the next section but one) the Poincaré group (isometries of the Lorentzian space $\mathcal{E}^{1,3}$). From what has been said already it should be clear that the translation parts of these groups present no particular problems; we shall therefore concentrate on their linear parts, which are $O(3)$ and $O(1,3)$ respectively.

The orientation-preserving elements of the orthogonal group $O(3)$ are just the rotations of Euclidean space, and so the subgroup $SO(3)$ is called the *rotation group*: it is with this that we shall be mainly concerned in this section. We shall often find it convenient to use vector notation.

The rotation group. A rotation of $\mathcal{E}^3$ is an orientation-preserving isometry which leaves a point of the space fixed. With respect to right-handed orthonormal affine coordinates based on the fixed point the rotation is therefore represented by a special orthogonal matrix, that is, an element of $SO(3)$. A familiar specific example of a rotation is the rotation about the x^3-axis given by the matrix

$$\begin{pmatrix} \cos t & -\sin t & 0 \\ \sin t & \cos t & 0 \\ 0 & 0 & 1 \end{pmatrix}.$$

The x^3-axis, which is left fixed by this transformation, is called the axis of the rotation, and the rotation is through angle t. Note that looking down the x^3-axis from the positive side towards the origin, with right-handed coordinates, one sees the rotation as counter-clockwise for t positive.

Exercise 21. Let $\mathbf{n}$ be a unit vector, with components n^a. Define a matrix $R = (R^a_b)$ by

$$R(\mathbf{x}) = \cos t\, \mathbf{x} + (1 - \cos t)(\mathbf{n} \cdot \mathbf{x})\mathbf{n} + \sin t\, \mathbf{n} \times \mathbf{x},$$

or in component form

$$R^a_b = \cos t\, \delta^a_b + (1 - \cos t)\delta_{bc} n^a n^c + \sin t\, \delta^{ac} \epsilon_{bcd} n^d.$$

(Here δ_{bc} are the components of the Euclidean scalar product in orthonormal coordinates, and therefore just the components of the identity matrix, and δ^{bc} likewise for the dual scalar product; they are included to ensure that the formula conforms with our range and summation conventions. The symbol ϵ_{abc} is the Levi-Civita alternating symbol defined in Section 2 of Chapter 4.) Show that R is a rotation which leaves fixed the line through the origin determined by the unit vector $\mathbf{n}$. □

This rotation is the rotation about the *axis* $\mathbf{n}$ (or more accurately the line determined by $\mathbf{n}$) through angle t.

Exercise 22. Show that rotation about $\mathbf{n}$ through angle s followed by rotation about $\mathbf{n}$ through angle t amounts to rotation about $\mathbf{n}$ through angle $s + t$. □

Exercise 23. Show that for any rotation R and any position vectors $\mathbf{x}$, $\mathbf{y}$, $R(\mathbf{x}\times\mathbf{y}) = R(\mathbf{x})\times R(\mathbf{y})$. Deduce that if S is rotation about $\mathbf{n}$ through angle t then RSR^{-1} is rotation about $R(\mathbf{n})$ through angle t. □

The rotations about a fixed axis constitute a one-parameter group of isometries, by Exercise 22. Slightly more generally, if R_t is the one-parameter group defined in Exercise 21 then $R_{\nu t}$ is also a one-parameter group, where ν is any constant; for $\nu \neq 0$ it is also a group of rotations about the axis $\mathbf{n}$, and (if t is thought of as the time) ν is the angular speed of rotation. Our aim now is to show that every one-parameter group of rotations is of this form, and that every rotation, except the identity, is a rotation about a unique axis and therefore lies on some one-parameter group of rotations. (The identity lies on every one-parameter group, of course.)

We show first that every rotation $R \neq I_3$ is a rotation about an axis. Since the rotation is a linear transformation of a vector space of odd dimension it must have at least one real eigenvalue. Let e be an eigenvalue and $\mathbf{n}$ a unit eigenvector belonging to it. Since R preserves the length of vectors, $R(\mathbf{n}) = e\mathbf{n}$ must also be a unit vector, and therefore $e = \pm 1$. Since R is an orthogonal transformation it must map the 2-plane orthogonal to $\mathbf{n}$ to itself (Exercise 5) and its restriction to this 2-plane may therefore be expressed in one of the forms exhibited in Exercise 3. If it takes the first form, that of a rotation of the 2-plane, then e must be 1 since R preserves orientation in $\mathcal{E}^3$; and $\mathbf{n}$ defines the axis of rotation. If not, there is a fixed line in the 2-plane, which is the required axis; $e = -1$ and the rotation is a rotation through π about this axis, which has the effect of reflecting the 2-plane. This result, that every rotation is a rotation about an axis, is known as Euler's theorem.

We now show that every continuous one-parameter group of rotations, not consisting merely of the identity transformation, is a group of rotations about a fixed common axis. By a continuous one-parameter group of rotations we mean a set of rotations $\{R_t\}$ depending on a parameter t, satisfying the one-parameter group property, and depending continuously on t in the sense that the matrix elements of R_t with respect to a fixed basis are continuous functions of t. We consider an element R_t of the one-parameter group, not being the identity: then by Euler's theorem it is a rotation about an axis $\mathbf{n}$. Now the pth power of a rotation is a rotation about the same axis through p times the angle, if it is not the identity. Consider, then, the rotation $R_{t/p}$. It cannot be the identity, and so is a rotation about some axis $\mathbf{n}'$. But $(R_{t/p})^p = R_t$; thus $\mathbf{n}'$ is an axis for R_t, $\mathbf{n}' = \mathbf{n}$, and $R_{t/p}$ and R_t are rotations about the same axis. It follows that for any rational number $r = p/q$, R_{rt} is a rotation about the same axis as R_t, since $R_{rt} = (R_{t/q})^p$. Since the one-parameter group is continuous, if R_{rt} is a rotation about axis $\mathbf{n}$ for every rational r then R_{kt} is a rotation about axis $\mathbf{n}$ for every real number k. The one-parameter group therefore consists of rotations about $\mathbf{n}$. Moreoever, if R_t is a rotation through angle s then R_{kt} is a rotation through angle ks, in the first place for rational k, and then by continuity for all real k. Thus the parameter t is proportional to the angle of rotation.

The Lie algebra of Killing fields of Euclidean 3-space. The infinitesimal generator of a one-parameter group of rotations about an axis is easily calculated by the methods of Section 2. For example, the generator X_3 of rotations about the x^3-axis (parametrised by angle) is $x^1\partial_2 - x^2\partial_1$.

Exercise 24. Show that the generators X_1, X_2 of the corresponding one-parameter groups of rotations about the x^1- and x^2-axes are $x^2\partial_3 - x^3\partial_2$ and $x^3\partial_1 - x^1\partial_3$ respectively. □

Exercise 25. Show that the brackets of the Killing fields X_1, X_2 and X_3 are given by

$$[X_2, X_3] = -X_1 \qquad [X_3, X_1] = -X_2 \qquad [X_1, X_2] = -X_3.$$

Let $T_a = \partial_a$, $a = 1, 2, 3$, be the generators of translations of $\mathcal{E}^3$ along the coordinate axes. Show that

$$[T_1, X_1] = 0 \qquad [T_1, X_2] = -T_3 \qquad [T_1, X_3] = T_2$$

and evaluate the corresponding brackets involving T_2 and T_3. □

The results of Exercise 25, together with the fact that all brackets of generators of translations are zero, specify the structure of the Lie algebra of Killing fields of $\mathcal{E}^3$. They may be conveniently written in the form

$$[X_a, X_b] = -\epsilon_{abc}\delta^{cd}X_d \qquad [T_a, X_b] = -\epsilon_{abc}\delta^{cd}T_d \qquad [T_a, T_b] = 0.$$

Angular velocity. The general rotation about a specified axis is given in Exercise 21. Making allowance for the fact that the parameter of a one-parameter group of rotations need not be the angle of rotation, but may be only proportional to it, we may write for the general one-parameter group of rotations

$$R_t(\mathbf{x}) = \cos\nu t\,\mathbf{x} + (1 - \cos\nu t)(\mathbf{n}\cdot\mathbf{x})\mathbf{n} + \sin\nu t\,\mathbf{n}\times\mathbf{x}.$$

Exercise 26. Show that the infinitesimal generator of this one-parameter group is $\nu n^a X_a$, where n^a are the components of $\mathbf{n}$ and X_a the generators of rotations about the coordinate axes (Exercise 24). Express this vector field in terms of the coordinate fields. □

We call the vector $\mathbf{w} = \nu\mathbf{n}$ the *angular velocity* of the one-parameter group of rotations, and $|\nu|$ its *angular speed*. The one-parameter group and its generator are completely determined by the angular velocity.

In terms of the basis Killing fields X_a the Killing field corresponding to the angular velocity vector $\mathbf{w}$ is $w^a X_a = \nu n^a X_a$. But in terms of the coordinate vector fields any infinitesimal generator of rotations corresponds to a skew-symmetric matrix A and takes the form $A^a_b x^b \partial_a$. We have therefore established a correspondence between vectors in $\mathcal{E}^3$ and skew-symmetric 3×3 matrices. The correspondence is given by

$$\mathbf{w} \longleftrightarrow \begin{pmatrix} 0 & -w^3 & w^2 \\ w^3 & 0 & -w^1 \\ -w^2 & w^1 & 0 \end{pmatrix}$$

or

$$A^a_b = \delta^{ac}\epsilon_{bcd}w^d.$$

Exercise 27. Verify that if W is the skew-symmetric matrix corresponding to the vector w then the map $\mathbf{w} \to W$ is a linear isomorphism; $W(\mathbf{x}) = \mathbf{w} \times \mathbf{x}$ for any vector $\mathbf{x}$; $[W_1, W_2]$ is the matrix corresponding to $\mathbf{w}_1 \times \mathbf{w}_2$; and $\exp(tW)$ is the one-parameter group of rotations from which we started. □

The dimension of the space is a key factor here, for only when $n = 3$ is the dimension of the space of $n \times n$ skew-symmetric matrices, $\frac{1}{2}n(n-1)$, equal to n.

Exercise 28. Deduce from the Jacobi identity for brackets of vector fields the vector identity

$$\mathbf{a} \times (\mathbf{b} \times \mathbf{c}) + \mathbf{b} \times (\mathbf{c} \times \mathbf{a}) + \mathbf{c} \times (\mathbf{a} \times \mathbf{b}) = 0.$$ □

The conclusions of Exercise 27 may be summarised by saying that the Lie algebra of infinitesimal rotations is isomorphic to the Lie algebra of Euclidean 3-vectors with the vector product as bracket operation. It follows also from Exercise 27 (2) that the velocity of a point $\mathbf{x}$ undergoing rotation with angular velocity $\mathbf{w}$ is given by

$$\frac{d\mathbf{x}}{dt} = \mathbf{w} \times \mathbf{x}.$$

This is a well-known formula from the kinematics of rigid bodies, for the case of rotation with constant angular speed about a fixed axis. We can pursue this line of argument to cover more general rotatory motion, as follows.

Consider the motion of a rigid body with one point fixed. The position of such a body at time t is determined by a rotation matrix $R(t)$, which may be defined as the rotation required to carry a right-handed set of axes fixed in space into coincidence with a similar set of axes fixed in the body, the space and body axes both being supposed to have their origins at the fixed point, and to be coincident initially. The one-parameter family of rotations $R(t)$ is not assumed to form a one-parameter group. We compute the velocity of a point fixed in the body. If the initial position of the point is given by the (constant) vector $\mathbf{x}_0$, then its position at time t is $\mathbf{x}(t) = R(t)\mathbf{x}_0$. Its velocity may be expressed in terms of its current position as follows:

$$\frac{d\mathbf{x}}{dt}(t) = \frac{dR}{dt}(t)\mathbf{x}_0 = \frac{dR}{dt}(t)R(t)^{-1}\mathbf{x}(t).$$

Set

$$W(t) = \frac{dR}{dt}(t)R(t)^{-1}.$$

Then $W(t)$ is skew-symmetric: using the orthogonality of $R(t)$ we have $R(t)^{-1} = R(t)^T$ and so

$$W(t)^T = R(t)\frac{dR^T}{dt}(t) = -\frac{dR}{dt}(t)R(t)^T = -W(t).$$

Thus $W(t)$ corresponds to a vector $\mathbf{w}(t)$, which we call the *instantaneous angular velocity* of rotation; and

$$\frac{d\mathbf{x}}{dt}(t) = \mathbf{w}(t) \times \mathbf{x}(t).$$

In effect, the definition of the instantaneous angular velocity at time t amounts to approximating the rotation $R(t+h)R(t)^{-1}$ by $\exp(hW(t))$, for small h. This is quite analogous to the definition of instantaneous (linear) velocity, in which one effectively approximates the difference between two translations of a one-parameter family, representing the position of a point in general motion, by a one-parameter group of translations, whose generator is the instantaneous velocity vector.

6. Parametrising Rotations

We have defined the general rotation in terms of its axis, specified by a unit vector, and its angle. The group of rotations is thus a three-dimensional object, in the sense that three parameters are required to fix a rotation, two for the direction of the unit vector along the axis and one for the angle. (The word "parameter" here is used to mean just a label or coordinate, without the additional meaning it has in the phrase "one-parameter group".) It is no coincidence that the dimension of the Lie algebra is the same as the dimension of the group in this sense: this relation will be explored in Chapter 12. For the present our concern is to find convenient and useful ways of parametrising the elements of the rotation group.

We could of course simply use the matrix entries to parametrise the elements of $SO(3)$. But they are not independent: the nine matrix entries are subject to six conditions (the columns of an orthogonal matrix are unit vectors and are pairwise orthogonal) and, in addition, to the condition that the determinant be positive. It is not clear how to solve the equations of condition so as to express all nine matrix entries in terms of just three of them.

The parametrisation in terms of axis and angle is not always the most useful, but there are several points of interest about it worth mentioning. Each pair $(\mathbf{n}, t)$ determines a rotation, where the unit vector $\mathbf{n}$ lies along the axis of the rotation and t is the angle of rotation. However, more than one pair $(\mathbf{n}, t)$ may correspond to one given rotation. Certainly, choices of t differing by an integer multiple of 2π, with the same $\mathbf{n}$, determine the same rotation. So do $(-\mathbf{n}, -t)$ and $(\mathbf{n}, t)$. And $(\mathbf{n}, 0)$ determines the same rotation for all $\mathbf{n}$, namely the identity. We may avoid much of this ambiguity by restricting the values of t to lie between 0 and π inclusive. Then each rotation, other than the identity or a rotation through π, corresponds to a unique pair $(\mathbf{n}, t)$ with $0 < t < \pi$. But it remains true that $(\mathbf{n}, 0)$ defines the identity for all $\mathbf{n}$; and also $(\mathbf{n}, \pi)$ and $(-\mathbf{n}, \pi)$ determine the same rotation.

This parametrisation gives a nice picture of the group $SO(3)$ as a whole. Consider, for each unit vector $\mathbf{n}$ and each t with $0 \leq t \leq \pi$, the point $t\mathbf{n}$ in $\mathcal{E}^3$. The set of such points makes up a solid sphere, or ball, of radius π. Each point in the interior of this ball determines a rotation and does so uniquely (the centre of the sphere corresponding to the identity). The only remaining ambiguity is that diametrically opposite points on the sphere bounding the ball determine the same rotation. The elements of $SO(3)$ are therefore in $1:1$ correspondence with the points of a ball in $\mathcal{E}^3$, the diametrically opposite points of whose boundary have been identified. In fact this correspondence is topological, in the sense that nearby points in the ball (allowing for the identifications) correspond to rotations which do not differ much in their effects. It should be apparent that because of the necessity of making these identifications the topology of the rotation group is not entirely trivial.

Euler angles. Another parametrisation of $SO(3)$ may be constructed from the fact that every rotation matrix may be obtained by multiplying matrices of rotations about just two of the axes. This is seen from the following construction. Suppose that R is a rotation which takes the standard basis vectors $\mathbf{e}_1, \mathbf{e}_2, \mathbf{e}_3$ to the vectors $\mathbf{e}'_1, \mathbf{e}'_2, \mathbf{e}'_3$ of a new right-handed orthonormal basis. Assume that $\mathbf{e}'_3 \neq \pm\mathbf{e}_3$ (the cases $\mathbf{e}'_3 = \pm\mathbf{e}_3$ may be dealt with separately). We decompose R into a succession of three rotations about coordinate axes as follows. Let R_3 be the rotation about $\mathbf{e}_3$ which brings $\mathbf{e}'_3$ into the $\mathbf{e}_1\mathbf{e}_3$-plane. Let R_2 be the rotation about $\mathbf{e}_2$ which takes $\mathbf{e}_3$ into $R_3\mathbf{e}'_3$. Then $R_3^{-1}R_2\mathbf{e}_3 = \mathbf{e}'_3 = R\mathbf{e}_3$ and so $R_2^{-1}R_3R$ is a rotation about $\mathbf{e}_3$, say $\hat{R}_3$. Thus

$$R = R_3^{-1}R_2\hat{R}_3$$

where R_3 (and therefore R_3^{-1}) and $\hat{R}_3$ are rotations about $\mathbf{e}_3$ and R_2 is a rotation about $\mathbf{e}_2$. Suppose that the angles of rotation of R_3^{-1}, R_2 and $\hat{R}_3$ are φ, ϑ and ψ respectively. Then the matrix representing R is given by

$$\begin{pmatrix} \cos\varphi & -\sin\varphi & 0 \\ \sin\varphi & \cos\varphi & 0 \\ 0 & 0 & 1 \end{pmatrix} \begin{pmatrix} \cos\vartheta & 0 & \sin\vartheta \\ 0 & 1 & 0 \\ -\sin\vartheta & 0 & \cos\vartheta \end{pmatrix} \begin{pmatrix} \cos\psi & -\sin\psi & 0 \\ \sin\psi & \cos\psi & 0 \\ 0 & 0 & 1 \end{pmatrix}$$

$$= \begin{pmatrix} \cos\varphi\cos\vartheta\cos\psi & -\cos\varphi\cos\vartheta\sin\psi & \cos\varphi\sin\vartheta \\ \quad -\sin\phi\sin\psi & \quad -\sin\varphi\cos\psi & \\ \sin\varphi\cos\vartheta\cos\psi & -\sin\varphi\cos\vartheta\sin\psi & \sin\varphi\sin\vartheta \\ \quad +\cos\varphi\sin\psi & \quad +\cos\varphi\cos\psi & \\ -\sin\vartheta\cos\psi & +\sin\vartheta\sin\psi & \cos\vartheta \end{pmatrix}.$$

The angles φ, ϑ, ψ are called the *Euler angles* of the rotation. The appropriate ranges of values of the Euler angles are

$$0 \leq \varphi < 2\pi \qquad 0 \leq \vartheta \leq \pi \qquad 0 \leq \psi < 2\pi.$$

The identity is given by $\varphi = \vartheta = \psi = 0$. The case where R is a rotation about $\mathbf{e}_3$ is covered by $\vartheta = 0$. The case when $\mathbf{e}'_3 = -\mathbf{e}_3$ is covered by $\vartheta = \pi$. The parametrisation is ambiguous in both these cases: apart from ϑ the angles are not uniquely determined.

Stereographic projection and the Cayley-Klein parameters. We next describe a parametrisation of $SO(3)$ by means of complex parameters, the Cayley-Klein parameters, which is of great importance. The general idea is that a rotation of $\mathcal{E}^3$ about a fixed point determines a rotation of any sphere with that point as centre. The points of the sphere may be made to correspond with the points of its equatorial plane by stereographic projection: then each rotation of the sphere induces a transformation of the plane. If the points of the plane are given complex coordinates, so that the plane is treated as an Argand plane, the transformations corresponding to rotations turn out to have a straightforward complex representation.

Without loss one may consider a unit sphere, since a rotation is determined by its action on a sphere of any radius. Orthonormal coordinates are taken in $\mathcal{E}^3$ with origin at the fixed point of the rotations. Let Σ denote the unit sphere with centre at the origin, let N $(0,0,1)$ and S $(0,0,-1)$ be its North and South poles respectively, and let Π be the equatorial plane $x^3 = 0$. Through any point P of the sphere (other than S) extend the line SP from the South pole until it intersects the plane Π. Let Q $(\xi^1, \xi^2, 0)$ be the point of intersection.

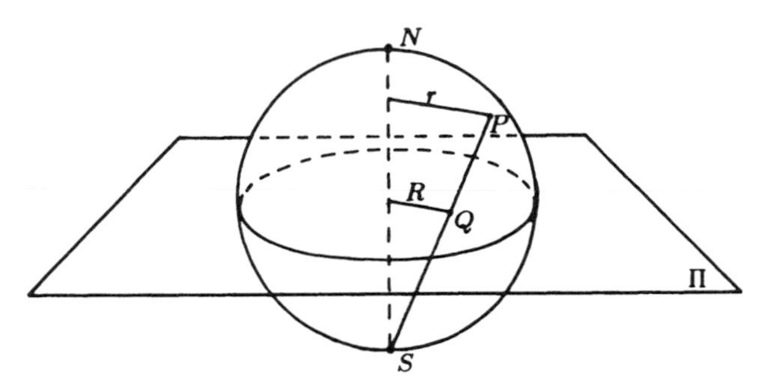

Fig. 1 Projection of a sphere onto its equatorial plane.

If r is the distance to P from the polar axis and R the distance to Q from the centre of the sphere then by similar triangles

$$\frac{R}{r} = \frac{1}{1+x^3}.$$

Thus
$$\xi^1 = \frac{x^1}{1+x^3} \qquad \xi^2 = \frac{x^2}{1+x^3}$$

and if we introduce the complex coordinate $\varsigma = \zeta^1 + i\zeta^2$ on Π then

$$\varsigma = \frac{x^1 + ix^2}{1+x^3}.$$

Exercise 29. Show that ς may also be written

$$\varsigma = \frac{1-x^3}{x^1 - ix^2}.$$

Show that

$$\varsigma\bar{\varsigma} = \frac{1-x^3}{1+x^3}$$

and deduce that

$$x^1 = \frac{\varsigma + \bar{\varsigma}}{1+\varsigma\bar{\varsigma}} \qquad x^2 = -i\left(\frac{\varsigma - \bar{\varsigma}}{1+\varsigma\bar{\varsigma}}\right) \qquad x^3 = \frac{1-\varsigma\bar{\varsigma}}{1+\varsigma\bar{\varsigma}}.$$ □

The map $\Sigma - S \to \Pi$ so defined is called *stereographic projection*. It associates a unique point Q on Π with each point P, other than S, on Σ. The South pole S may be regarded as being sent to infinity by the projection. With this convention, stereographic projection is a bijective map from Σ to the extended complex plane $\Pi \cup \{\infty\} = \Pi^\star$.

We note in passing that one may regard stereographic projection as a means of specifying coordinates for the points of the sphere; though one point, the South pole, is not assigned coordinates by this construction.

If the sphere is now rotated the projections of its points on Π^* will also move. Thus any rotation induces a transformation of Π^*. We shall derive this transformation.

First of all, rotate the sphere through an angle φ about the x^3-axis. Then $x^1 + ix^2 \mapsto e^{i\varphi}(x^1 + ix^2)$, while x^3 is unchanged: so the induced transformation of ς is simply

$$\varsigma \mapsto e^{i\varphi}\varsigma.$$

Consider next a rotation about the x^2-axis through an angle ϑ. This is not quite so straightforward. The effect of the rotation is to send $x^3 + ix^1$ to $e^{i\vartheta}(x^3 + ix^1)$ and leave x^2 unchanged. We set

$$\eta = \frac{x^3 + ix^1}{1 + x^2},$$

and then $\eta \mapsto e^{i\vartheta}\eta$. From the expressions for x^1, x^2, x^3 in Exercise 29 we obtain

$$\eta = \frac{1 + i(\varsigma + \bar{\varsigma}) - \varsigma\bar{\varsigma}}{1 - i(\varsigma - \bar{\varsigma}) + \varsigma\bar{\varsigma}} = \frac{(1 + i\varsigma)(1 + i\bar{\varsigma})}{(1 - i\varsigma)(1 + i\bar{\varsigma})} = \frac{(1 + i\varsigma)}{(1 - i\varsigma)}$$

whence

$$\varsigma = i\left(\frac{1 - \eta}{1 + \eta}\right).$$

The transformation of Π^* is therefore given by

$$\varsigma \mapsto i\left(\frac{1 - e^{i\vartheta}\eta}{1 + e^{i\vartheta}\eta}\right) = i\left(\frac{(1 - i\varsigma) - e^{i\vartheta}(1 + i\varsigma)}{(1 - i\varsigma) + e^{i\vartheta}(1 + i\varsigma)}\right)$$
$$= \frac{(e^{i\vartheta} + 1)\varsigma - i(e^{i\vartheta} - 1)}{i(e^{i\vartheta} - 1)\varsigma + (e^{i\vartheta} + 1)} = \frac{(\cos\frac{1}{2}\vartheta)\varsigma + \sin\frac{1}{2}\vartheta}{-(\sin\frac{1}{2}\vartheta)\varsigma + \cos\frac{1}{2}\vartheta}.$$

Exercise 30. Show that the transformation of Π^* corresponding to a rotation about the x^1-axis through angle χ is

$$\varsigma \mapsto \frac{(\cos\frac{1}{2}\chi)\varsigma - i\sin\frac{1}{2}\chi}{-i(\sin\frac{1}{2}\chi)\varsigma + \cos\frac{1}{2}\chi}.$$ □

A transformation of the (extended) complex plane of the form

$$\varsigma \mapsto \frac{a\varsigma + b}{c\varsigma + d},$$

where a, b, c, d are complex constants with $ad - bc \neq 0$, is called a *fractional linear transformation*. The condition $ad - bc \neq 0$ ensures that the transformation is a bijective map of the extended plane onto itself. The transformations of the complex plane obtained by stereographic projection from rotations about the x^1- and x^2-axes are manifestly of this form, but so in fact is the transformation obtained from rotation about the x^3-axis, which may be rewritten, for later convenience,

$$\varsigma \mapsto \frac{e^{i\varphi/2}\varsigma + 0}{0 \cdot \varsigma + e^{-i\varphi/2}}.$$

Thus a rotation about any coordinate axis induces a fractional linear transformation of Π^*. Now the transformation induced by the composition of two rotations will be just the composition of the transformations induced by the rotations separately. So we must next consider the composition of fractional linear transformations.

Exercise 31. Show that the composition of two fractional linear transformations is again a fractional linear transformation. □

Every rotation may be expressed as the product of rotations about coordinate axes. Thus to every rotation there corresponds a fractional linear transformation $\varsigma \mapsto (a\varsigma + b)/(c\varsigma + d)$. Its coefficients a, b, c, d are initially determined only up to multiplication by a complex constant: they will be normalised by demanding that $ad - bc = 1$. It will be observed that our expressions for the transformations corresponding to rotations about the coordinate axes already satisfy this condition.

Exercise 32. By composing the appropriate fractional linear transformations for coordinate axis rotations, show that a general rotation with Euler angles φ, ϑ, ψ induces a fractional linear transformation whose normalised coefficients a, b, c, d are given by

$$a = e^{i(\varphi+\psi)/2} \cos \tfrac{1}{2}\vartheta \qquad b = e^{i(\varphi-\psi)/2} \sin \tfrac{1}{2}\vartheta$$
$$c = -e^{-i(\varphi-\psi)/2} \sin \tfrac{1}{2}\vartheta \qquad d = e^{-i(\varphi+\psi)/2} \cos \tfrac{1}{2}\vartheta.$$ □

Exercise 33. Observe that the coefficients in Exercise 32 satisfy $d = \bar{a}$, $c = -\bar{b}$, as well as $ad - bc = 1$. Deduce that $a\bar{a} + b\bar{b} = |a|^2 + |b|^2 = 1$. Show that, conversely, any pair of complex numbers a, b such that $|a|^2 + |b|^2 = 1$ may be expressed in the form

$$a = e^{i(\varphi+\psi)/2} \cos \tfrac{1}{2}\vartheta \qquad b = e^{i(\varphi-\psi)/2} \sin \tfrac{1}{2}\vartheta$$

with $0 \leq \varphi < 2\pi$, $0 \leq \vartheta \leq \pi$, $0 \leq \psi < 2\pi$, which determine therefore the Euler angles for some rotation. Show directly, by using the expressions for x^1, x^2, x^3 in terms of ς given in Exercise 29, that the transformation of Σ determined by the fractional linear transformation $\varsigma \mapsto (a\varsigma + b)/(-\bar{b}\varsigma + \bar{a})$ with $|a|^2 + |b|^2 = 1$ has the matrix

$$\begin{pmatrix} p^2 - q^2 - r^2 + s^2 & -2(pq + rs) & 2(pr - qs) \\ 2(pq - rs) & p^2 - q^2 + r^2 - s^2 & 2(ps + qr) \\ -2(pr + qs) & 2(qr - ps) & p^2 + q^2 - r^2 - s^2 \end{pmatrix}$$

where $a = p + iq$, $b = r + is$ (and so $p^2 + q^2 + r^2 + s^2 = 1$). Confirm that this is orthogonal and has determinant $+1$, and is therefore a rotation. □

Thus to every rotation there corresponds a pair of complex numbers a, b such that $|a|^2 + |b|^2 = 1$, and to every such pair of complex numbers there corresponds a rotation. Complex numbers a, b determined in this way are called *Cayley-Klein parameters* for the rotation. The extent to which Cayley-Klein parametrisation is ambiguous still has to be determined: this will be one of the aims of the following discussion.

It will have been apparent, from Exercise 31, that composition of fractional linear transformations is related to matrix multiplication. In fact we may replace any fractional linear transformation by a 2×2 complex matrix, through the introduction of projective coordinates. We set $\varsigma = \varsigma^2/\varsigma^1$. This leaves ς^1 and ς^2 undetermined up to a common complex factor, but this is a convenience, not a nuisance. A fractional linear transformation will be recovered if it is assumed that the complex vector $Z = (\varsigma^1, \varsigma^2)^T$ transforms by matrix multiplication:

$$Z \mapsto UZ \qquad U = \begin{pmatrix} a & b \\ c & d \end{pmatrix}.$$

We require that the entries in U satisfy $ad - bc = 1$, so that $\det U = 1$, and U is a unimodular, or special, matrix. With this restriction it is easily seen that U and U' determine the same fractional linear transformation if and only if $U' = \pm U$. Furthermore, the matrix product $U_1 U_2$ represents the composite of the fractional linear transformations corresponding to U_1 and U_2.

The further conditions on a, b, c, d given in Exercise 33 reduce the set of matrices of interest to those of the form

$$\begin{pmatrix} a & b \\ -\bar{b} & \bar{a} \end{pmatrix} \qquad \text{where } |a|^2 + |b|^2 = 1.$$

Such a matrix U is necessarily *unitary*: that is to say, it satisfies $UU^\dagger = U^\dagger U = I_2$, where $U^\dagger$ is the complex conjugate transpose of U. In fact

$$UU^\dagger = \begin{pmatrix} a & b \\ -\bar{b} & \bar{a} \end{pmatrix} \begin{pmatrix} \bar{a} & -b \\ \bar{b} & a \end{pmatrix} = \begin{pmatrix} a\bar{a} + b\bar{b} & 0 \\ 0 & a\bar{a} + b\bar{b} \end{pmatrix}.$$

Thus U belongs to the group $SU(2)$ of special (or unimodular) unitary 2×2 matrices.

Exercise 34. Show that, conversely, every element of $SU(2)$ takes this form. □

The argument so far has shown that each element of $SU(2)$ determines a rotation (Exercise 33), and every rotation may be so obtained (Exercise 32). It has been hinted further that the association of rotations and elements of $SU(2)$ preserves multiplication, in other words that the map $SU(2) \to SO(3)$ is actually a homomorphism: but this remains to be finally established. One could do it by brute force, using the formula in Exercise 33, but fortunately there is a more civilised way of proceeding.

Corresponding to the transformation of Z by an $SU(2)$ matrix, namely $Z \mapsto UZ$, there is a transformation $Z^\dagger \mapsto Z^\dagger U^\dagger$ of its complex conjugate transpose $Z^\dagger = (\bar{\varsigma}^1, \bar{\varsigma}^2)$. Now $Z^\dagger Z = \varsigma^1\bar{\varsigma}^1 + \varsigma^2\bar{\varsigma}^2$, a real number, and under the transformation by U, $Z^\dagger Z \mapsto Z^\dagger U^\dagger U Z$, so this number is unchanged. On the other hand,

$$ZZ^\dagger = \begin{pmatrix} \varsigma^1\bar{\varsigma}^1 & \varsigma^1\bar{\varsigma}^2 \\ \bar{\varsigma}^1\varsigma^2 & \varsigma^2\bar{\varsigma}^2 \end{pmatrix}$$

is a hermitian matrix, that is, it is equal to its complex conjugate transpose. Under the transformation by U, $ZZ^\dagger \mapsto U(ZZ^\dagger)U^\dagger$. From the defining relations for stereographic projection (Exercise 29) we obtain (with $\varsigma = \varsigma^2/\varsigma^1$)

$$x^1 = \frac{\bar{\varsigma}^1\varsigma^2 + \varsigma^1\bar{\varsigma}^2}{\varsigma^1\bar{\varsigma}^1 + \varsigma^2\bar{\varsigma}^2} \qquad x^2 = -i\left(\frac{\bar{\varsigma}^1\varsigma^2 - \varsigma^1\bar{\varsigma}^2}{\varsigma^1\bar{\varsigma}^1 + \varsigma^2\bar{\varsigma}^2}\right) \qquad x^3 = \frac{\varsigma^1\bar{\varsigma}^1 - \varsigma^2\bar{\varsigma}^2}{\varsigma^1\bar{\varsigma}^1 + \varsigma^2\bar{\varsigma}^2}$$

from which there follow

$$\bar{\varsigma}^1\varsigma^2 = \tfrac{1}{2}Z^\dagger Z(x^1 + ix^2)$$

$$\varsigma^1\bar{\varsigma}^1 = \tfrac{1}{2}Z^\dagger Z(1 + x^3) \qquad \varsigma^2\bar{\varsigma}^2 = \tfrac{1}{2}Z^\dagger Z(1 - x^3).$$

Thus

$$\frac{1}{Z^\dagger Z}ZZ^\dagger = \frac{1}{2}\begin{pmatrix} 1 + x^3 & x^1 - ix^2 \\ x^1 + ix^2 & 1 - x^3 \end{pmatrix}.$$

So a point $\mathbf{x} = (x^1, x^2, x^3)$ on the unit sphere Σ determines a hermitian matrix, which we shall denote $\sigma(\mathbf{x})$.

Exercise 35. Show that $\sigma(\mathbf{x})$, as well as being hermitian, has determinant 0 and trace 1. Show that σ is a bijective map of Σ onto the set of hermitian 2×2 matrices with determinant 0 and trace 1. □

Exercise 36. Let h be any hermitian 2×2 matrix with determinant zero and trace 1, and let U be any element of $SU(2)$. Show that $UhU^\dagger$ has these same properties. □

It follows that, for any $\mathbf{x} \in \Sigma$ and $U \in SU(2)$, the matrix $U\sigma(\mathbf{x})U^\dagger$ represents a point on Σ. The map $\mathbf{x} \mapsto \sigma^{-1}(U\sigma(\mathbf{x})U^\dagger)$ is just the rotation of the sphere determined by U, say R_U. Then for any two elements U_1, U_2 of $SU(2)$ we have

$$R_{U_1U_2}(\mathbf{x}) = \sigma^{-1}(U_1U_2\sigma(\mathbf{x})U_2^\dagger U_1^\dagger) = \sigma^{-1}(U_1\sigma(R_{U_2}(\mathbf{x}))U_1^\dagger) = R_{U_1}R_{U_2}(\mathbf{x})$$

so that

$$R_{U_1U_2} = R_{U_1}R_{U_2}$$

and the map $U \mapsto R_U$ is therefore a homomorphism of the groups $SU(2)$ and $SO(3)$.

The question, which $SU(2)$ matrices correspond to a given rotation, or equivalently, what is the kernel of the homomorphism above, has yet to be answered. To answer it, suppose that R_U is the identity rotation. Then $U\sigma(\mathbf{x})U^\dagger = \sigma(\mathbf{x})$ for all $\mathbf{x} \in \Sigma$. Multiplying by U on the right one obtains $U\sigma(\mathbf{x}) = \sigma(\mathbf{x})U$, whence, considering either an arbitrary $\mathbf{x}$ or three specific linearly independent ones, one quickly finds $U = \pm I_2$. Thus U and U' determine the same rotation if and only if $U'U^{-1} = \pm I_2$, that is, $U' = \pm U$. This ambiguity of sign cannot be avoided. It is already present at the stage of passing from $SU(2)$ to fractional linear transformations. Again, the matrix $\mathrm{diag}(e^{i\varphi/2}, e^{-i\varphi/2})$ determines a rotation around the x^3-axis through angle φ, but as φ increases steadily from 0 to 2π the matrix changes smoothly from I_2 to $-I_2$, even though rotation through 2π is indistiguishable from the identity.

The kernel of a homomorphism of groups is a normal subgroup of the domain group. The kernel in this case consists of the two matrices $\pm I_2$ and is thus isomorphic to $\mathbf{Z}_2$, the cyclic group of order 2; and $SO(3)$ is isomorphic to the quotient group $SU(2)/\{\pm I_2\}$.

Finally, observe that it is easy to identify $SU(2)$ as a topological space: each element is uniquely determined by a pair of complex numbers $a = p + iq$, $b = r + is$, with $|a|^2 + |b|^2 = p^2 + q^2 + r^2 + s^2 = 1$. Thus the elements of $SU(2)$ are in $1:1$ correspondence with the points of the unit 3-sphere, the set of points distant 1 from a fixed point of $\mathcal{E}^4$. To obtain $SO(3)$ we have to identify diametrically opposite points on this 3-sphere. Alternatively we may restrict attention to one hemisphere, say that with $p \geq 0$: but it will still be necessary to identify opposite points on the boundary $p = 0$. By projecting onto the hyperplane $p = 0$ one sees that the 3-hemisphere is equivalent to a ball in $\mathcal{E}^3$, diametrically opposite points of whose boundary have to be identified; thus we recover the topological picture of $SO(3)$ described before.

7. The Lorentz Group

A *Lorentz transformation* of $\mathcal{E}^{1,3}$ is an isometry which leaves a point of the space fixed. With respect to orthonormal affine coordinates based on the fixed point the

Lorentz transformation is therefore represented by a matrix which is orthogonal with respect to the scalar product of signature $(1,3)$. Thus a Lorentz transformation is represented by an element of $O(1,3)$. In special relativity this group is usually denoted L and called the *Lorentz group*, and we follow this practice here.

We shall mainly be concerned with orientation-preserving Lorentz transformations, that is, with elements of $SO(1,3)$. Such transformations are said to be *proper*, and $SO(1,3)$ is called the *proper Lorentz group* and denoted L_+. There is a further specialisation of Lorentz transformations concerned with time-sense. We have distinguished (Chapter 7, Section 2)

$$\left\{\begin{array}{l}\text{timelike}\\ \text{null}\\ \text{spacelike}\end{array}\right\} \quad \text{vectors } v \text{, for which} \quad \left\{\begin{array}{l}g(v,v) > 0\\ g(v,v) = 0\\ g(v,v) < 0\end{array}\right\}$$

(recall that g has the matrix $\operatorname{diag}(-1,-1,-1,+1)$ in orthonormal coordinates). Since Lorentz transformations preserve g, they preserve the timelike, null or spacelike character of vectors. In particular, any Lorentz transformation preserves the null cone, as a whole; it may however interchange the future and the past. In fact if v and w are timelike vectors both pointing to the future or both pointing to the past then $g(v,w) > 0$, while if one points to the future and the other to the past then $g(v,w) < 0$ (Chapter 7, Exercise 9). Since a Lorentz transformation λ preserves g, if $g(v,w) > 0$ then $g(\lambda(v),\lambda(w)) > 0$. Thus a Lorentz transformation either preserves future-pointing vectors and past-pointing vectors separately, or it interchanges the whole future with the whole past. A Lorentz transformation which preserves the future and the past separately (maps future-pointing timelike vectors to future-pointing timelike vectors) is called *time-preserving* or *orthochronous*, while one which interchanges future and past is called time-reversing or antichronous. The orthochronous Lorentz transformations constitute a subgroup of L, denoted $L^\uparrow$ and called the *orthochronous Lorentz group*.

The Lorentz transformations which are both time- and orientation-preserving form a subgroup of L called the *restricted Lorentz group*. This group is the intersection of L_+ and $L^\uparrow$ and is denoted $L_+^\uparrow$. It plays a central role in this section.

Exercise 37. Show that $\lambda \in L^\uparrow$ if and only if $\lambda_4^4 > 0$. □

Exercise 38. Show that each of the following matrices defines a Lorentz transformation:

$$O = \begin{pmatrix} -1 & 0 & 0 & 0\\ 0 & -1 & 0 & 0\\ 0 & 0 & -1 & 0\\ 0 & 0 & 0 & 1\end{pmatrix} \quad T = \begin{pmatrix} -1 & 0 & 0 & 0\\ 0 & -1 & 0 & 0\\ 0 & 0 & -1 & 0\\ 0 & 0 & 0 & -1\end{pmatrix} \quad B = \begin{pmatrix} 1 & 0 & 0 & 0\\ 0 & 1 & 0 & 0\\ 0 & 0 & 1 & 0\\ 0 & 0 & 0 & -1\end{pmatrix}.$$

Show that O is orientation-reversing but time-preserving; that T is time-reversing but orientation-preserving; and that B reverses both time and orientation. Show that together with the identity I_4 they form a group of four elements with the multiplication rules $O^2 = T^2 = B^2 = I_4$; $OT = TO = B$, $TB = BT = O$, $BO = OB = T$. Show that $L_+^\uparrow$ is a normal subgroup of L and that the quotient group $L/L_+^\uparrow$ is the four element group just defined. □

Every rotation in the spacelike 3-plane $x^4 = 0$ is a Lorentz transformation (in $L_+^\uparrow$). More generally, given any timelike vector, any rotation in its orthogonal 3-

plane is a Lorentz transformation. Such a Lorentz transformation leaves fixed the timelike vector and a spacelike vector orthogonal to it, namely any one on the axis of rotation. It therefore leaves pointwise fixed the timelike 2-plane spanned by these vectors, and so it leaves fixed the two independent null directions which this 2-plane contains. Conversely any Lorentz transformation which fixes two independent null directions induces an orthogonal transformation in the orthogonal spacelike 2-plane, and if the transformation is to be proper this must be a rotation.

As an example of a more distinctively "Lorentz" Lorentz transformation, consider the transformation whose matrix is

$$\begin{pmatrix} 1 & 0 & 0 & 0 \\ 0 & 1 & 0 & 0 \\ 0 & 0 & \cosh t & \sinh t \\ 0 & 0 & \sinh t & \cosh t \end{pmatrix}.$$

This leaves invariant the timelike 2-plane spanned by e_3 and e_4, and also the orthogonal spacelike 2-plane spanned by e_1 and e_2, acting as the identity in the latter. If we set $v = -\tanh t$ then the transformation in the 2-plane spanned by e_3 and e_4 is given by

$$\acute{x}^3 = \frac{1}{\sqrt{1-v^2}}(x^3 - vx^4) \qquad \acute{x}^4 = \frac{1}{\sqrt{1-v^2}}(-vx^3 + x^4).$$

These are the equations relating the coordinates of intertial observers in special relativity in relative motion along their x^3-axes with (constant) relative speed v. The Lorentz transformation whose matrix is displayed above is called a boost in the x^3x^4-plane. More generally, any Lorentz transformation (in $L_+^\uparrow$) which leaves fixed every vector in a spacelike 2-plane is called a *boost*.

Exercise 39. Show that by suitable choice of t any unit future pointing timelike vector in the x^3x^4-plane may be obtained from e_4 by applying the standard boost whose matrix is given above. Deduce that any any timelike vector may be transformed by a boost into any other of the same magnitude and time-sense. □

Exercise 40. Show that the two null vectors $k = (0,0,1,1)^T$ and $l = (0,0,-1,1)^T$ in the x^3x^4-plane transform under the boost given above by $k \mapsto e^t k$, $l \mapsto e^{-t} l$. Show that for any boost there is a pair of independent null vectors k, l which transform in the same way. □

Both a boost and a rotation have a pair of independent null eigenvectors (in the case of a rotation, each with eigenvalue 1). Conversely, any element of $L_+^\uparrow$ having a pair of independent null eigenvectors must consist of a boost in the timelike 2-plane they span, and a rotation in the orthogonal spacelike 2-plane (it must certainly map this spacelike 2-plane to itself, and must therefore be a rotation of it). We call such a transformation a *boost plus rotation*. However, this does not exhaust all the possibilities for elements of $L_+^\uparrow$. Every element of $L_+^\uparrow$ must leave at least one null direction fixed. The reason for this is essentially topological. A time-preserving Lorentz transformation maps the future null cone to itself: the projection of the future null cone from the origin onto the 3-plane $x^4 = 1$ (say) is a sphere in that plane, and so the Lorentz transformation induces a transformation of the sphere. When the Lorentz transformation belongs to $L_+^\uparrow$ this is an orientation-preserving diffeomorphism, and it is known that such a transformation of the sphere must have at least one fixed point. Any fixed point determines a fixed null direction of the Lorentz transformation. There are elements of $L_+^\uparrow$ which leave just one null direction fixed, as we now show.

Suppose that $\lambda \in L_+^\uparrow$ has the null vector k as an eigenvector, with $\lambda(k) = Ak$, $A > 0$. Then λ maps the 3-plane orthogonal to k to itself: this is a null 3-plane, which contains k, and we may choose within it a pair of unit orthogonal spacelike vectors r and s, which are also orthogonal to k. Then $\lambda(r)$ and $\lambda(s)$ are linear combinations of k, r and s, and using the fact that λ preserves scalar products we find that

$$\lambda(r) = Bk + \cos\vartheta r + \sin\vartheta s$$
$$\lambda(s) = Ck - \sin\vartheta r + \cos\vartheta s$$

for some B, C and ϑ.

Exercise 41. The vectors r, s are not uniquely determined but may always be changed by the addition of a multiple of k whilst retaining their orthonormality properties. Show that in general $\acute{r} = r + ak$, $\acute{s} = s + bk$ may be chosen so that λ leaves the spacelike 2-plane spanned by $\acute{r}$ and $\acute{s}$ invariant (and acts as rotation through ϑ in it), and in this case λ is a boost plus rotation. But show that this cannot be done if $A = 1$ and $\vartheta = 0$. □

We concentrate therefore on the case $\lambda(k) = k$, $\lambda(r) = Bk + r$, $\lambda(s) = Ck + s$. To complete the description of λ we introduce a further null vector l, orthogonal to both r and s but independent of k; we shall scale l so that $g(k,l) = 1$.

Exercise 42. Show that l must transform by λ according to

$$\lambda(l) = l + \tfrac{1}{2}(B^2 + C^2)k + Br + Cs.$$

Show that when B and C are not both zero, λ has no other null eigenvector than k and deduce that it cannot be a boost plus rotation. □

Transformations of this kind are called *null rotations* about k.

Exercise 43. Show that null rotations about a fixed null vector k form a subgroup of $L_+^\uparrow$, which is commutative, and is isomorphic to $\mathbf{R}^2$. Deduce that every null rotation lies on a one-parameter group of null rotations. □

We have shown that an element of $L_+^\uparrow$ has at least one null eigenvector: if it has exactly one then it is a null rotation about that null vector; if two (or more) it is a boost plus rotation, possibly with one component being the identity, unless it is the identity itself. Furthermore, since every rotation lies on a one-parameter group of rotations, every boost lies on a one-parameter group of boosts (as follows in effect from Exercises 4 and 6) and every null rotation lies on a one-parameter group of null rotations, it follows that every element of $L_+^\uparrow$ lies on a one-parameter subgroup.

The Lie algebra of Lorentzian Killing fields. The Lie algebra of Killing fields of $\mathcal{E}^{1,3}$ is 10-dimensional. A basis for it may be made up as follows: three generators of rotations

$$X_1 = x^2\partial_3 - x^3\partial_2 \qquad X_2 = x^3\partial_1 - x^1\partial_3 \qquad X_3 = x^1\partial_2 - x^2\partial_1$$

three generators of boosts

$$Y_1 = x^1\partial_4 + x^4\partial_1 \qquad Y_2 = x^2\partial_4 + x^4\partial_2 \qquad Y_3 = x^3\partial_4 + x^4\partial_3$$

and four translations $T_\alpha = \partial_\alpha$, $\alpha = 1,2,3,4$. The brackets of generators of rotations and of translations along the space axes are the same as for the Euclidean case.

Exercise 44. Show that

$$[X_1, Y_1] = 0 \qquad [X_1, Y_2] = -Y_3 \qquad [X_1, Y_3] = Y_2$$
$$[Y_2, Y_3] = X_1 \qquad [Y_3, Y_1] = X_2 \qquad [Y_1, Y_2] = X_3$$
$$[T_1, Y_1] = T_4 \qquad [T_1, Y_2] = [T_1, Y_3] = 0$$
$$[T_4, X_1] = [T_4, X_2] = [T_4, X_3] = 0 \qquad [T_4, Y_1] = T_1$$

and compute the remaining brackets. □

These results may be summarised as follows $(a, b, c = 1, 2, 3)$

$$[X_a, X_b] = -\epsilon_{abc}\delta^{cd}X_d \quad [Y_a, Y_b] = \epsilon_{abc}\delta^{cd}X_d \quad [X_a, Y_b] = -\epsilon_{abc}\delta^{cd}Y_d$$
$$[T_a, X_b] = -\epsilon_{abc}\delta^{cd}T_d \quad [T_a, Y_b] = \delta_{ab}T_4 \quad [T_4, X_a] = 0 \quad [T_4, Y_a] = T_a$$
$$[T_\alpha, T_\beta] = 0.$$

8. The Celestial Sphere

We now extend to the restricted Lorentz group $L_+^\uparrow$ the ideas leading to the establishing of the homomorphism $SU(2) \to SO(3)$, with the help of the celestial sphere.

To construct the celestial sphere, imagine that you are looking at the night sky around you. You may locate each star by marking its apparent direction on a transparent sphere with yourself at the centre, as if you were surrounded by a planetarium. This sphere we call the *celestial sphere*. Let $\{e_\alpha\}$ be an orthonormal basis whose timelike member e_4 is your 4-velocity, and let k^α be the components of a future pointing null vector: so that $(-k^\alpha)$ might represent, for example, the momentum of a photon arriving from a particular star. Then the direction of the star image in your rest frame will be given by the vector (k^a) (note that α ranges from 1 to 4, a from 1 to 3). We choose to scale (k^α) so that $k^4 = 1$; then

$$(k^1)^2 + (k^2)^2 + (k^3)^2 = 1,$$

so that the point (k^a) lies on the unit sphere. Pursuing the ideas of the previous section, we associate with the null vector $k = (k^\alpha)$ the hermitian 2×2 matrix

$$\sigma(k) = \tfrac{1}{2}\begin{pmatrix} k^4 + k^3 & k^1 - ik^2 \\ k^1 + ik^2 & k^4 - k^3 \end{pmatrix}.$$

In the case in which $k^4 = 1$, this is precisely the same process as we used in Section 4, and $\sigma(k)$ has determinant zero and trace 1. But we may extend the idea a little by dropping the restriction that $k^4 = 1$: then σ is a bijective map between the null cone and the set of hermitian 2×2 matrices with determinant zero.

If now the orthonormal basis is rotated, keeping the timelike member unchanged, the effect of the rotation on the matrix $\sigma(k)$ will be given by the appropriate $SU(2)$ matrix, as described in the previous section. A boost in the x^3x^4

2-plane will transform the components of vectors according to

$$(k^\alpha) \mapsto (k^1, k^2, \cosh t\, k^3 + \sinh t\, k^4, \sinh t\, k^3 + \cosh t\, k^4)$$

and the corresponding transformation of $\sigma(k)$ may be written

$$\sigma(k) \mapsto \tfrac{1}{2}\begin{pmatrix} e^t(k^4 + k^3) & k^1 - ik^2 \\ k^1 + ik^2 & e^{-t}(k^4 - k^3) \end{pmatrix} = S\sigma(k)S^\dagger$$

where S is the unimodular (but not unitary) matrix

$$\begin{pmatrix} e^{t/2} & 0 \\ 0 & e^{-t/2} \end{pmatrix}.$$

But it is only to be expected that in extending the discussion of the previous section from $SO(3)$ to $L_+^\uparrow$ we shall have to go outside the group $SU(2)$.

The restricted Lorentz group and SL(2,C). We now extend the map σ one stage further, by removing the restriction that its argument be null. Let σ be the map $\mathcal{E}^{1,3} \to \mathcal{H}$, the space of all hermitian 2×2 matrices, by

$$\sigma(x) = \begin{pmatrix} x^4 + x^3 & x^1 - ix^2 \\ x^1 + ix^2 & x^4 - x^3 \end{pmatrix}$$

where x^α are the coordinates of x with respect to an orthonormal coordinate system (the factor $\frac{1}{2}$ no longer has any significance so we drop it). Then

$$\det \sigma(x) = g(x,x) \qquad\qquad \operatorname{tr} \sigma(x) = 2x^4.$$

Now if S is any unimodular matrix and h any hermitian one then
(1) $ShS^\dagger$ is hermitian
(2) the map $\mathcal{H} \to \mathcal{H}$ by $h \mapsto ShS^\dagger$ is linear
(3) $\det(ShS^\dagger) = \det h$
the third of these being due to the multiplicative property of determinants and the fact that $\det S = 1$. Thus the map λ_S defined by

$$\lambda_S(x) = \sigma^{-1}(S\sigma(x)S^\dagger)$$

is affine, leaves the origin fixed and preserves norms of vectors: it is therefore a Lorentz transformation. We show that it is actually an element of $L^\uparrow$. For this it is enough to show that the image of just one future-pointing timelike vector is future pointing. We take the vector with components $(0,0,0,1)$, for which the corresponding hermitian matrix is the unit matrix. We have therefore to calculate the trace of $SS^\dagger$ for any unimodular matrix S. This is easily found to be $|a|^2 + |b|^2 + |c|^2 + |d|^2$, if

$$S = \begin{pmatrix} a & b \\ c & d \end{pmatrix},$$

and since this is positive, $\lambda_S \in L^\uparrow$. A tedious calculation shows that $\lambda_S \in L_+$ also, so that $\lambda_S \in L_+^\uparrow$. Moreover, for any two unimodular matrices S_1, S_2,

$$\lambda_{S_1S_2}(x) = \sigma^{-1}(S_1S_2\sigma(x)S_2^\dagger S_1^\dagger) = \sigma^{-1}(S_1\sigma(\lambda_{S_2}(x))S_1^\dagger) = \lambda_{S_1}(\lambda_{S_2}(x))$$

and so

$$\lambda_{S_1S_2} = \lambda_{S_1} \circ \lambda_{S_2}.$$

Thus the map from $SL(2,\mathbf{C})$, the group of unimodular 2×2 complex matrices, to $L_+^\uparrow$, given by $S \mapsto \lambda_S$, is a homomorphism.

We shall now show that this homomorphism is surjective. To do so we take advantage of the fact that every rotation in the $x^1x^2x^3$-space is the image of some S in $SL(2,\mathbf{C})$ (in fact in $SU(2)$), and every boost in the x^3x^4-plane is the image of an S in $SL(2,\mathbf{C})$, namely the diagonal matrix given in the previous subsection. We shall show that every restricted Lorentz transformation may be written as the product of matrices of rotations in the $x^1x^2x^3$-space and of a boost in the x^3x^4-plane.

Let λ be an element of $L_+^\uparrow$ and $\{\acute{e}_\alpha\}$ the images of the basis vectors $\{e_\alpha\}$ under λ, so that the $\acute{e}_\alpha$ form an oriented orthonormal basis with $\acute{e}_4$ a future-pointing timelike vector. There is a rotation R_1 leaving e_4 fixed such that $R_1(\acute{e}_4)$ lies in the x^3x^4-plane. There is a boost B in the x^3x^4-plane such that $B(e_4) = R_1(\acute{e}_4)$ (Exercise 39). Then $\mu = R_1^{-1}B$ satisfies $\mu(e_4) = \acute{e}_4$: thus $\mu^{-1}(\lambda(e_4)) = e_4$, and so $\mu^{-1}\lambda$ is a rotation leaving e_4 fixed, say $\mu^{-1}\lambda = R_2$. Finally,

$$\lambda = \mu R_2 = R_1^{-1}BR_2.$$

Here R_1^{-1} and R_2 are rotations in the $x^1x^2x^3$-space and B a boost in the x^3x^4-plane. There are therefore $SU(2)$ matrices U_1, U_2 and a diagonal $SL(2,\mathbf{C})$ matrix D such that

$$\lambda_{U_1} = R_1^{-1} \qquad \lambda_{U_2} = R_2 \qquad \lambda_D = B,$$

whence

$$\lambda_{U_1DU_2} = \lambda.$$

The map $S \mapsto \lambda_S$ is a surjective homomorphism. Its kernel comprises those $S \in SL(2,\mathbf{C})$ for which $ShS^\dagger = h$ for all hermitian h. In particular, such S must satisfy $SI_2S^\dagger = I_2$; it must therefore be unitary, and so by arguments given earlier $S = \pm I_2$. Once again the kernel of the homomorphism is $\mathbf{Z}_2$, and we have established the isomorphism of $L_+^\uparrow$ with $SL(2,\mathbf{C})/\{\pm I_2\}$.

The Lie algebras of SU(2) and SL(2,C). We now show how to find the Lie algebras of $SU(2)$ and $SL(2,\mathbf{C})$, acting on the complex vector space $\mathbf{C}^2$ whose elements $Z = (\varsigma^1, \varsigma^2)^T$ were introduced in Section 6.

The indices A, B will range and sum over 1, 2. We shall use "coordinate vector fields" $\partial_A = \partial/\partial\varsigma^A$ on $\mathbf{C}^2$, which operate formally in just the same way as the ordinary affine coordinate vector fields on a real affine space, so long as the functions on which they act are functions only of the ς^A, and not of their complex conjugates. Then any one-parameter group of 2×2 complex matrices λ_t acting on $\mathbf{C}^2$ has for its generator the vector field $A^B_C\varsigma^C\partial_B$ in the usual way, where $A = (A^B_C) = d/dt(\lambda_t)(0)$. If λ_t is a one-parameter group of unimodular matrices then by differentiating the condition $\det\lambda_t = 1$ and setting $t = 0$ one obtains

$$\operatorname{tr} A = A^1_1 + A^2_2 = 0.$$

Thus the coefficient matrix of the generator of a one-parameter group of unimodular transformations is trace-free: $A^B_B = 0$, but A is otherwise arbitrary. It may be shown that every such vector field generates a one-parameter subgroup of $SL(2,\mathbf{C})$.

The conditions that λ_t be also unitary are that $(\lambda_t)^2_2 = (\bar{\lambda}_t)^1_1$ and $(\lambda_t)^2_1 = -(\bar{\lambda}_t)^1_2$, so that the generator of a one-parameter subgroup of $SU(2)$ must have in addition $A^2_2 = \bar{A}^1_1$ and $A^2_1 = -\bar{A}^1_2$. These conditions may be summed up as follows:

$$A^\dagger = -A \qquad\qquad \operatorname{tr} A = 0.$$

Thus A must be anti-hermitian as well as trace free if $A^B_C \varsigma^C \partial_B$ is to be a generator of a one-parameter subgroup of $SU(2)$. Again, it may be shown that every such vector field does generate a one-parameter subgroup of $SU(2)$.

Exercise 45. Show that the vector fields X_1, X_2, X_3 whose coefficient matrices are

$$\begin{pmatrix} 0 & -i/2 \\ -i/2 & 0 \end{pmatrix} \qquad \begin{pmatrix} 0 & 1/2 \\ -1/2 & 0 \end{pmatrix} \qquad \begin{pmatrix} i/2 & 0 \\ 0 & -i/2 \end{pmatrix}$$

form a basis for the Lie algebra of infinitesimal generators of $SU(2)$ (over the reals, so that every element of the Lie algebra is uniquely expressible as a linear combination of these with real coefficients). Show that these basis vector fields have precisely the same bracket relations as the identically named basis elements for the Lie algebra of $SO(3)$ (Exercises 24, 25). □

Exercise 46. Show that the vector fields X_1, X_2, X_3 and $Y_1 = iX_1$, $Y_2 = iX_2$, $Y_3 = iX_3$ form a basis for the Lie algebra of infinitesimal generators of $SL(2,\mathbf{C})$, and that these satisfy the same bracket relations as the identically named basis elements for the Lie algebra of $L^\uparrow_+$ (Exercise 44). □

There is therefore a bijective correspondence between the Lie algebras of $SL(2,\mathbf{C})$ and $L^\uparrow_+$ which preserves brackets; and likewise for the Lie algebras of $SU(2)$ and $SO(3)$. Such a correspondence is called a *Lie algebra isomorphism*. It is interesting to observe that the Lie algebras are isomorphic even though the groups are not, but only "nearly" so (isomorphic up to a finite subgroup). These groups and algebras are useful examples to bear in mind for the general discussion of the relation between Lie groups and their Lie algebras in Chapter 12.

Summary of Chapter 8

An isometry of an affine metric space $\mathcal{A}$ is a smooth map $\phi\colon \mathcal{A} \to \mathcal{A}$ such that $g(\phi_* u, \phi_* v) = g(u,v)$: it therefore preserves the metric. An isometry of an affine metric space is necessarily an affine map. Isometries form a group, whose linear part may be identified with the orthogonal group $O(p, n-p)$ of matrices satisfying $M^T G M = G$, where G is the diagonal matrix $\operatorname{diag}(+1, +1, \dots, -1, -1, \dots)$ of appropriate signature representing the scalar product in an orthonormal basis. Every translation is an isometry. An isometry preserves orientation if its linear part has determinant $+1$ (the only alternative value is -1); the group of orthogonal matrices with determinant $+1$ is denoted $SO(p, n-p)$.

The infinitesimal generators of one-parameter groups of isometries are called infinitesimal isometries or Killing fields. In orthonormal affine coordinates the vector field $(A^a_b x^b + P^a)\partial_a$ is an infinitesimal isometry if and only if $A = (A^a_b)$ is skew-symmetric in the sense that $A^T G + GA = 0$. Infinitesimal isometries are solutions X of Killing's equation, which may be written in several forms: $\mathcal{L}_X g = 0$; $X(g(V,W)) = g([X,V],W) + g(V,[X,W])$; $g(\nabla_V X, W) + g(V, \nabla_W X) = 0$. In the first case the Lie derivative is defined by extension of the definition of the Lie derivative of a form: $\mathcal{L}_X g = d/dt(\phi_t{}^* g)(0)$ where ϕ is the flow of X. This definition applies more generally, to covariant tensor fields (fields of multilinear functions on the tangent spaces to $\mathcal{A}$). The Killing fields (infinitesimal isometries) form a Lie algebra, that is, a finite-dimensional vector space closed under bracket. The dimension of the algebra is $\frac{1}{2}n(n+1)$ where $n = \dim \mathcal{A}$.

A conformal transformation ϕ satisfies $g(\phi_* v, \phi_* w) = \kappa g(v, w)$ for some positive function κ. Infinitesimal conformal transformations or conformal Killing fields are generators of flows of conformal transformations. They are solutions of $\mathcal{L}_X g = \rho g$ for some function ρ, and also form a Lie algebra, except in dimension 2; they are not necessarily affine.

The Euclidean group is the isometry group of $\mathcal{E}^3$. Its orientation-preserving linear part consists of rotations, each of which leaves a line fixed, its axis. Rotations about a given axis parametrised by a chosen multiple of the angle of rotation form a one-parameter group; every one-parameter subgroup of $O(3)$ is of this form, and every element of $SO(3)$ lies on a one-parameter subgroup. The generators X_1, X_2, X_3 of rotations about the coordinate axes satisfy the bracket relations $[X_2, X_3] = -X_1$ and its cyclical variants. There is a bijective correspondence between generators of rotations and 3-vectors in which the bracket goes over to the vector product. The vector corresponding to a particular one-parameter group is the angular velocity of the rotation, and points along the axis.

Each rotation may be parametrised by a unit vector along the axis, and the angle of rotation: the rotations collectively form a 3-dimensional space which can be pictured as a solid 3-sphere with the diametrically opposite points of its surface identified. Rotations may also be parametrised by three angles of rotation about coordinate axes, the Euler angles. Finally, rotations may be parametrised by two complex numbers a, b satisfying $|a|^2 + |b|^2 = 1$, the Cayley-Klein parameters. This parametrisation is arrived at *via* the stereographic projection of the unit sphere onto its equatorial plane, which allows one to correlate rotations of the sphere and certain fractional linear transformations of the plane. In the end this procedure amounts to establishing a homomorphism of $SU(2)$ (special unitary 2×2 complex matrices) onto $SO(3)$, whose kernel is the two element group $\{\pm I_2\}$.

The Poincaré group is the isometry group of $\mathcal{E}^{1,3}$. Its linear part is called the Lorentz group L, and the subgroup of L of orientation- and time-sense-preserving transformations is $L_+^\uparrow$, the restricted Lorentz group. There are essentially three types of restricted Lorentz transformation: rotations in a spacelike 2-plane; boosts in a timelike 2-plane; and null rotations. A basis of infinitesimal Lorentz transformations consists of three generators of rotations in the x^2x^3-, x^3x^1- and x^1x^2-planes, and three generators of boosts in the x^1x^4-, x^2x^4- and x^3x^4-planes.

The homomorphism $SU(2) \to SO(3)$ may be extended to a homomorphism $SL(2,\mathbf{C}) \to L_+^\uparrow$ which is again surjective and has kernel $\{\pm I_2\}$. The Lie algebras of $SU(2)$ and $SL(2,\mathbf{C})$ are isomorphic to those of $SO(3)$ and $L_+^\uparrow$ respectively.

9. GEOMETRY OF SURFACES

This chapter should be viewed as a point of transition between the considerations of affine spaces of the first half of the book and those of the more general spaces—differentiable manifolds—of the second. The surfaces under consideration are those smooth 2-dimensional surfaces, sensible to sight and touch, of 3-dimensional Euclidean space with which everyone is familiar: sphere, cylinder, ellipsoid ... In the first instance the metrical properties of such surfaces are deduced from those of the surrounding space. One of the main geometrical tasks is to formulate a definition and measure of the curvature of a surface. One such measure is the Gaussian curvature; Gauss, for whom it is named, discovered that it is in fact an intrinsic property of the surface, which is to say that it can be calculated in terms of measurements carried out entirely within the surface and without reference to the surrounding space. This is a most important result, because it renders possible the definition and study of surfaces in the abstract and, by a rather obvious process of generalisation to higher dimensions, of so-called Riemannian and pseudo-Riemannian manifolds, of which the space-times of general relativity are examples.

We shall show in this chapter how the machinery of earlier chapters is used to study the differential geometry of 2-surfaces in Euclidean 3-space, and so pave the way to the study of manifolds in later chapters.

1. Surfaces

In earlier chapters we have used two methods of representing a surface: as a level surface of a smooth function (Section 4 of Chapter 2), and by means of coordinates, as in the case of stereographic coordinates for the sphere (Section 6 of Chapter 8; the discussion of submanifolds in Chapter 6 provides another and more general example). For the purposes of the present chapter the use of coordinates is the more convenient method. We shall describe the assignment of coordinates to the points of a surface in terms of a smooth map, as we did for submanifolds in Chapter 6. Now, however, we suppose that the map ϕ in question is defined on $\mathbf{R}^2$ or some open subset of it; and we suppose further that orthonormal coordinates have been chosen, once for all, in the codomain $\mathcal{E}^3$. Thus ϕ will generally be thought of in terms of its coordinate presentation.

We shall require that a coordinate map ϕ have the property that the induced map ϕ_* is injective (as a linear map of tangent spaces) at every point of the domain of ϕ. This requirement is designed to eliminate from consideration points where the surface may fail to be smooth. For example, one might use the map

$$\phi\colon (\xi^1, \xi^2) \mapsto (\xi^1 \cos \xi^2, \xi^1 \sin \xi^2, \xi^1)$$

to assign coordinates to the points of the cone $(x^1)^2 + (x^2)^2 - (x^3)^2 = 0$. At its vertex, the origin, this cone evidently fails to be a smooth surface. The induced map ϕ_* has the matrix representation

$$\begin{pmatrix} \cos\xi^2 & -\xi^1\sin\xi^2 \\ \sin\xi^2 & \xi^1\cos\xi^2 \\ 1 & 0 \end{pmatrix},$$

which is the Jacobian matrix of the coordinate pesentation of ϕ. It is clear that ϕ_* fails to be injective when $\xi^1 = 0$, that is, at those points which are mapped to the vertex. Unfortunately failure of ϕ_* to be injective does not necessarily indicate failure of the image of ϕ in $\mathcal{E}^3$ to be a smooth surface. Consider for example the map

$$\phi: (\xi^1, \xi^2) \mapsto (\sin\xi^1\cos\xi^2, \sin\xi^1\sin\xi^2, \cos\xi^1),$$

which assigns coordinates to the points of the unit sphere $(x^1)^2 + (x^2)^2 + (x^3)^2 = 1$: here (ξ^1, ξ^2) are *polar coordinates* for the sphere, derived from spherical polar coordinates for $\mathcal{E}^3$ (Chapter 2, Exercise 24). The Jacobian matrix of the coordinate presentation of ϕ is now

$$\begin{pmatrix} \cos\xi^1\cos\xi^2 & -\sin\xi^1\sin\xi^2 \\ \cos\xi^1\sin\xi^2 & \sin\xi^1\cos\xi^2 \\ -\sin\xi^1 & 0 \end{pmatrix},$$

from which it is clear that ϕ_* fails to be injective when $\xi^1 = 0$ in this case also: here the points in $\mathbf{R}^2$ with $\xi^1 = 0$ are those which are mapped to the North pole $(1,0,0)$. In this latter case the fault clearly lies with the coordinates, not with the nature of the subset of $\mathcal{E}^3$ with which we are dealing.

In the case of the cone no coordinate map can be found, in any neighbourhood of the vertex, whose induced map is injective at points corresponding to the vertex: in effect the existence of such a coordinate map would imply that the cone had a unique tangent plane at its vertex. On the other hand, it is easy to find a coordinate map onto a neighbourhood of the North pole of the sphere whose induced map is injective at the corresponding points of $\mathbf{R}^2$, by using polar coordinates based on some other point as pole, for example, or stereographic coordinates. What is clearly not so easy (and is in fact impossible) is to find a single coordinate map for the whole sphere whose induced map is always injective. Thus in defining a surface by means of coordinates one must demand that the induced map of the coordinate map be injective, to avoid the possibility of points like the vertex of a cone; but one must then allow for the fact that more than one coordinate system may be needed to cover the whole surface.

A subset $\mathcal{S}$ of $\mathcal{E}^3$ is called a *surface* if around each of its points there may be found an open set $\mathcal{O}$ in $\mathcal{E}^3$ such that $\mathcal{O} \cap \mathcal{S}$ is the image of an open set $\mathcal{P}$ in $\mathbf{R}^2$ by a smooth map $\phi: \mathcal{P} \to \mathcal{E}^3$ for which ϕ_* is injective at each point of $\mathcal{P}$. This definition is at the same time a generalisation and a specialisation of the definition of a submanifold given in Chapter 6: here we allow for the necessity of using several coordinate systems to cover a surface, but restrict the dimensions in question. Such a map ϕ will be called a *parametrisation* of $\mathcal{O} \cap \mathcal{S}$ (as before), or a *local parametrisation* of $\mathcal{S}$.

The fact that a parametrisation ϕ has injective induced map means that ϕ is itself locally injective. However, as the example

$$(\xi^1,\xi^2) \mapsto (\sin\xi^1\cos\xi^2, \sin\xi^1\sin\xi^2, \cos\xi^1)$$

with $\mathcal{P} = \{(\xi^1,\xi^2) \in \mathbf{R}^2 \mid 0 < \xi^1 < \pi\}$ and $\mathcal{O} = \{(x^1,x^2,x^3) \in \mathcal{E}^3 \mid x^1,x^2 \neq 0\}$ shows, a parametrisation need not be injective on the whole of its domain. In order that the parametrisation have the desirable property that different coordinates label different points of the surface, it may be necessary to restrict its domain. For any surface $\mathcal{S}$ a family of parametrisations $\phi\colon \mathcal{P} \to \mathcal{E}^3$, $\psi\colon \mathcal{Q} \to \mathcal{E}^3$ ... may be found, each of which is injective, such that the sets $\phi(\mathcal{P})$, $\psi(\mathcal{Q})$, ... together cover $\mathcal{S}$. In the case of the sphere, for example, such a family, containing just two injective parametrisations, may be constructed on the principle of polar coordinates, as follows:

$$\mathcal{P} = \mathcal{Q} = \{(\xi^1,\xi^2) \in \mathbf{R}^2 \mid 0 < \xi^1 < \pi,\ 0 < \xi^2 < 2\pi\}$$
$$\phi(\xi^1,\xi^2) = (\sin\xi^1\cos\xi^2, \sin\xi^1\sin\xi^2, \cos\xi^1)$$
$$\psi(\xi^1,\xi^2) = (-\sin\xi^1\cos\xi^2, \cos\xi^1, \sin\xi^1\sin\xi^2).$$

But this is by no means the only way of injectively parametrising the sphere. Alternatively, one could use stereographic projection from North and South poles, for example.

Exercise 1. By using the formulae of Chapter 8, Section 6, show that the stereographic parametrisations are given by $\mathcal{P} = \mathcal{Q} = \mathbf{R}^2$,

$$\phi(\xi^1,\xi^2) = (2\xi^1, 2\xi^2, -1 + (\xi^1)^2 + (\xi^2)^2)/(1 + (\xi^1)^2 + (\xi^2)^2)$$
$$\psi(\xi^1,\xi^2) = (2\xi^1, 2\xi^2, 1 - (\xi^1)^2 - (\xi^2)^2)/(1 + (\xi^1)^2 + (\xi^2)^2).$$

The points on the sphere other than the North and South poles each have two sets of stereographic coordinates. One may therefore define a map of $\mathbf{R}^2 - \{(0,0)\}$ to itself by mapping (ξ^1,ξ^2) to the North pole stereographic coordinates of the point whose South pole stereographic coordinates are (ξ^1,ξ^2). Show that this map is given by $(\xi^1,\xi^2) \mapsto (\xi^1,\xi^2)/((\xi^1)^2 + (\xi^2)^2)$, and observe that it is smooth. □

Exercise 2. Show that on the sphere the point whose standard polar coordinates are (ξ^1,ξ^2) has North pole stereographic coordinates $(\cot\frac{1}{2}\xi^1\cos\xi^2, \cot\frac{1}{2}\xi^1\sin\xi^2)$. □

Again, perpendicular projection from any plane through the centre of the sphere onto the sphere may be used to construct two maps of the interior of the unit circle (in the plane) into $\mathcal{E}^3$ whose images are the two hemisperes into which the sphere is divided by the plane. In the case of the equatorial plane these maps are given by

$$(\xi^1,\xi^2) \mapsto \left(\xi^1, \xi^2, \pm\sqrt{1 - (\xi^1)^2 - (\xi^2)^2}\right).$$

The images of the interior of the unit circle under these two maps are the hemispheres with $x^3 > 0$ and $x^3 < 0$ respectively, and the maps are parametrisations of these hemispheres; the equator itself is excluded, however. But by using the six parametrisations based on the three coordinate planes in $\mathcal{E}^3$ the sphere is completely covered.

These examples reveal three points of general significance.

(1) Two different injective parametrisations covering parts of the same surface provide distinct sets of coordinates for the points belonging to the intersection of

their images; the transformation between these coordinates is defined by a smooth bijective map between open subsets of $\mathbf{R}^2$.

(2) The last construction shows how parametrisations may be found to cover the level surface of a smooth function F on $\mathcal{E}^3$: in general, if (x_0^1, x_0^2, x_0^3) is a point of the level surface at which $\partial F/\partial x^3 \neq 0$, say, then the level surface may be represented in the form $x^3 = f(x^1, x^2)$ near (x_0^1, x_0^2, x_0^3), and then $(\xi^1, \xi^2) \mapsto (\xi^1, \xi^2, f(\xi^1, \xi^2))$ provides the required parametrisation. Such a procedure will always work, for one or other of the coordinates, provided that the partial derivatives of F do not all vanish simultaneously at any point of the level surface, that is, provided dF is never zero. Thus the level surfaces of a smooth function F are surfaces indeed, provided that dF is not zero.

(3) There are many different local parametrisations covering parts of a surface, and therefore many different ways of assigning coordinates to the points of the surface: none is to be preferred to any other, except perhaps (as in the case of the sphere) by custom or symmetry; in this respect a surface is quite different from the affine space in which it lies. It is desirable, therefore, as far as possible to use methods which do not depend on a particular choice of coordinates, at least for general theoretical work, though specific calculations will usually require specific coordinates.

2. Differential Geometry on a Surface

Tangent and cotangent spaces. At each point x on a surface S there is defined a 2-dimensional subspace of $T_x\mathcal{E}^3$ consisting of those tangent vectors which are tangent to curves lying in the surface; this is the *tangent space* to the surface at x, which we denote T_xS. Its dual T_x^*S is the *cotangent space* to the surface at x.

Elements of T_xS may be thought of as vectors in a plane in $\mathcal{E}^3$ touching the surface at x. Alternatively they may be regarded as differential operators (directional derivatives) which act on functions specified on the surface, say by restriction; these operators satisfy the linearity and Leibniz rules. A local parametrisation of S defines local coordinate vectors at each point in its image: they are the tangent vectors to the coordinate curves through the point, or equally the images of the coordinate vectors in $\mathbf{R}^2$ by the linear map of tangent vectors induced by the parametrisation. These coordinate vectors form a basis for the tangent space to S. A parametrisation also defines coordinate functions on the surface whose differentials give the dual basis of the cotangent space. We shall denote the coordinate functions by ξ^1, ξ^2 and the coordinate vector fields and differentials by $\partial_a = \partial/\partial\xi^a$ and $d\xi^a$ in the usual way: here a ranges and sums over 1, 2. Each tangent vector may thus be expressed either as a linear combination of $\partial/\partial\xi^1$ and $\partial/\partial\xi^2$, or (thinking of it as a tangent vector to $\mathcal{E}^3$) as a linear combination of $\partial/\partial x^1$, $\partial/\partial x^2$ and $\partial/\partial x^3$, the orthonormal coordinate vectors in $\mathcal{E}^3$.

Exercise 3. Show that the tangent vector $\partial/\partial x^1 + \partial/\partial x^2$ at $(0,0,1)$ in $\mathcal{E}^3$ is tangent to the unit sphere, and that its representation with respect to the stereographic coordinate vectors (based on the North pole) is $\frac{1}{2}(\partial/\partial\xi^1 + \partial/\partial\xi^2)$. □

Exercise 4. Show that the tangent space at a point x on a level surface of a function F consists of those vectors $v \in T_x\mathcal{E}^3$ such that $\langle v, dF\rangle = 0$. □

The Euclidean scalar product in $\mathcal{E}^3$ defines a scalar product on each tangent space $T_x S$ by restriction. This scalar product, or metric, g, may be used to calculate lengths of, and angles between, vectors tangent to the surface, and may be used to raise and lower indices, all without reference to the ambient space.

Classically, a metric would be expressed in the form

$$ds^2 = E(d\xi^1)^2 + 2F d\xi^1 d\xi^2 + G(d\xi^2)^2 = g_{ab}d\xi^a d\xi^b.$$

The main point here is that $g_{ab} = g(\partial/\partial\xi^a, \partial/\partial\xi^b)$ are the components of the metric with respect to the coordinate vectors, which will not in general be orthonormal. On the other hand it will certainly be possible in the neighbourhood of any point to find vector fields tangent to the surface which are orthonormal: but they will not necessarily be coordinate vector fields.

The various operations on vector fields and forms which we have introduced earlier may be applied to vector fields and forms on a surface, that is to say, vector fields and forms whose values at each point of the surface are elements of the tangent space to the surface at that point, the cotangent space to the surface, or its second exterior power (there are no non-zero p-forms for $p > 2$). In particular, a vector field V on a surface generates a flow of transformations of the surface into itself, whose induced linear maps map tangent spaces to the surface to other tangent spaces to the surface. Thus the Lie derivative of one vector field on the surface by another is again a vector field on the surface. Alternatively, the bracket of two vector fields on the surface, regarded as a differential operator on functions on the surface, defines another vector field on the surface. Again, if a vector field V is given in $\mathcal{E}^3$ in a neighbourhood of a surface, which happens to be tangent to the surface at points on it, and if $\phi: \mathcal{P} \to \mathcal{E}^3$ is a parametrisation, then there is a vector field on $\mathcal{P}$ to which V is ϕ-related. Since the brackets of ϕ-related vector fields are ϕ-related (Chapter 3, Section 10) it follows that the bracket of two vector fields in $\mathcal{E}^3$ tangent to the surface is again tangent to the surface.

The exterior derivative of a 1-form θ on the surface is given by

$$d\theta(U,V) = U\langle V,\theta\rangle - V\langle U,\theta\rangle - \langle [U,V],\theta\rangle,$$

where U, V are vector fields on the surface. It defines a 2-form on the surface.

An important example of a vector field in $\mathcal{E}^3$, specified on a surface S, which is not however a vector field on the surface in the sense just described, is furnished by a normal field. A vector field N defined on the surface such that, at each point x, N_x is orthogonal to $T_x S$ (considered as a subspace of $T_x\mathcal{E}^3$) is called a *normal* field; and if it is of unit length, a unit normal field. At each point of S there are two unit normal vectors, which point in opposite directions. Whether a consistent choice can be made to form a unit normal field all over the surface depends on whether the surface is orientable: a familiar example in which this cannot be done is the Möbius band.

Exercise 5. Show that if a surface S admits a global unit normal field N then $N \lrcorner \Omega$ restricted to the surface is nowhere vanishing, where Ω is the volume 3-form in $\mathcal{E}^3$. Show that $N \lrcorner \Omega$ is in fact a volume 2-form determined by the metric on S. Show conversely that if there is a nowhere vanishing 2-form on S then S admits a global unit normal field. □

Exercise 6. Show that with respect to North pole stereographic coordinates the metric on the sphere is given by

$$ds^2 = 4((d\xi^1)^2 + (d\xi^2)^2)/(1 + (\xi^1)^2 + (\xi^2)^2)^2.$$

Find vector fields proportional to $\partial/\partial\xi^1$ and $\partial/\partial\xi^2$ which are unit, and compute their bracket. Find the volume 2-form, in terms of $d\xi^1$ and $d\xi^2$, with respect to which $(\partial/\partial\xi^1, \partial/\partial\xi^2)$ is positively oriented. Find the 1-forms obtained by lowering the indices on the two unit vector fields and compute their exterior derivatives, expressing the answers as multiples of the volume form. □

Exercise 7. Show that on a level surface of a function F whose differential does not vanish on the surface grad F is a normal field. □

3. Curvature

The curvature of a surface is made manifest by the way the normal changes its direction as one moves from point to point. A surface will usually curve by different amounts, and possibly in different senses, in different directions. This idea of curvature being measured by the change in the unit normal, and being direction dependent, is captured by the definition we shall now develop.

On a surface S, with N one of the two (local) unit normal fields, the covariant derivative $\nabla_v N$ with respect to vectors v tangent to the surface has several important properties. First, since the covariant derivative (computed according to the rules of covariant differentiation in $\mathcal{E}^3$) respects the Euclidean scalar product,

$$g(\nabla_v N, N) = \tfrac{1}{2} v(g(N, N)) = 0$$

because N is unit, and so $\nabla_v N$, being orthogonal to N, is tangent to the surface. Now the map $v \mapsto \nabla_v N$ is linear; and therefore for each $x \in S$ there is defined by this means a linear map of $T_x S$ into itself. Finally, this map is symmetric with respect to the surface metric, in the sense that for any $v, w \in T_x S$

$$g(\nabla_v N, w) = g(v, \nabla_w N).$$

In order to show that this is the case we shall have to make use of the fact that for any two vector fields tangent to S, say V, W, the bracket $[V, W]$ is also tangent to S, and to relate the bracket to the covariant derivative we must deal with vector fields defined not just on the surface but in an open set in $\mathcal{E}^3$ about the point x in the surface. The construction of suitable vector fields is left to the reader in the following exercise.

Exercise 8. Show that given $v, w \in T_x S$ there are vector fields V, W defined on an open set $\mathcal{O}$ containing x in $\mathcal{E}^3$ such that V and W are tangent to S on S and take the values v, w at x, as follows. Let ϕ be a parametrisation of a region of S about x. Extend ϕ to a map $\hat{\phi}$ of an open subset of $\mathbf{R}^3$ into $\mathcal{E}^3$ by

$$\hat{\phi}(\xi^1, \xi^2, \xi^3) = \phi(\xi^1, \xi^2) + \xi^3 N$$

where N is the normal at $\phi(\xi^1, \xi^2)$. Show that $\hat{\phi}$ is smooth and that at points for which $\xi^3 = 0$ its induced map $\hat{\phi}_*$ is non-singular, and deduce that about each such point (whose image lies on S) there is an open set in $\mathbf{R}^3$ on which $\hat{\phi}$ is injective. By the assumption that

ϕ is a parametrisation there are vectors v_0, w_0 in $T_{(\xi^1,\xi^2)}\mathbf{R}^2$ such that $\phi_* v_0 = v$, $\phi_* w_0 = w$ (where $\phi(\xi^1, \xi^2) = x$). Let V_0, W_0 be the constant vector fields on $\mathcal{E}^3$ which are everywhere parallel to v_0, w_0, considered as vectors tangent to the $\xi^1\xi^2$-plane at $(\xi^1, \xi^2, 0)$. Show that $\hat{\phi}_* V_0$ and $\hat{\phi}_* W_0$ are well-defined local vector fields on $\mathcal{E}^3$ with the required properties. □

With vector fields V, W constructed as in the exercise and with N extended into $\mathcal{E}^3$ by defining it as $\hat{\phi}_*(\partial/\partial\xi^3)$ we have

$$\begin{aligned} g(\nabla_V N, W) - g(V, \nabla_W N) &\\ &= V\big(g(N,W)\big) - W\big(g(N,V)\big) - g(N, \nabla_V W - \nabla_W V) \\ &= V\big(g(N,W)\big) - W\big(g(N,V)\big) - g\big(N, [V,W]\big). \end{aligned}$$

On the surface, $V\big(g(N,W)\big)$ depends only on the values of $g(N,W)$ on $\mathcal{S}$ since V is tangent to $\mathcal{S}$. But $g(N,W) = 0$ on $\mathcal{S}$, and so $V\big(g(N,W)\big) = 0$ there. Likewise, $W\big(g(N,V)\big) = 0$ on $\mathcal{S}$. Finally, $g\big(N,[V,W]\big) = 0$ on $\mathcal{S}$ because $[V,W]$ is tangent to $\mathcal{S}$ (actually, $[V,W] = 0$ by construction). Thus at $x \in \mathcal{S}$

$$g(\nabla_v N, w) = g(v, \nabla_w N).$$

The linear map $T_x\mathcal{S} \to T_x\mathcal{S}$ by $v \mapsto \nabla_v N$ could reasonably be called the curvature map; it is in fact called the *Weingarten map*.

Exercise 9. Compute the Weingarten map at the origin for the surface given in orthonormal coordinates by $x^3 = a_1(x^1)^2 + a_2(x^2)^2$ (an elliptic paraboloid for $a_1a_2 > 0$, a hyperbolic paraboloid for $a_1a_2 < 0$). Express the result as a matrix with respect to an orthonormal basis for the tangent plane to the surface, whose vectors lie in the directions of the x^1- and x^2-axes. □

Exercise 10. Show that the normal component of $\nabla_V W$ (where V, W are any two vector fields tangent to $\mathcal{S}$) is $-g(\nabla_V N, W)$. □

The curvature properties of a surface at a point are defined by the algebraic invariants of the Weingarten map. Its determinant is the *Gaussian curvature* K and half its trace the *mean curvature* H of the surface at the point. Since the Weingarten map is symmetric with respect to g and therefore will be represented by a symmetric matrix with respect to an orthonormal basis for $T_x\mathcal{S}$ it has real eigenvalues, and eigenvectors corresponding to distinct eigenvalues are orthogonal. These eigenvalues are called the *principal curvatures* and the corresponding eigenvectors the *principal directions* at the point. The elliptic paraboloid of Exercise 9 is an example of a surface with positive Gaussian curvature and (provided $a_1 \neq a_2$) distinct principal curvatures, whose principal directions at the origin are the x^1- and x^2-axes. The hyperbolic paraboloid has negative Gaussian curvature, and the same principal directions. If the Weingarten map is a multiple of the identity the point is called an *umbilic*: this occurs when $a_1 = a_2$ in Exercise 9, and at every point of the sphere. If the Weingarten map is zero the point is a *parabolic umbilic*, or *planar* point (thus the Gaussian curvature may be zero at a point without the point being planar: all points of a cylinder have zero Gaussian curvature without being planar).

Exercise 11. Show that the origin is a planar point of the "monkey saddle" surface $x^3 = x^1(x^1 - \sqrt{3}x^2)(x^1 + \sqrt{3}x^2)$. □

4. Surface Geometry using Exterior Forms

We have defined the curvature in terms of vector fields on the surface; it is also possible to discuss the geometry of a surface in terms of forms. On the surface, choose a pair of 1-forms θ^1, θ^2 which are orthonormal (that is, obtained by lowering the indices on a pair of orthonormal vector fields V_1, V_2 on the surface, using the metric on the surface). Now $\{V_1, V_2, V_3 = N\}$ is an orthonormal basis for vector fields on $\mathcal{E}^3$ defined on the surface, which can be extended into an open set about the surface as described in Exercise 8. The 1-forms θ^1, θ^2, θ^3on $\mathcal{E}^3$ dual to these satisfy the structure equations

$$d\theta^\alpha + \omega^\alpha_\beta \wedge \theta^\beta = 0, \qquad \alpha, \beta = 1, 2, 3$$

where the connection forms ω^α_β, defined by

$$\nabla_U V_\beta = \langle U, \omega^\alpha_\beta \rangle V_\alpha,$$

satisfy

$$\omega^\gamma_\beta \delta_{\alpha\gamma} + \omega^\gamma_\alpha \delta_{\beta\gamma} = 0$$

(see Section 7 of Chapter 5 and Exercise 52 of Chapter 7). Consider these structure equations on the surface (so that the vector arguments are restricted to be vectors tangent to the surface). Since θ^3 vanishes if its argument is tangent to the surface we have

$$d\theta^1 + \omega^1_2 \wedge \theta^2 = 0 \qquad d\theta^2 + \omega^2_1 \wedge \theta^1 = 0$$

because of the skew-symmetry of ω^α_β; also, $\omega^2_1 + \omega^1_2 = 0$. We set $\omega = \omega^2_1$ and call it the *connection form* of the surface for the orthonormal basis of 1-forms $\{\theta^1, \theta^2\}$; then

$$d\theta^1 = \omega \wedge \theta^2 \qquad d\theta^2 = -\omega \wedge \theta^1.$$

These are the *first structure equations* of the surface.

Exercise 12. The connection form may be defined as follows. Since V_1, V_2 are orthonormal, for any vector v tangent to S the tangential component of $\nabla_v V_1$ is in the direction of V_2, and that of $\nabla_v V_2$ is in the direction of V_1. Define 1-forms ω_1, ω_2 on S as follows: $\langle v, \omega_1 \rangle = g(\nabla_v V_1, V_2)$, $\langle v, \omega_2 \rangle = g(V_1, \nabla_v V_2)$. Thus ω_1, ω_2 measure the components of $\nabla_v V_1$, $\nabla_v V_2$ tangent to S. Show that $\omega_1 = -\omega_2 = \omega$. Deduce the first structure equations directly. □

Since θ^3 vanishes if its argument is tangent to the surface, so does $d\theta^3$ if both of its arguments are tangent to the surface; thus on the surface the *symmetry condition*

$$\omega^3_1 \wedge \theta^1 + \omega^3_2 \wedge \theta^2 = 0$$

holds. Now

$$\nabla_v V_3 = \langle v, \omega^1_3 \rangle V_1 + \langle v, \omega^2_3 \rangle V_2 = -\langle v, \omega^3_1 \rangle V_1 - \langle v, \omega^3_2 \rangle V_2,$$

and on S, $V_3 = N$: thus ω^1_3 and ω^2_3 are related to the Weingarten map. In fact if we set $\langle V_a, \omega^b_3 \rangle = W^b_a$ then (W^b_a) is the matrix of the Weingarten map with respect to the orthonormal basis $\{V_a\}$. The symmetry condition then corresponds to the symmetry of this matrix. Moreover $\omega_3^a = W_b^a\theta^b$, from which it follows that the Gaussian curvature K and mean curvature H are given by

$$\omega_3^1 \wedge \omega_3^2 = \det(W_b^a)\theta^1 \wedge \theta^2 = K\theta^1 \wedge \theta^2$$
$$\omega_3^1 \wedge \theta^2 + \theta^1 \wedge \omega_3^2 = \operatorname{tr}(W_b^a)\theta^1 \wedge \theta^2 = 2H\theta^1 \wedge \theta^2.$$

The connection forms ω_β^α for the local orthonormal basis $\{V_\alpha\}$ in $\mathcal{E}^3$ satisfy

$$d\omega_\beta^\alpha + \omega_\gamma^\alpha \wedge \omega_\beta^\gamma = 0$$

as we showed in Section 7 of Chapter 5. Specialising again to the surface one obtains first *Gauss's equation*

$$d\omega = -\omega_3^1 \wedge \omega_3^2,$$

and then the *Codazzi equations*

$$d\omega_3^1 = \omega \wedge \omega_3^2 \qquad\qquad d\omega_3^2 = -\omega \wedge \omega_3^1.$$

Finally, from Gauss's equation and the equation for the Gaussian curvature there follows the *second structure equation* for the surface:

$$d\omega = -K\theta^1 \wedge \theta^2.$$

5. The Levi-Civita Connection

It is interesting to compare the two structure equations for a surface with those for Euclidean space of the same dimension, $\mathcal{E}^2$. In each case there is a single connection form ω and the first structure equations are the same for both:

$$d\theta^1 = \omega \wedge \theta^2 \qquad\qquad d\theta^2 = -\omega \wedge \theta^1.$$

The second structure equation for $\mathcal{E}^2$ is

$$d\omega = 0$$

while that for the surface is

$$d\omega = -K\theta^1 \wedge \theta^2.$$

This confirms that $\mathcal{E}^2$, which may after all be considered as a surface in $\mathcal{E}^3$, has zero Gaussian curvature, as one might expect. More suggestive is the identity of the first structure equations. Exercise 12 showed that the connection form ω may be defined in terms of the components of covariant derivatives tangential to the surface. In the case of $\mathcal{E}^2$, considered as a surface in $\mathcal{E}^3$, covariant differentiation of a vector field on $\mathcal{E}^2$ along a vector tangent to $\mathcal{E}^2$ is the same whether the operation is carried out with respect to parallelism in $\mathcal{E}^2$ or in $\mathcal{E}^3$. In the case of a surface we have been forced up to now to rely on the parallelism in $\mathcal{E}^3$ to compute covariant derivatives even when the direction of differentiation and the differentiated vector field were both tangent to the surface. The fact that this process produces the same first structure equations as those of $\mathcal{E}^2$ suggests that it may be possible to introduce a concept of parallelism and an operation of covariant differentiation in the surface, enjoying some of the properties of parallelism and covariant differentiation in $\mathcal{E}^2$, from which the first structure equations would follow. That this generalisation of the concept of parallelism can be made was discovered by Levi-Civita and the structure obtained is named after him as the Levi-Civita connection of the surface. The clue to the construction is contained in Exercise 12. Before we explain it in detail, however, we shall describe a thought experiment which suggests that one at least of the properties of parallelism in $\mathcal{E}^2$ will have to be renounced.

The example concerns parallelism on the sphere. The great circles on a sphere (the intersections of the sphere with planes through its centre) are, like the straight lines in $\mathcal{E}^2$, curves of minimal length, and therefore constitute a possible generalisation of straight lines. There are two properties of parallelism in $\mathcal{E}^2$ concerned with straight lines which we seek also to generalise: first, the tangent vectors to an affinely parametrised straight line are parallel along it; and secondly, vectors specified at different points on the line, and given to be parallel, will all make the same angle with it. The difficulty which arises in the case of the sphere becomes clear if one attempts to endow parallel vectors along great circles with these properties. Consider for example two great circles through the North and South poles of the sphere, say the Greenwich meridian and the meridian at 90° W; and consider the vector at the North pole which is the initial tangent vector to the Greenwich meridian, say v_N. Then so far as the Greenwich meridian is concerned, the vector at the South pole parallel to v_N is again the tangent to the Greenwich meridian. But the rule of constant angles, applied to the other great circle, produces at the South pole a different vector parallel to v_N—in fact the negative of the first. And clearly, by choosing other meridians, one could obtain vectors parallel to v_N, according to these criteria, in all possible directions at the South pole.

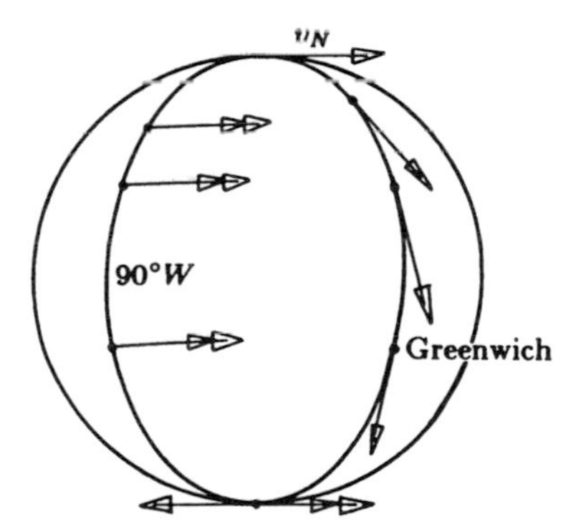

Fig. 1 Parallel vectors on a sphere.

The major difference between Levi-Civita parallelism on surfaces and parallelism in $\mathcal{E}^2$, which is so convincingly demonstrated by this example, is that parallelism on a surface is path-dependent: that is to say, there is no longer any substance to the statement that vectors tangent to a surface at different points are parallel; one may only ask whether or not vectors tangent to the surface, specified along a curve in it, are parallel. Despite this apparent defect, this concept of parallelism and the operation of covariant differentiation deduced from it are enormously useful, and indeed the dependence of parallelism on the path is related to the curvature of the surface in a very interesting way.

Now to the definitions. We say that a vector field V along a curve σ in a surface $\mathcal{S}$, which is tangent to $\mathcal{S}$ at each point $\sigma(t)$, is *parallel* along σ in the sense of Levi-Civita if $\nabla_{\dot\sigma(t)}V$ is normal to $\mathcal{S}$ for every t in the domain of σ (where $\dot\sigma(t)$ is the vector tangent to σ at $\sigma(t)$). We define a rule of covariant differentiation associated with Levi-Civita parallelism as follows: the *covariant derivative* $\hat\nabla_v W$ of the vector field W tangent to $\mathcal{S}$ by the vector v tangent to $\mathcal{S}$ is the component of $\nabla_v W$ tangent to $\mathcal{S}$ (where ∇ denotes the covariant derivative operator in $\mathcal{E}^3$, as before). This law of parallelism, with its associated covariant differentiation, is called the *Levi-Civita connection* on $\mathcal{S}$. The definitions are motivated by Exercise 12; we have still to show that they reproduce most, although not all, of the properties of parallelism and the covariant derivative in $\mathcal{E}^2$, or indeed $\mathcal{E}^n$.

The covariant derivative. It is convenient to begin by examining the properties of the covariant derivative operator $\hat\nabla$. It will be recalled (Chapter 3, Section 11 and Chapter 7, Section 6) that the corresponding operator ∇ in $\mathcal{E}^n$ has the following properties:

(1) $\nabla_{U+V}W = \nabla_U W + \nabla_V W$
(2) $\nabla_{fV}W = f\nabla_V W$
(3) $\nabla_U(V + W) = \nabla_U V + \nabla_U W$
(4) $\nabla_V(fW) = f\nabla_V W + (Vf)W$
(5) $U(g(V,W)) = g(\nabla_U V, W) + g(V, \nabla_U W)$
(6) $\nabla_V W - \nabla_W V = [V,W]$
(7) $\nabla_U\nabla_V W - \nabla_V\nabla_U W = \nabla_{[U,V]}W$

for any vector fields U, V, W and function f on $\mathcal{E}^n$. The first four of these properties are concerned with the linearity (or otherwise) of ∇: in particular, the contrasting effects of multiplication of the arguments by functions exhibited in (2) and (4) are distinctive features of covariant differentiation. Property (5) is a consequence of the fact that parallel translation preserves scalar products (and therefore lengths and angles). The last two properties express the interrelationships of covariant differentiation and the bracket operation on vector fields. Furthermore, properties (1)–(6) of this list uniquely determine the covariant derivative operator in $\mathcal{E}^n$, since it follows that

$$g(\nabla_U V, W) = \tfrac{1}{2}\{U(g(V,W)) + V(g(U,W)) - W(g(U,V)) + g([U,V],W) - g([U,W],V) - g(U,[V,W])\}$$

(Exercise 50 of Chapter 7), a formula which defines $\nabla_U V$ in terms of the directional derivative and bracket once g is given. Note that property (7), the second order commutation relation, is not involved in this determination; it is a consequence of the fact that coordinates, namely affine coordinates, may be found in $\mathcal{E}^n$ with respect to which affine the metric has constant components, and that parallel vector fields have constant components in terms of these coordinates; this property is therefore special.

The Levi-Civita covariant derivative on a surface enjoys all of properties (1)–(6), but not in general property (7). If N is a unit normal vector field on the surface then

$$\hat{\nabla}_U V = \nabla_U V - g(\nabla_U V, N)N$$

for any vector fields U, V tangent to the surface. From this it follows that (for example) property (4) holds for $\hat{\nabla}$ as a consequence of its truth for ∇: for if f is any function on $\mathcal{S}$, and F any function defined in a neighbourhood of a point of $\mathcal{S}$ in $\mathcal{E}^3$ which agrees with f on $\mathcal{S}$, then (on $\mathcal{S}$)

$$\begin{aligned}\hat{\nabla}_V(fW) &= \nabla_V(FW) - g(\nabla_V(FW), N)N \\ &= F\nabla_V W + (VF)W - Fg(\nabla_V W, N)N - (VF)g(W,N)N \\ &= f\nabla_V W + (Vf)W - fg(\nabla_V W, N)N \qquad \text{on } \mathcal{S} \\ &= f\hat{\nabla}_V W + (Vf)W.\end{aligned}$$

Exercise 13. Prove properties (1), (2) and (3) similarly. □

Exercise 14. Prove property (5). □

Property (6) for $\hat{\nabla}$ follows from the same property for ∇ (applied to any extensions of the vector fields V and W to $\mathcal{E}^3$) and the fact that, since V and W are tangent to $\mathcal{S}$, so is $[V,W]$.

Exercise 15. Show that for any vector fields U, V, W tangent to $\mathcal{S}$,

$$\begin{aligned}g(\hat{\nabla}_U V, W) = \tfrac{1}{2}\{&U(g(V,W)) + V(g(U,W)) - W(g(U,V)) \\ &+ g([U,V],W) - g([U,W],V) - g(U,[V,W])\}.\end{aligned}$$

(Since the arguments of g are all vector fields tangent to $\mathcal{S}$, g here represents the metric on the surface). □

Exercise 16. Suppose that $V = V^a\partial/\partial\xi^a$, $W = W^a\partial/\partial\xi^a$ with respect to coordinates (ξ^1, ξ^2) on the surface. Show that

$$\hat{\nabla}_V W = V^c\left(\frac{\partial W^a}{\partial\xi^c} + \Gamma^a_{bc}W^b\right)\frac{\partial}{\partial\xi^a},$$

where the coefficients Γ^a_{bc} are defined in terms of the metric components g_{ab} by

$$\Gamma^a_{bc} = \frac{1}{2}g^{ad}\left(\frac{\partial g_{cd}}{\partial\xi^b} + \frac{\partial g_{bd}}{\partial\xi^c} - \frac{\partial g_{bc}}{\partial\xi^d}\right).$$ □

Exercise 17. Let $\{U_1, U_2\}$ be an orthonormal basis of vector fields on $\mathcal{S}$ with dual orthonormal basis of 1-forms $\{\theta^1, \theta^2\}$. Show that for every vector v tangent to $\mathcal{S}$, $v \mapsto g(\hat{\nabla}_v U_1, U_2)$ defines a 1-form ω on $\mathcal{S}$, the connection form of the Levi-Civita connection. Show that $g(\hat{\nabla}_v U_2, U_1) = -g(\hat{\nabla}_v U_1, U_2)$; deduce that

$$\hat{\nabla}_V W = V\langle W, \theta^a\rangle U_a + \langle W, \theta^1\rangle\langle V, \omega\rangle U_2 - \langle W, \theta^2\rangle\langle V, \omega\rangle U_1$$

and rederive the first structure equations. □

The Levi-Civita covariant derivative makes sense in several forms: when V and W are vector fields tangent to $\mathcal{S}$ then $\hat{\nabla}_V W$ is a vector field tangent to $\mathcal{S}$; when v is a vector tangent to $\mathcal{S}$ at x and W a vector field defined in a neighbourhood of x in $\mathcal{S}$ and tangent to $\mathcal{S}$ then $\hat{\nabla}_v W$ is well-defined and is a vector tangent to $\mathcal{S}$ at x; when σ is a curve in $\mathcal{S}$ and W a vector field defined along σ and tangent to $\mathcal{S}$ then $\hat{\nabla}_{\dot\sigma} W$ is a vector field on σ tangent to $\mathcal{S}$. This last construction, which corresponds to what we called the absolute derivative in Chapter 2, is useful in considering parallelism in the sense of Levi-Civita, to which we now turn.

Parallelism. A vector field V along a curve σ in $\mathcal{S}$, which is tangent to $\mathcal{S}$, is *parallel* along σ if $\hat{\nabla}_{\dot{\sigma}}V = 0$, in other words if $\nabla_{\dot{\sigma}}V$ is normal to $\mathcal{S}$. From Exercise 16 it appears that the components of a parallel vector field satisfy a pair of linear first order ordinary differential equations

$$\frac{dV^a}{dt} + \Gamma^a_{bc}V^b\frac{d\sigma^c}{dt} = 0.$$

Any solution (V^1, V^2) of these equations defines a parallel vector field along σ. From the properties of such systems of differential equations it follows that for a given tangent vector v to $\mathcal{S}$ at a given point $\sigma(t_1)$ of σ there is a unique parallel vector field along σ coincident with v at $\sigma(t_1)$. Using this result we define a map from (say) $T_{\sigma(t_1)}\mathcal{S}$ to $T_{\sigma(t_2)}\mathcal{S}$, called *parallel translation along* σ, denoted Υ, as follows: given $v \in T_{\sigma(t_1)}\mathcal{S}$, $\Upsilon(v)$ is the value at $\sigma(t_2)$ of the parallel vector field along σ whose initial value (at $\sigma(t_1)$) is v. It follows from the properties of linear differential equations that any linear combination of parallel vector fields, with constant coefficients, is parallel and so Υ is linear; that the only parallel vector field which takes the value 0 anywhere is the zero field and so Υ is injective; and thus by dimensionality that Υ is an isomorphism. Thus, just as in the case of parallelism in $\mathcal{E}^n$, Levi-Civita parallelism defines an isomorphism of distinct tangent spaces; and by property (5) of the covariant derivative, parallel translation preserves lengths and angles. However, there is this important difference, that on a surface parallel translation depends on the path joining the two points in question.

Exercise 18. Examine the effect of a reparametrisation of σ and confirm that parallel translation depends on the path, not the curve. □

Exercise 19. Compute the equations of parallel translation on a sphere in terms of spherical polar coordinates and confirm the correctness of the description of parallel translation along meridians given above. □

Geodesics. An autoparallel curve on a surface, that is to say, a curve whose tangent vector is parallelly transported along itself, is called a *geodesic*. The coordinate functions of a geodesic γ satisfy the second-order non-linear differential equations

$$\frac{d^2\gamma^a}{dt^2} + \Gamma^a_{bc}\frac{d\gamma^b}{dt}\frac{d\gamma^c}{dt} = 0.$$

It follows that there is a unique geodesic through a given point of the surface with given initial tangent vector.

Exercise 20. Show that (in terms of the covariant derivative operator in the ambient space $\mathcal{E}^3$) a curve γ in a surface is a geodesic if and only if its acceleration $\nabla_{\dot{\gamma}}\dot{\gamma}$ is normal to the surface, so that the geodesics are the paths of particles constrained to move on the surface by smooth constraints and under the influence of no other forces. □

Exercise 21. Show that a plane curve in a surface is a geodesic if and only if it is the intersection of the surface with a plane everywhere normal to it, and confirm that the great circles are geodesics on the sphere. To show that these are the only geodesics, write down the geodesic equations using polar coordinates for the sphere (Section 1) and deduce that the functions A, B, C given by

$$A = -\dot{\xi}^1 \sin \xi^2 - \dot{\xi}^2 \sin \xi^1 \cos \xi^1 \cos \xi^2$$
$$B = \dot{\xi}^1 \cos \xi^2 - \dot{\xi}^2 \sin \xi^1 \cos \xi^1 \sin \xi^2$$
$$C = \dot{\xi}^2 \sin^2 \xi^1$$

are constants along any geodesic; show that $A \sin \xi^1 \cos \xi^2 + B \sin \xi^1 \sin \xi^2 + C \cos \xi^1 = 0$, and infer that the geodesic lies in the plane through the centre of the sphere perpendicular to the vector whose orthogonal components are (A, B, C). □

Exercise 22. Deduce from the fact that parallel translation preserves lengths that the tangent vector to a geodesic has constant length, and that therefore two geodesics with the same path differ by an affine reparametrisation $s \mapsto as + b$ (a, b constant). Show that any other reparametrisation of a geodesic produces a curve which satisfies $\hat{\nabla}_{\dot{\gamma}}\dot{\gamma} = \kappa\dot{\gamma}$ where κ is some function along the curve; and that conversely any curve whose tangent vector satisfies this equation is a reparametrisation of a geodesic. □

6. Connection and Curvature

We return now to the consideration of the final property of the covariant derivative in $\mathcal{E}^n$, the second order commutation relation

$$\nabla_U \nabla_V W - \nabla_V \nabla_U W = \nabla_{[U,V]} W,$$

and describe how this is modified on a surface $\mathcal{S}$. By using the definition of $\hat{\nabla}$ we may express $\nabla_V W$ (where V and W are tangent to $\mathcal{S}$) in the form

$$\nabla_V W = \hat{\nabla}_V W + g(\nabla_V W, N)N = \hat{\nabla}_V W - g(W, \nabla_V N)N$$

since $g(W, N) = 0$. Notice that $\nabla_V N$ is the vector field obtained by applying the Weingarten map to V. By covariantly differentiating again we obtain

$$\begin{aligned} 0 &= \nabla_U \nabla_V W - \nabla_V \nabla_U W - \nabla_{[U,V]} W \\ &= \hat{\nabla}_U \hat{\nabla}_V W - \hat{\nabla}_V \hat{\nabla}_U W - \hat{\nabla}_{[U,V]} W \\ &\quad + g(W, \nabla_U N)\nabla_V N - g(W, \nabla_V N)\nabla_U N, \end{aligned}$$

where in eliminating the remaining terms we have used the fact that (for example) $g(\hat{\nabla}_U W, \nabla_V N) = g(\nabla_U W, \nabla_V N)$ because $\nabla_V N$ is tangential and $\nabla_U W - \hat{\nabla}_U W$ is normal to $\mathcal{S}$. Thus

$$\hat{\nabla}_U \hat{\nabla}_V W - \hat{\nabla}_V \hat{\nabla}_U W - \hat{\nabla}_{[U,V]} W = g(W, \nabla_V N)\nabla_U N - g(W, \nabla_U N)\nabla_V N.$$

The right hand side of this equation is formed from the Weingarten map; the fact that it is skew symmetric in U and V suggests (since the dimension is 2) that it should be expressible in terms of the determinant of the Weingarten map, that is, the Gaussian curvature K.

Exercise 23. Show that

$$g(W, \nabla_V N)\nabla_U N - g(W, \nabla_U N)\nabla_V N = K\big(g(V, W)U - g(U, W)V\big).$$ □

Exercise 24. Show that the equation

$$\hat{\nabla}_U \hat{\nabla}_V W - \hat{\nabla}_V \hat{\nabla}_U W - \hat{\nabla}_{[U,V]} W = K\big(g(V, W)U - g(U, W)V\big)$$

is equivalent to the second structure equation. □

The results of this calculation may be summed up by saying that the Gaussian curvature measures the non-commutativity of second covariant derivatives.

7. Tensor Fields

The calculation above reveals another unexpected and important fact about $\hat{\nabla}_U\hat{\nabla}_V W - \hat{\nabla}_V\hat{\nabla}_U W - \hat{\nabla}_{[U,V]}W$: though it appears on the face of it that this particular combination involves derivatives of the vector fields U, V and W, it is clear from the right hand side of the equation in Exercise 24 that in fact it depends only on their values. That is to say, if we write $R(U,V)W$ to denote this expression (a notation which reflects the similar roles of U and V and the different role of W in it) then the value of the vector field $R(U,V)W$ at a point x in S depends only on the values of U, V and W at x. This observation is confirmed when $R(U,V)W$ is calculated in terms of the connection coefficients Γ^a_{bc} relative to some coordinate system on S.

Exercise 25. Show that $R(U,V)W = R^a{}_{bcd}W^bU^cV^d\partial_a$ where

$$R^a{}_{bcd} = \frac{\partial\Gamma^a_{bd}}{\partial\xi^c} - \frac{\partial\Gamma^a_{bc}}{\partial\xi^d} + \Gamma^a_{ce}\Gamma^e_{bd} - \Gamma^a_{de}\Gamma^e_{bc}.$$ □

The object R defined in this way is an example of a tensor field. To be precise, a *tensor* at a point x in a surface S is a map from the r-fold Cartesian product of T_xS into either $\mathbf{R}$ or T_xS, as appropriate, which is multilinear; it is said to be of *type* $(0,r)$ or $(1,r)$ respectively. A *tensor field* is a choice of tensor at each point of S, of the same type everywhere, which is smooth in the sense that its components with respect to coordinate vector fields are smooth (local) functions. It is multilinear over the functions on S. Thus R is a tensor field of type $(1,3)$. Other tensor fields already introduced are g (type $(0,2)$) and any form (type $(0,p)$ if it is a p-form). We have also made use of the identity map of tangent spaces, represented by the Kronecker delta δ^a_b; this is a tensor field of type $(1,1)$. Tensor fields of type $(0,p)$ are said to be covariant, of valence p (Chapter 8, Section 3).

Actually, the definition given above, although it generalises the definition of Chapter 8, still does not cover the most general kind of tensor, even allowing for the limitation on dimension due to the fact that we are here concerned only with surfaces: more general kinds of tensor will be discussed in a later chapter.

Exercise 26. Show that a type $(1,1)$ tensor field may be regarded as a field of linear maps of tangent spaces. The Weingarten map is an example of a type $(1,1)$ tensor field on a surface. Show that if T is a type $(1,1)$ tensor field then $(v,w) \mapsto g(T(v),w)$ is a type $(0,2)$ tensor field, obtained by lowering an index on T; and that in the case of the Weingarten map this tensor field is symmetric. □

The symmetric type $(0,2)$ tensor field constructed in this way from the Weingarten map is called the *second fundamental form* of the surface.

As we pointed out above, it is not immediately obvious that R is a tensor field, especially since $(U,V) \mapsto \nabla_U V$ does not define a tensor field (its value at a point depends on the derivatives of the components of V). The non-tensorial nature of the covariant derivative shows up in the effect of multiplying V by a function:

$$\nabla_U(fV) = f\nabla_U V + (Uf)V.$$

In order for a map of vector fields on S to be a tensor field it must be multilinear, not just over the reals, but over the algebra of smooth functions $\mathcal{F}(S)$. That is to say, in order to be a tensor field, a map $T\colon \mathcal{X}(\mathcal{S})^r \to \mathcal{F}(\mathcal{S})$ or $\mathcal{X}(\mathcal{S})$ (here $\mathcal{X}(\mathcal{S})$ is the set of smooth vector fields on $\mathcal{S}$, a module over $\mathcal{F}(\mathcal{S})$) must satisfy the multilinearity conditions

$$T(V_1, V_2, \ldots, V_k + V_k', \ldots, V_r) = T(V_1, V_2, \ldots, V_k, \ldots, V_r) + T(V_1, V_2, \ldots, V_k', \ldots V_r)$$
$$T(V_1, V_2, \ldots, fV_k, \ldots, V_r) = fT(V_1, V_2, \ldots, V_k, \ldots, V_r)$$

for $V_1, V_2, \ldots, V_k, V_k', \ldots, V_r \in \mathcal{X}(\mathcal{S})$ and $f \in \mathcal{F}(\mathcal{S})$. The covariant derivative satisfies the first of these, but not the second. Both conditions are required, however, in order that it be possible to express T in terms of components. (To be precise, we should explain here that we are working essentially in a coordinate patch: there are some technical problems which have to be overcome before the following arguments work globally; we postpone discussing them until later.) Suppose, for definiteness, that T is of type $(1,2)$. Then, in terms of any basis $\{U_a\}$ of vector fields we may set

$$T(U_a, U_b) = T^c{}_{ab} U_c$$

where the functions $T^c{}_{ab}$ are the *components* of T with respect to the basis. So far we have merely exploited the fact that $T(U_a, U_b) \in \mathcal{X}(\mathcal{S})$ and $\{U_a\}$ is a vector field basis. But if T is a tensor field then it must be the case that for any vector fields $V = V^a U_a$, $W = W^a U_a$, where V^a, W^a are functions,

$$T(V, W) = T^c{}_{ab} V^a W^b U_c,$$

and this requires the multilinearity property asserted above.

It is clear that R, defined by

$$R(U,V)W = \hat{\nabla}_U \hat{\nabla}_V W - \hat{\nabla}_V \hat{\nabla}_U W - \hat{\nabla}_{[U,V]} W,$$

is additive in each of its arguments (that is to say, it satisfies the first of the multilinearity conditions). We can show directly, without appealing to its expression in terms of the Weingarten map, that R is a tensor field (of type $(1,3)$) by showing that it also satisfies the second of these conditions. That it does so follows from numbers (2) and (4) of the properties of a connection, and the rule $[fU, V] = f[U,V] - (Vf)U$ for the bracket. Thus

$$\begin{aligned} R(fU,V)W &= \hat{\nabla}_{fU} \hat{\nabla}_V W - \hat{\nabla}_V \hat{\nabla}_{fU} W - \hat{\nabla}_{[fU,V]} W \\ &= f\hat{\nabla}_U \hat{\nabla}_V W - \hat{\nabla}_V (f\hat{\nabla}_U W) - f\hat{\nabla}_{[U,V]} W + (Vf)\hat{\nabla}_U W \\ &= f\hat{\nabla}_U \hat{\nabla}_V W - f\hat{\nabla}_V \hat{\nabla}_U W - (Vf)\hat{\nabla}_U W - f\hat{\nabla}_{[U,V]} W + (Vf)\hat{\nabla}_U W \\ &= fR(U,V)W. \end{aligned}$$

The argument for $R(U, fV)W$ is essentially the same.

Exercise 27. Show that $R(U,V)fW = fR(U,V)W$. □

This confirms that R is a tensor field.

Exercise 28. Show that, for any operator ∇ which satisfies properties (1)–(4) of a connection, the map $(U,V) \mapsto \nabla_U V - \nabla_V U - [U,V]$ is a tensor field of type $(1,2)$. □

In the case of parallelism in an affine space the tensor field defined in this exercise is zero; however, it is possible to construct covariant differentiation operators for which it is non-zero, a matter discussed in Chapter 11.

We have shown how to construct a connection—that is to say, a concept of parallelism and an associated covariant derivative—on a surface, which satisfies most of the properties of the corresponding structure in $\mathcal{E}^n$ but is path-dependent. It has been shown that this connection, though defined initially in terms of normal and tangential components, is uniquely determined by the metric (Exercises 15-17) and is therefore in principle definable in terms of operations carried out entirely within the surface, and not referring to the ambient Euclidean space, once the metric is known; it is therefore said to be *intrinsic*. It has also been shown that second covariant derivatives do not commute (even when allowance is made for the possibility that the vector fields involved do not commute), but rather that the commutator of second covariant derivatives defines a type $(1,3)$ tensor field, which we call the *curvature tensor* of the surface. This tensor field is defined intrinsically, but it has been shown to be equivalent to the Gaussian curvature, which is not on the face of it intrinsic since it is defined in terms of the normal field. This important fact, that there is an intrinsic measure of curvature, was regarded by Gauss, who discovered it and called it his *Theorema Egregium*, as one of his most significant results. It opens the way to a study of curved spaces in the abstract, initiated by Riemann. Generalisations of matters concerned with connections and their curvature tensors will occupy much of the rest of this book.

Exercise 29. Define a type $(0,4)$ tensor field (also denoted by R) by $R(U_1,U_2,V_1,V_2) = g(R(U_1,U_2)V_1,V_2)$. Show that

$$R(U_2,U_1,V_1,V_2) = R(U_1,U_2,V_2,V_1) = -R(U_1,U_2,V_1,V_2).$$

Show that it follows from these identities and the fact that the surface has dimension 2 that with respect to a basis of vector fields on the surface the tensor R has just one independent non-vanishing component, say R_{1212}, all others being either 0 or determined in terms of this one. Show that R_{1212} takes the same value with respect to any orthonormal basis of a given orientation, and deduce that if K is the common value then

$$R(U_1,U_2,V_1,V_2) = K\big(g(U_1,V_1)g(U_2,V_2) - g(U_1,V_2)g(U_2,V_1)\big)$$

(compare Exercise 24). □

8. Abstract Surfaces

The previous sections were concerned with identifying features of the geometry of a surface which are intrinsic in the sense that they depend on operations carried out in the surface itself without reference to the ambient space (once the basic notions of topology, differentiability and measurement have been fixed). It is natural to take this process one stage further and define a surface in the abstract, without reference to an ambient space, thus finally stripping away any dependence on $\mathcal{E}^3$ at all. What we are left with is the most general kind of 2-dimensional space on which "vector calculus" can be carried out and which has a Euclidean-like measure of lengths and angles; it is an example of a Riemannian differentiable manifold. At least one further generalisation is possible—to other dimensions—and since the whole process is discussed in detail in the next chapter we shall be somewhat brief here.

The basic requirement of an abstract surface is that its points may be labelled by coordinates. The discussions of surfaces in $\mathcal{E}^3$ and of curvilinear coordinates in affine spaces have indicated that it is unreasonable to expect that in general one coordinate system can be found to cover the whole surface, that all partial coordinate systems should be regarded as equally satisfactory, and that different coordinate systems should be related by smooth coordinate transformations. The part of the definition of an abstract surface which is concerned with its topological and analytical properties uses these three features of coordinate systems. There is, however, one small technical problem to overcome: as the reader may have noticed there is a discrepancy between the way we described coordinates for affine space in Chapter 1 and the way we described coordinates for surfaces in $\mathcal{E}^3$ in this chapter. In either case a coordinate system is a bijective map between a subset of the affine space or surface and an open subset of $\mathbf{R}^n$ (where $n = 2$ for the surface). However, for an affine space we regarded the map as going from the space to $\mathbf{R}^n$ (assigning a point to its coordinates), whereas for a surface we have regarded the map as going in the opposite direction, from $\mathbf{R}^2$ to the surface (attaching a coordinate label to each point). Though each possibility has its advantages, the former is generally preferred for the abstract definition (mainly because then the coordinates may be regarded as functions on the space) and so this change is made in the following definition.

An *abstract surface* $\mathcal{S}$ is a set of points which may be given coordinates in the following sense. A (local) *coordinate chart* for $\mathcal{S}$ is a bijective map ψ of a subset of $\mathcal{S}$ (a *coordinate patch*) onto an open subset of $\mathbf{R}^2$. Two local coordinate charts ψ_1, ψ_2 defined on different patches $\mathcal{P}_1$, $\mathcal{P}_2$ are *smoothly related* provided that both of the *coordinate transformation* maps, $\psi_1 \circ \psi_2^{-1}: \psi_2(\mathcal{P}_1 \cap \mathcal{P}_2) \to \psi_1(\mathcal{P}_1 \cap \mathcal{P}_2)$ and its inverse $\psi_2 \circ \psi_1^{-1}$, are smooth maps of open sets in $\mathbf{R}^2$. An *atlas* for $\mathcal{S}$ is a collection of coordinate charts and patches which cover $\mathcal{S}$, in the sense that each point of $\mathcal{S}$ belongs to some patch, the charts on each pair of patches being smoothly related.

Such notions as the smoothness of functions on an abstract surface, or of maps between abstract surfaces, may be defined in terms of the smoothness of their coordinate presentations, much as in Chapter 1. At each point on an abstract surface one may define the tangent space, whose elements must be regarded as differential operators on smooth functions, which satisfy the linearity and Leibniz rules, since there is no internal or external affine structure to allow one to use the limiting vector definition. The cotangent space is defined to be the dual of the tangent space. Thus vector fields, forms and tensor fields in general may be defined on an abstract surface and the usual operations (bracket, Lie derivative, exterior product, exterior derivative ...) may be carried out. Details are given in Chapter 10.

The metric structure of an abstract surface is provided by the assumption that there is a symmetric type $(0,2)$ tensor field singled out, which has the property that, regarded as a bilinear form on each tangent space, it is positive-definite. This metric tensor field may be used to define lengths of vectors and angles between them, lengths of curves, raising and lowering of indices and the other operations associated with a Euclidean metric. In particular there is a unique connection satisfying the first six properties set out in Section 5, defined just as in Exercises 15–17. This connection will be associated with a law of parallel translation which is in general path-dependent, and will have a curvature associated with it, so that the commutator of second covariant derivatives will not be zero but will define a curvature tensor field. This is finally what is meant by the statement that Gaussian curvature is intrinsic.

A surface in $\mathcal{E}^3$ may be thought of as a realisation of an abstract surface $\mathcal{S}$ by means of a smooth map or "imbedding" $\mathcal{S} \to \mathcal{E}^3$ such that the metric induced on $\mathcal{S}$ from $\mathcal{E}^3$ is the same as that which is there already. The non-intrinsic geometrical properties of a surface in $\mathcal{E}^3$, such as its mean curvature, are then properties of the imbedding. Thus the plane $\mathcal{E}^2$ may be considered as an abstract surface and has zero Gaussian curvature; it may be imbedded in $\mathcal{E}^3$ in many different ways, for example as a plane (zero mean curvature), or as a cylinder of radius r (mean curvature $1/2r$).

Summary of Chapter 9

The term "surface" may be applied either to a suitable subset of $\mathcal{E}^3$, or to a suitable object in the abstract. In either case, what distinguishes a surface is the existence of local coordinate systems such that each point may be labelled by two coordinates, and such that coordinate transformations, which may be regarded as maps between open subsets of $\mathbf{R}^2$, are smooth. A tangent vector at a point on a surface is an $\mathbf{R}$-valued linear operator on smooth functions on the surface which satisfies Leibniz's rule. All the usual objects—vector fields, their flows, their brackets, exterior 1- and 2-forms—may be defined on a surface. A surface in $\mathcal{E}^3$ inherits a metric from the Euclidean metric of the ambient space.

The Weingarten map corresponding to a unit normal field N on a surface in $\mathcal{E}^3$ is the linear map of the tangent space defined by $v \mapsto \nabla_v N$. It is symmetric (with respect to the metric); it describes the way the surface curves in different directions, and its invariants are the curvatures of the surface: the determinant is the Gaussian curvature K and the trace is twice the mean curvature H. The curvature may also be represented in terms of an orthonormal basis of 1-forms $\{\theta^1, \theta^2\}$ on the surface: in particular, $d\theta^1 = \omega \wedge \theta^2$ and $d\theta^2 = -\omega \wedge \theta^1$ where ω is the corresponding connection 1-form, and $d\omega = -K\theta^1 \wedge \theta^2$; these are the first and second structure equations for the surface.

There is on a surface a uniquely defined covariant derivative operator $\hat{\nabla}$, with corresponding law of parallel translation, which enjoys almost all of the properties of the connection in Euclidean space, except for two major differences: parallel translation is path-dependent, and covariant derivative operators do not commute, even when they are differentiating in the directions of commuting vector fields. This connection is called the Levi-Civita connection of the surface, and may best be defined by the formula

$$g(\hat{\nabla}_U V, W) = \tfrac{1}{2}\{U(g(V,W)) + V(g(U,W)) - W(g(U,V)) + g([U,V],W) - g([U,W],V) - g(U,[V,W])\}$$

where U, V, W are any vector fields on the surface and g is its metric. The non-commutativity of covariant derivatives is related to the Gaussian curvature by

$$\hat{\nabla}_U \hat{\nabla}_V W - \hat{\nabla}_V \hat{\nabla}_U W - \hat{\nabla}_{[U,V]} W = K(g(V,W)U - g(U,W)V).$$

This establishes that the left-hand side, denoted $R(U,V)W$, is a tensor field, that is, depends multilinearly on its arguments over $\mathcal{F}(S)$; it is called the curvature tensor. It shows also that the Gaussian curvature is an intrinsic object, that is, that it may be defined in terms of the surface metric alone (unlike the mean curvature, which depends also on the realisation of the surface in Euclidean space). In other words, both the Levi-Civita conection and the Gaussian curvature may be defined for any abstract surface which has a metric, that is, a symmetric positive-definite type $(0,2)$ tensor field, defined on it, without the necessity of conceiving of the surface as imbedded in $\mathcal{E}^3$.

10. MANIFOLDS

The treatment of surfaces in Chapter 9 shows that restricting oneself to affine spaces and subspaces would be a limitation on geometrical thinking much too severe to be tolerable. The more general idea foreshadowed there and developed in the rest of this book is that of a manifold, a space which locally resembles an affine space but globally may be quite different. Manifolds are used in Lagrangian and Hamiltonian mechanics, where configuration space and phase space are among the relevant examples, in general relativity theory, where space-time is a manifold and no longer an affine space, and in the theory of groups: the rotation and Lorentz groups considered in Chapter 8 are among those which may to advantage be considered to be manifolds.

In this chapter we define manifolds and maps between them, and go on to explain what modifications must be made to the ideas introduced, in the affine space context, in Chapters 1 to 7 to adapt these ideas to manifolds.

1. Manifolds Defined

We begin with two examples. Like the sphere, dealt with in the previous chapter, these examples lack the property possessed by affine spaces that one may label points with a single coordinate system so that

(1) nearby points have nearby coordinates, and

(2) every point has unique coordinates.

On the other hand, as in the case of the sphere, it is possible to choose for each of these examples a set of coordinate systems such that

(1) nearby points have nearby coordinates in at least one coordinate system, and

(2) every point has unique coordinates in each system which covers it.

Our first example is the configuration space of a double pendulum, which is shown in the figure on the next page. A rod AB turns about an axis at A. A second rod BC turns about a parallel axis fixed to the first rod at B, so that the whole system moves in a plane. A mass is attached to the second rod at C. The joints are supposed to be arranged so that B can turn in a complete circle around A and C can turn in a complete circle around B. Thus the configuration of the system may be specified by giving a point on each of two circles; in order that nearby positions of the system should always be represented by nearby points in configuration space it is necessary that the two circles be configured into a torus. Suppose for example that the position of the rod AB is represented by an angular coordinate ϑ which runs from 0 to 2π, that of BC by another angular coordinate φ which also runs from 0 to 2π. Then configurations for which $\vartheta = 2\pi$ (and φ takes any specified value) must be identified with those for which $\vartheta = 0$ (and φ takes the same value), since these values of ϑ represent the same positions of the rod AB. In the same way, configurations for which $\varphi = 2\pi$ must be identified with those for which $\varphi = 0$. In the figure the identifications are indicated by arrows. Imagining the resulting rectangle made out of indiarubber one may join the identified edges and the result is a torus.

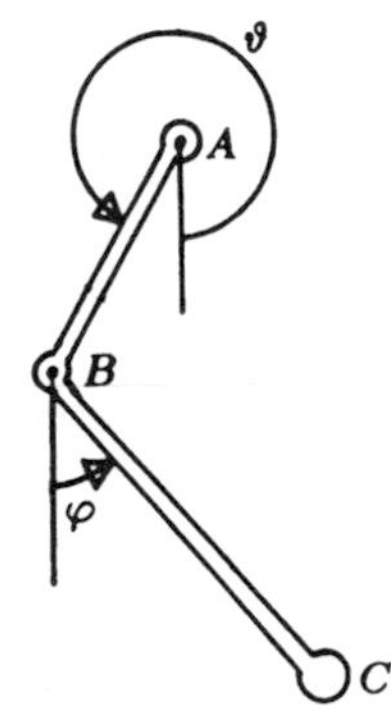

Fig. 1 A double pendulum.

The problem of introducing coordinates on the torus is similar to that which arises for the circle and the sphere; a fairly straightforward covering by four coordinate patches is shown in the next figure.

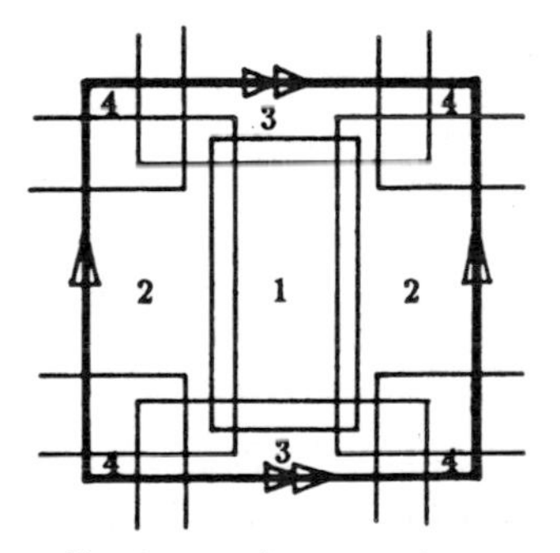

Fig. 2 Coordinate patches on a torus.

Our second example is the configuration space of a rigid body turning about a fixed point. Any given position of the body may be specified by the (unique) rotation which takes it from some standard position to the given position. Thus the configurations of the body are in 1 : 1 correspondence with the elements of the rotation group $SO(3)$, and two rotations should be considered nearby if they yield nearby configurations of the body. The simplest parametrisation of the rotation group for the inspection of the group as a whole is the one whereby rotation through angle t about axis with direction $\mathbf{n}$ (a unit vector) is represented by the vector $t\mathbf{n}$. In this way every rotation is represented by at least one point inside or on the surface of the sphere of radius π. However, diametrically opposite points on the surface represent the same position of the body, achieved by rotations through π in opposite senses, and must therefore be identified with one another. We cannot use these parameters as coordinates if we wish to preserve the uniqeness of labelling enjoyed by points in an affine space.

These two simple examples taken from mechanics, together with the examples already discussed in Chapter 9, point the need for a definition of a kind of space more general than affine space, in which nevertheless the use of coordinates remains a possibility.

Charts and manifolds. A manifold is a set in which it makes sense to introduce coordinates, at least locally. In this respect it behaves locally like an affine space, except that there are no preferred systems of coordinates comparable to affine coordinates. The definition may conveniently be reached in two stages, the first establishing the topological properties, the second, differentiability.

Let $\mathcal{M}$ be a topological space. If every point of $\mathcal{M}$ has a neighbourhood homeomorphic to an open subset of $\mathbf{R}^m$, and if furthermore it is a Hausdorff space with a countable basis, then $\mathcal{M}$ is called a *topological manifold* of dimension m (or *topological m-manifold*). The latter restrictions prohibit various more or less bizarre constructions which would otherwise qualify as manifolds. However, this is a somewhat technical matter which need not cause undue concern.

More structure is required in order that the usual operations of the calculus should be possible. This structure may be introduced by specifying admissible coordinate systems. Let $\mathcal{M}$ be a topological m-manifold. A *chart* on $\mathcal{M}$ comprises an open set $\mathcal{P}$ of $\mathcal{M}$, called a *coordinate patch*, and a map $\psi\colon \mathcal{P} \to \mathbf{R}^m$ which is a homeomorphism of $\mathcal{P}$ onto an open subset of $\mathbf{R}^m$. If x lies in $\mathcal{P}$, then the pair $(\mathcal{P}, \psi)$ is called a chart around x. The definition of a topological manifold guarantees the existence of a chart around each point.

The map ψ is used to assign coordinates to points of $\mathcal{P}$ in exactly the same way as in an affine space (Chapter 2, Section 6): Π^a denotes the projection of $\mathbf{R}^m$ on its ath factor; the coordinate functions on $\mathcal{P}$ are the functions

$$x^a = \Pi^a \circ \psi\colon \mathcal{P} \to \mathbf{R} \qquad a = 1, 2, \ldots, m.$$

We need next to establish a criterion of mutual consistency of coordinate systems by specifying conditions to be satisfied when two charts overlap. It is at this point that the concept of differentiability, or smoothness, is introduced into the structure. Suppose that $(\mathcal{P}_1, \psi_1)$ and $(\mathcal{P}_2, \psi_2)$ are two charts on $\mathcal{M}$, with overlapping coordinate patches. In the overlap $\mathcal{P}_1 \cap \mathcal{P}_2$ two maps to $\mathbf{R}^m$ are specified. Since these maps are homeomorphisms they are invertible, and therefore maps between open subsets of $\mathbf{R}^m$ may be specified by

$$\chi = \psi_2 \circ \psi_1^{-1}\colon \psi_1(\mathcal{P}_1 \cap \mathcal{P}_2) \to \psi_2(\mathcal{P}_1 \cap \mathcal{P}_2)$$
$$\chi^{-1} = \psi_1 \circ \psi_2^{-1}\colon \psi_2(\mathcal{P}_1 \cap \mathcal{P}_2) \to \psi_1(\mathcal{P}_1 \cap \mathcal{P}_2).$$

The question of smoothness is now reduced to consideration of the maps χ and χ^{-1}. Precisely the same conditions will be imposed as in the affine case, namely that χ and χ^{-1} are both C^∞, or smooth, which is to say C^k for every k. This means that the functions relating the coordinates in two overlapping patches may

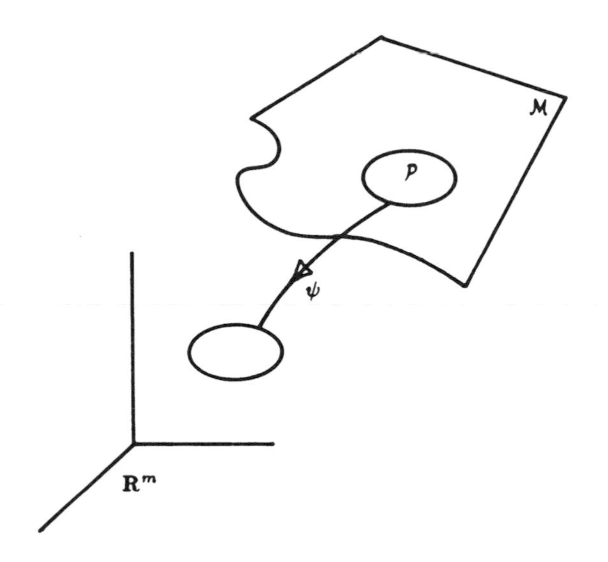

Fig. 3 A chart around a point in a manifold.

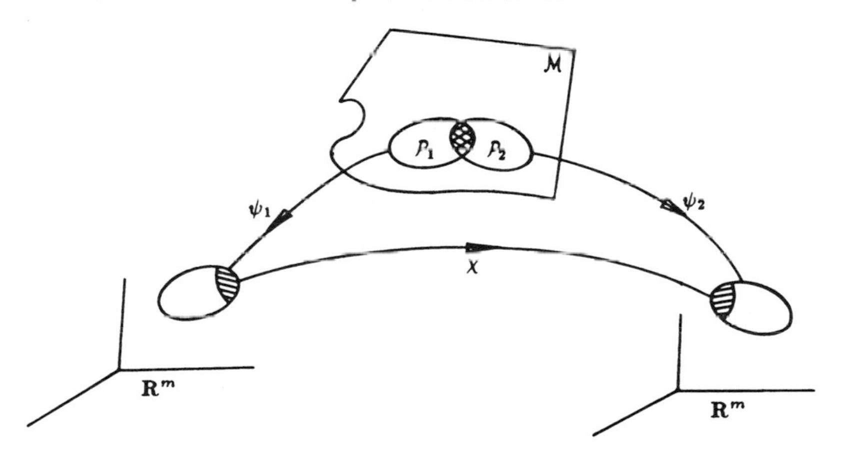

Fig. 4 Two overlapping patches in $\mathcal{M}$ and a change of coordinates.

be differentiated any number of times. The difference between this and the affine case is that here all coordinate systems are curvilinear, none having priority over others, and all of equal status, only the relations between them being restricted. Pairs of charts related in this way are said to be *smoothly related*. It is convenient to say that two charts are smoothly related also if their domains do not intersect.

Since smooth functions of smooth functions are smooth functions, the composition of smooth maps yields a smooth map. Consequently it makes sense to allow all charts which are smoothly related. This is the mathematical realisation of the physical idea that all coordinate systems are equally good. The inclusion of all such charts is the point of the following definitions.

Let $\mathcal{M}$ be a topological manifold. A *smooth atlas* for $\mathcal{M}$ is a collection of pairwise smoothly related charts whose coordinate patches cover $\mathcal{M}$. Thus every point of $\mathcal{M}$ must lie in some patch of the atlas, and thereby acquire coordinates, and where two sets of coordinates are both in operation they must be smoothly related. An atlas is called *complete* if it is not a proper subcollection of any other atlas; this means that there is no chart, smoothly related to all the charts in the atlas, which is not itself already in the atlas. Any atlas may be completed by adding to it all the charts not already in it which are smoothly related to those in it. There is no question of enumerating all the charts in a complete atlas. The point is merely that any coordinate system which is related to those of an atlas by smooth coordinate transformations is as admissible as those already in the atlas.

An m-dimensional topological manifold $\mathcal{M}$, together with a complete atlas, is called an m-dimensional C^∞, or smooth, *differentiable manifold*. A complete atlas is also sometimes called a differentiable structure for $\mathcal{M}$.

Suppose that $(\mathcal{P}_1, \psi_1)$ and $(\mathcal{P}_2, \psi_2)$ are two charts on an m-dimensional smooth manifold $\mathcal{M}$ with overlapping coordinate patches. We may use the map χ and its inverse to express the relation between the coordinates belonging to $\mathcal{P}_1$ and to $\mathcal{P}_2$, as follows. If $x_1^a = \Pi^a \circ \psi_1$ and $x_2^a = \Pi^a \circ \psi_2$ denote the coordinate functions on $\mathcal{P}_1$ and $\mathcal{P}_2$ respectively then

$$x_2^a = \chi^a(x_1^b) \qquad x_1^a = (\chi^{-1})^a(x_2^b).$$

The invertibility of χ implies that its Jacobian matrix and the Jacobian matrix of χ^{-1} are inverses of each other. This is often expressed in the form

$$\frac{\partial x_2^a}{\partial x_1^c}\frac{\partial x_1^c}{\partial x_2^b} = \delta_b^a \qquad \frac{\partial x_1^a}{\partial x_2^c}\frac{\partial x_2^c}{\partial x_1^b} = \delta_b^a.$$

Examples of manifolds. We conclude this section by giving some more examples of manifolds and atlases. It is enough to give one atlas, which may in principle always be completed.

(1) Any open subset $\mathcal{P}$ of $\mathbf{R}^m$ is a manifold, with an atlas consisting of one chart $(\mathcal{P}, \psi)$, with ψ the identity map on $\mathbf{R}^m$ restricted to $\mathcal{P}$. This example is enough for almost all work in tensor calculus, except where integration is involved. In particular, the whole of $\mathbf{R}^m$ may be regarded as a manifold, $m = 1, 2, \ldots$

(2) Any affine space, with an atlas consisting of one chart of affine coordinates, is a manifold. In this example the affine structure is exploited only to construct the coordinate chart, and apart from that it may be ignored. Permissible curvilinear coordinates on affine space are precisely those coordinates which are smoothly related to affine coordinates.

(3) The m-sphere

$$\mathcal{S}^m = \{\, (x^1, x^2, \ldots, x^{m+1}) \in \mathcal{E}^{m+1} \mid \sum_{k=1}^{m+1} (x^k)^2 = 1 \,\}$$

is a manifold. At least two charts are needed. Stereographic projection may be generalised from 3 to m dimensions to yield the required charts, one by projection from each pole $(0,0,\dots,0,\pm 1)$.

(4) The Cartesian product of two manifolds may be made into a manifold. Let $\mathcal{M}$ and $\mathcal{N}$ be manifolds; then $\mathcal{M} \times \mathcal{N}$ is the set of pairs (x, y) with $x \in \mathcal{M}$ and $y \in \mathcal{N}$. If $\{(\mathcal{P}_\alpha, \psi_\alpha)\}$ is an atlas for $\mathcal{M}$ and $\{(\mathcal{Q}_a, \phi_a)\}$ is an atlas for $\mathcal{N}$ then $\{(\mathcal{P}_\alpha \times \mathcal{Q}_a, \Psi_{\alpha a})\}$ is an atlas for $\mathcal{M} \times \mathcal{N}$, where $\Psi_{\alpha a}\colon \mathcal{P} \times \mathcal{Q} \to \mathbf{R}^{m+n}$ is given by $\Psi_{\alpha a}(x, y) = \big(\psi_\alpha(x), \phi_a(y)\big)$.

(5) Let $\mathcal{M}$ be a manifold, and let $\mathcal{O}$ be an open subset of $\mathcal{M}$. Let $\{(\mathcal{P}_\alpha, \psi_\alpha)\}$ be an atlas for $\mathcal{M}$. Then $\{(\mathcal{P}_\alpha \cap \mathcal{O}, \psi_\alpha|_{\mathcal{P}_\alpha \cap \mathcal{O}})\}$ is an atlas for $\mathcal{O}$, as manifold. It follows, in particular, that if $(\mathcal{P}, \psi)$ is a chart on $\mathcal{M}$ then $\mathcal{P}$ may be regarded as a manifold with $\{(\mathcal{P}, \psi)\}$ as an atlas, consisting of a single chart.

(6) Let $M_n(\mathbf{R})$ denote the set of $n \times n$ matrices with real entries, and for any $A \in M_n(\mathbf{R})$ let $x^a_b(A)$ denote the entry in the ath row and bth column of A. Then $(x^1_1, x^1_2, \dots, x^1_n, x^2_1, \dots, x^n_n)$ may be taken as coordinate functions on $M_n(\mathbf{R})$, making it into an n^2-dimensional manifold. If the singular matrices, defined by the polynomial equation $\det\big(x^a_b(A)\big) = 0$, are removed what remains is an open subset which may also be made into an n^2-dimensional manifold.

2. Maps of Manifolds

A number of examples of maps between manifolds have already appeared in various contexts: affine maps, assignment of coordinates by a chart, parametrisation of surfaces in Euclidean space are all special cases of maps between manifolds. The introduction of a geometrical structure such as a vector field may also be described by a map, as will appear later on. In fact, almost anything worth talking about may be described by a map. The definition and exploitation of maps depend on the possibility of using coordinate charts both to give explicit form to maps and to carry out computations.

The question at once arises, what maps should be allowed which are in a reasonable sense compatible with the manifold structure. The answer is conveniently formulated in terms of coordinate presentations. Let $\mathcal{M}$ and $\mathcal{N}$ be (smooth) manifolds, not necessarily of the same dimension, and let $\phi\colon \mathcal{M} \to \mathcal{N}$ be a map. Let $(\mathcal{P}, \psi)$ and $(\mathcal{Q}, \xi)$ be charts on $\mathcal{M}$ and $\mathcal{N}$ respectively, chosen so that the overlap $\phi(\mathcal{P}) \cap \mathcal{Q}$ is not empty. This is always possible, whatever $\mathcal{P}$ may be, because every point of $\phi(\mathcal{P})$ lies in some patch of $\mathcal{N}$. Let the coordinate functions in these charts be x^a on $\mathcal{P} \subset \mathcal{M}$ $(a = 1, 2, \dots, m = \dim \mathcal{M})$ and y^α on $\mathcal{Q} \subset \mathcal{N}$ $(\alpha = 1, 2, \dots, n = \dim \mathcal{N})$. The *coordinate presentation* of the map ϕ, with respect to these charts, is the map $\xi \circ \phi \circ \psi^{-1}$ which, on its domain of definition,

goes up from $\mathbf{R}^m$ to $\mathcal{M}$: ψ^{-1}

then goes across from $\mathcal{M}$ to $\mathcal{N}$: ϕ

then goes down from $\mathcal{N}$ to $\mathbf{R}^n$: ξ.

The coordinate presentation $\xi \circ \phi \circ \psi^{-1}$ of ϕ is therefore a map of a subset of $\mathbf{R}^m$ into $\mathbf{R}^n$. To be precise, the domain of $\xi \circ \phi \circ \psi^{-1}$ is $\psi(\phi^{-1}(\phi(\mathcal{P}) \cap \mathcal{Q}))$, which is an open subset of $\mathbf{R}^m$, and its image is $\xi\big(\phi(\mathcal{P}) \cap \mathcal{Q}\big)$, which is an open subset of $\mathbf{R}^n$.

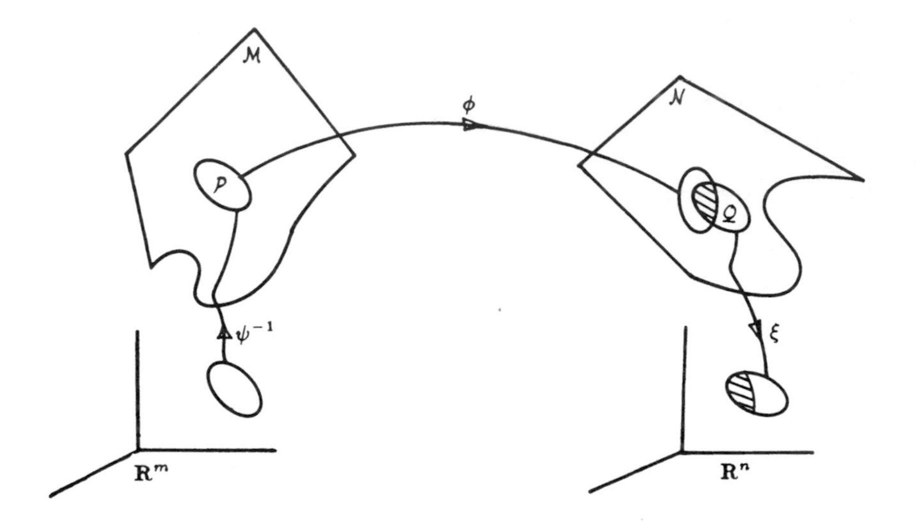

Fig. 5 Charts for a map of manifolds.

A coordinate presentation of a map of manifolds is therefore a map of open subsets of real number spaces.

The n functions on $\psi(\phi^{-1}(\phi(\mathcal{P}) \cap \mathcal{Q})) \subset \mathbf{R}^m$ which give the coordinates y^α of $\phi(x)$ in terms of the coordinates x^a of $x \in \mathcal{P}$ will be written ϕ^α, so that the coordinate presentation of ϕ may be expressed in the form

$$y^\alpha = \phi^\alpha(x^a).$$

A map is called *smooth* if its presentations in coordinates are given by smooth functions, for all charts of complete atlases for both domain and codomain manifold. Most maps of interest in applications are smooth; and in the rest of this book, maps may be assumed to be smooth, or it may be assumed that smoothness can be proved from other assumptions made about them. Fortunately it is not necessary to examine the presentation of a map in every pair of coordinate charts of complete atlases for both its domain $\mathcal{M}$ and codomain $\mathcal{N}$ to determine whether it is smooth. Provided that a map has a smooth coordinate presentation with respect to enough charts on $\mathcal{M}$ to cover it, and enough charts on $\mathcal{N}$ to cover it (that is, with respect to an atlas for $\mathcal{M}$ and one for $\mathcal{N}$), it will be smooth: for its coordinate presentation with respect to any charts smoothly related to those of the two atlases will clearly also be smooth.

An important special case of a map of manifolds is a smooth bijective map with a smooth inverse; such a map is called a *diffeomorphism*, and two manifolds connected by a diffeomorphism are said to be *diffeomorphic*. From the differential-geometric point of view, diffeomorphic manifolds not distinguished by some other structure are effectively the same. Notice, by the way, that a map ϕ can be smooth and invertible without having a smooth inverse; the map $\mathbf{R} \to \mathbf{R}$ by $x \mapsto x^3$, whose inverse is not smooth at 0 (Chapter 2, Exercise 19), is a simple example.

Affine transformations of an affine space are diffeomorphisms of the affine space to itself. Rotations of a sphere about its centre are diffeomorphisms of the sphere to itself. According to Examples 3 and 5 the sphere with either North or South pole deleted is a manifold. On this manifold, stereographic projection to the plane is a diffeomorphism. However, the sphere as a whole is not diffeomorphic to the plane.

Immersions and imbeddings. If $\phi\colon \mathcal{M} \to \mathcal{N}$ is a smooth map, if $(\mathcal{P}, \psi)$ is a chart about some point $x \in \mathcal{M}$, and if $(\mathcal{Q}, \xi)$ is a chart about $\phi(x) \in \mathcal{N}$, we may form the Jacobian matrix $(\partial\phi^\alpha/\partial x^a)$ of ϕ, or strictly of its coordinate presentation. The Jacobian matrix of ϕ changes when the coordinates are changed; but it does so by pre- and post-multiplication by non-singular matrices, namely the Jacobian matrices of the coordinate transformations in $\mathbf{R}^n$ and $\mathbf{R}^m$ respectively. It follows that the rank of the Jacobian matrix of ϕ at any point is independent of a change of coordinates and is therefore a property of ϕ itself: we call it the *rank* of ϕ at the point.

A smooth map whose rank does not vary from point to point is rather easier to deal with than one whose rank does vary. It follows from the results of Chapter 6, Section 4 that if a smooth map ϕ has constant rank k on $\mathcal{M}$ then coordinate charts may always be found on $\mathcal{M}$ and $\mathcal{N}$ with respect to which the coordinate presentation of ϕ is given by

$$\begin{gathered}\phi^1(x^a) = x^1 \quad \phi^2(x^a) = x^2 \quad \dots \quad \phi^k(x^a) = x^k \\ \phi^{k+1}(x^a) = \phi^{k+2}(x^a) = \cdots = \phi^n(x^a) = 0.\end{gathered}$$

Two particular extreme cases stand out. First, when $k = n \leq m$, the coordinate presentation of ϕ corresponds to projection of $\mathbf{R}^m$ onto the first n factors. We call a smooth map whose rank is everywhere equal to the dimension of its codomain a *submersion*. At the other extreme, when $k = m \leq n$, the coordinate presentation of ϕ corresponds to the injection of $\mathbf{R}^m$ into $\mathbf{R}^n$ as a coordinate m-plane. We call a smooth map whose rank is everywhere equal to the dimension of its domain an *immersion*.

It is clear that an immersion is locally injective: no two points of $\mathcal{M}$ lying in a coordinate neighbourhood in which ϕ has the coordinate presentation given above can have the same image. However, an immersion need not be injective globally. Moreover, an immersion may have other undesirable global features: for example, the curve whose image is shown in the figure is an immersion of $\mathbf{R}$ in $\mathbf{R}^2$, but its image approaches a point of itself asymptotically. This is a topological peculiarity of the map; the point of the following definition is to exclude such possibilities. A smooth map is an *imbedding* if it is an injective immersion and is a homeomorphism onto its image.

Submanifolds. If $\phi\colon \mathcal{M} \to \mathcal{N}$ is an immersion then in the special coordinates described above the image $\phi(\mathcal{M}) \subset \mathcal{N}$ is represented locally by a coordinate m-plane, and the first m of the coordinates on $\mathcal{N}$ serve as coordinates for it. It is therefore appropriate to consider $\phi(\mathcal{M})$ as, locally, a submanifold of $\mathcal{N}$. We say

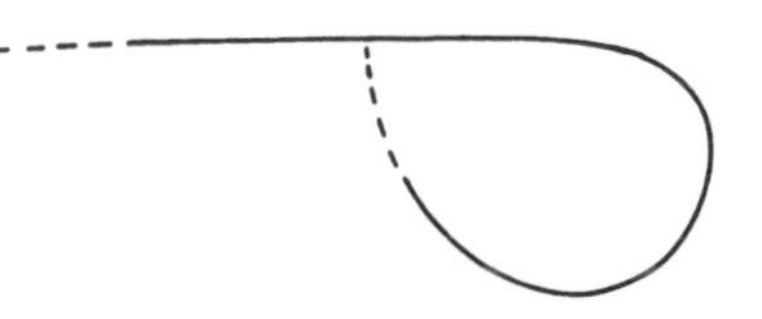

Fig. 6 An immersion whose image approaches itself.

that a subset of $\mathcal{N}$ which is the image of an immersion $\mathcal{M} \to \mathcal{N}$ is an *immersed submanifold* of $\mathcal{N}$, while a subset of $\mathcal{N}$ which is the image of an imbedding is an *imbedded submanifold*, or simply a *submanifold* of $\mathcal{N}$.

Suppose that $\mathcal{S}$ is a subset of a smooth manifold $\mathcal{N}$ with the property that about each point in $\mathcal{S}$ there is a chart of $\mathcal{N}$ such that the part of $\mathcal{S}$ covered by the chart coincides with the coordinate m-plane $y^{m+1} = y^{m+2} = \cdots = y^n = 0$. Then the restrictions of these charts to $\mathcal{S}$ define on it the structure of a smooth manifold of dimension m, and the injection $\mathcal{S} \to \mathcal{N}$, which maps each point of $\mathcal{S}$, considered as a differentiable manifold in its own right, to the same point regarded as a point of the differentiable manifold $\mathcal{N}$, is an immersion. Thus $\mathcal{S}$ is an immersed submanifold of $\mathcal{N}$.

In particular, if $f^{m+1}, f^{m+2}, \ldots, f^n$ are smooth functions on $\mathcal{N}$, then the subset $\mathcal{S}$ of $\mathcal{N}$ on which they simultaneously vanish is an immersed submanifold of $\mathcal{N}$, provided that the differentials $df^{m+1}, df^{m+2}, \ldots, df^n$ are linearly independent everywhere on $\mathcal{S}$. For when this is the case the matrix of partial derivatives of the coordinate representatives of the f^i $(i = m+1, m+2, \ldots, n)$ with respect to any coordinates (z^α) on $\mathcal{N}$ has rank $n - m$, and so without loss of generality it may be assumed that the $(n-m) \times (n-m)$ matrix $(\partial f^i / \partial z^j)$ is non-singular. It then follows that if

$$y^1 = z^1, y^2 = z^2, \ldots, y^m = z^m, y^{m+1} = f^{m+1}(z^\alpha), \ldots, y^n = f^n(z^\alpha)$$

then the y^α form a coordinate system with respect to which $\mathcal{S}$ is given by

$$y^{m+1} = y^{m+2} = \cdots y^n = 0.$$

Exercise 1. Show that, more generally, if $\phi\colon \mathcal{N}_1 \to \mathcal{N}_2$ is a submersion then the inverse image of any point of $\mathcal{N}_2$ is an immersed submanifold of $\mathcal{N}_1$. □

3. Curves and Functions

In this section we generalise to the context of manifolds the ideas of curve and function introduced in Chapter 2, Section 1 and used continually since then.

Curves. Let $\mathcal{M}$ be a smooth manifold. A *curve* in $\mathcal{M}$ is a map $\mathbf{R} \to \mathcal{M}$, or a map $I \to \mathcal{M}$ where I is an open interval of $\mathbf{R}$. A curve is smooth if it is defined by a smooth map of manifolds. It may be helpful, however, to give the definition of smoothness for a curve in detail.

Let $\sigma\colon I \to \mathcal{M}$ be a curve in $\mathcal{M}$ (I may be the whole of $\mathbf{R}$). The curve σ is said to be smooth on a subinterval J of I if there is a chart $(\mathcal{P}, \psi)$ of $\mathcal{M}$ such that $\sigma(J)$ lies in $\mathcal{P}$ and the *coordinate presentation* of σ, $(\sigma^a) = (x^a \circ \sigma)\colon J \to \mathbf{R}^m$, is given by smooth functions. If $\sigma(J)$ lies in two overlapping coordinate patches and σ is smooth in one chart then it will also be smooth in the other, because of the assumed smoothness of coordinate changes. Since the whole of $\mathcal{M}$ is covered by charts, so is the whole of $\sigma(I)$, and σ is called *smooth* (without further qualification) if its domain is covered by overlapping intervals in each of which it is smooth. This definition depends on the differentiable structure of $\mathcal{M}$ but not on the choice of particular charts.

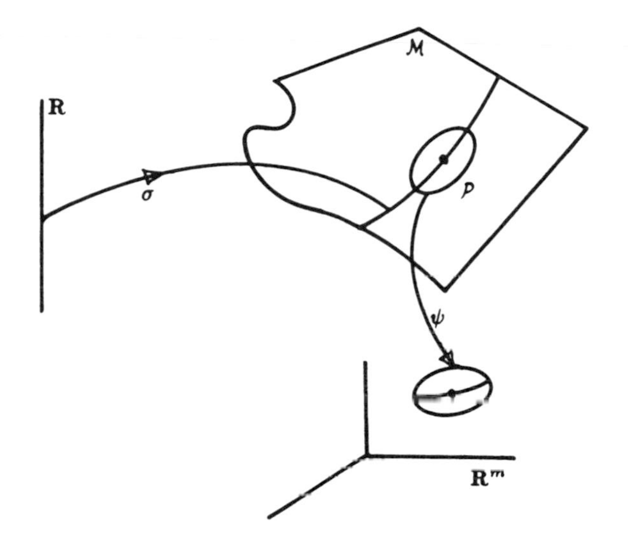

Fig. 7 Coordinate presentation of a curve.

No fresh ideas are needed in order to generalise to manifolds the definitions of constant curve, or reparametrisation, which are therefore not repeated here (see Chapter 2, Section 1).

Functions. The definition of smoothness for a function on a manifold follows a similar pattern. A (real) function on $\mathcal{M}$ is a map $f: \mathcal{M} \to \mathbf{R}$.

As in the affine case, there is a problem of notation for functions, which is resolved in the same way. We distinguish a function from its coordinate presentation, but now the latter cannot be given all over the manifold at once unless there happens to exist a global coordinate system. If $(\mathcal{P}, \psi)$ is a chart on $\mathcal{M}$ with coordinates (x^a), then the *coordinate presentation* of a function f in this chart is the map

$$f^x = f \circ \psi^{-1}: \psi(\mathcal{P}) \subset \mathbf{R}^m \to \mathbf{R}.$$

The coordinate presentation may be distinguished by an index identifying the coordinates which are being used, as before.

If the function f^x is a smooth function of m variables then f is called smooth in $\mathcal{P}$. Because charts are smoothly related, if f is smooth in $\mathcal{P}$, it is also smooth in

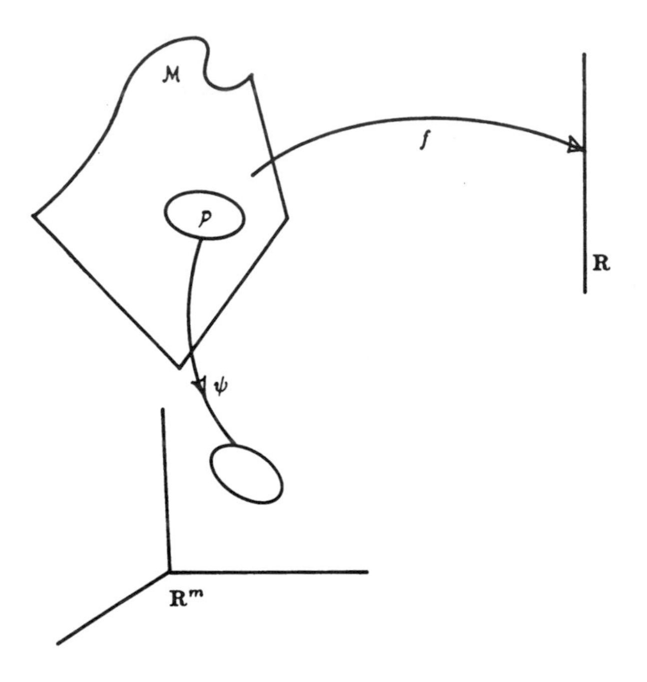

Fig. 8 Coordinate presentation of a function.

the overlap with $\mathcal{P}$ of the patch $\mathcal{P}'$ of any other chart $(\mathcal{P}', \psi')$. A function which is smooth on all the charts of an atlas is called a *smooth function*. Thus a smooth function on $\mathcal{M}$ is a map $\mathcal{M} \to \mathbf{R}$ which is smooth as a map of manifolds. As in the affine case, we shall deal always with smooth functions.

The smooth functions on $\mathcal{M}$ form an algebra, which we shall denote $\mathcal{F}(\mathcal{M})$.

One useful consequence of our choice of smoothness condition is the existence of so-called *bump functions*. Let $\mathcal{O}$ be an open subset of a smooth manifold $\mathcal{M}$ and x a point of $\mathcal{O}$. Then one can find a smooth function b such that $b(x) = 1$ but $b(y) = 0$ for $y \notin \mathcal{O}$. The construction is based on the smooth function h on $\mathbf{R}$ described in Chapter 2, Exercise 4 which takes the value 1 on some closed interval containing a given point of $\mathbf{R}$ but vanishes outside a larger open interval. Let $\mathcal{P}$ be a coordinate patch about x, with coordinates chosen so that x is at the origin; it is convenient to think of the coordinate space as being Euclidean. Let R be a positive number sufficiently small that the open ball of radius R lies in the coordinate neighbourhood and the corresponding subset of $\mathcal{P}$ is contained in $\mathcal{O}$. Define b as follows: for $y \notin \mathcal{P}$, $b(y) = 0$; the coordinate presentation of b on $\mathcal{P}$ is given by $b^x(x^1, x^2, \ldots, x^m) = h(r)$, where r is the usual radial coordinate and h the smooth real function which takes the value 1 on $[-R/2, R/2]$ say and vanishes outside $(-R, R)$. Then b has the required properties: it is a bump function at x.

Tangent vectors. There are some manifolds, such as affine spaces, on which the idea of a displacement vector makes sense. However, on an arbitrary manifold without appropriate additional structure there is nothing analogous to a displacement vector. For instance, the differences between the coordinates of two arbitrarily chosen points may be altered by a coordinate change in a completely arbitrary way without any effect on the coordinates of two others, and there is no longer any reason to prefer one coordinate system over another. Consequently the definition of a tangent vector as a limit of displacement vectors along chords is not amenable to generalisation to the case of manifolds.

The directional derivative definition, on the other hand, can be generalised quite readily. In the first place, if σ is a smooth curve in $\mathcal{M}$ and f is a smooth function on $\mathcal{M}$, then $f \circ \sigma$ is a smooth real function of one variable and (exactly as in the affine case) the derivative $d/ds(f \circ \sigma)(0)$ represents the rate of change of the function along the curve. In any coordinates (x^a) of a chart around the point $x = \sigma(0)$

$$\frac{d}{ds}(f \circ \sigma)(0) = \frac{d}{ds}(f^x(\sigma^a))(0) = \frac{\partial f^x}{\partial x^b}\dot{\sigma}^b(0),$$

the partial derivatives in the last expression being evaluated at $\left(x^a(x)\right)$.

In the affine case, the derivative of a function along a curve is the same for any two curves with the same tangent vector at a chosen point, and all such curves constitute an equivalence class defined by and defining the tangent vector. Moreover, the directional derivative so defined has the properties

$$v(af + bg) = avf + bvg \qquad \text{linearity}$$

$$v(fg) = (vf)g(x) + f(x)(vg) \qquad \text{Leibniz}$$

for all real a, b and all smooth functions f and g. Conversely any map of functions which enjoys these properties fixes a unique tangent vector v at x.

We now turn this around, following the revised definition for the affine case given in Chapter 2, Section 2, and define a *tangent vector* at a point x of a smooth manifold $\mathcal{M}$ to be an operator on smooth functions on $\mathcal{M}$ which is linear and satisfies Leibniz's rule as set out above.

We shall denote the set of tangent vectors to $\mathcal{M}$ at x by $T_x\mathcal{M}$. To make this set into a vector space we define the linear combination $av + bw$ by

$$(av + bw)f = avf + bwf$$

where $v, w \in T_x\mathcal{M}$, $a, b \in \mathbf{R}$ and $f \in \mathcal{F}(\mathcal{M})$. The linear space $T_x\mathcal{M}$ is called the *tangent space* to $\mathcal{M}$ at x.

The partial differentiation operators $\partial_a = \partial/\partial x^a$ with respect to the coordinates x^a in any chart around X are tangent vectors, whose action on functions is given by

$$f \mapsto \partial_a f = \frac{\partial f^x}{\partial x^a}(x^b),$$

the partial derivative being evaluated at the coordinates (x^b) of x. We shall show that these coordinate tangent vectors form a basis for the tangent space $T_x\mathcal{M}$, which is therefore of (finite) dimension m, the dimension of the manifold itself. To do so we shall need to use the following fact of analysis: if F is any smooth function on $\mathbf{R}^m$ then there are smooth functions F_a such that

$$F(\xi^c) = F(0) + \xi^a F_a(\xi^c).$$

This follows from the fact that

$$\int_0^1 \frac{d}{dt}\{F(t\xi^c)\}dt = F(\xi^c) - F(0);$$

for

$$\int_0^1 \frac{d}{dt}\{F(t\xi^c)\}dt = \xi^a \int_0^1 \frac{\partial F}{\partial \xi^a}(t\xi^c)dt$$

and the integral is a smooth function for each a. Note that

$$\frac{\partial F}{\partial \xi^a}(0) = F_a(0).$$

Suppose that v is any tangent vector at $x \in \mathcal{M}$, so that v is a linear operator on smooth functions which obeys Leibniz's rule. We show first that such an operator can be made to act on smooth local functions defined near x, so as still to satisfy linearity and Leibniz's rule. Let f be a smooth local function defined near x. We call a smooth globally defined function F an extension of f if F and f agree on some neighbourhood of x (contained in the domain of f). Extensions of a local function may always be found, for example by multiplying it by a suitable bump function. We propose to define the operation of a tangent vector v at x on a local function f by setting $vf = vF$ where F is an extension of f: but this will make sense only if vF has the same value regardless of the extension chosen. If F_1 and F_2 are two extensions of f then $F_1 - F_2$ vanishes on a neighbourhood $\mathcal{O}$ of x. Let b be a bump function at x, which vanishes outside some open set whose closure is contained in $\mathcal{O}$. Then $b(F_1 - F_2)$ is identically zero, while $b(x) = 1$. By linearity of the operator v

$$v\big(b(F_1 - F_2)\big) = v(0) = 0,$$

while by Leibniz's rule

$$\begin{aligned} v\big(b(F_1 - F_2)\big) &= v(b)\big(F_1(x) - F_2(x)\big) + b(x)\big(v(F_1) - v(F_2)\big) \\ &= v(F_1) - v(F_2). \end{aligned}$$

Thus $v(F_1) = v(F_2)$, and the operation of v on local functions is well-defined.

Now let x^a be the coordinate functions of a coordinate chart about x, chosen so that $\big(x^a(x)\big) = 0$, so that x is at the origin of coordinates. Then we may define m numbers v^a by $v^a = v(x^a)$. We shall show that, with respect to these coordinates, $v = v^a \partial_a$.

First of all, it follows from Leibniz's rule that for any function f

$$vf = v(1 \cdot f) = v(1)f(x) + 1 \cdot vf$$

and therefore $v(1) = 0$. Then by linearity, for any $c \in \mathbf{R}$

$$vc = v(c \cdot 1) = cv(1) = 0.$$

Thus v certainly vanishes on constants, as one would expect. Now for any smooth function f, by the analytical result above there are local smooth functions f_a such that $f = f(x) + f_a x^a$, where $f_a(x) = \partial_a f(x)$. Then since $f(x)$ is a constant (the value of f at the fixed point x)

$$vf = (vf_a)x^a(x) + f_a(x)v(x^a) = v^a \partial_a f(x)$$

as required. The coordinate vectors ∂_a certainly span $T_x\mathcal{M}$; they are clearly linearly independent (since $\partial_a x^b = \delta_a^b$); they therefore form a basis for $T_x\mathcal{M}$.

Note that in this argument essential use is made of the fact that the functions concerned are smooth: bump functions are required, and only when f is smooth will the corresponding f_a also be smooth.

Exercise 2. Show that if x^a and $\acute{x}^a$ are coordinates in two charts around x then $\acute{\partial}_a = (\partial x^b/\partial \acute{x}^a)\partial_b$. □

Exercise 3. Show that the derivative along a curve σ defines a tangent vector, whose components with respect to a coordinate basis are $\dot{\sigma}^a(s)$. □

In the tensor calculus a tangent vector is usually defined by giving the components in one coordinate chart and asserting that under change of chart they transform by the formula given in Exercise 2. Many modern authors, on the other hand, define a tangent vector to be the equivalence class of curves which yields the appropriate directional derivative. These definitions are both equivalent to the one given above, and in any particular context one should employ whichever is the most convenient.

Level surfaces and covectors. The ideas of level surfaces and linear forms may be taken over from the affine case essentially unchanged. If f is a function on $\mathcal{M}$ and c is a real number then the set of points $f^{-1}(c)$ at which f takes the value c, if not empty, is called a *level surface* of f. If σ is a curve lying in $f^{-1}(c)$ then $d/ds(f \circ \sigma) = 0$ at any point of σ, so that the tangent vectors to curves in the level surface are those which satisfy $vf = 0$.

For fixed f and x the map $T_x\mathcal{M} \to \mathbf{R}$ by $v \mapsto vf$ defines a linear form df on $T_x\mathcal{M}$ called the *differential* of f. The vector space of linear forms on $T_x\mathcal{M}$ is called the *cotangent space* at x and denoted $T_x^*\mathcal{M}$, and the linear forms are often called *cotangent vectors* or *covectors*. The differential of f, provided it is not zero, determines the tangent space to a level surface of f, as the subspace of $T_x\mathcal{M}$ of codimension 1 consisting of the tangent vectors which it annihilates. If $df = 0$ at x, on the other hand, the level surface may not have a well-defined tangent space at x.

The idea of a tangent hyperplane does not survive generalisation from the affine case.

As in the affine case, the covectors dx^a, the differentials of the coordinate functions, constitute a basis for $T_x^*\mathcal{M}$ dual to the basis $\{\partial_a\}$ of $T_x\mathcal{M}$. Any element α of $T_x^*\mathcal{M}$ may therefore be written uniquely in the form $\alpha_a dx^a$, where $\alpha_a = \langle \partial_a, \alpha \rangle$. In particular, since vf may always be expressed as $vf = v^a \partial f^x/\partial x^a$ in a local chart about x, we have

$$df = \frac{\partial f^x}{\partial x^a} dx^a = \langle \partial_a, df \rangle dx^a,$$

so that the differential of a function is effectively the same as the total differential encountered in elementary calculus; but in the absence of a metric, df is to be distinguished from the gradient vector, as before.

Exercise 4. Show that if x^a and $\acute{x}^a$ are coordinates in two charts about x then $d\acute{x}^a = (\partial\acute{x}^a/\partial x^b)dx^b$. □

In the tensor calculus a covector is usually defined by giving the components in one coordinate chart together with the above transformation formula.

Induced maps. Let $\phi\colon \mathcal{M} \to \mathcal{N}$ be a smooth map of manifolds. Then ϕ may be used to move objects back and forth between $\mathcal{M}$ and $\mathcal{N}$, in particular, curves and tangent vectors from $\mathcal{M}$ to $\mathcal{N}$, functions and covectors from $\mathcal{N}$ to $\mathcal{M}$.

Let σ be a curve in $\mathcal{M}$. Composing σ with ϕ, one obtains a curve $\phi \circ \sigma$ in $\mathcal{N}$, called the curve induced by ϕ from σ.

Now let h be a function on $\mathcal{N}$. Composing ϕ with h, one obtains a function $h \circ \phi$ on $\mathcal{M}$, called the function induced by ϕ from h.

Exercise 5. Show that if ϕ, σ and h are smooth, so are $\phi \circ \sigma$ and $h \circ \phi$. □

Exercise 6. Show that the map $\mathcal{F}(\mathcal{N}) \to \mathcal{F}(\mathcal{M})$ by $h \mapsto h \circ \phi$ is an algebra homomorphism; that is, it preserves sums and products of functions. □

Notice that ϕ induces a curve in $\mathcal{N}$ from a curve in $\mathcal{M}$, but induces a function on $\mathcal{M}$ from a function on $\mathcal{N}$. Thus curves go cogrediently (in the same direction as ϕ) while functions go cotragrediently (in the opposite direction). This is just what one would expect from the affine case.

It is straightforward, and very useful, to extend induced maps from curves and functions to vectors and covectors.

Let $\phi\colon \mathcal{M} \to \mathcal{N}$ be a smooth map and v a tangent vector to $\mathcal{M}$ at x. An operator $\phi_* v\colon \mathcal{F}(\mathcal{N}) \to \mathbf{R}$ is defined by

$$(\phi_* v)h = v(h \circ \phi) \qquad \text{for all } h \in \mathcal{F}(\mathcal{N}).$$

This operator is a tangent vector at $\phi(x)$, because if f and g are functions on $\mathcal{N}$ and a and b any numbers then

$$(af + bg) \circ \phi = a(f \circ \phi) + b(g \circ \phi) \quad \text{and} \quad (fg) \circ \phi = (f \circ \phi)(g \circ \phi),$$

(Exercise 6) so that

$$(\phi_* v)(af + bg) = a(\phi_* v)f + b(\phi_* v)g$$

and

$$(\phi_* v)(fg) = ((\phi_* v)f)g(\phi(x)) + f(\phi(x))((\phi_* v)g).$$

The tangent vector $\phi_* v$ so defined is said to be *induced* from v by ϕ, and $\phi_*\colon T_x\mathcal{M} \to T_{\phi(x)}\mathcal{N}$ by $v \mapsto \phi_* v$ is called the *induced map of tangent vectors*.

Exercise 7. Show that if v is a tangent vector at some point to a curve σ in $\mathcal{M}$ then $\phi_* v$ is the tangent vector to the induced curve in $\mathcal{N}$ at the corresponding point. □

Exercise 8. Show that ϕ_* is a linear map of vector spaces. □

Exercise 9. Show that a smooth map ϕ is an immersion if and only if ϕ_* is everywhere injective, and a submersion if and only if ϕ_* is everywhere surjective. □

All this may be turned around and used to induce covectors, instead of vectors. However covectors, like functions, go in the opposite direction to ϕ. Let α be a covector at $\phi(x) \in \mathcal{N}$. A linear form $\phi^*\alpha$ on $T_x\mathcal{M}$, hence a covector at $x \in \mathcal{M}$, is defined as follows:

$$\langle v, \phi^*\alpha \rangle = \langle \phi_* v, \alpha \rangle \qquad \text{for all } v \in T_x\mathcal{M}.$$

Thus $\phi^*: T^*_{\phi(x)}\mathcal{N} \to T^*_x\mathcal{M}$ is just the adjoint of the linear map $\phi_*: T_x\mathcal{M} \to T^*_{\phi(x)}\mathcal{N}$, and is therefore a linear map of cotangent spaces. It is called the *induced map of covectors*.

Exercise 10. Show that for any $h \in \mathcal{F}(\mathcal{N})$, $\phi^*(dh) = d(h \circ \phi)$. □

Exercise 11. Show that if $\kappa: \mathcal{K} \to \mathcal{M}$ and $\mu: \mathcal{M} \to \mathcal{N}$ are smooth maps of manifolds then $(\mu \circ \kappa)_* = \mu_* \circ \kappa_*$ and $(\mu \circ \kappa)^* = \kappa^* \circ \mu^*$. □

It is essential to be able to compute induced maps in terms of coordinates.

Exercise 12. The coordinate presentation of the map ϕ is $y^\alpha = \phi^\alpha(x^a)$, where (x^a) are local coordinates around $x \in \mathcal{M}$ and y^α are local coordinates around $\phi(x)$ in $\mathcal{N}$. Show that if $v = v^a\partial_a \in T_x\mathcal{M}$ then $\phi_* v = v^a(\partial_a\phi^\beta)\partial_\beta$, while if $\alpha = \alpha_\beta dy^\beta \in T^*_{\phi(x)}\mathcal{N}$ then $\phi^*\alpha = \alpha_\beta(\partial_a\phi^\beta)dx^a$. □

4. Vector Fields and Flows

In this section we continue the generalisation to the context of manifolds of ideas introduced earlier in the affine case; we deal now with the one-parameter groups, vector fields, flows and congruences introduced originally in Chapter 3.

One-parameter groups of diffeomorphisms. Recall (Section 2) that a diffeomorphism of a manifold $\mathcal{M}$ (to itself) is a smooth map ϕ with a smooth inverse. Combining this with the idea of a one-parameter group (Chapter 3, Section 2) we define a *one-parameter group of diffeomorphisms* of $\mathcal{M}$ to be a smooth map $\phi: \mathbf{R} \times \mathcal{M} \to \mathcal{M}$ with the properties

(1) for each t, $\phi_t: \mathcal{M} \to \mathcal{M}$ by $x \mapsto \phi_t(x) = \phi(t, x)$ is a diffeomorphism

(2) $\phi(0, x) = x$ for all $x \in \mathcal{M}$

(3) for all $x \in \mathcal{M}$ and all $s, t \in \mathbf{R}$, $\phi(s, \phi(t, x)) = \phi(s + t, x)$.

This definition is based on the construction in Exercise 5 of Chapter 3; it incorporates a convenient smoothness condition. Because of (2) and (3), the diffeomorphisms ϕ_t satisfy the conditions $\phi_0 = \mathrm{id}_\mathcal{M}$ and $\phi_s \circ \phi_t = \phi_{s+t}$. The set of all diffeomorphisms of $\mathcal{M}$ forms, under composition, a group $\operatorname{diff}\mathcal{M}$ and the map $t \mapsto \phi_t$ is a homomorphism of groups $\mathbf{R} \to \operatorname{diff}\mathcal{M}$, the real line being considered as an additive group (compare Exercise 6 of Chapter 3).

Let x be any point of $\mathcal{M}$. The set of points into which x is mapped by ϕ_t as t varies is a smooth curve σ_x called the *orbit* of x under ϕ_t and given by $\sigma_x(t) = \phi_t(x)$. If y lies on the orbit of x, so that $y = \sigma_x(s) = \phi_s(x)$ for some $s \in \mathbf{R}$, then $\sigma_y(t) = \phi_t(\phi_s(x)) = \phi_{s+t}(x) = \sigma_x(s+t)$, and so the curves σ_x and σ_y are congruent: they differ only by a change of origin of the parameter (Chapter 2, Section 1). Every point of $\mathcal{M}$ lies on the orbits of a congruent set, and no two congruent sets intersect. As in the affine case, a collection of curves on $\mathcal{M}$, such that each point lies on the curves of a congruent set and no two curves from distinct congruent sets intersect, is called a *congruence of curves*. A unique tangent vector at each point of $\mathcal{M}$ may be associated with a one-parameter group or a congruence, namely the tangent vector to the orbit or congruent set through that point. A choice of a tangent vector at each point of $\mathcal{M}$ is called a *vector field* on $\mathcal{M}$; thus a unique vector field is associated with any congruence or one-parameter group. This vector field is called the (*infinitesimal*) *generator* of the one-parameter group.

A vector field V is smooth if the function Vf is smooth whenever the function f is smooth.

Exercise 13. Show that a vector field is smooth if and only if its components in the charts of any atlas are smooth functions. Derive the transformation law for the components of a smooth vector field under a coordinate transformation (compare Exercise 2, where the rule for a coordinate vector is given). □

The vector fields on $\mathcal{M}$ form a module over the algebra of smooth functions $\mathcal{F}(\mathcal{M})$; we shall denote this module $\mathcal{X}(\mathcal{M})$. Unlike the affine case, where the existence of global coordinates ensures that there is a basis for the module of vector fields, $\mathcal{X}(\mathcal{M})$ need have no basis: the most that can be said is that any point has a neighbourhood on which local vector fields are defined which form a basis of the tangent space at each point in the neighbourhood. The coordinate fields on any coordinate patch provide an example of such a local basis.

Flows. As the examples given in the affine case show, there are vector fields on manifolds which do not generate one-parameter groups, and the more general concept of a flow introduced in the affine case may be generalised to encompass these examples.

A *flow* (or *local group of local transformations*) on a manifold $\mathcal{M}$ is a smooth map $\phi\colon \mathcal{D} \to \mathcal{M}$, where $\mathcal{D}$ is an open subset of $\mathbf{R} \times \mathcal{M}$ which contains $\{0\} \times \mathcal{M}$, such that $\phi(0, x) = x$ for each $x \in \mathcal{M}$ and that $\phi(s, \phi(t, x)) = \phi(s + t, x)$ whenever both sides are meaningful. As before, ϕ_t is given by $\phi_t(x) = \phi(t, x)$. Orbits, changes of origin, congruent sets, congruences and generating vector fields are defined exactly as in the affine case (Chapter 3, Section 3).

A unique vector field is associated with every flow, and the point of the definition, of course, is that a unique flow is associated with every vector field. The only novelty here is that the integral curves have to be pieced together as one moves from one coordinate chart to another. This tedious process yields the following result. Let V be a smooth vector field on $\mathcal{M}$. Then there exists an open subset $\mathcal{D}$ of $\mathbf{R} \times \mathcal{M}$ containing $\{0\} \times \mathcal{M}$ and a flow ϕ on $\mathcal{D}$ having V as generator. With this result one may retrieve the diagram drawn for the affine case:

$$
\begin{array}{ccc}
 & \text{vector field} & \\
\nearrow & & \searrow \\
\text{flow} & \longleftrightarrow & \text{congruence}
\end{array}
$$

The implications expressed in this diagram are that whenever one of the three constructions is given existence of the other two is assured.

Lie transport and the Lie derivative. The finite version of Lie transport developed in the affine case in Chapter 3, Section 5 depends on the existence of displacement vectors and cannot be generalised to manifolds. However, the infinitesimal version generalises without significant alteration, and leads to the definition of the Lie derivative. Let V be a vector field on a manifold $\mathcal{M}$ and ϕ the flow which it generates; let W be a vector field given on an integral curve σ of V, not necessarily maximal. We may without loss of generality assume that $x = \sigma(0)$. From the definition of a flow it follows that there is an open neighbourhood $\mathcal{O}$ of x and an open interval I containing 0 for which $\phi_t(y)$ is determined for every $y \in \mathcal{O}$ and $t \in I$. For $t \in I$ the vector $\phi_{-t*}W(t)$ may be constructed. The *Lie derivative* of W at x is the vector

$$\lim_{t\to 0} \frac{1}{t}\left(\phi_{-t*}W(t) - W(0)\right) = \frac{d}{dt}\left(\phi_{-t*}W(t)\right)(0).$$

By carrying out the same construction at each point of the curve, a vector field $\mathcal{L}_V W$ is obtained. In many applications ϕ is a one-parameter group, not merely a flow, and W is defined all over $\mathcal{M}$, and then the precautions about domains of definition are unnecessary.

If a vector w is given at one point, a vector field W may be constructed along the orbit of that point by $W(t) = \phi_{t*}w$. A vector extended to a vector field in this way is said to be *Lie transported* along the flow. The vector field W then has the property $\mathcal{L}_V W = 0$ wherever it is specified. Conversely, a vector field W for which this condition holds must have been obtained by Lie transport.

The Lie derivative has, as in the affine case, the properties

$$\mathcal{L}_U(aV + bW) = a\mathcal{L}_U V + b\mathcal{L}_U W \qquad U, V, W \in \mathcal{X}(\mathcal{M}), a, b \in \mathbf{R}$$

$$\mathcal{L}_V(fW) = f\mathcal{L}_V W + (Vf)W \qquad f \in \mathcal{F}(\mathcal{M}).$$

The Lie derivative of a covector field α is defined, as in the affine case, by

$$\mathcal{L}_V\alpha(0) = \lim_{t\to 0}\left(\phi_t{}^*\alpha(t) - \alpha(0)\right)$$

and the equations

$$V\langle W, \alpha\rangle = \langle \mathcal{L}_V W, \alpha\rangle + \langle W, \mathcal{L}_V\alpha\rangle$$

$$\mathcal{L}_V(df) = d(Vf)$$

are also satisfied. The arguments leading to these results do not depend on the affine structure and may be repeated word for word in any coordinate chart.

The bracket of two vector fields, considered as operators on functions obeying the linearity and Leibniz rules, is their commutator as before; it is related to the Lie derivative in the same way as in the affine case. Thus

$$\mathcal{L}_V W = [V, W] = -\mathcal{L}_W V$$

$$[\mathcal{L}_V, \mathcal{L}_W] = \mathcal{L}_{[V,W]}.$$

The argument given to establish the first of these in the affine case, in Sections 7 and 9 of Chapter 3, may be easily adapted to the more general situation.

The definition of Ψ-relatedness of vector fields given in Chapter 3, Section 10 extends without obvious change to the case where Ψ is a smooth map of manifolds; and the argument that shows that the brackets of Ψ-related vector fields share that property goes through unchanged.

5. Tensor Fields

In this section we make some generalisations which do more than merely adapt to manifolds ideas already developed in the affine space case: we introduce a general idea of tensor fields, which encompasses the exterior forms introduced in Chapter 4, the bilinear forms and metrics of Chapter 7, the covariant tensor fields of Chapter 8 and the curvature tensor of Chapter 9. We begin with the algebraic foundations of the concept.

Let $\mathcal{V}_1, \mathcal{V}_2, \ldots, \mathcal{V}_r$ be vector spaces (over $\mathbf{R}$). A map

$$T: \mathcal{V}_1 \times \mathcal{V}_2 \times \cdots \times \mathcal{V}_r \to \mathbf{R}$$

is *multilinear* if it is linear in each factor, that is , if

$$\begin{aligned} &T(v_1, v_2, \ldots, cv_k + c'v'_k, \ldots, v_r) \\ &\qquad = cT(v_1, v_2, \ldots, v_k, \ldots, v_r) + c'T(v_1, v_2, \ldots, v'_k, \ldots, v_r) \end{aligned}$$

for $k = 1, 2, \ldots, r$. Now let $\mathcal{V}$ be a fixed vector space (over $\mathbf{R}$), and let each of $\mathcal{V}_1, \mathcal{V}_2, \ldots, \mathcal{V}_r$ be either $\mathcal{V}^*$ or $\mathcal{V}$, so that there are altogether p copies of $\mathcal{V}^*$ and q copies of $\mathcal{V}$ (in some order), with $p+q = r$. A multilinear map $T: \mathcal{V}_1 \times \mathcal{V}_2 \times \cdots \times \mathcal{V}_r \to \mathbf{R}$ is then called a *tensor* of *type*, or *valence*, (p, q) over $\mathcal{V}$. It may happen that all the copies of $\mathcal{V}^*$ appear first and all the copies of $\mathcal{V}$ afterwards, in the Cartesian product, so that

$$T: \underbrace{\mathcal{V}^* \times \cdots \mathcal{V}^*}_{p \text{ copies}} \times \underbrace{\mathcal{V} \times \cdots \mathcal{V}}_{q \text{ copies}} \to \mathbf{R},$$

but this need not be the case. If $\{e_a\}$ is a basis for $\mathcal{V}$ and $\{\theta^a\}$ the dual basis for $\mathcal{V}^*$, then T is determined completely by its action on each sequence chosen from these—in the case just mentioned, by

$$T^{a_1 a_2 \ldots a_p}{}_{b_1 b_2 \ldots b_q} = T\left(\theta^{a_1}, \theta^{a_2}, \ldots, \theta^{a_p}, e_{b_1}, e_{b_2}, \ldots, e_{b_q}\right)$$

where all the as and bs range over $1, 2, \ldots, m = \dim \mathcal{V}$. These m^r numbers are called the *components* of T with respect to the chosen bases.

A tensor of type $(1, r)$ as defined here is a multilinear map $\mathcal{V}^* \times \mathcal{V} \times \cdots \times \mathcal{V} \to \mathbf{R}$. Thus if T is a tensor of type $(1, r)$ and $v_1, v_2, \ldots, v_r$ are any fixed elements of $\mathcal{V}$, then the map $\mathcal{V}^* \to \mathbf{R}$ by $\alpha \mapsto T(\alpha, v_1, v_2, \ldots, v_r)$ is linear; it is thus an element of $(\mathcal{V}^*)^*$, the space dual to $\mathcal{V}^*$. But this is just $\mathcal{V}$. There is thus an alternative interpretation of a tensor of type $(1, r)$: it is a multilinear map of $\mathcal{V} \times \mathcal{V} \times \cdots \times \mathcal{V} \to \mathcal{V}$, where there are r copies of $\mathcal{V}$ in the domain.

Exercise 14. Show conversely how an r-fold multilinear map $\mathcal{V} \times \mathcal{V} \times \cdots \times \mathcal{V} \to \mathcal{V}$ may be used to define a tensor of type $(1, r)$ in the original sense. □

Thus the curvature tensor at a point of a surface, as defined in Chapter 9, is a tensor of type $(1, 3)$ according to this alternative definition.

Notice that the type counts the number of copies of $\mathcal{V}$ and $\mathcal{V}^*$ which appear, but not their order, so that a trilinear map $\mathcal{V} \times \mathcal{V} \times \mathcal{V}^* \to \mathbf{R}$ and a trilinear map $\mathcal{V} \times \mathcal{V}^* \times \mathcal{V} \to \mathbf{R}$ are of the same type $(1,2)$, for example.

The tensors of a given type, with the factors of the Cartesian product in a given order, form a vector space, with addition defined by

$$(c_1T_1 + c_2T_2)(v_1, v_2, \ldots, v_r) = c_1T_1(v_1, v_2, \ldots, v_r) + c_2T_2(v_1, v_2, \ldots, v_r),$$

where the argument v_k belongs to the kth factor of the Cartesian product, be it $\mathcal{V}$ or $\mathcal{V}^*$.

Exercise 15. Show that the dimension of a space of tensors of type (p,q) is m^r where $m = \dim \mathcal{V}$ and $r = p + q$. □

There is a natural isomorphism between tensors of the same type with the factors in different orders, defined by rearranging the factors. For example, if $T: \mathcal{V} \times \mathcal{V} \times \mathcal{V}^* \to \mathbf{R}$ is a tensor of type $(1,2)$, then a tensor T', of the same type but mapping $\mathcal{V} \times \mathcal{V}^* \times \mathcal{V} \to \mathbf{R}$, may be defined by $T'(v_1, \alpha, v_2) = T(v_1, v_2, \alpha)$ for all $v_1, v_2 \in \mathcal{V}$ and all $\alpha \in \mathcal{V}^*$. Tensors which are related by such a rearrangement are called *isomers*.

All these constructions may be carried out at a point x of a manifold $\mathcal{M}$, taking $T_x\mathcal{M}$ for $\mathcal{V}$ and $T_x^*\mathcal{M}$ for $\mathcal{V}^*$. A *tensor* of type (p,q) at $x \in \mathcal{M}$ is a multilinear map $\mathcal{V}_1 \times \mathcal{V}_2 \times \cdots \times \mathcal{V}_r \to \mathbf{R}$, where each $\mathcal{V}_k$ is either $T_x^*\mathcal{M}$ or $T_x\mathcal{M}$, there are p of the former and q of the latter, and $r = p + q$. In practice, a tensor at a single point of a manifold in isolation is not of much interest: one is far more likely to have to deal with tensor fields. A *tensor field* on $\mathcal{M}$ is a choice of tensor at each point of $\mathcal{M}$, of a fixed type and order of factors in the argument. From a tensor field of type (p,q) and any p 1-forms and q vector fields one may build a function by the obvious pointwise construction; the tensor field is said to be smooth if the resulting function is smooth for every choice of smooth 1-form and vector field arguments. One may also define components of a tensor field with respect to any local basis of vector fields and dual basis of 1-forms, simply by carrying out the construction of components described above in the vector space context pointwise; these components will be local functions, smooth if the tensor field is. We shall have to deal only with smooth tensor fields.

Exercise 16. Show that under a change of coordinates the components of a tensor field with respect to a coordinate basis transform linearly and homogeneously: for example

$$\hat{T}^a{}_{bc} = \frac{\partial \hat{x}^a}{\partial x^d} \frac{\partial x^e}{\partial \hat{x}^b} \frac{\partial x^f}{\partial \hat{x}^c} T^d{}_{ef}. \qquad \square$$

This property of tensor fields is the defining property used in tensor calculus.

We have carried on the description of tensor fields so far by starting from a pointwise definition. However, it may happen that a tensor field is defined in the first instance as a map of vector fields and 1-forms: the curvature tensor introduced in Chapter 9 is a case in point. In such a case it is useful to have a test of whether or not a given object is tensorial which deals directly with the object as it is defined. Now a tensor field of type (p,q) on $\mathcal{M}$ defines a map $\Xi_1 \times \Xi_2 \times \cdots \times \Xi_r \to \mathcal{F}(\mathcal{M})$ where each Ξ_k is either $\mathcal{X}^*(\mathcal{M})$, the $\mathcal{F}(\mathcal{M})$ module of covector fields or 1-forms, or $\mathcal{X}(\mathcal{M})$, and there are p of the former and q of the latter in some order, with $r = p + q$. Moreover this map is $\mathcal{F}(\mathcal{M})$-multilinear. It is this property of $\mathcal{F}(\mathcal{M})$-multilinearity which provides the required test. For example if α is a given 1-form then $(V, W) \mapsto V\langle W, \alpha\rangle - W\langle V, \alpha\rangle - \langle [V, W], \alpha\rangle$ is a tensor field of type $(0, 2)$, namely the 2-form $d\alpha$; but in contrast $(V, W) \mapsto V\langle W, \alpha\rangle - W\langle V, \alpha\rangle$ is not a tensor field, because $V\langle fW, \alpha\rangle - fW\langle V, \alpha\rangle = f\{V\langle W, \alpha\rangle - W\langle V, \alpha\rangle\} + (Vf)\langle W, \alpha\rangle$ and the final term spoils the $\mathcal{F}(\mathcal{M})$-multilinearity (compare Chapter 5, Exercise 13 and the subsequent discussion). Thus $\mathcal{F}(\mathcal{M})$-multilinearity gives one a filter for the removal of objects which are not tensor fields. Its function is to guarantee that a tensor field can be specified by giving its components with respect to any basis. There is however a technical difficulty which arises in the present case which has not arisen before. This is that the $\mathcal{F}(\mathcal{M})$-multilinearity property involves only globally defined vector fields and 1-forms, while components are expressed in terms of local fields. We should like to be able to say that an $\mathcal{F}(\mathcal{M})$-multilinear map $\Xi_1 \times \Xi_2 \times \cdots \times \Xi_r \to \mathcal{F}(\mathcal{M})$, where each Ξ_k is either $\mathcal{X}^*(\mathcal{M})$ or $\mathcal{X}(\mathcal{M})$, defines a tensor field; but we have no reason to suppose, on the face of it, that such a map would make sense when its arguments are only locally defined; and this would be an essential step in reconstructing an **R**-multilinear map of tangent and cotangent spaces.

This problem does not arise for affine spaces (because of the existence of global coordinate fields); nor does it occur for the curvature tensor on a surface (because this can be expressed in terms of the metric in a way which makes its tensorial nature clear). In fact it is true for a C^∞ manifold that an $\mathcal{F}(\mathcal{M})$-multilinear map $\Xi_1 \times \Xi_2 \times \cdots \times \Xi_r \to \mathcal{F}(\mathcal{M})$, where each Ξ_k is either $\mathcal{X}(\mathcal{M})$ or $\mathcal{X}^*(\mathcal{M})$, defines a tensor field; we shall now give an indication of the proof.

For ease of exposition we shall consider a specific case: type $(1, 1)$ tensor fields. We consider therefore a map T from $\mathcal{X}(\mathcal{M})$ to itself which satisfies the conditions

$$T(f_1 V_1 + f_2 V_2) = f_1 T(V_1) + f_2 T(V_2)$$

for all $f_1, f_2 \in \mathcal{F}(\mathcal{M})$, $V_1, V_2 \in \mathcal{X}(\mathcal{M})$. It is not at all clear that it makes sense to talk of "the value of T at some point $x \in \mathcal{M}$"; but this is what we must establish to show that T is indeed a type $(1, 1)$ tensor field, according to the pointwise definition. Now the value of $T(V)$ at x is well-defined, for any $V \in \mathcal{X}(\mathcal{M})$ and any $x \in \mathcal{M}$, since $T(V)$ is just a vector field. The question to be faced is this: is the value of $T(V)$ at a point determined by the value of V at that point? Or does it depend on the values of V at other points, as would be the case if $T(V)$ were, in some sense, a derivative of V? We shall show that the linearity condition implies that $T(V)_x$ is completely determined by V_x.

We show first that T is well-defined on local vector fields. The argument is similar to the one given already in the discussion of tangent vectors in Section 3. Every local vector field may be extended to a global one by multiplying it by a suitable bump function. The definition of T may thus be extended to apply when its argument is a local vector field, but this will make sense only if the result does not depend on how the local vector field is extended. We must therefore show that $T(V)_x$ depends only on the values of V in some neighbourhood of x. Suppose first that V is zero on some neighbourhood $\mathcal{O}$ of x. Let $\tilde{\mathcal{O}}$ be an open set containing

x whose closure is contained in $\mathcal{O}$, and let b be a bump function which takes the value 1 at x and which vanishes outside $\bar{\mathcal{O}}$. Then the vector field bV is identically zero, and since by its assumed linearity T maps the zero vector field to the zero vector field, $T(bV) = 0$. Thus $bT(V) = 0$, and $b(x)T(V)_x = 0$; but $b(x) = 1$, and therefore $T(V)_x = 0$. For given $\mathcal{O}$, this conclusion holds for every $x \in \mathcal{O}$: and so if V is zero on $\mathcal{O}$, so is $T(V)$. It follows that if $V_1 = V_2$ on an open set $\mathcal{O}$ then $T(V_1) = T(V_2)$ on $\mathcal{O}$, and therefore that T is well-defined on local vector fields. To evaluate T on a local vector field V whose domain is the open set $\mathcal{O}$ we may choose any global vector field $\bar{V}$ which agrees with V near $x \in \mathcal{O}$ and set $T(V)_x = T(\bar{V})_x$. Moreover, T satisfies the same linearity conditions whether its arguments are local vector fields and functions or global ones. So suppose that $\{U_a\}$ is a basis of local vector fields on some open set $\mathcal{O}$. Then

$$T(U_a) = T_a^b U_b$$

for some local functions T_a^b on $\mathcal{O}$; and for any V, local or global,

$$T(V) = V^a T_a^b U_b \qquad \text{where } V = V^a U_a.$$

It follows that $T(V)_x$ depends only on the value of V at x—and depends linearly on it. Thus T is a type $(1,1)$ tensor field.

Exercise 17. Let A be a type $(1,1)$ tensor field, and for any vector fields V, W set

$$N_A(V,W) = A^2([V,W]) + [A(V),A(W)] - A([A(V),W]) - A([V,A(W)]),$$

where $A^2 = A \circ A$. Show that N_A is a type $(1,2)$ tensor field. (Assume that the $\mathcal{F}(\mathcal{M})$-multilinearity condition works in general.) □

A tensor field of type $(p,0)$ is called *contravariant* and a tensor field of type $(0,q)$ is called *covariant*. The nomenclature is unfortunate, because under maps of manifolds covariant tensor fields, like exterior forms which they include as a special case, map contragrediently, while contravariant tensor fields, like vector fields, need not map at all, but if they do, map cogrediently. For example if T is a tensor field of type $(0,q)$ on a manifold $\mathcal{N}$ and $\phi: \mathcal{M} \to \mathcal{N}$ is a smooth map of manifolds then one may define a tensor field ϕ^*T of type $(0,q)$ on $\mathcal{M}$ by

$$(\phi^*T)_x(v_1, v_2, \ldots, v_q) = T_{\phi(x)}(\phi_* v_1, \phi_* v_2, \ldots, \phi_* v_q).$$

If ϕ is a diffeomorphism then tensors of any type map in both directions; for example if T is of type $(1,2)$ then

$$(\phi^*T)_x(\alpha, v, w) = T_{\phi(x)}(\phi^{-1*}\alpha, \phi_* v, \phi_* w),$$

and so on. This makes it possible to define the Lie derivative of a tensor of any type along a flow: if ϕ is a flow with generator V, and T is a tensor field, then $\mathcal{L}_V T$ is a tensor field of the same type defined by

$$\mathcal{L}_V T = \lim_{t \to 0} \frac{1}{t}(\phi_t^* T - T).$$

Exercise 18. Show that if T is a tensor field of type (p,q) (whose 1-form arguments come first, for convenience) and if $\theta^1, \theta^2, \dots, \theta^p$ and $W_1, W_2, \dots, W_q$ are 1-forms and vector fields, then

$$\begin{aligned}(\mathcal{L}_V T)(\theta^1, \theta^2, \dots, \theta^p, W_1, W_2, \dots, W_q) \\ = V(T(\theta^1, \theta^2, \dots, \theta^p, W_1, W_2, \dots, W_q)) \\ - \sum_{j=1}^{p} T(\theta^1, \dots, \mathcal{L}_V \theta^j, \dots, \theta^p, W_1, W_2, \dots, W_q) \\ - \sum_{k=1}^{q} T(\theta^1, \theta^2, \dots, \theta^p, W_1, \dots, [V, W_k], \dots, W_q).\end{aligned}$$

Derive an expression for the components of $\mathcal{L}_V T$ in a coordinate basis. □

6. Exterior Algebra and Calculus

The exterior algebra and differential calculus of exterior forms developed for affine spaces in Chapters 4 and 5 goes over essentially without alteration to the context of manifolds, as we shall now show.

A p-form on a manifold is a tensor field of type $(0,p)$ which is alternating in its arguments, which is to say, changes sign if any two of them are interchanged. The set of p-forms on $\mathcal{M}$ is denoted $\bigwedge^p \mathcal{M}^*$.

The operations of exterior algebra, including exterior and interior multiplication, are carried out pointwise, just as in the affine case. If ω is a p-form and χ a q-form then, for any vector fields $V_1, V_2, \dots, V_{p+q}$,

$$\begin{aligned}(\omega \wedge \chi)(V_1, V_2, \dots, V_{p+q}) \\ = \frac{1}{p!\,q!} \sum_{\pi} \epsilon(\pi) \omega(V_{\pi(1)}, V_{\pi(2)}, \dots, V_{\pi(p)}) \chi(V_{\pi(p+1)}, V_{\pi(p+2)}, \dots, V_{\pi(p+q)})\end{aligned}$$

the sum being over all permutations π of $(1, 2, \dots, p+q)$. The exterior product is distributive and associative, and

$$\chi \wedge \omega = (-1)^{pq} \omega \wedge \chi.$$

With these operations defined, the set of all forms (of all degrees) forms an algebra over $\mathcal{F}(\mathcal{M})$, the exterior algebra of forms on $\mathcal{M}$, denoted $\bigwedge \mathcal{M}^*$.

The interior product of a p-form ω and a vector field V is the $(p-1)$-form $V \lrcorner \omega$ defined by

$$(V \lrcorner \omega)(V_1, V_2, \dots, V_{p-1}) = \omega(V, V_1, V_2, \dots, V_{p-1}).$$

In any coordinate chart a p-form ω may be written

$$\omega = \frac{1}{p!} \omega_{a_1 a_2 \dots a_p} dx^{a_1} \wedge dx^{a_2} \wedge \cdots \wedge dx^{a_p}$$

where

$$\omega_{a_1 a_2 \dots a_p} = \omega(\partial_{a_1}, \partial_{a_2}, \dots, \partial_{a_p}) = \omega_{[a_1 a_2 \dots a_p]}.$$

A smooth map $\phi \colon \mathcal{M} \to \mathcal{N}$ induces a map $\phi^* \colon \bigwedge \mathcal{N}^* \to \bigwedge \mathcal{M}^*$, defined as for tensors, which preserves degree. The map ϕ^* is $\mathcal{F}(\mathcal{M})$-linear, and preserves exterior products.

The exterior derivative $d\colon \bigwedge^p \mathcal{M}^* \to \bigwedge^{p+1} \mathcal{M}^*$ is defined by

$$\begin{aligned} d\omega(V_1, V_2, \dots, V_{p+1}) &= \sum_{r=1}^{p+1} (-1)^{r+1} V_r(\omega(V_1, \dots \widehat{V_r} \dots, V_{p+1})) \\ &\quad + \sum_{1 \le r < s \le p+1} (-1)^{r+s} \omega([V_r, V_s], V_1, \dots \widehat{V_r} \dots \widehat{V_s} \dots, V_{p+1}). \end{aligned}$$

The exterior derivative is $\mathbf{R}$-linear and satisfies

$$\begin{aligned} d(f\omega) &= f\,d\omega + df \wedge \omega \\ d(\omega \wedge \chi) &= d\omega \wedge \chi + (-1)^p \omega \wedge d\chi \qquad \omega \in \textstyle\bigwedge^p \mathcal{M}^* \\ d^2 &= 0 \\ \phi^*(d\omega) &= d(\phi^*\omega), \end{aligned}$$

all exactly as in the affine case.

The Lie derivative $\mathcal{L}_V\omega$ of a p-form ω along a vector field V is defined as for a tensor field, and satisfies

$$(\mathcal{L}_V\omega)(W_1, W_2, \dots, W_p) = V(\omega(W_1, W_2, \dots, W_p)) - \sum_{r=1}^{p} \omega(W_1, \dots, [V, W_r], \dots, W_p).$$

Moreover, the Lie derivative is $\mathbf{R}$-linear in V and ω, and

$$\begin{aligned} \mathcal{L}_V(f\omega) &= f\mathcal{L}_V\omega + (Vf)\omega \\ \mathcal{L}_V(\omega \wedge \chi) &= \mathcal{L}_V\omega \wedge \chi + \omega \wedge \mathcal{L}_V\chi \\ \mathcal{L}_V(d\omega) &= d(\mathcal{L}_V\omega) \\ \mathcal{L}_{fV}\omega &= f\mathcal{L}_V\omega + df \wedge (V \lrcorner \omega) \\ \mathcal{L}_V\mathcal{L}_W\omega - \mathcal{L}_W\mathcal{L}_V\omega &= \mathcal{L}_{[V,W]}\omega \\ \mathcal{L}_V\omega &= V \lrcorner d\omega + d(V \lrcorner \omega), \end{aligned}$$

again all as in the affine case.

Closed and exact forms. As before, a form ω such that $d\omega = 0$ is said to be *closed*, and one which is an exterior derivative, $\omega = d\chi$, is said to be *exact*. An exact form is necessarily closed, since $d^2 = 0$. A closed form is locally exact, in the following sense: every point has a neighbourhood on which is defined a local form χ such that $\omega = d\chi$. To see this it is enough to realise that a coordinate chart about the point x such that x is at the origin of coordinates and the corresponding open subset of $\mathbf{R}^m$ is star-shaped with respect to the origin (Chapter 5, Exercise 34), will serve, because the argument given in the affine case will then apply.

However, a closed form on a manifold need not be globally exact. A simple example of a closed but non-exact form is furnished by the 1-form

$$\alpha = \frac{-x^2 dx^1 + x^1 dx^2}{(x^1)^2 + (x^2)^2}$$

on $\mathbf{R}^2 - \{(0,0)\}$. This 1-form is in fact the Cartesian expression for what is often written $d\vartheta$, where ϑ is the polar angle: but this does not mean that α is exact, because the polar angle does not define a smooth function on $\mathbf{R}^2 - \{(0,0)\}$ (compare Chapter 5, Exercise 32).

For a slightly more complicated example, consider the 2-form on the unit sphere in $\mathcal{E}^3$ which is given, with respect to stereographic coordinates based on the North pole, by

$$\omega = \frac{dx^1 \wedge dx^2}{\left(1 + (x^1)^2 + (x^2)^2\right)^2}.$$

That this does define a smooth global 2-form on the sphere is easily checked by transforming to stereographic coordinates based on the South pole (the required coordinate transformation is given in Chapter 9, Exercise 1); it turns out that ω has the same expression with respect to either coordinates, and so in particular is smooth at the North pole as well as everywhere else. It is a closed 2-form (as every 2-form on a 2-dimensional manifold must be). A 1-form on the coordinate chart corresponding to North pole stereographic coordinates, whose exterior derivative is the given 2-form, is easily found (by the method of Chapter 5, Section 8, or by guesswork): for example

$$\frac{-x^2 dx^1 + x^1 dx^2}{2\left(1 + (x^1)^2 + (x^2)^2\right)}.$$

Any other such 1-form must differ from this by an exact form. The 1-form which has the same expression as this in South pole stereographic coordinates has ω for its exterior derivative on that coordinate chart. When it is transformed to North pole stereographic coordinates it is found to differ from the 1-form given above by

$$\frac{-x^2 dx^1 + x^1 dx^2}{2\left((x^1)^2 + (x^2)^2\right)},$$

this 1-form being defined on $\mathbf{R}^2 - \{(0,0)\}$, corresponding to the region of overlap of the two coordinate charts. Had the original 2-form ω on the sphere been exact, this 1-form would also have been exact; but (as follows from the first example) it is not exact (though it is closed). It follows that the original 2-form is not exact. (The 2-form ω is, apart from a constant multiple, the volume 2-form on the sphere induced from the standard volume in $\mathcal{E}^3$).

Whether or not there exist on a manifold closed forms which are not (globally) exact depends on the global topological properties of the manifold. It is a most interesting question, but one we shall not pursue further here.

7. Frobenius's Theorem

A distribution on a manifold $\mathcal{M}$ is an assignment, to each point x of $\mathcal{M}$, of a subspace of $T_x\mathcal{M}$. It is to be assumed that all the subspaces thus defined have the same dimension, and that they vary smoothly from point to point. This latter requirement may be expressed in two equivalent ways: first, by requiring that every point of $\mathcal{M}$ has a neighbourhood in which smooth local vector fields $V_1, V_2, \ldots, V_k$ may be found which form at each point in the neighbourhood a basis for the subspace distinguished by the distribution (whose dimension is therefore k); alternatively, by requiring the existence about each point of $\mathcal{M}$ of $m - k$ independent smooth local 1-forms $\theta^{k+1}, \theta^{k+2}, \ldots, \theta^m$ which form a basis at each point for the space of constraint forms for the distinguished subspace.

An immersed submanifold of $\mathcal{M}$ is said to be an *integral submanifold* of a distribution $\mathcal{D}$ if, at each point x on it, its tangent space coincides with the subspace $\mathcal{D}_x$ of $T_x\mathcal{M}$. The necessary and sufficient conditions for a distribution to be integrable (that is, for local integral submanifolds of it to exist through every point of $\mathcal{M}$) are those given in Chapter 6. These have two alternative formulations: in terms of a local vector field basis,

$$[V_\alpha, V_\beta] = f_{\alpha\beta}^{\gamma} V_\gamma \qquad \alpha, \beta, \gamma = 1, 2, \ldots, k$$

for some functions $f_{\alpha\beta}^{\gamma}$; or in terms of a local basis for constraint 1-forms,

$$d\theta^\rho = \lambda_\sigma^\rho \wedge \theta^\sigma \qquad \rho, \sigma = k+1, k+2, \ldots, m$$

for some 1-forms λ_σ^ρ.

When a distribution is integrable, the existence of local integral submanifolds is assured by the argument of Chapter 6. In fact, $\mathcal{M}$ may be covered by coordinate charts in each of which the coordinates are such that the submanifolds $x^{k+1} =$ constant, $x^{k+2} =$ constant, $\ldots, x^m =$ constant, are integral submanifolds of the distribution. The question remains how to piece these local integral submanifolds together to form maximal integral submanifolds. The problem is best solved by approaching it from a slightly different direction. We call a smooth curve in $\mathcal{M}$ whose tangent vector everywhere belongs to $\mathcal{D}$ an *integral curve* of $\mathcal{D}$; more generally, a continuous curve in $\mathcal{M}$ which is made up of a finite number of smooth segments each of which is an integral curve of $\mathcal{D}$ we call a *piecewise integral* curve of $\mathcal{D}$. For a given point $x \in \mathcal{M}$ we define the *leaf* of $\mathcal{D}$ through x, $L(x)$, to be the set of all points of $\mathcal{M}$ accessible from x along piecewise integral curves of $\mathcal{D}$. These concepts may be defined regardless of whether or not $\mathcal{D}$ is integrable: but when it is integrable each leaf $L(x)$ is an immersed submanifold of $\mathcal{M}$, and is an integral submanifold of $\mathcal{D}$ which is maximal in the sense that any other connected integral submanifold of $\mathcal{D}$ through x must be contained in it. In fact, suppose that y is any point of $L(x)$, and consider a coordinate chart about y whose coordinates are adapted to the integral submanifolds of the distribution in the way described above. Then every point in the corresponding patch which lies on the submanifold given by $x^{k+1} = x^{k+1}(y)$, $x^{k+2} = x^{k+2}(y), \ldots, x^m = x^m(y)$ can be joined to x by a piecewise integral curve of $\mathcal{D}$ (*via* y), and therefore itself belongs to $L(x)$; conversely every point of $L(x)$ which lies in the patch and which can be joined to y by an integral curve of $\mathcal{D}$ lying in the patch belongs to this submanifold. The first k coordinates of the chart then serve to define a coordinate system about y on $L(x)$: it may be shown that $L(x)$ acquires in this way the structure of a smooth manifold, and is an immersed submanifold of $\mathcal{M}$.

Now suppose given any connected integral submanifold $\mathcal{S}$ of $\mathcal{D}$ through x. Any connected manifold is pathwise connected: that is, any point y of $\mathcal{S}$ may be joined to x by a piecewise smooth curve lying in $\mathcal{S}$. But the smooth segments of this curve must then be integral curves of $\mathcal{D}$; thus y may be joined to x by a piecewise integral curve of $\mathcal{D}$ and is therefore in $L(x)$. Thus $L(x)$ contains $\mathcal{S}$, and is maximal.

However, the leaf of an integrable distribution may still have somewhat complicated structure (in relation to $\mathcal{M}$) even in simple-looking cases. The following example is standard. Consider the 2-torus, regarded as before as a square in $\mathbf{R}^2$ with opposite edges identified. Define a 1-dimensional distribution on the torus by taking for vector field basis the single vector field $cos\varphi\partial_1 + \sin\varphi\partial_2$ where φ is some constant. The integral submanifolds are given by $x^1 \sin\varphi - x^2 \cos\varphi = $ constant (where x^1 and x^2 take their values modulo 1). Now if $\tan\varphi$ is rational then each leaf is a closed path on the torus and is an imbedded submanifold. But if not, it can be shown that the leaf of any point, while it does not return to that point, comes back repeatedly arbitrarily close to it; and in fact the leaf is a dense subset of the torus, that is, it intersects every neighbourhood of every point of the torus.

8. Metrics on Manifolds

In this section we introduce the idea of a metric on an arbitrary manifold. This is simultaneously a generalisation of the idea of an affine metric, introduced in Chapter 7, Section 5, and of the idea of a metric on an abstract surface, introduced in Chapter 9, Section 8.

An affine metric on an affine space $\mathcal{A}$ is a symmetric bilinear form, that is, a symmetric tensor field of type $(0,2)$, constructed on each tangent space to $\mathcal{A}$ from a scalar product on the underlying vector space by identifying the tangent space with it. It is distinguished by the fact that its value on any two everywhere parallel vector fields is constant, or in other words, that in an affine coordinate system its components are all constants.

A metric on an abstract surface is a positive-definite symmetric tensor of type $(0,2)$ assigned on the surface. It is more general than an affine metric in that there need not be even a coordinate patch in which the components are all constants, but less general in that it is only defined for dimension 2 and is restricted in its signature.

The generalisation to arbitrary manifolds gives up the constancy of the affine case and gives up the restriction on dimension and signature of the abstract surface case.

Let $\mathcal{M}$ be a manifold of dimension m. A *metric* on $\mathcal{M}$ is a non-degenerate symmetric tensor field g of type $(0,2)$ on $\mathcal{M}$, which is to say, an $\mathcal{F}(\mathcal{M})$-bilinear map $\mathcal{X}(\mathcal{M}) \times \mathcal{X}(\mathcal{M}) \to \mathcal{F}(\mathcal{M})$ such that $g(W,V) = g(V,W)$ (as functions on $\mathcal{M}$) for any vector fields V, W, and that $g(V,W) = 0$ for all W only when $V = 0$. Assignment of a metric on $\mathcal{M}$ is equivalent to assignment of a scalar product in each tangent space, with the proviso that if V and W are smooth vector fields then $g(V,W)$ is a smooth function.

The remarks about signature in Chapter 7, Section 1 (which deals with scalar products on vector spaces) apply to each tangent space; in particular, g is called *positive-definite* if $g_x(v,v) > 0$ whenever $v \in T_x\mathcal{M} \neq 0$. If g is positive-definite then there is

(1) a basis $\{V_a\}$ for vector fields on each coordinate patch such that $g(V_a, V_b) = \delta_{ab}$ all over the patch, and

(2) a coordinate chart around any point such that $g(\partial_a, \partial_b) = \delta_{ab}$ at that point; but

(3) in general no neighbourhood of any point throughout which $g(\partial_a, \partial_b) = \delta_{ab}$ in any chart.

The possibility of finding a local orthonormal basis of vector fields as given in (1) arises from the fact that a system of vector fields can constitute a basis for tangent vectors at each point without reducing to a coordinate basis in any chart.

A positive-definite metric is often called a *Riemannian metric*, *Riemannian structure* or *Euclidean structure* on $\mathcal{M}$, and $\mathcal{M}$, endowed with a Riemannian metric, is often called a *Riemannian manifold*.

If g is of signature $(r, m - r)$ then there is

(1) a basis $\{V_a\}$ for vector fields on each coordinate patch such that $g(V_a, V_b) = \eta_{ab}$ all over the patch, where

$$\eta_{ab} = \begin{cases} +1 & \text{for } a = b = 1, 2, \ldots, r \\ -1 & \text{for } a = b = r+1, r+2, \ldots, m \\ 0 & \text{for } a \neq b. \end{cases}$$

(2) a coordinate chart around any point such that $g(\partial_a, \partial_b) = \eta_{ab}$ at the point; but

(3) in general no neighbourhood throughout which $g(\partial_a, \partial_b) = \eta_{ab}$, in any chart.

A non-singular, but not positive-definite, metric is often called a *pseudo-Riemannian metric* or *pseudo-Riemannian structure* on $\mathcal{M}$, and $\mathcal{M}$, endowed with a pseudo-Riemannian metric, is often called a *pseudo-Riemannian manifold*.

If $r = 1$, which is the case in classical gravitation theory and other applications, then g is often called a *hyperbolic normal* or *Lorentzian metric* or a *Lorentz structure* on $\mathcal{M}$. The distinction between timelike, spacelike and null vectors and directions may be taken over unaltered from the affine case, and readily extended to vector fields.

A metric has components $g_{ab} = g(\partial_a, \partial_b)$ and may be written

$$ds^2 = g_{ab} dx^a dx^b.$$

The operations of raising and lowering indices described in Chapter 7 may be applied in a Riemannian or pseudo-Riemannian manifold; they are carried out in essentially the same way as before. In particular, a function f may be used to define a vector field, its *gradient*, by the rule $g(\operatorname{grad} f, V) = Vf$ for every vector field f.

We may also raise and lower indices on tensor fields. For example, if A is a type $(1,1)$ tensor field then $(U, V) \mapsto g\big(A(U), V\big)$ defines a type $(0,2)$ tensor field; and conversely, given any type $(0,2)$ tensor field B there is a type $(1,1)$ tensor field A such that $B(U, V) = g\big(A(U), V\big)$. Then A is obtained by raising an index on B (caution: there are two indices to choose from), and conversely B is obtained by lowering an index on A.

A type $(0,2)$ tensor field is symmetric if its value is unchanged by interchange of its arguments. A type $(1,1)$ tensor field is *symmetric* with respect to g if the corresponding field with lowered index is symmetric.

Isometries. If $\phi\colon \mathcal{M} \to \mathcal{N}$ is a map of manifolds and $\hat{g}$ is a metric tensor on $\mathcal{N}$ then $\phi^*\hat{g}$, defined by $(\phi^*\hat{g})_x(v,w) = \hat{g}_{\phi(x)}(\phi_* v, \phi_* w)$ for all $v, w \in T_x\mathcal{M}$, is a symmetric type $(0,2)$ tensor field on $\mathcal{M}$.

Exercise 19. Show, by means of an example, that $\phi^*\hat{g}$ is not necessarily a metric on $\mathcal{M}$. □

If there is already a metric tensor on $\mathcal{M}$, say g, and $\phi^*\hat{g} = g$ then ϕ is called an *isometry*. An example of this is the realisation of an abstract surface by an isometric imbedding described in Chapter 9, Section 8.

One important case is that in which $\mathcal{N} = \mathcal{M}$ and the map belongs to a one-parameter group or flow. Thus a flow ϕ is an *isometric flow* of a manifold $\mathcal{M}$ with metric g if $g_{\phi_t(x)}(\phi_{t*}v, \phi_{t*}w) = g_x(v,w)$ whenever the left-hand side is defined, and if an isometric flow is actually a one-parameter group it is called a *one-parameter group of isometries*. Locally it makes no difference whether one is dealing with a flow or a one-parameter group. If ϕ is any flow, with generator X, then it follows from the definition of the Lie derivative of a tensor field that the condition for ϕ to be an isometric flow is $\mathcal{L}_X g = 0$. The components of this equation in a coordinate basis are called *Killing's equations*, (or the equation itself may be called Killing's equation) and the solutions are called Killing vector fields of g.

Exercise 20. Show that Killing's equations are

$$X^c \partial_c g_{ab} + g_{ac}\partial_b X^c + g_{bc}\partial_a X^c = 0. \qquad \square$$

Exercise 21. Find three linearly independent solutions of Killing's equations for the metric of the unit sphere specified in polar coordinates ϑ, φ by $ds^2 = d\vartheta^2 + \sin^2\vartheta d\varphi^2$. (We have written ϑ, φ here instead of ξ^1, ξ^2 which we used when we introduced polar coordinates for the sphere in Chapter 9, Section 1.) Verify that these solutions are $\mathbf{R}$-linear combinations of the restriction to the unit sphere of the generators of rotations of $\mathcal{E}^3$. □

Exercise 22. Show that if X and Y are Killing vector fields then so are $kX + lY$, for $k, l \in \mathbf{R}$, and $[X,Y]$. □

It is known that the Killing fields constitute a (finite-dimensional) Lie algebra.

Conformal structures. Roughly speaking, a conformal transformation is a map which preserves the metric up to a scalar factor (in general variable). The importance of this idea for applications is that in the case of a pseudo-Riemannian structure such a transformation leaves the null cones unaltered.

Let $\mathcal{M}$ and $\mathcal{N}$ be manifolds, endowed with metrics g and $\hat{g}$ respectively. A smooth map $\phi\colon \mathcal{M} \to \mathcal{N}$ induces the symmetric tensor field $\phi^*\hat{g}$ on $\mathcal{M}$, and ϕ is said to be *conformal* if

$$\phi^*\hat{g} = \kappa g$$

where κ is a positive function on $\mathcal{M}$. In particular, if κ is constant, ϕ is called *homothetic*; if $\kappa = 1$ we regain the case of an isometry, already discussed.

Of greatest interest is the case in which $\mathcal{N} = \mathcal{M}$ and $\hat{g} = g$: a diffeomorphism ϕ of $\mathcal{M}$ which is conformal is called a *conformal transformation* of $\mathcal{M}$ in this case.

Exercise 23. Show that the conformal transformations of a manifold form a group. □

A related notion is that of a *conformal change* or *conformal rescaling*: here we deal with one manifold, which carries two metric structures, say g and $\hat{g}$, which are conformally related: $\hat{g} = \kappa g$, where κ is a positive function, as before. Two metric structures which are related by a conformal change are called *conformally equivalent*.

Exercise 24. Show that conformal equivalence is an equivalence relation. □

An equivalence class of conformally equivalent metric structures on $\mathcal{M}$ is called a *conformal structure* on $\mathcal{M}$. Specifying a conformal structure amounts to specifying a metric structure up to a positive scalar factor. A property common to all members of an equivalence class is called *conformally invariant*.

Exercise 25. Show that the definition of conformal transformation passes to the quotient to apply to conformal structures. □

Remark first of all that the ratio of scalar products is well-defined: if g and $\hat{g}$ are conformally equivalent metrics of Lorentz signature and $u_1, u_2, v_1, v_2 \in T_x\mathcal{M}$ then

$$\frac{\hat{g}_x(u_1,u_2)}{\hat{g}_x(v_1,v_2)} = \frac{g_x(u_1,u_2)}{g_x(v_1,v_2)}$$

(one must of course avoid $g_x(v_1,v_2) = 0$). Whether a vector is timelike, null or spacelike is conformally invariant:

$$\left\{\begin{array}{l} \hat{g}_x(v,v) > 0 \\ \hat{g}_x(v,v) = 0 \\ \hat{g}_x(v,v) < 0 \end{array}\right\} \iff \left\{\begin{array}{l} g_x(v,v) > 0 \\ g_x(v,v) = 0 \\ g_x(v,v) < 0 \end{array}\right\} \iff v \text{ is } \left\{\begin{array}{l} \text{timelike} \\ \text{null} \\ \text{spacelike} \end{array}\right\}$$

(the signature is chosen as in Chapter 7, Section 2). Orthogonality (between any pair of vectors u, v) is defined by $\hat{g}_x(u,v) = g_x(u,v) = 0$. The angle ϑ between spacelike vectors u, v is well-defined:

$$\cos\vartheta = \frac{\hat{g}_x(u,v)}{\sqrt{\hat{g}_x(u,u)\hat{g}_x(v,v)}} = \frac{g_x(u,v)}{\sqrt{g_x(u,u)g_x(v,v)}}.$$

A hypersurface (submanifold of dimension $n-1$) is called

$$\left\{\begin{array}{l} \text{spacelike} \\ \text{null} \\ \text{timelike} \end{array}\right\} \text{ according as its normal is } \left\{\begin{array}{l} \text{timelike} \\ \text{null} \\ \text{spacelike.} \end{array}\right\}$$

All these properties may in fact be defined entirely in terms of the null cone.

Summary of Chapter 10

A manifold is a set in which it makes sense to introduce coordinates locally; it differs from an affine space in that there need not necessarily be any globally defined coordinates. A topological manifold of dimension m is a topological space each point of which lies in an open subset which is homeomorphic to an open subset of $\mathbf{R}^m$; such an open subset, together with the homeomorphism, is called a chart of the manifold. If the homeomorphisms of open subsets of $\mathbf{R}^m$ which represent coordinate transformations on the overlap of two charts are smooth then the charts are smoothly related. A (smooth) atlas is a collection of pairwise smoothly related charts covering the manifold; it is complete if it is not a proper subcollection of any other atlas. A differentiable manifold is a topological manifold with a complete atlas.

Maps of manifolds are smooth if smoothly represented with respect to charts in domain and codomain. An immersion is a smooth map of manifolds whose Jacobian matrix (with respect to any charts) has rank equal to the dimension of its domain; a submersion is a smooth map whose rank is equal to the dimension of its codomain. A subset of a manifold which is the image of another manifold by an immersion is called an immersed submanifold; it is an imbedded submanifold (or just a submanifold) if the immersion is in addition injective and a homeomorphism onto its image. A curve is a smooth map of (an open interval) of $\mathbf{R}$ into a manifold; a function is a smooth map of the manifold to $\mathbf{R}$. With these definitions practically all of the concepts from Chapters 1–7 which are not specifically tied to the affine structure of an affine space may be generalised to apply to manifolds. We mention only those points where caution is necessary in making the generalisation, or where some new element is introduced.

In defining a tangent vector it is necessary to use the directional derivative definition, since in general the idea of a displacement vector makes no sense in a manifold. It remains true that the space of tangent vectors at a point of an m-dimensional smooth manifold is an m-dimensional real vector space, though to prove it requires the use of a technical lemma from analysis. Cotangent vectors are defined as duals of tangent vectors, as before.

The integral curves of a vector field, whose existence in any chart is assured by the existence theorem for systems of ordinary differential equations, must be pieced together as one moves from chart to chart. One obtains thereby a unique maximal integral curve which passes (with parameter value 0) through a given point of the manifold. But a vector field may generate only a flow, and not a one-parameter group, as is known from the affine case.

A tensor at a point x of a manifold $\mathcal{M}$ is an $\mathbf{R}$-multilinear map of the Cartesian product of p copies of $T_x^*\mathcal{M}$ and q copies of $T_x\mathcal{M}$, in some order, to $\mathbf{R}$. Such a tensor is said to be of type (p,q). A tensor field on a manifold is a smooth assignment to each point of a tensor at that point; or equivalently, an $\mathcal{F}(\mathcal{M})$-multilinear map of 1-forms and vector fields to $\mathcal{F}(\mathcal{M})$. (Strictly speaking, proof of this equivalence requires a technical analytic result to allow one to replace global fields by local fields as arguments.) If there are p 1-form and q vector field arguments the tensor field is said to be of type (p,q), in accordance with the pointwise definition. A tensor field of type $(1,r)$ may be regarded, alternatively, as an r-fold $\mathcal{F}(\mathcal{M})$-multilinear map $\mathcal{X}(\mathcal{M})^r \to \mathcal{X}(\mathcal{M})$, where $\mathcal{X}(\mathcal{M})$ is the $\mathcal{F}(\mathcal{M})$-module of vector fields on $\mathcal{M}$.

The algebra and calculus of exterior forms (which are special kinds of tensor fields) follows much the same pattern as in an affine space, except that a closed form is not necessarily exact (though it is locally exact).

The results of Frobenius's theorem on the integrability of distributions apply also as in the case of an affine space, though again there is a technical difficulty in piecing together local integral submanifolds (defined on coordinate patches) to form maximal ones.

A metric on a manifold, of whatever signature, is defined in much the same way as a metric on an affine space, but there will not generally be a coordinate chart in which its components are constants. Two metrics are conformally related if one is a scalar multiple of the other by a positive function.

Notes to Chapter 10

1. Hausdorff spaces with countable bases. A Hausdorff space is one in which any two distinct points lie in disjoint open sets. A topology has a countable base if the sets in its base may be put into 1 : 1 correspondence with the positive integers. For details of these ideas see for example Kelley [1955], and for their relevance to the definition of a manifold for example Brickell and Clark [1970].

2. Closed and exact forms. An exact form is necessarily closed. A closed form is locally exact, that is, may be written as an exterior derivative in a neighbourhood of each point of its domain of definition. The example given at the end of Section 6 shows that there are closed forms which are not everywhere exact. Call two forms equivalent if they differ by an exact form. The number of inequivalent closed p-forms on a manifold is determined by the topological properties of the manifold. We do not attempt to explain this: to make the statement precise would take us too far afield. An elementary treatment, with many applications, has been given by Flanders [1963]. A more advanced version is in Warner [1971]; de Rham [1955] is the standard classic.

11. CONNECTIONS

The one major item of discussion from Chapters 1 to 9 which has not been generalised so as to apply in a differentiable manifold is the idea of a connection, that is, of parallelism and the associated operation of covariant differentiation. This is the subject of the present chapter.

It may be recalled that in an affine space it makes sense to say whether two vectors defined at different points are parallel, because they may be compared with the help of the natural identification of tangent spaces at different points. On a surface, on the other hand, no such straightforward comparison of tangent vectors at different points is possible; there is however a plausible and workable generalisation from affine spaces to surfaces in which the criterion of parallelism depends on a choice of path between the points where the tangent vectors are located. Though the covariant differentiation operator associated with this path-dependent parallelism satisfies the first order commutation relation of affine covariant differentiation, $\nabla_U V - \nabla_V U - [U,V] = 0$, it fails to satisfy the second order one, $\nabla_U \nabla_V W - \nabla_V \nabla_U W - \nabla_{[U,V]} W = 0$ in general; and indeed its failure to do so is intimately related to the curvature of the surface.

In generalising these notions further, from surfaces to arbitrary differentiable manifolds, we have to allow for the arbitrariness of dimension; we have to develop the theory without assuming the existence of a metric in the first instance (though we shall consider that important specialisation in due course); and we have to allow for the possibility that not even the first order commutation relation survives.

As an illustration of the last point we describe a rather natural definition of parallelism on the surface of the Earth which, unlike Levi-Civita parallelism, satisfies the second order commutation relation but not the first. This construction is due to É. Cartan. Imagine the Earth to be a perfect sphere with the North pole of polar coordinates placed at the North magnetic pole. Then at each point on the surface a compass needle would point along the meridian, and a navigator might therefore choose to call vectors at different points parallel if they had the same lengths and made equal angles, in the same sense, with the meridians at the points in question. This definition would of course break down at the poles themselves, which will therefore be left out of the following argument. On the rest of the sphere parallelism of vectors at different points is thereby defined in a way independent of any path between the points. It follows that the second order commutation relation is satisfied. In the usual polar coordinates ϑ, φ on the sphere the vector fields $\partial/\partial\vartheta$ and $(\sin\vartheta)^{-1}\partial/\partial\varphi$ are parallel fields; if the first order commutation relation held they would therefore commute, but evidently $[\partial/\partial\vartheta, (\sin\vartheta)^{-1}\partial/\partial\varphi] \neq 0$. Cartan described this failure to commute in terms of a construction which he called the "torsion" of the connection; we shall define torsion in Section 3 below.

Most of the techniques for handling general connections used in this chapter are simple generalisations of techniques introduced in the discussion of specific cases in previous chapters. We develop the theory of connections, starting from ideas of parallelism, first by vector methods, then by methods of exterior calculus. After a discussion of geodesics and exponential maps, we describe the Levi-Civita connection in a pseudo-Riemannian manifold.

1. Parallelism and Connections on Manifolds

Following the lead suggested by surface theory, we shall define parallelism on a manifold with respect to a path. It will be recalled that a path in a manifold $\mathcal{M}$ is a curve freed from its parametrisation. The essential notion of parallelism is that one should be able to identify the tangent spaces at any two points, once given a path joining them. This identification should preserve the linear structure of the tangent spaces. Vectors at points x and y, which are identified in this way, are parallel with respect to the given path. One would expect that if z is a point on the path intermediate between x and y then vectors at x and y will be parallel if and only if they are both parallel to the same vector at z.

We therefore define a *rule of parallel transport* along a path as a collection of non-singular linear maps $\Upsilon_{y,x}: T_x\mathcal{M} \to T_y\mathcal{M}$, one for every pair of points x, y on the path, such that for any point z on the path

$$\Upsilon_{y,z} \circ \Upsilon_{z,x} = \Upsilon_{y,x}.$$

It follows from this that $\Upsilon_{x,x}$ is the identity on $T_x\mathcal{M}$, and $\Upsilon_{x,y} = \left(\Upsilon_{y,x}\right)^{-1}$.

Pedantry would require that $\Upsilon_{y,x}$ be labelled by the path as well as the points on it, since the possibility that the map depends on the path is an essential feature of this construction, but so long as only one path is being considered at a time this dependence will be understood rather than inscribed in a yet more cumbersome notation.

If a rule of parallel transport is given for each path of $\mathcal{M}$, one says that a rule of parallel transport is given in $\mathcal{M}$. We assume that if one path is a subset of another, then the rule of parallel transport on the subset is that obtained by restriction.

A vector field given along a path is called a *parallel field* along the path, or said to be *parallelly transported*, if it may be obtained by parallel transport from a vector given at some point of the path; thus W is a parallel field if $W_y = \Upsilon_{y,x}W_x$ for each y and some x on the path.

Some conditions of smoothness have to be imposed; we shall do this in full when we define the covariant derivative below. For the present we suppose merely that parallel fields are smooth.

Exercise 1. Let V and W be parallel fields along a given path obtained by parallel transport of V_x and W_x from a point x along the path. Show that for any real a and b, $aV + bW$ is a parallel field obtained by parallel transport of $aV_x + bW_x$ from x. □

Exercise 2. Let x be a point chosen on a given path and w a vector given at x. Construct a vector field W along the path by parallel transport of w. Show that if z is some point on the path distinct from x, $\acute{w} = \Upsilon_{z,x}w$, and $\acute{W}$ is a vector field constructed by parallel transport of $\acute{w}$ along the path, then $\acute{W}$ and W coincide. □

Exercise 3. With the help of a parallelly transported basis, show that if W is a smooth field (not necessarily a parallel field) given along a path and if vectors $w(y)$ are defined in $T_x\mathcal{M}$ by $w(y) = \Upsilon_{x,y}W_y$ then in any smooth parametrisation of the path $w(y)$ depends smoothly on the parameter, and therefore defines (the path of) a smooth curve in $T_x\mathcal{M}$. □

The ideas of parallel transport and parallel field may be extended immediately from vectors to subspaces of the tangent spaces along a path: if a subspace H_x of $T_x\mathcal{M}$ is given, one may define a field of subspaces along a given path through x by parallelly transporting the vectors in H_x, and thereby constructing subspaces $H_y = \{\, \Upsilon_{y,x}v \mid v \in H_x \,\}$. The field of subspaces obtained in this way is said to be parallel along the path.

Parallel transport may also be extended to covectors in a straightforward way: a non-singular linear map $\Upsilon^*_{y,x}: T^*_x\mathcal{M} \to T^*_y\mathcal{M}$ is defined by $\langle w, \Upsilon^*_{y,x}\alpha\rangle = \langle \Upsilon_{x,y}w, \alpha\rangle$ for each $\alpha \in T^*_x\mathcal{M}$ and for all $w \in T_y\mathcal{M}$. This rule ensures that parallel transport preserves pairings.

As in the affine case one may employ parallel transport to construct along a curve an absolute derivative of a vector field which is not necessarily parallel: the result is another vector field along the curve. Let W be a vector field defined along a curve σ. Let $W(t)$ denote the value of W at $\sigma(t)$ and let $W(t+\delta)_{\|}$ be the vector at $\sigma(t)$ obtained by parallelly transporting $W(t+\delta)$ along the path defined by σ from $\sigma(t+\delta)$ to $\sigma(t)$: $W(t+\delta)_{\|} = \Upsilon_{\sigma(t),\sigma(t+\delta)}W(t+\delta)$. The *absolute derivative* of W along σ at $\sigma(t)$ is

$$\frac{DW}{Dt}(t) = \lim_{\delta\to 0}\frac{1}{\delta}\left(W(t+\delta)_{\|} - W(t)\right) = \frac{d}{ds}\left\{\Upsilon_{\sigma(t),\sigma(s)}W(s)\right\}_{s=t}.$$

Exercise 4. Show that if W, W_1 and W_2 are vector fields and f a function, all given on σ, then

$$\frac{D}{Dt}(W_1 + W_2) = \frac{DW_1}{Dt} + \frac{DW_2}{Dt} \qquad \frac{D}{Dt}(fW) = f\frac{DW}{Dt} + \frac{df}{dt}W.$$

Deduce that if $\{V_a\}$ is a basis of vector fields along σ and if $W = W^aV_a$ then

$$\frac{DW}{Dt} = W^a\frac{DV_a}{Dt} + \frac{dW^a}{dt}V_a.$$ □

Exercise 5. Show from the definition that if a vector field W is parallel along a path then $DW/Dt = 0$ for any curve with that path as image. Deduce that if W is smooth (but not necessarily parallel) then DW/Dt is smooth along the path. □

Exercise 6. Show that the absolute derivative satisfies the chain rule for reparametrisation (compare Chapter 2, Exercise 43): if $\hat\sigma = \sigma \circ h$ is a reparametrisation, and $\hat W(s)$ denotes the value of W at $\hat\sigma(s) = \sigma(h(s))$, then $D\hat W/Ds = \dot h(DW/Dt)\circ h$. □

We prove now that if $DW/Dt = 0$ along a path then W is a parallel field. Fix a point x on the path; take a curve σ which defines the path, such that $x = \sigma(0)$.

Observe first that

$$\begin{aligned}\frac{d}{dt}\{\Upsilon_{x,\sigma(t)}W(t)\} &= \frac{d}{ds}\{\Upsilon_{x,\sigma(t+s)}W(t+s)\}_{s=0}\\ &= \frac{d}{ds}\{\Upsilon_{x,\sigma(t)}\circ\Upsilon_{\sigma(t),\sigma(t+s)}W(t+s)\}_{s=0}\\ &= \Upsilon_{x,\sigma(t)}\frac{d}{ds}\{\Upsilon_{\sigma(t),\sigma(t+s)}W(t+s)\}_{s=0}\\ &= \Upsilon_{x,\sigma(t)}\frac{DW}{Dt}(t).\end{aligned}$$

Thus if $DW/Dt = 0$ in some interval about x then $\Upsilon_{x,\sigma(t)}W(t)$ is a constant vector in $T_x\mathcal{M}$, say w, and so $W(t) = \Upsilon_{\sigma(t),x}w$ is parallel.

2. Covariant Differentiation

In affine space, and on a surface, the absolute derivative at a point depends only on the tangent vector to the curve, and not on the choice of a particular curve with that tangent vector. Moreover, it depends linearly on the tangent vector. Examples may be constructed to show that the rules of parallel translation outlined above are not in themselves sufficient to ensure that this is the case. Nevertheless these are important properties of the absolute derivative, which we seek to generalise, and they must therefore be the subject of additional assumptions.

We assume that the rule of parallel transport in question satisfies the following additional conditions:

(1) the absolute derivative along a curve depends, at any point, only on the tangent vector to the curve, in the following sense: if σ and τ are curves such that $\sigma(0) = \tau(0) = x$ and $\dot\sigma(0) = \dot\tau(0)$, and if W is a vector field given near x, then the absolute derivatives of W at x along σ and τ are equal

(2) given a vector field W defined near x the map $T_x\mathcal{M} \to T_x\mathcal{M}$ which takes each $u \in T_x\mathcal{M}$ to the absolute derivative of W at x, along any curve whose tangent vector at x is u, is a linear map.

Such a rule of parallel transport is said to determine a *linear connection* on $\mathcal{M}$. This term refers equally to the rule of parallelism, to the associated absolute derivative, or to the covariant derivative operator now to be defined.

According to assumption (1) above we may unambiguously associate, with each $u \in T_x\mathcal{M}$ and each vector field W defined near x, an element $\nabla_u W$ of $T_x\mathcal{M}$ by $\nabla_u W = DW/Dt(0)$, where the absolute derivative is taken along any curve σ such that $\sigma(0) = x$ and $\dot\sigma(0) = u$. We call $\nabla_u W$ the *covariant derivative* of W along u.

Using the covariant derivative one may construct from two local vector fields U and W with the same domain a further local vector field $\nabla_U W$ whose value at any point x where U and W are defined is $\nabla_{U_x} W$. To complete the definition of a smooth linear connection, we require $\nabla_U W$ to be smooth for every smooth U and W.

The covariant derivative has the properties

(1) $\nabla_{U+V}W = \nabla_U W + \nabla_V W$

(2) $\nabla_{fU}W = f\nabla_U W$

(3) $\nabla_U(V+W) = \nabla_U V + \nabla_U W$
(4) $\nabla_U(fW) = f\nabla_U W + (Uf)W$.

Here U, V, W are locally defined smooth vector fields and f a locally defined smooth function. The first two properties follow from the linearity assumption (2) above, the second two from properties of parallel translation. These properties correspond to the first four of those listed for the covariant derivative in Chapter 3, Section 11 and Chapter 9, Section 5.

Exercise 7. Show that a covariant derivative operator (that is, an operator on local vector fields satisfying the properties given above) defines a rule of parallel translation along paths, by using the fact that $\nabla_{\dot\sigma} W = 0$ is a set of linear first order differential equations for the components of W. □

Exercise 8. Extend the operation of covariant differentiation to 1-forms along the lines set out in Exercise 45 of Chapter 2. □

It is a straightforward matter to express covariant derivatives in terms of an arbitrary local basis for vector fields, and in particular in terms of a coordinate basis. Let $\{U_a\}$ be a local basis of vector fields on $\mathcal{M}$. Then the covariant derivatives $\nabla_{U_c} U_b$ may be expressed in terms of the basis itself, say

$$\nabla_{U_c} U_b = \gamma^a_{bc} U_a.$$

The functions γ^a_{bc} are the *coefficients of linear connection* with respect to the local basis. From the properties of ∇ set out above it follows that if $V = V^a U_a$ and $W = W^a U_a$ then

$$\nabla_V W = V^c(U_c(W^a) + \gamma^a_{bc} W^b) U_a.$$

In particular, if $\{U_a\} = \{\partial_a\}$ is a coordinate basis then

$$\nabla_V W = V^c(\partial_c W^a + \Gamma^a_{bc} W^b)\partial_a$$

where, as is customary, we have written Γ^a_{bc} for the coefficients of connection with respect to the coordinate basis:

$$\nabla_{\partial_c}\partial_b = \Gamma^a_{bc}\partial_a.$$

The term in parentheses in the expression for $\nabla_V W$ is often abbreviated to $W^a{}_{\|c}$: thus $\nabla_V W = (\nabla_V W)^a \partial_a$ where $(\nabla_V W)^a = W^a{}_{\|c} V^c$ and $W^a{}_{\|c} = \partial_c W^a + \Gamma^a_{bc} W^b$.

Exercise 9. If $\{\acute U_a\}$ is another local basis, where $\acute U_a = A^b_a U_b$ with (A^b_a) a non-singular matrix of functions, then the coefficients of connection $\acute\gamma^a_{bc}$ corresponding to $\{\acute U_a\}$ are given in terms of those corresponding to $\{U_a\}$, γ^a_{bc}, by

$$A^a_f \acute\gamma^f_{bc} = \acute U_c(A^a_b) + \gamma^a_{de} A^d_b A^e_c.$$

In particular, if $\{U_a\}$, $\{\acute U_a\}$ are coordinate bases, so that (A^b_a) is the inverse of the Jacobian matrix of the coordinate transformation, then

$$\acute\Gamma^a_{bc} = \frac{\partial^2 x^d}{\partial \acute x^b \partial \acute x^c}\frac{\partial \acute x^a}{\partial x^d} + \Gamma^d_{ef}\frac{\partial \acute x^a}{\partial x^d}\frac{\partial x^e}{\partial \acute x^b}\frac{\partial x^f}{\partial \acute x^c}.$$ □

Exercise 10. Show that if $\{\theta^a\}$ is the local basis of 1-forms dual to $\{U_a\}$ and if $\alpha = \alpha_a\theta^a$ then

$$\nabla_V\alpha = V^c(U_c(\alpha_b) - \gamma^a_{bc}\alpha_a)\theta^b$$

and in particular if $U_a = \partial_a$ and $\theta^a = dx^a$ then $\nabla_V\alpha = V^c\alpha_{b\|c}dx^b$, where $\alpha_{b\|c} = \partial_c\alpha_b - \Gamma^a_{bc}\alpha_a$. □

Exercise 11. Extend covariant differentiation to tensors of type $(0,2)$ by requiring that it act as a derivation. □

3. Torsion and Curvature

For a general choice of connection on a manifold there is no reason to suppose that either of the commutation rules

$$\nabla_U V - \nabla_V U - [U,V] = 0$$
$$\nabla_U \nabla_V W - \nabla_V \nabla_U W - \nabla_{[U,V]} W = 0$$

need be satisfied. However, the left hand sides of these equations, when not zero, define quantities which are helpful in the geometrical interpretation of the connection and the associated rule of parallel transport, as is clear from Chapter 9 and the introductory remarks to this chapter, and they are of great importance (especially the second) in physical applications. We therefore define, for a given connection, its *torsion* $T\colon \mathcal{X}(\mathcal{M}) \times \mathcal{X}(\mathcal{M}) \to \mathcal{X}(\mathcal{M})$ and *curvature* $R\colon \mathcal{X}(\mathcal{M}) \times \mathcal{X}(\mathcal{M}) \times \mathcal{X}(\mathcal{M}) \to \mathcal{X}(\mathcal{M})$ by

$$T(U,V) = \nabla_U V - \nabla_V U - [U,V]$$
$$R(U,V)W = \nabla_U \nabla_V W - \nabla_V \nabla_U W - \nabla_{[U,V]} W.$$

Just as for a surface, the curvature is a type $(1,3)$ tensor field; and the torsion is a type $(1,2)$ tensor field (see Exercise 28 of Chapter 9).

Exercise 12. Show that T and R satisfy the necessary linearity properties over $\mathcal{F}(\mathcal{M})$ to be tensor fields. Show that if the coefficients of the connection with respect to a basis $\{U_a\}$ of vector fields are γ^a_{bc} then the components of T and R, defined by $T(U_b,U_c) = T^a_{bc} U_a$ and $R(U_c,U_d)U_b = R^a{}_{bcd} U_a$, are given by

$$T^a_{bc} = \gamma^a_{cb} - \gamma^a_{bc} - C^a_{bc}$$
$$R^a{}_{bcd} = U_c(\gamma^a_{bd}) - U_d(\gamma^a_{bc}) + \gamma^e_{bd}\gamma^a_{ec} - \gamma^e_{bc}\gamma^a_{ed} - C^e_{cd}\gamma^a_{be}$$

where $[U_b,U_c] = C^a_{bc} U_a$. Thus in particular when $\{U_a\}$ is a coordinate basis

$$T^a_{bc} = \Gamma^a_{cb} - \Gamma^a_{bc}$$
$$R^a{}_{bcd} = \partial_c \Gamma^a_{bd} - \partial_d \Gamma^a_{bc} + \Gamma^c_{bd}\Gamma^a_{ec} - \Gamma^e_{bc}\Gamma^a_{ed}.$$ □

Exercise 13. Show that $T(V,U) = -T(U,V)$ and $R(V,U)W = -R(U,V)W$. □

If the torsion of a connection vanishes, the connection is called *symmetric*. Most connections met with in practice are symmetric; however, there are occasions where non-symmetric connections occur naturally in geometrical circumstances, and there have been attempts, none entirely satisfactory, to exploit the possibility of torsion in physical theories.

Exercise 14. A manifold may be equipped with two (or more) connections. Show that if ∇ and $\hat{\nabla}$ are connections on $\mathcal{M}$, and if U and V are vector fields, then $D(U,V) = \hat{\nabla}_U V - \nabla_U V$ depends $\mathcal{F}(\mathcal{M})$-linearly on V as well as on U, and infer that D so defined is a type $(1,2)$ tensor field. Show that, conversely, if D is any type $(1,2)$ tensor field and ∇ is a connection then $\hat{\nabla}$ defined by $\hat{\nabla}_U V = \nabla_U V + D(U,V)$ is also a connection. Show that if the connection coefficients of ∇ and $\hat{\nabla}$ with respect to some basis are γ^a_{bc} and $\hat{\gamma}^a_{bc}$ respectively then the components of D with respect to that basis are $\hat{\gamma}^a_{cb} - \gamma^a_{cb}$. Show that the torsions of ∇ and $\hat{\nabla}$, denoted by T and $\hat{T}$ respectively, are related by

$\hat{T}(U,V) = T(U,V) + D(U,V) - D(V,U)$. Thus the torsions are the same if and only if D is symmetric. Show that, on the other hand, if ∇ is given and D is chosen to be $-\frac{1}{2}T$ then $\hat{\nabla}$ is symmetric. □

Covariant derivatives of tensor fields. The torsion and curvature tensors of a linear connection satisfy certain identities which are known collectively as the Bianchi identities. We shall shortly derive these identities, but since they involve the covariant derivatives of the torsion and curvature we must first explain how to define the covariant derivative of a tensor field. We shall concentrate on tensor fields of type $(0,p)$ and $(1,p)$, though the same principles apply to tensor fields of any type.

The covariant differentiation operator is extended to tensor fields on the same principles as it was extended from vector fields to 1-forms, and to tensor fields of type $(0,2)$ in Exercise 11: that is, by ensuring that appropriate forms of the linearity and Leibniz rules of differentiation apply, which is to say, that it acts as a derivation.

Suppose for definiteness that S is a type $(1,2)$ tensor field. Then for any vector fields V, W, $S(V,W)$ is a vector field whose covariant derivative with respect to U, say, may be formed. In order that ∇_U should act as a derivation, this covariant derivative should be expressible in the form

$$\nabla_U\big(S(V,W)\big) = (\nabla_U S)(V,W) + S(\nabla_U V,W) + S(V,\nabla_U W).$$

By making $(\nabla_U S)(V,W)$ the subject of this relation, we obtain the appropriate definition of the covariant derivative of S:

$$(\nabla_U S)(V,W) = \nabla_U\big(S(V,W)\big) - S(\nabla_U V,W) - S(V,\nabla_U W).$$

Suppose, more generally, that S is a type (p,q) tensor field with $p = 0$ or 1. In order to treat the cases $p = 0$ and $p = 1$ on the same footing it is convenient here to extend the operation of covariant differentiation to functions also by the rule $\nabla_U f = Uf$. Then the *covariant derivative* $\nabla_U S$ is defined by

$$(\nabla_U S)(V_1,V_2,\ldots,V_q) = \nabla_U\big(S(V_1,V_2,\ldots,V_q)\big) - \sum_{k=1}^{q} S(V_1,V_2,\ldots,\nabla_U V_k,\ldots,V_q).$$

The value of this definition is that, as well as respecting the rules of differentiation, it also preserves the tensorial nature of S.

Exercise 15. Confirm that $\nabla_U S$ is a tensor field of the same type as S. □

Exercise 16. Show that with respect to coordinates

$$(\nabla_U S)(\partial_{a_1},\partial_{a_2},\ldots,\partial_{a_q}) = U^a\left(\partial_a S_{a_1a_2\ldots a_q} - \sum_{k=1}^{q}\Gamma^b_{a_k a} S_{a_1a_2\ldots b\ldots a_q}\right)$$

if S is of type $(0,q)$, and

$$(\nabla_U S)(\partial_{a_1},\partial_{a_2},\ldots,\partial_{a_q}) = U^a\left(\partial_a S^b{}_{a_1a_2\ldots a_q} + \Gamma^b_{ca} S^c{}_{a_1a_2\ldots a_q} - \sum_{k=1}^{q}\Gamma^c_{a_k a} S^b{}_{a_1a_2\ldots c\ldots a_q}\right)\partial_b$$

if it is of type $(1,q)$. □

The coefficient $\partial_a S_{a_1 a_2 \ldots a_q} - \sum_{k=1}^{q} \Gamma_{a_k a}^{b} S_{a_1 a_2 \ldots b \ldots a_q}$ is usually abbreviated to $S_{a_1 a_2 \ldots a_q \| a}$, while

$$\partial_a S^b{}_{a_1 a_2 \ldots a_q} + \Gamma^b_{ca} S^c{}_{a_1 a_2 \ldots a_q} - \sum_{k=1}^{q} \Gamma_{a_k a}^{c} S^b{}_{a_1 a_2 \ldots c \ldots a_q}$$

is usually written $S^b{}_{a_1 a_2 \ldots a_q \| a}$. These quantities are called the *components* of the covariant derivatives of the tensor fields. Another common notation for components of covariant derivatives is $\nabla_a S_{a_1 a_2 \ldots a_q}$ for $S_{a_1 a_2 \ldots a_q \| a}$ and $\nabla_a S^b{}_{a_1 a_2 \ldots a_q}$ for $S^b{}_{a_1 a_2 \ldots a_q \| a}$. Some authors write ; for $\|$ and a few write , but this last practice is not recommended since the same notation is often used for the partial derivative and confusion could be disastrous.

Exercise 17. Show that the identity type $(1,1)$ tensor field I defined by $I(U) = U$ for all vector fields U (whose components with respect to any basis are those of the Kronecker δ) has vanishing covariant derivative with respect to any linear connection. □

Exercise 18. Show that for any 1-form θ

$$\langle T(U,V), \theta \rangle = d\theta(U,V) + \langle U, \nabla_V \theta \rangle - \langle V, \nabla_U \theta \rangle. \quad \square$$

Exercise 19. For a given vector field W the map $\mathcal{X}(\mathcal{M}) \to \mathcal{X}(\mathcal{M})$ by $V \mapsto \nabla_V W$ is a type $(1,1)$ tensor field, which may be denoted A_W. Show that

$$(\nabla_U A_W)(V) - (\nabla_V A_W)(U) = R(U,V)W - A_W(T(U,V)).$$

Show that if $W = W^a \partial_a$ then A_W has components $W^a{}_{\|b}$, and deduce that for a symmetric connection

$$W^a{}_{\|bc} - W^a{}_{\|cb} = R^a{}_{dcb} W^d.$$

Here $W^a{}_{\|bc}$ has been written for $W^a{}_{\|b\|c}$. □

Exercise 20. Given a type (p,q) tensor field S ($p = 0$ or 1) set

$$\nabla S(V, V_1, V_2, \ldots, V_q) = (\nabla_V S)(V_1, V_2, \ldots, V_q).$$

Show that ∇S is a tensor field of type $(p, q+1)$. □

Exercise 21. If A is a type $(1,1)$ tensor field and S a type $(1,r)$ tensor field, one may define a new type $(1,r)$ tensor field $A(S)$ by

$$A(S)(V_1, V_2, \ldots, V_r) = A(S(V_1, V_2, \ldots, V_r)) - \sum_{k=1}^{r} S(V_1, V_2, \ldots, A(V_k), \ldots, V_r)$$

(adapting the model provided by the covariant derivative of a tensor field). Using the fact that $W \mapsto R(U,V)W$, for fixed U, V, is a type $(1,1)$ tensor field, show that

$$\nabla_U \nabla_V S - \nabla_V \nabla_U S - \nabla_{[U,V]} S = R(U,V)(S). \quad \square$$

Exercise 22. Propose a definition of the covariant derivative of a type (p,q) tensor field for general p. Ensure that the covariantly differentiated vector tensor field is a tensor field of the same type. Find an expression for the components of the covariantly differentiated field. □

The Bianchi identities. The Bianchi identities are obtained by taking covariant derivatives of T and of R. We deal with the case of a symmetric connection, leaving the more general case to the exercises. In the absence of torsion, by covariantly differentiating the first symmetry relation we obtain

$$\nabla_U \nabla_V W - \nabla_U \nabla_W V - \nabla_U [V,W] = 0.$$

Permuting the arguments U, V, W cyclically and adding the results one obtains

$$\begin{aligned}\nabla_U \nabla_V W - \nabla_U \nabla_W V - \nabla_U [V, W] \\ + \nabla_V \nabla_W U - \nabla_V \nabla_U W - \nabla_V [W, U] \\ + \nabla_W \nabla_U V - \nabla_W \nabla_V U - \nabla_W [U, V] = 0.\end{aligned}$$

Substituting $\nabla_U \nabla_V W - \nabla_V \nabla_U W = R(U, V)W + \nabla_{[U,V]} W$ from the definition of the curvature, and the equations obtained from this by permuting the arguments cyclically, one is left with

$$\begin{aligned}R(U, V)W + \nabla_{[U,V]} W - \nabla_W [U, V] \\ + R(V, W)U + \nabla_{[V,W]} U - \nabla_U [V, W] \\ + R(W, U)V + \nabla_{[W,U]} V - \nabla_V [W, U] = 0.\end{aligned}$$

In the absence of torsion the second and third terms in each line altogether cancel out by the Jacobi identity, leaving

$$R(U, V)W + R(V, W)U + R(W, U)V = 0.$$

This is the *Ricci identity* or *first Bianchi identity* for a symmetric connection.

Exercise 23. Show that for an arbitrary linear connection (with torsion)

$$\begin{aligned}&R(U, V)W + R(V, W)U + R(W, U)V \\ &\quad = (\nabla_U T)(V, W) + (\nabla_V T)(W, U) + (\nabla_W T)(U, V) \\ &\qquad + T(T(V, W), U) + T(T(W, U), V) + T(T(U, V), W).\end{aligned}$$ □

This is the more general form of the first Bianchi identity.

Exercise 24. Show that for a symmetric connection the components of the curvature tensor with respect to a coordinate basis satisfy

$$R^d{}_{cba} = -R^d{}_{cab}$$
$$R^d{}_{abc} + R^d{}_{cab} + R^d{}_{bca} = 0.$$ □

If for every pair of vector fields U and V the torsion $T(U, V)$ is a linear combination of U and V then the connection is called *semi-symmetric*.

Exercise 25. Show that the "magnetic" connection on the sphere described in the introduction to this chapter is semi-symmetric. □

Exercise 26. Show that for a semi-symmetric connection there is a 1-form τ such that $T(U, V) = \langle V, \tau \rangle U - \langle U, \tau \rangle V$, and show that with respect to a coordinate basis $\tau = (m-1)^{-1} T^a_{ab} \, dx^b$ where T^a_{bc} are the components of T and $m = \dim \mathcal{M}$. Show that

$$(\nabla_U T)(V, W) = \langle W, \nabla_U \tau \rangle V - \langle V, \nabla_U \tau \rangle W$$

and that the first Bianchi identity becomes

$$R(U, V)W + R(V, W)U + R(W, U)V = -d\tau(U, V)W - d\tau(V, W)U - d\tau(W, U)V.$$ □

The covariant derivative of the curvature is given by

$$\begin{aligned}(\nabla_U R)(V, W)W' &= \nabla_U \nabla_V \nabla_W W' - \nabla_U \nabla_W \nabla_V W' - \nabla_U \nabla_{[V,W]} W' \\ &\quad - R(\nabla_U V, W)W' - R(V, \nabla_U W)W' - R(V, W)\nabla_U W'.\end{aligned}$$

Permuting the arguments U, V and W cyclically and adding the results one obtains

$$\begin{aligned}(\nabla_U R)(V,W)W' + (\nabla_V R)(W,U)W' + (\nabla_W R)(U,V)W' \\ = R([V,W],U)W' + R([W,U],V)W' + R([U,V],W)W' \\ -R(\nabla_V W,U)W' - R(\nabla_W U,V)W' - R(\nabla_U V,W)W' \\ +R(\nabla_W V,U)W' + R(\nabla_U W,V)W' + R(\nabla_V U,W)W' \\ +\nabla_{[[V,W],U]}W' + \nabla_{[[W,U],V]}W' + \nabla_{[[U,V],W]}W'.\end{aligned}$$

By Jacobi's identity, the sum of the terms in the last line is zero. In the absence of torsion the terms in the preceding three lines cancel in threes, leaving the *second Bianchi identity* for a symmetric connection,

$$(\nabla_U R)(V,W)W' + (\nabla_V R)(W,U)W' + (\nabla_W R)(U,V)W' = 0.$$

Note that the argument W' is unaffected by any of the rearrangements of the other vector fields. Accordingly, it is often convenient to regard $R(U,V)$ as defining a type $(1,1)$ tensor field which is alternating in the two vector field arguments U and V. The notation has anticipated this point of view. With this understanding the second Bianchi identity for a symmetric connection may be written

$$(\nabla_U R)(V,W) + (\nabla_V R)(W,U) + (\nabla_W R)(U,V) = 0.$$

Exercise 27. Show that for a symmetric connection the second Bianchi identity may be written

$$R^e{}_{dab\|c} + R^e{}_{dbc\|a} + R^e{}_{dca\|b} = 0$$

in terms of components with respect to a coordinate basis. □

Exercise 28. Show that for an arbitrary linear connection (with torsion)

$$\begin{aligned}(\nabla_U R)(V,W) + (\nabla_V R)(W,U) + (\nabla_W R)(U,V) \\ = R(U,T(V,W)) + R(V,T(W,U)) + R(W,T(U,V)).\end{aligned}$$ □

This is the more general form of the second Bianchi identity.

4. Connection Forms and Structure Equations

The formalism for covariant differentiation developed in the last section, with the accent very much on vector fields, has a dual version, in terms of exterior forms, which also has a variety of applications. It is described in this section. In particular, we shall examine the consequences of the vanishing of the curvature of a connection, which it is simpler to do using forms. The basic concepts have been introduced in the context of the natural connection on affine space in Chapter 5, Section 7. In fact the task for this section is essentially to repeat the argument given there but making allowance for the possibility of non-vanishing torsion and curvature.

Let $\{U_a\}$ be a local basis of vector fields on a manifold $\mathcal{M}$ with connection ∇, and let $\{\theta^a\}$ be the dual local basis of 1-forms. The *connection forms* ω^a_b corresponding to these bases are defined by

$$\langle V, \omega^a_b \rangle = \langle \nabla_V U_b, \theta^a \rangle$$

for an arbitrary vector field V. That this equation does define a 1-form (locally, with the same domain as the U_a and θ^a) follows from the linearity properties of ∇_V with respect to V.

Exercise 29. Show that $\omega^a_b = \gamma^a_{bc}\theta^c$, where γ^a_{bc} are the coefficients of the connection with respect to the basis $\{U_a\}$. □

Exercise 30. Show that the operation of covariant differentiation of a vector field $W = W^a U_a$ may be expressed in terms of the connection forms as follows:

$$\nabla_V W = (V(W^a) + W^b \langle V, \omega^a_b \rangle) U_a,$$

or equivalently

$$\nabla_V W = (V\langle W, \theta^a \rangle + \langle W, \theta^b \rangle \langle V, \omega^a_b \rangle) U_a. \qquad \square$$

From this second expression for $\nabla_V W$ in Exercise 30 we may express the torsion of the connection in terms of θ^a and ω^a_b, as follows:

$$\begin{aligned}
\langle T(V,W), \theta^a \rangle &= \langle \nabla_V W, \theta^a \rangle - \langle \nabla_W V, \theta^a \rangle - \langle [V,W], \theta^a \rangle \\
&= V\langle W, \theta^a \rangle + \langle W, \theta^b \rangle \langle V, \omega^a_b \rangle \\
&\quad - W\langle V, \theta^a \rangle - \langle V, \theta^b \rangle \langle W, \omega^a_b \rangle - \langle [V,W], \theta^a \rangle \\
&= d\theta^a(V,W) + (\omega^a_b \wedge \theta^b)(V,W).
\end{aligned}$$

Introducing the *torsion 2-forms* Θ^a by

$$\Theta^a(V,W) = \langle T(V,W), \theta^a \rangle,$$

we write the last formula entirely in terms of forms:

$$d\theta^a + \omega^a_b \wedge \theta^b = \Theta^a.$$

These are *Cartan's first structure equations.*

There is another set of structure equations involving the curvature. In order to simplify its derivation we note that from $\langle U_b, \theta^a \rangle = \delta^a_b$ it follows that $\nabla_V \theta^a = -\langle V, \omega^a_b \rangle \theta^b$, and so

$$\begin{aligned}
\langle \nabla_V \nabla_W U_b, \theta^a \rangle &= V\langle \nabla_W U_b, \theta^a \rangle - \langle \nabla_W U_b, \nabla_V \theta^a \rangle \\
&= V\langle W, \omega^a_b \rangle + \langle V, \omega^a_c \rangle \langle W, \omega^c_b \rangle.
\end{aligned}$$

From this it follows that

$$\begin{aligned}
\langle R(V,W)U_b, \theta^a \rangle &= V\langle W, \omega^a_b \rangle + \langle V, \omega^a_c \rangle \langle W, \omega^c_b \rangle \\
&\quad - W\langle V, \omega^a_b \rangle - \langle W, \omega^a_c \rangle \langle V, \omega^c_b \rangle - \langle [V,W], \omega^a_b \rangle \\
&= d\omega^a_b(V,W) + (\omega^a_c \wedge \omega^c_b)(V,W).
\end{aligned}$$

The *curvature 2-forms* Ω^a_b of the connection with respect to the given basis are the defined by

$$\Omega^a_b(V,W) = \langle R(V,W)U_b, \theta^a \rangle.$$

The last derived formula may be written entirely in terms of forms:

$$d\omega^a_b + \omega^a_c \wedge \omega^c_b = \Omega^a_b.$$

These are *Cartan's second structure equations.*

There are thus at least three different ways to describe connections, torsion and curvature: abstract, tensor-analytical and exterior-analytical. To make the distinctions clearer we compare the alternative definitions of torsion:

$$T(V,W) = \nabla_V W - \nabla_W V - [V,W] \qquad \text{abstract}$$

$$T^a_{bc} = \Gamma^a_{cb} - \Gamma^a_{bc} \qquad \text{tensor-analytical}$$

$$\Theta^a = d\theta^a + \omega^a_b \wedge \theta^b \qquad \text{exterior-analytical.}$$

The three are essentially equivalent. Each has its advantages, and is to be preferred in certain contexts, and all are used in the literature.

Exercise 31. Make the same comparison for curvature. □

The Bianchi identities. On taking the exterior derivatives of the structure equations, and substituting for $d\theta^a$ and $d\omega^a_b$ from them, one finds that

$$d\Theta^a + \omega^a_b \wedge \Theta^b = \Omega^a_b \wedge \theta^b$$

$$d\Omega^a_b + \omega^a_c \wedge \Omega^c_b - \Omega^a_c \wedge \omega^c_b = 0.$$

These are equivalent to the first and second Bianchi identities respectively.

Exercise 32. Confirm this equivalence. □

In this version the Bianchi identities may be seen as consequences of the fact that $d^2 = 0$. They are relations between 3-forms, which explains why in the tensor version the identities involve cyclic sums; it also explains why the Bianchi identities are vacuous on a 2-dimensional manifold and therefore play no part in surface geometry.

Change of basis.

Exercise 33. Suppose that a new local basis of vector fields $\{\hat{U}_a\}$ is chosen, so that $\hat{U}_a = A^b_a U_b$, where the A^b_a are local smooth functions whose values at each point are the entries in a non-singular matrix. Show that the connection quantities associated with the new basis are related to those associated with the old as follows:

$$\hat{\theta}^a = (A^{-1})^a_b \theta^b \qquad \hat{\omega}^a_b = (A^{-1})^a_c dA^c_b + (A^{-1})^a_c \omega^c_d A^d_b$$

$$\hat{\Theta}^a = (A^{-1})^a_b \Theta^b \qquad \hat{\Omega}^a_b = (A^{-1})^a_c \Omega^c_d A^d_b.$$

Confirm that the transformation rule for connection forms is equivalent to the one for connection coefficients given in Exercise 9. □

The transformation rules for the torsion and curvature forms are therefore straightforward, as their tensorial character demands; while the non-tensorial nature of the connection coefficients shows up in the exterior derivative term in the transformation rule for the connection forms.

5. Vector- and Matrix-Valued Forms

The reader may have noticed that the disposition of the indices in all the equations deduced in the previous section is exactly what one would expect in matrix multiplication either of a vector by a matrix or of two matrices. In fact, it is possible to combine exterior calculus and matrix algebra so as to express these equations even more economically.

The idea is to regard θ^a (for instance) as the ath entry in a vector of 1-forms; and ω^a_b as the (a,b) entry in a matrix of 1-forms. More exactly, we define a *vector-valued 1-form* θ as a map from $\mathcal{X}(\mathcal{M})$ to the space of column vectors of size k, say, with entries from $\mathcal{F}(\mathcal{M})$, satisfying the tensorial requirements for a 1-form:

$$\theta(V+W) = \theta(V) + \theta(W) \qquad V,W \in \mathcal{X}(\mathcal{M})$$
$$\theta(fV) = f\theta(V) \qquad f \in \mathcal{F}(\mathcal{M}).$$

Thus for any V, $\theta(V) = \langle V, \theta\rangle$ is an element of $(\mathcal{F}(\mathcal{M}))^k$, considered as a column vector.

Evidently the definition extends to give a definition of a *vector-valued p-form*, which must satisfy the usual conditions of multilinearity and skew-symmetry. Again, a *matrix-valued p-form* is an $\mathcal{F}(\mathcal{M})$-multilinear alternating map from $(\mathcal{X}(\mathcal{M}))^p$ to the space of $k \times k$ matrices with entries from $\mathcal{F}(\mathcal{M})$. In the applications we have in mind here, k will be m, the dimension of the manifold; but this is not an essential part of the definition. Each component of a vector- or matrix-valued p-form is a p-form in the ordinary sense. Moreover, if ν is a vector-valued p-form and $V_1, V_2, \ldots, V_p$ are vector fields then $\nu(V_1, V_2, \ldots, V_p)$ is a column vector of functions, or a *vector-valued function*; and similarly for a matrix-valued p-form.

Matrix multiplication of vector- or matrix-valued forms by matrix- valued functions (0-forms) is quite straightforward: if (say) θ is a vector-valued 1-form and A is a matrix-valued function then $A\theta$ is the vector-valued 1-form such that $(A\theta)(W)$ is the vector obtained by multiplying the vector $\theta(W)$ by the matrix A: that is to say, $(A\theta)(W) = A(\theta(W))$ for all $W \in \mathcal{X}(\mathcal{M})$. Thus the transformation rules under a change of basis given in Exercise 33 may be written

$$\hat{\theta} = A^{-1}\theta \qquad \hat{\omega} = A^{-1}dA + A^{-1}\omega A$$
$$\hat{\Theta} = A^{-1}\Theta \qquad \hat{\Omega} = A^{-1}\Omega A.$$

Here A^{-1} is the matrix-valued function whose value at a point is the inverse of the matrix defined by A, which is assumed non-singular, and dA is the matrix-valued 1-form obtained by taking the exterior derivative of each entry of A.

Exercise 34. Show that if A is a matrix-valued function then the definition of dA may be couched in terms entirely similar to those used in the definition of the exterior derivative of an ordinary function, that is, $\langle V, dA\rangle = V(A)$, $V \in \mathcal{X}(\mathcal{M})$. Show also that (for example) if θ is a vector-valued 1-form then the vector-valued 2-form $d\theta$ given by $d\theta(V,W) = V\langle W,\theta\rangle - W\langle V,\theta\rangle - \langle [V,W],\theta\rangle$ has for its components just the exterior derivatives of the components of θ. Conclude that exterior calculus extends to vector- and matrix-valued forms with no formal change at the theoretical level, and by simply operating on components at the practical level. □

The only complications—and surprises—in this scheme occur when one has to combine exterior and matrix multiplication in forming the products of (say) two matrix-valued forms. The process is in fact quite straightforward: at the practical level one simply multiplies the two matrices together in the usual way, but remembers to combine the elements using exterior multiplication. But the non-commutativity of the two kinds of multiplication involved may lead to the frustration of expectations based on either. Thus for example the matrix exterior product of a matrix-valued 1-form with itself is not necessarily zero: in fact if ω is a matrix-valued 1-form then for $V, W \in \mathcal{X}(\mathcal{M})$

$$(\omega \wedge \omega)(V, W) = \langle V \omega\rangle\langle W, \omega\rangle - \langle W, \omega\rangle\langle V, \omega\rangle.$$

Matrix multiplication is implied on the right hand side. With this understanding, we write the structure equations in the form

$$d\theta + \omega \wedge \theta = \Theta$$
$$d\omega + \omega \wedge \omega = \Omega,$$

the first being an equation between vector-valued 2-forms, the second an equation between matrix-valued 2-forms.

Exercise 35. Show that the Bianchi equations may be written

$$d\Theta + \omega \wedge \Theta = \Omega \wedge \theta$$
$$d\Omega + \omega \wedge \Omega - \Omega \wedge \omega = 0.$$ □

Exercise 36. Let α and β be matrix-valued 1-forms; define $[\alpha \wedge \beta]$ by

$$[\alpha \wedge \beta](V, W) = [\langle V, \alpha\rangle, \langle W, \beta\rangle] - [\langle W, \alpha\rangle, \langle V, \beta\rangle]$$

for any pair of vector fields V, W, where the square brackets on the right signify the commutator of matrices. Show that $[\alpha \wedge \beta]$ is a matrix-valued 1-form, and that both terms are necessary on the right for this to be so. Show that $[\beta \wedge \alpha] = [\alpha \wedge \beta]$. □

Exercise 37. Show that the second structure equation and second Bianchi identity may be written

$$d\omega + \tfrac{1}{2}[\omega \wedge \omega] = \Omega \qquad d\Omega + [\omega \wedge \Omega] = 0.$$ □

Exercise 38. Suppose that $\theta^{p+1}, \theta^{p+2}, \ldots, \theta^m$ are 1-forms, which define, by constraint, a distribution. Combine them into an $(m-p)$-vector-valued 1-form θ. Show that the conditions of Frobenius's theorem may be written $d\theta = \lambda \wedge \theta$ where λ is an $(m-p) \times (m-p)$-matrix-valued 1-form. □

6. Vanishing Curvature and Torsion

The curvature and torsion of the natural connection on an affine space both vanish. The coordinate vector fields for an affine coordinate system have two significant properties from this point of view. In the first place, they are parallel vector fields, so that their covariant derivatives in any direction vanish; and secondly they are coordinate vector fields, so their brackets vanish. It follows from the first of these properties that the curvature vanishes, and from the second that the torsion does.

In this section we shall show the converse: that if both curvature and torsion vanish then locally at least the manifold admits a coordinate system which is affine-like, in the sense that its coordinate vector fields are parallel.

We consider first, however, the case of a manifold with a connection whose curvature vanishes, without making any assumptions about the torsion. A sufficient condition for this to occur is that there should exist a local basis of vector fields which are parallel, in the sense of having vanishing covariant derivatives as before, but now not necessarily a coordinate basis. In fact, with respect to such a basis the connection forms are all zero, and so the curvature vanishes (and the torsion forms are exact although not necessarily zero). A connection with the property that there exists about each point a local basis of parallel vector fields is called a *complete parallelism*; thus the curvature of a complete parallelism is zero. We shall show that the converse is true. We shall do so by using the structure equations, but we shall not be able to identify *a priori* which local basis of vector fields is likely to be parallel. Even in the case of an affine space it is not apparent, when a basis of vector fields other than an affine coordinate basis is used, that the curvature vanishes. The strategy of the proof is to start with an arbitrary basis of vector fields and seek a transformation to a parallel one, using the vanishing of the curvature to show that this is possible. It is worthwhile therefore to examine first the effect of using a non-parallel basis for a complete parallelism.

We shall use vector- and matrix-valued forms, and the notation explained immediately before Exercise 34, with the old basis being non-parallel and the new one parallel. Then $\hat{\omega} = 0$ and so $\omega = -(dA)A^{-1}$.

Exercise 39. Show that one may equally well write $\omega = AdA^{-1}$, and check that $d\omega + \omega \wedge \omega = 0$. □

It is to be expected that if $\hat{\Omega} = 0$ then $\Omega = 0$, since both the transformation rule, and the tensorial nature of the curvature tensor, demand it. However, the role of the curvature may be seen in another light. We shall rewrite $\omega = -(dA)A^{-1}$ as $dA + \omega A = 0$. We regard this as an equation to find A, the transformation from the non-parallel to a parallel basis. Taking the exterior derivative of this equation will provide integrability conditions (it amounts to the standard device of differentiating again and using the symmetry of second partial derivatives). These conditions are

$$(d\omega)A - \omega \wedge dA = 0,$$

which gives, on substituting for dA,

$$(d\omega + \omega \wedge \omega)A = 0,$$

that is

$$\Omega A = 0.$$

Thus the vanishing of the curvature may be thought of as an integrability condition for the equation $d\omega + \omega A = 0$.

We shall now show that when the curvature vanishes this equation is indeed integrable. The argument is based on Frobenius's Theorem in the form version, this time for matrix-valued forms.

Consider the manifold $\mathcal{O} \times M_m(\mathbf{R})$, where $M_m(\mathbf{R})$ is the manifold of $m \times m$ matrices (Chapter 10, Section 1, Example 6) and $\mathcal{O}$ is the open set of $\mathcal{M}$ on which a local basis of vector fields is defined (and $m = \dim \mathcal{M}$). Let X be the matrix-valued function on $\mathcal{O} \times M_m(\mathbf{R})$ whose entries are just the coordinate functions on $M_m(\mathbf{R})$. Then we define on $\mathcal{O} \times M_m(\mathbf{R})$ a matrix-valued 1-form

$$\mu = dX + \omega X.$$

Strictly speaking one should distinguish between forms on $\mathcal{O}$ and $M_m(\mathbf{R})$ and their pull-backs to $\mathcal{O} \times M_m(\mathbf{R})$ by projection; however, no confusion should arise from this abuse of notation. The matrix-valued 1-form μ defines a distribution on $\mathcal{O} \times M_m(\mathbf{R})$ of dimension m, by constraint.

We shall test the entries in μ to see whether they satisfy the conditions of Frobenius's theorem. Now

$$\begin{aligned} d\mu &= (d\omega)X - \omega \wedge dX \\ &= (d\omega)X - \omega \wedge (\mu - \omega X) \\ &= (d\omega + \omega \wedge \omega)X - \omega \wedge \mu, \end{aligned}$$

and so when the curvature vanishes

$$d\mu = -\omega \wedge \mu.$$

This amounts to the conditions of Frobenius's theorem. Integral submanifolds of the distribution defined by μ will be of dimension m. Provided they are transverse to the $M_m(\mathbf{R})$ factor they will be expressible in the form $X = A(x)$ where A is a matrix-valued function on $\mathcal{O}$ (and $x \in \mathcal{O}$). It is clear from the form of μ that its integral submanifolds are nowhere tangent to the $M_m(\mathbf{R})$ factor, since μ reduces to dX on any vector tangent to the $M_m(\mathbf{R})$ factor, and the entries of dX are linearly independent. Choose a point $x_0 \in \mathcal{O}$ and consider the integral submanifold through $(x_0, I_m) \in \mathcal{O} \times M_m(\mathbf{R})$, where I_m is the identity matrix. The integral submanifold through (x_0, I_m) is thus defined by a matrix-valued function A such that $A(x_0) = I_m$, and thus A is non-singular on an open neighbourhood of x_0. Since μ vanishes on an integral submanifold, A satisfies $dA + \omega A = 0$. Thus the local basis obtained by transforming by A is parallel. It follows that if a manifold has a connection with vanishing curvature, this connection is a complete parallelism.

Exercise 40. Repeat the argument with the indices in evidence, to confirm that nothing is lost by using matrix-valued forms. □

If, as well as the curvature being zero, the torsion is zero then the vector fields of a parallel basis commute, and so local coordinates may be found with the parallel fields as coordinate basis fields (Chapter 6, Section 4). These coordinates are then affine-like, as described above. Under these circumstances the connection is said to be *flat*.

Exercise 41. Consider, in the Euclidean plane $\mathcal{E}^2$, the two orthonormal vector fields

$$U_1 = \cos\vartheta\partial_1 + \sin\vartheta\partial_2 \qquad U_2 = -\sin\vartheta\partial_1 + \cos\vartheta\partial_2$$

with respect to orthonormal coordinates, where ϑ is a smooth function on $\mathcal{E}^2$. Suppose that U_1 and U_2 are given to be parallel vector fields. Show that the torsion of the resulting complete parallelism satisfies $T(U_1, U_2) = (U_1\vartheta)U_1 + (U_2\vartheta)U_2$, so that the torsion vanishes if and only if ϑ is constant, in which case the basis $\{U_1, U_2\}$ is obtained from the coordinate basis by a fixed rotation. □

7. Geodesics

One may generalise most conveniently the idea of a straight line from affine spaces to manifolds with connection by using the property of a straight line that its (1-dimensional) tangent spaces are parallel along it. In a manifold with connection a path whose tangent spaces are parallel along it is called a *geodesic*. On an affinely parametrised straight line, in affine space, the tangent vectors form a parallel field; generalising, one calls a curve whose tangent vectors form a parallel field with respect to the given connection an *affinely parametrised geodesic*. Thus on any curve γ whose image is a geodesic path, $D\dot{\gamma}/Dt$ is a multiple of $\dot{\gamma}$, and on an affinely parametrised geodesic $D\dot{\gamma}/Dt = 0$.

Exercise 42. Show that any geodesic may be affinely parametrised. □

In view of this result, it is seldom important to maintain the distinction between a geodesic (path) and an affinely parametrised geodesic (curve) and we shall use "geodesic" to mean "affinely parametrised geodesic" unless the context demands a distinction.

Exercise 43. Show that, just as in the case of a surface (Chapter 9, Section 5), the equations of a geodesic in local coordinates are

$$\frac{d^2\gamma^a}{dt^2} + \Gamma^a_{bc}\frac{d\gamma^b}{dt}\frac{d\gamma^c}{dt} = f\frac{d\gamma^a}{dt}$$

for some real function f, and that $f = 0$ if and only if the geodesic is affinely parametrised. □

Exercise 44. Show that two different affine parametrisations of the same geodesic (path) can differ only by an affine reparametrisation $t \mapsto at + b$ with a and b constant and $a \neq 0$. □

It follows from Exercise 43 that geodesics are indifferent to torsion: two connections with the same symmetric part have the same geodesics, irrespective of their torsions. Moreover, since the geodesic equations are second order ordinary differential equations, there is a unique geodesic with given initial point and given initial tangent vector: that is to say, given $x \in \mathcal{M}$ and $v \in T_x\mathcal{M}$ there is a unique (affinely parametrised) geodesic γ such that $\gamma(0) = x$ and $\dot{\gamma}(0) = v$. However, the existence theorem for solutions of systems of ordinary differential equations guarantees a solution only in some neighbourhood of 0 in $\mathbf{R}$; although for given initial conditions such solution elements may be smoothly pieced together to form a geodesic of maximal domain, there is no guarantee that the maximal domain will be the whole of $\mathbf{R}$. A manifold with connection every one of whose geodesics may be extended to the whole of $\mathbf{R}$ is said to be *geodesically complete*. An example of a geodesically incomplete manifold is obtained by removing a single point from the geodesically complete manifold $\mathbf{R}^m$. This device may appear somewhat artificial, but in any case where the connection is determined by a positive definite metric this is essentially the only way of introducing incompleteness. For the space-time of general relativity theory, on the other hand, geodesic completeness is incompatible with other, physically reasonable, conditions.

Since geodesics have been mentioned in the same context as integral curves of a vector field, it would be as well to emphasise that the totality of geodesics on a manifold does not form a congruence of curves, any more than the totality of straight lines in affine space does, because there are many geodesics through each point of the manifold. This is not to say that one never considers congruences of curves each of which is a geodesic—or equivalently, vector fields tangent to geodesic congruences. The models provided by the set of (affinely parametrised) straight lines in affine space, and the set of (affinely parametrised) great circles of the sphere, are good guides here. The advantages of working with a congruence consisting of all geodesics at once may be recovered by relocating the geodesics in another, larger manifold: this will be explained in Chapter 13.

On an affine space the straight lines are represented, in affine coordinates, by affine functions. On a general manifold with connection it is not possible to find a coordinate system, even locally, in which the geodesics are represented in this way; but for each point it is possible to find a local coordinate system with respect to which the geodesics through that point are affinely represented. Such coordinates are called normal coordinates about the point in question. Construction of normal coordinates is based on the properties of geodesics, and is best presented in terms of a map known as the exponential map.

Before defining the exponential map we must point out another property of geodesics. We consider geodesics with initial point a fixed point, say x, of $\mathcal{M}$. Then each tangent vector $v \in T_x\mathcal{M}$ uniquely determines a geodesic, which we shall denote γ_v: it is fixed by the conditions $\gamma_v(0) = x$, $\dot\gamma_v(0) = v$. It follows from Exercise 43 that for $k \in \mathbf{R}$, $k \neq 0$, γ_{kv} is obtained by reparametrising γ_v by $t \mapsto kt$; that is to say $\gamma_{kv}(t) = \gamma_v(kt)$. Now given any $v \neq 0$, γ_v is defined on some open interval $(-\varepsilon, \varepsilon)$. Thus γ_{kv} is defined on $(-\varepsilon/|k|, \varepsilon/|k|)$. By choosing $|k|$ sufficiently small it may be ensured that $\gamma_{kv}(t)$ is defined for $t = 1$. Moreover, $\gamma_0(t) = x$ for all t, and so γ_0 is defined on $\mathbf{R}$. In fact, there is an open neighbourhood $\mathcal{O}$ of $0 \in T_x\mathcal{M}$ such that for each $v \in \mathcal{O}$, $\gamma_v(1)$ is defined. (When the manifold is complete, we can take for $\mathcal{O}$ the whole of $T_x\mathcal{M}$.)

The exponential map. The *exponential map* $\exp\colon \mathcal{O} \subset T_x\mathcal{M} \to \mathcal{M}$ is defined by

$$\exp(v) = \gamma_v(1).$$

It is a smooth map, since the geodesic equations are of a type whose solutions depend smoothly on their initial conditions. Moreover, $\exp(0) = x$. The exponential map associates with each direction in $T_x\mathcal{M}$ a segment of a geodesic which starts off from x in that direction; it is designed, by fixing the parameter value at 1 but varying the initial tangent vector, to be injective, at least near 0. We shall prove, by showing that the induced map $\exp_*$ is non-singular at 0, that exp is a diffeomorphism of some neighbourhood of 0, possibly smaller than $\mathcal{O}$, in $T_x\mathcal{M}$, with a neighbourhood of x in $\mathcal{M}$. As is apparent from the example of the sphere, where all the geodesics through any one point intersect again in the antipodal point, one cannot expect exp to be a diffeomorphism in the large; in general, non-vanishing curvature of a connection may cause focusing of its geodesics, in which case the exponential map will be only a local diffeomorphism.

We shall now compute $\exp_*\colon T_0(T_x\mathcal{M}) \to T_x\mathcal{M}$. We shall identify the tangent space at 0 to the vector space $T_x\mathcal{M}$ with $T_x\mathcal{M}$ itself in the usual way: the element $v \in T_x\mathcal{M}$ is thought of as the tangent vector at $s = 0$ to the curve $s \mapsto sv$ in $T_x\mathcal{M}$. Then $\exp_*(v)$ is the tangent vector at $s = 0$ to the image curve $s \mapsto \exp(sv)$ in $\mathcal{M}$. But

$$\exp(sv) = \gamma_{sv}(1) = \gamma_v(s),$$

whose tangent vector at $s = 0$ is just v. Thus

$$\exp_*(v) = v$$

and therefore $\exp_*$ is just the identity map of $T_x\mathcal{M}$. This is certainly non-singular. It follows from the inverse function theorem that exp has a smooth inverse defined on a neighbourhood of x in $\mathcal{M}$, and therefore defines a diffeomorphism of a neighbourhood of 0 in $T_x\mathcal{M}$ with a neighbourhood of x in $\mathcal{M}$.

Exercise 45. The purpose of this exercise is to show that ordinary exponentiation of real numbers is an example of an exponential map in the above sense. Writing $\delta_t(x) = xe^t$, show that δ_t is a one-parameter group of transformations of the positive real numbers $\mathbf{R}^+$ with generator $x\partial/\partial x$. Next, show that there is a unique connection on $\mathbf{R}^+$ with respect to which this generator is a parallel vector field, and that this connection is given by

$$\nabla_{\partial/\partial x}\left(\frac{\partial}{\partial x}\right) = -\frac{1}{x}\frac{\partial}{\partial x}$$

(this is not, of course, the usual connection on $\mathbf{R}$). Show that the geodesic γ of this connection with initial conditions $\gamma(0) = 1$, $\dot\gamma(0) = k$ is given by $\gamma(t) = e^{kt}$. Infer that the exponential map $\exp\colon \mathbf{R} \to \mathbf{R}^+$ based at $1 \in \mathbf{R}^+$ is given by ordinary exponentiation. □

Normal coordinates. The exponential map may be used to define a coordinate system near x, since by choosing a basis one may identify $T_x\mathcal{M}$ with $\mathbf{R}^m$. Coordinates obtained in this way are called *normal coordinates* about x. Thus if y is a point sufficiently close to x, its normal coordinates are the components, with respect to the chosen basis of $T_x\mathcal{M}$, of the vector v such that $\gamma_v(1) = y$. Now the geodesics in $\mathcal{M}$ through x are the images by the exponential map of radial straight lines in $T_x\mathcal{M}$, and are therefore represented in the form $x^a = v^a t$ in terms of normal coordinates (x^a). Therefore $d^2\gamma/dt^2 = 0$ at x on each geodesic, and inspection of the geodesic equation reveals that the connection coefficients for a normal coordinate system must satisfy at the point x the relation

$$\Gamma^a_{bc} + \Gamma^a_{cb} = 0.$$

This property is much used in tensor calculus to prove tensor identities involving the curvature tensor, since it simplifies the expression for the curvature tensor at x. This is especially convenient when the connection is symmetric, since then $\Gamma^a_{bc}(x) = 0$ and so (for example) at x

$$R^a{}_{bcd} = \partial_c\Gamma^a_{bd} - \partial_d\Gamma^a_{bc},$$

from which the Ricci identity is immediate.

Exercise 46. Without assuming that the connection is symmetric, prove the full first Bianchi identity by this method. □

Exercise 47. Given any type $(0,p)$ tensor field K, one defines its covariant differential ∇K to be the type $(0,p+1)$ tensor field given by $(\nabla K)(V,V_1,V_2,\ldots,V_p) = \nabla_V K(V_1,V_2,\ldots,V_p)$ (Exercise 20). Show that if ∇ is symmetric and ω is a p-form then $d\omega$ is the alternating part of $\nabla\omega$. □

A normal coordinate neighbourhood is star-shaped with respect to its origin x: each y in the neighbourhood may be joined to x by a geodesic segment lying entirely within the neighbourhood. In terms of normal coordinates this segment looks as if it were an affine line. This makes normal coordinate neighbourhoods convenient for topological arguments: for example, on a normal neighbourhood every closed form is exact. It is possible, by more sophisticated arguments, to prove the existence of neighbourhoods with the stronger property of convexity: a *convex normal neighbourhood* is one in which each pair of points may be joined by a geodesic segment lying within the neighbourhood.

8. Affine Maps and Transformations

We shall now consider transformations of a manifold with connection, and maps between two manifolds with connection, which generalise affine transformations of, and affine maps between, affine spaces in the sense that they preserve parallelism.

A map of manifolds with connection which preserves parallelism is called an affine map. To be precise, a smooth map $\phi: \mathcal{M} \to \hat{\mathcal{M}}$ is an *affine map* if, for each path in $\mathcal{M}$ and each pair of points x, y on the path,

$$\hat{\Upsilon}_{\phi(y),\phi(x)} \circ \phi_{*x} = \phi_{*y} \circ \Upsilon_{y,x}$$

where Υ and $\hat{\Upsilon}$ are respectively the rules of parallel transport along the path in $\mathcal{M}$ and along its image by ϕ in $\hat{\mathcal{M}}$. As usual, ϕ_* denotes the map of tangent vectors induced by ϕ, and the condition is that the induced map "intertwine" parallel transport on corresponding paths in the two manifolds.

Exercise 48. Show that a curve $\gamma: I \to \mathcal{M}$ is an affinely parametrised geodesic if and only if it is an affine map, where I has the affine connection it inherits as an open subset of $\mathbf{R}$. □

Exercise 49. Show that the composition of affine maps is affine. □

In particular, if $\phi: \mathcal{M} \to \hat{\mathcal{M}}$ is affine and if γ is a geodesic in $\mathcal{M}$ then $\phi \circ \gamma$ is a geodesic in $\hat{\mathcal{M}}$. It follows that if $\exp: T_x\mathcal{M} \to \mathcal{M}$ is the exponential map at $x \in \mathcal{M}$ and $\widehat{\exp}: T_{\phi(x)}\hat{\mathcal{M}} \to \hat{\mathcal{M}}$ is the exponential map at the image point $\phi(x) \in \hat{\mathcal{M}}$ then

$$\phi \circ \exp = \widehat{\exp} \circ \phi_{*x}.$$

Thus with respect to normal coordinates ϕ is represented by a linear map.

Affine maps also preserve covariant differentiation, as one might expect. Suppose that $\mathcal{M}$ and $\hat{\mathcal{M}}$ are manifolds with connections ∇ and $\hat{\nabla}$ respectively. Recall that a vector field $\hat{V}$ on $\hat{\mathcal{M}}$ is said to be ϕ-related to a vector field V on $\mathcal{M}$ if for all $x \in \mathcal{M}$, $\phi_{*x}V_x = \hat{V}_{\phi(x)}$. If $\hat{V}$ is ϕ-related to V and $\hat{W}$ is ϕ-related to W, and if ϕ is affine, then $\hat{\nabla}_{\hat{V}}\hat{W}$ is ϕ-related to $\nabla_V W$. In fact, for $v \in T_x\mathcal{M}$ and W any vector field along a curve σ such that $\sigma(0) = x$ and $\dot{\sigma}(0) = v$,

$$\begin{aligned}
\hat{\nabla}_{\phi_* v}(\phi_* W) &= \frac{d}{dt}\left(\hat{\Upsilon}_{\phi(x),\phi(\sigma(t))}\phi_* W_{\sigma(t)}\right)_{t=0} \\
&= \phi_*\left(\frac{d}{dt}\left(\Upsilon_{x,\sigma(t)}W_{\sigma(t)}\right)_{t=0}\right) \\
&= \phi_*\left(\nabla_v W\right).
\end{aligned}$$

Exercise 50. From the assertions in the preceding paragraph, deduce that if T and $\hat{T}$ are the torsion tensors and R and $\hat{R}$ the curvature tensors of ∇ and $\hat{\nabla}$ and if ϕ is affine then $\hat{T}(\hat{V}, \hat{W})$ is ϕ-related to $T(V, W)$, and $\hat{R}(\hat{U}, \hat{V})\hat{W}$ is ϕ-related to $R(U, V)W$. □

Affine transformations. Suppose a connection to have been given on $\mathcal{M}$, once for all. An affine map of $\mathcal{M}$ to itself which is a diffeomorphism is called an *affine transformation* of $\mathcal{M}$.

The best strategy for studying structure-preserving maps of a manifold to itself is usually to investigate the generators of one-parameter groups of such maps. This was the case for volume-preserving maps Chapter 5, Section 5) and for isometries (Chapter 8, Section 2), and it is also the case for affine transformations. A one-parameter group ϕ_t of diffeomorphisms of $\mathcal{M}$ is called *affine* if ϕ_t is an affine transformation of $\mathcal{M}$ for each t. The generator X of a one-parameter group of affine transformations is called an *infinitesimal affine transformation*. We shall derive the conditions satisfied by an infinitesimal affine transformation. Since the corresponding one-parameter group consists of affine transformations, for every pair of vector fields V, W

$$\phi_{t*}(\nabla_V W) = \nabla_{(\phi_{t*}V)}\phi_{t*}W = \nabla_{(\phi_{t*}V-V)}\phi_{t*}W + \nabla_V(\phi_{t*}W - W) + \nabla_V W$$

so that

$$\phi_{t*}(\nabla_V W) - \nabla_V W = \nabla_{(\phi_{t*}V-V)}\phi_{t*}W + \nabla_V(\phi_{t*}W - W).$$

Dividing by t and taking the limit as $t \to 0$ one obtains

$$\mathcal{L}_X \nabla_V W = \nabla_{[X,V]} W + \nabla_V \mathcal{L}_X W.$$

It is also true, conversely, that if this condition holds then X is an infinitesimal affine transformation, except that there is in general no guarantee that X will generate a full one-parameter group; however it will generate a flow of local affine transformations.

Exercise 51. Show that if X is an infinitesimal affine transformation then $\mathcal{L}_X \nabla_V - \nabla_V \mathcal{L}_X = \nabla_{[X,V]}$ for every vector field V whenever the operators apply either to a function or to a vector field; and deduce that the equation is true, as an equation between operators, when applied to any tensor field. □

Exercise 52. Show that any linear combination, with constant coefficients, of infinitesimal affine transformations is again an infinitesimal affine transformation, and that the bracket of two infinitesimal affine transformations is again one. □

Exercise 53. Show that the condition for $X = X^a\partial_a$ to be an infinitesimal affine transformation, when expressed in terms of local coordinates, is

$$\frac{\partial^2 X^c}{\partial x^a \partial x^b} + \frac{\partial X^d}{\partial x^a}\Gamma^c_{db} + \frac{\partial X^d}{\partial x^b}\Gamma^c_{ad} - \frac{\partial X^c}{\partial x^d}\Gamma^d_{ab} + X^d \frac{\partial \Gamma^c_{ab}}{\partial x^d} = 0. \qquad \square$$

Exercise 54. Show that if X is an infinitesimal affine transformation then $\mathcal{L}_X T = 0$ and $\mathcal{L}_X R = 0$. □

Exercise 55. Let $\mathcal{M}$ be a manifold with connection ∇, and let ϕ be a diffeomorphism of $\mathcal{M}$, not assumed affine. For any vector fields V, W on $\mathcal{M}$ set $\nabla^\phi_V W = \phi_*^{-1}(\nabla_{\phi_* V}\phi_* W)$. Show that ∇^ϕ is a connection on $\mathcal{M}$. Show that for any one-parameter group of diffeomorphisms ϕ_t, not assumed affine, of $\mathcal{M}$ the tensor field $D = d/dt(\nabla^{\phi_t} - \nabla)_{t=0}$ satisfies

$$D(V,W) = \mathcal{L}_X \nabla_V W - \nabla_V \mathcal{L}_X W - \nabla_{[X,V]} W,$$

where X is the infinitesimal generator of ϕ_t; and so deduce again the condition for X to be an infinitesimal affine transformation. □

By Exercise 52 the set of infinitesimal affine transformations is a subspace (over $\mathbf{R}$) of the space of all vector fields on $\mathcal{M}$ and is closed under bracket. We now show that it is a finite dimensional subspace, an important property which it shares with the space of infinitesimal isometries of a Riemannian or pseudo-Riemannian manifold, for example; thus, like a space of infinitesimal isometries, it is a Lie algebra. (There are, however, many examples of spaces of infinitesimal generators of transformations preserving some geometrical structure which are not finite dimensional: volume preserving transformations, and symplectic transformations—the "isometries" of a closed, non-degenerate 2-form—are examples.)

The infinitesimal affine transformations of an m-dimensional affine space constitute an extreme case. They consist of the vector fields which are linear (but not necessarily homogeneous) when expressed in terms of affine coordinates. They certainly form a linear space, whose dimension is just m^2 (dimension of the homogeneous vector fields) plus m (dimension of the constant vector fields). It may be anticipated, therefore, that the dimension of the space of infinitesimal affine transformations of a manifold $\mathcal{M}$ with connection, which we shall denote $\mathcal{A}(\mathcal{M})$, is at most $m^2 + m$ where $m = \dim \mathcal{M}$. This we shall now prove; and we shall also show that if $\dim \mathcal{A}(\mathcal{M}) = m^2 + m$ then the connection on $\mathcal{M}$ is flat, so that it looks locally at least just like the connection on affine space.

The proof is based on the fact that, if $\phi\colon \mathcal{M} \to \mathcal{M}$ is an affine transformation which leaves a point $x \in \mathcal{M}$ fixed, and if exp is the exponential map at x, then

$$\phi \circ \exp = \exp \circ \phi_{*x},$$

where ϕ_{*x} is a linear map of $T_x\mathcal{M}$ to itself. Thus ϕ is determined, at least so far as its transformation of a normal coordinate neighbourhood is concerned, by the linear map ϕ_{*x}, that is, by an $m \times m$ matrix. In particular, if $\phi(x) = x$ and ϕ_{*x} is the identity on $T_x\mathcal{M}$ then ϕ is the identity on a neighbourhood of x.

Exercise 56. Show that if X is the generator of a one-parameter group ϕ_t and $X_x = 0$, so that $\phi_t(x) = x$ for all t, then $\phi_{t*}\colon T_x\mathcal{M} \to T_x\mathcal{M}$ is the exponential of the linear map whose matrix with respect to a coordinate basis of $T_x\mathcal{M}$ is $(\partial X^a/\partial x^b)(x)$, where $X = X^a\partial_a$. Deduce that if X is an infinitesimal generator of affine transformations, and if $X_x = 0$, then in terms of normal coordinates at x, $X = X^a_b x^b \partial_a$ where $X^a_b = (\partial X^a \partial/x_b)(x)$. □

Suppose that $X = X^a\partial_a \in \mathcal{A}(\mathcal{M})$. If both X^a and $(\partial X^a/\partial x^b)$ vanish at a point in one coordinate system they do in all. By the exercise, if $X^a(x) = 0$ and $(\partial X^a/\partial x^b)(x) = 0$ in terms of normal coordinates then X vanishes on the normal coordinate neighbourhood. Let $\mathcal{O}$ be the set of points at which both (X^a) and $(\partial X^a/\partial x^b)$ vanish. Then $\mathcal{O}$ is open, because if $x \in \mathcal{O}$ then X vanishes on a normal neighbourhood of x; but $\mathcal{O}$ is defined by equations and is therefore also closed. In a connected space a subset which is both open and closed is either empty or the whole space. Thus if $\mathcal{M}$ is connected it follows that any $X \in \mathcal{A}$ which vanishes simultaneously with the matrix of partial derivatives of its components must be identically zero.

Choose now a point $x \in \mathcal{M}$ and a normal coordinate system based at x; and define a map $\mathcal{A}(\mathcal{M}) \to \mathbf{R}^m \oplus M_m(\mathbf{R})$ (the space of $m \times m$ matrices over $\mathbf{R}$) by $X \mapsto \big(X^a(x), (\partial X^a/\partial x^b)(x)\big)$, where $X = X^a\partial_a$. This is a linear map. It follows from the result in the previous paragraph that its kernel is the zero vector field. Thus the map is injective and so the dimension of $\mathcal{A}(\mathcal{M})$ must be finite and cannot exceed that of $\mathbf{R}^m \oplus M_m(\mathbf{R})$, which is $m + m^2$.

Suppose that $\dim \mathcal{A}(\mathcal{M}) = m + m^2$. Then, in particular, every vector field in a normal neighbourhood of $x \in \mathcal{M}$ which is linear with respect to normal coordinates is an infinitesimal affine transformation. Thus the equation obtained from that in Exercise 53 by taking $X^a = \delta^a_b x^b$ holds throughout the normal neighbourhood: it is

$$\Gamma^c_{ab} + x^d \frac{\partial \Gamma^c_{ab}}{\partial x^d} = 0.$$

It follows that $\Gamma^c_{ab} = 0$ at x, so the torsion certainly vanishes there; and by differentiating with respect to x^d, that $\partial \Gamma^c_{ab}/\partial x^d = 0$ at x, so that the curvature also vanishes at x. This applies at each point of $\mathcal{M}$, and so the connection is flat.

9. The Levi-Civita Connection on a (Pseudo-)Riemannian Manifold

In Section 5 of Chapter 9 we showed that there is a connection on a surface, defined initially in terms of the extrinsic geometry of the surface, but in fact uniquely determined by the properties

(1) that it is symmetric—it has no torsion; and

(2) that parallel transport preserves the metric.

By a simple generalisation of the arguments there we show now that there is a unique symmetric metric-preserving connection for any (pseudo-)Riemannian manifold, the Levi-Civita connection.

Let $\mathcal{M}$ be a manifold with metric g of any signature. The condition that parallel transport along a path from x to y preserve the metric properties of the manifold—that is, preserve all scalar products—is that

$$g_y(\Upsilon_{y,x}v, \Upsilon_{y,x}w) = g_x(v, w)$$

for every $v, w \in T_x\mathcal{M}$. Equivalently, this condition may be written

$$\frac{d}{dt}(g(V,W)) = 0$$

whenever V and W are parallel vector fields along a curve. For arbitrary vector fields V and W given along a curve, the same condition is

$$\frac{d}{dt}(g(V,W)) = g\left(\frac{DV}{Dt}, W\right) + g\left(V, \frac{DW}{Dt}\right).$$

In terms of covariant derivatives this condition may be rewritten

$$U(g(V,W)) = g(\nabla_U V, W) + g(V, \nabla_U W)$$

for any $U, V, W \in \mathcal{X}(\mathcal{M})$. Because ∇_U is a derivation, this amounts to

$$\nabla_U g = 0.$$

The two conditions satisfied by a symmetric connection which preserves the metric may therefore be written

$$\nabla_U V - \nabla_V U = [U, V]$$
$$\nabla_U g = 0 \qquad \text{for all } U, V \in \mathcal{X}(\mathcal{M}).$$

Exercise 57. Show that these conditions entail the relation

$$g(\nabla_U V, W) = \tfrac{1}{2}\{U(g(V,W)) + V(g(U,W)) - W(g(U,V)) + g([U,V],W) - g([U,W],V) - g(U,[V,W])\}.$$ □

Exercise 58. Show that in local coordinates the connection coefficients of a symmetric metric-preserving connection are given by

$$\Gamma^a_{bc} = \tfrac{1}{2} g^{ad}\left(\frac{\partial g_{cd}}{\partial x^b} + \frac{\partial g_{bd}}{\partial x^c} - \frac{\partial g_{bc}}{\partial x^d}\right).$$ □

Since no covariant derivatives appear on the right hand side of the relation stated in Exercise 57, this relation serves to define $\nabla_U V$. To establish that ∇ so defined is actually a connection it is necessary to check that it satisfies the conditions

(1) $\nabla_{U+V} W = \nabla_U W + \nabla_V W$
(2) $\nabla_{fU} V = f \nabla_U V$
(3) $\nabla_U (V + W) = \nabla_U V + \nabla_U W$
(4) $\nabla_V (fW) = f\nabla_V W + (Vf)W$.

which are satisfied by all linear connections. For example,

$$\begin{aligned} g(\nabla_{fU} V, W) &= fg(\nabla_U V, W) + \tfrac{1}{2}(Vf)g(U,W) - \tfrac{1}{2}(Wf)g(U,V) \\ &\quad - \tfrac{1}{2}g((Vf)U, W) + \tfrac{1}{2}g((Wf)U, V) \\ &= fg(\nabla_U V, W), \end{aligned}$$

which establishes property (2). Furthermore, $\nabla_U g = 0$, because

$$\begin{aligned} &g(\nabla_U V, W) + g(V, \nabla_U W) \\ &= U(g(V,W)) + \tfrac{1}{2}\{V(g(U,W)) - W(g(U,V)) \\ &\quad + g([U,V],W) - g([U,W],V) - g(U,[V,W]) \\ &\quad + W(g(U,V)) - V(g(U,W)) \\ &\quad + g([U,W],V) - g([U,V],W) - g(U,[W,V])\} \\ &= U(g(V,W)). \end{aligned}$$

Exercise 59. Complete the verification that ∇ defined in Exercise 57 is a symmetric connection. □

Exercise 60. By considering their transformation properties, show that

$$\Gamma^a_{bc} = \frac{1}{2} g^{ad}(\partial_b g_{cd} + \partial_c g_{bd} - \partial_d g_{bc})$$

are the coefficients of a connection in a coordinate basis. □

The uniqueness of the connection defined in Exercise 57 is an immediate consequence of the formula given there, since two connections ∇, $\hat{\nabla}$ both satisfying the conditions immediately preceeding the exercise must have $g(\nabla_U V, W) = g(\hat{\nabla}_U V, W)$ for all U, V, W and so be identical. The formula in Exercise 57 therefore defines a unique symmetric metric-preserving connection: this is the *Levi-Civita connection* for g.

The coefficients of connection in a coordinate basis displayed in Exercise 58 are called the *Christoffel symbols of the second kind*. They are often written $\Gamma^a_{bc} = \left\{ {a \atop b\,c} \right\} = g^{ad}[bc,d]$, where $[bc,d] = \frac{1}{2}(\partial_b g_{cd} + \partial_c g_{bd} - \partial_d g_{bc})$ are the *Christoffel symbols of the first kind*.

Exercise 61. The curvature tensor of a Levi-Civita connection satisfies the identities
$$g(R(U_1,U_2)V_1,V_2) = -g(R(U_1,U_2)V_2,V_1)$$
and
$$g(R(V_1,V_2)U_1,U_2) = g(R(U_1,U_2)V_1,V_2).$$
Verify these identities with the help of normal coordinates. □

Exercise 62. Show that the curvature tensor of a Levi-Civita connection on an m-dimensional manifold has $\frac{1}{12}m^2(m^2-1)$ linearly independent components. □

Exercise 63. Write out Lagrange's equations $d/dt(\partial L/\partial \dot{x}^a) - \partial L/\partial x^a = 0$ for the Lagrangian $L(x^a,\dot{x}^a) = \frac{1}{2}g_{bc}(x^a)\dot{x}^b\dot{x}^c$ and show that they are linear combinations of the geodesic equations for the Levi-Civita connection obtained from the metric with components g_{bc}. □

This gives a quick way of computing the Levi-Civita connection, since finding Lagrange's equations is simpler than evaluating the Christoffel symbols from scratch.

Exercise 64. Find the equations of geodesics for the (spherically symmetric static) metric
$$ds^2 = e^{2\nu}dt^2 - e^{2\lambda}dr^2 - r^2(d\vartheta^2 + \sin^2\vartheta d\varphi^2),$$
where ν, λ are functions of r alone. Writing x^1,x^2,x^3,x^4 for r,ϑ,φ,t respectively, show that the non-zero Christoffel symbols are
$$\Gamma^1_{11} = \lambda' \quad \Gamma^1_{22} = -re^{-2\lambda} \quad \Gamma^1_{33} = -r\sin^2\vartheta e^{-2\lambda} \quad \Gamma^1_{44} = \nu' e^{2(\nu-\lambda)}$$
$$\Gamma^2_{12} = \Gamma^2_{21} = \frac{1}{r} \quad \Gamma^2_{33} = -\sin\vartheta\cos\vartheta$$
$$\Gamma^3_{13} = \Gamma^3_{31} = \frac{1}{r} \quad \Gamma^3_{23} = \Gamma^3_{32} = \cot\vartheta$$
$$\Gamma^4_{14} = \Gamma^4_{41} = \nu'$$
(where $\lambda' = d\lambda/dr$, $\nu' = d\nu/dr$). Compute the components of the curvature tensor in an orthonormal frame whose vectors point in the coordinate directions. □

Exercise 65. Show that the components g^{ab} of the inverse or dual metric satisfy
$$\frac{\partial g^{ab}}{\partial x^c} + g^{ad}\Gamma^b_{dc} + g^{bd}\Gamma^a_{dc} = 0.$$
Deduce that for any type $(1,p)$ tensor field S
$$g^{bc}S^a{}_{bc\ldots e\|f} = (g^{bc}S^a{}_{bc\ldots e})_{\|f}.$$
□

The Ricci tensor and curvature scalar. From the curvature tensor of a Levi-Civita connection we construct a type $(1,1)$ tensor called the Ricci tensor and a function called the curvature scalar.

The trace of the map $U \mapsto R(\,\cdot\,,U)\,\cdot$ is a type $(0,2)$ tensor field, given explicitly by $(V,W) \mapsto \langle R(V,U_c)W,\theta^c\rangle$ where $\{U_c\}$ is any local vector field basis and $\{\theta^c\}$ the dual 1-form basis. This tensor field is symmetric, for
$$\langle R(V,U_c)W,\theta^c\rangle - \langle R(W,U_c)V,\theta^c\rangle = \langle R(V,W)U_c,\theta^c\rangle$$
by the first Bianchi identity; but $R(V,W)$ is skew-symmetric with respect to g (Exercise 61) and its trace is therefore zero (Chapter 8, Exercise 8). We define a type $(1,1)$ tensor field R^*, the *Ricci tensor*, by
$$g(V,R^*(W)) = \langle R(V,U_c)W,\theta^c\rangle.$$
Then R^* is symmetric with respect to g. Its trace $\rho = \langle R^*(U_c),\theta^c\rangle$ is the *curvature scalar* of the connection.

Exercise 66. Show that the components of $R^\star$ are given by $R^{\star a}_{\ b} = g^{ac}R^d_{\ cbd}$. Show that $\rho = R^{\star c}_{\ c}$. □

Exercise 67. For a surface, $R(U,V)W = K\big(g(V,W)U - g(U,W)V\big)$, where K is the Gaussian curvature. Compute the Ricci tensor and curvature in terms of K. □

Exercise 68. By taking suitable contractions of the second Bianchi identity show that the Ricci tensor and curvature satisfy

$$R^{\star a}_{\ b\|a} - \frac{1}{2}\rho_{\|a} = 0.$$ □

The tensor $R^\star - \frac{1}{2}\rho I$ is called the *Einstein tensor*. In the field equations of general relativity the Einstein tensor is identified as a multiple of the energy-momentum tensor of matter. The property of the Einstein tensor exhibited in the last exercise is an important factor in this identification.

Exercise 69. Let A be a type $(1,1)$ tensor which is symmetric with respect to g; set

$$R_A(U,V)W = -g\big(A(V),W\big)U - g(V,W)A(U) + g\big(A(U),W\big)V + g(U,W)A(V).$$

Show that R_A has all the algebraic symmetries of the curvature of a Levi-Civita connection. Show that the corresponding "Ricci tensor" and "curvature scalar" are given in terms of A by

$$R^\star_A = (m-2)A + (\operatorname{tr} A)I \qquad \rho_A = 2(m-1)\operatorname{tr} A$$

and deduce that

$$R_A(U,V)W + \frac{1}{(m-2)}\big(g(R^\star_A(V),W)U + g(V,W)R^\star_A(U) - g(R^\star_A(U),W)V - g(U,W)R^\star_A(V)\big)$$
$$- \frac{\rho}{(m-1)(m-2)}\big(g(V,W)U - g(U,W)V\big) = 0.$$ □

10. Conformal Geometry

The idea of a conformal rescaling of a (pseudo-) Riemannian metric was introduced in the last chapter, Section 8. We now describe some of the effects on the associated connection of making a conformal rescaling of a metric, with a view particularly to picking out geometric objects which are unaffected.

The first point we make is that a non-homothetic conformal transformation does not preserve the connection, nor the geodesics, although in the pseudo-Riemannian case it does preserve the null geodesics, meanwhile generally altering the affine parameters on them. Suppose that $\hat{g}$ is a metric obtained from g by a conformal rescaling. It is convenient to set $\hat{g} = e^{2\sigma}g$. Then the Levi-Civita connection $\hat{\nabla}$ for $\hat{g}$ is given by

$$\begin{aligned}\hat{g}\left(\hat{\nabla}_U V, W\right) &= \tfrac{1}{2}\{U\left(\hat{g}(V,W)\right) + V\left(\hat{g}(U,W)\right) - W\left(\hat{g}(U,V)\right)\\ &\qquad + \hat{g}([U,V],W) - \hat{g}([U,W],V) - \hat{g}(U,[V,W])\}\\ &= \hat{g}(\nabla_U V,W) + (U\sigma)\hat{g}(V,W) + (V\sigma)\hat{g}(U,W) - (W\sigma)\hat{g}(U,V)\\ &= \hat{g}\big(\nabla_U V + (U\sigma)V + (V\sigma)U - g(U,V)\operatorname{grad}\sigma, W\big)\end{aligned}$$

so that

$$\hat{\nabla}_U V = \nabla_U V + (U\sigma)V + (V\sigma)U - g(U,V)\operatorname{grad}\sigma.$$

Now a vector field V is geodesic tangent with respect to ∇ if and only if $\nabla_V V$ is a multiple of V, and in particular the geodesics are affinely parametrised if and only if $\nabla_V V = 0$. With respect to $\hat{\nabla}$,

$$\hat{\nabla}_V V = \nabla_V V + 2(V\sigma)V - g(V,V)\operatorname{grad}\sigma.$$

Because of the last term, a conformal change cannot in general preserve geodesics. However, if V is null, then

$$\hat{\nabla}_V V = \nabla_V V + 2(V\sigma)V,$$

so that $\hat{\nabla}_V V$ is still proportional to V. But it need not vanish, so that the affine property of V is spoilt, unless it happens that $V\sigma = 0$ (which cannot be true for all null geodesics unless σ is constant). Thus a conformal change preserves null geodesics but not affine parametrisation.

The other important fact is the invariance under conformal change of the *Weyl conformal curvature tensor* C of a Riemannian or pseudo-Riemannian structure on a manifold of dimension $m > 2$, defined in the following exercise.

Exercise 70. By using the formula for the Levi-Civita connection $\hat{\nabla}$ given above, show that the curvature tensors R and $\hat{R}$ of the connections ∇ and $\hat{\nabla}$ are related by $\hat{R} = R + R_A$, where R_A is the tensor defined in Exercise 69 and A is the type $(1,1)$ tensor given by

$$A(V) = \nabla_V \operatorname{grad}\sigma - (V\sigma)\operatorname{grad}\sigma - \tfrac{1}{2}g(\operatorname{grad}\sigma, \operatorname{grad}\sigma)V.$$

Deduce that if

$$\begin{aligned} C(U,V)W &= R(U,V)W \\ &+ \frac{1}{(m-2)}\left(g(R^*(V),W)U + g(V,W)R^*(U) - g(R^*(U),W)V - g(U,W)R^*(V)\right) \\ &\qquad - \frac{\rho}{(m-1)(m-2)}\left(g(V,W)U - g(U,W)V\right) \end{aligned}$$

and $\hat{C}$ is the corresponding tensor constructed from $\hat{R}$ then $\hat{C} = C$. □

The tensor C is the *Weyl conformal curvature tensor* of the metric g: the result of the exercise is that it is conformally invariant, that is, unchanged by a conformal rescaling. The Weyl tensor has zero trace on every pair of indices, as well as the algebraic symmetries of the curvature tensor. There has been an extensive search for other conformal invariants, and various algorithms for constructing them are known, but there is no known procedure for finding all those which may be constructed in a space of given dimension.

Summary of Chapter 11

A rule of parallel transport along a path in a manifold $\mathcal{M}$ is a collection of non-singular linear maps $\Upsilon_{y,x}\colon T_x\mathcal{M} \to T_y\mathcal{M}$, for every pair of points x, y on the path, such that $\Upsilon_{y,z} \circ \Upsilon_{z,x} = \Upsilon_{y,x}$ for every point z on the path. A rule of parallel transport in $\mathcal{M}$ is fixed when a rule is given along each path in it. A vector field W is parallel along a path if for any x and each y on the path $W_y = \Upsilon_{y,x}W_x$.

If σ is a curve through x and W is a vector field specified along σ then the absolute derivative of W along σ at $x = \sigma(0)$ is $DW/Dt = d/ds\{\Upsilon_{x,\sigma(s)}W(s)\}_{s=0}$.

A linear connection ∇ is the assignment, to each $x \in \mathcal{M}$ and to each vector field W defined near x, of a linear map $\nabla W\colon T_x\mathcal{M} \to T_x\mathcal{M}$ such that for any curve

σ through x the absolute derivative of W along σ is $\nabla_{\dot\sigma}W$. For any $v \in T_x\mathcal{M}$, $\nabla_v W$ is called the covariant derivative of W with respect to v. In the common domain of vector fields V and W, $\nabla_V W$ is the vector field with the same domain defined by $(\nabla_V W)_x = \nabla_{V_x} W$. It is assumed that if V and W are smooth then $\nabla_V W$ is smooth. The covariant derivative has the properties

$$\nabla_{(U+V)}W = \nabla_U W + \nabla_V W \qquad \nabla_U(V+W) = \nabla_U V + \nabla_U W$$
$$\nabla_{fV}W = f\nabla_V W \qquad \nabla_V(fW) = f\nabla_V W + (Vf)W.$$

A linear connection gives rise to a rule of parallel transport determined by solution of the equations $\nabla_{\dot\sigma}W = 0$. Let $\{U_a\}$ be a local basis for vector fields. The coefficients of linear connection with respect to this basis are the functions γ^a_{bc} defined by $\nabla_{U_c}U_b = \gamma^a_{bc}U_a$. If $\{U_a\}$ is a coordinate basis then Γ^a_{bc} is written for the connection coefficients, and $\nabla_V W = V^b W^a{}_{\|b}\partial_a$, where $W^a{}_{\|b} = \partial_b W^a + \Gamma^a_{cb}W^c$.

The covariant derivative of a function is its directional derivative, and the covariant derivative of a 1-form is defined so that the covariant derivative of a pairing is a derivation: $\langle W, \nabla_V\alpha\rangle = V\langle W,\alpha\rangle - \langle\nabla_V W,\alpha\rangle$. The covariant derivative of a tensor field of type (p,q) is also defined so that it is a derivation.

The torsion of a linear connection is the type $(1,2)$ tensor field T and the curvature the type $(1,3)$ tensor field R defined by

$$T(U,V) = \nabla_U V - \nabla_V U - [U,V]$$
$$R(U,V)W = \nabla_U\nabla_V W - \nabla_V\nabla_U W - \nabla_{[U,V]}W.$$

The connection is called symmetric if its torsion vanishes. Relative to a coordinate basis T and R are given by

$$T^a_{bc} = \Gamma^a_{cb} - \Gamma^a_{bc}$$
$$R^a{}_{bcd} = \partial_c\Gamma^a_{bd} - \partial_d\Gamma^a_{bc} + \Gamma^e_{bd}\Gamma^a_{ec} - \Gamma^e_{bc}\Gamma^a_{ed}.$$

The curvature and torsion satisfy

$$T(V,U) = -T(U,V) \qquad R(V,U)W = -R(U,V)W.$$

For a symmetric connection the curvature satisfies the Bianchi identities

$$R(U,V)W + R(V,W)U + R(W,U)V = 0$$
$$(\nabla_U R)(V,W) + (\nabla_V R)(W,U) + (\nabla_W R)(U,V) = 0.$$

Relative to a coordinate basis, for a symmetric connection,

$$R^d{}_{cab} = -R^d{}_{cba}$$
$$R^d{}_{cab} + R^d{}_{bca} + R^d{}_{abc} = 0$$
$$R^d{}_{eab\|c} + R^d{}_{eca\|b} + R^d{}_{ebc\|a} = 0.$$

More general versions of the Bianchi identities, involving the torsion, hold for non-symmetric connections.

Let $\{\theta^a\}$ be the basis for 1-forms dual to the local basis $\{U_a\}$ for vector fields. The connection forms ω^a_b are defined by $\langle V, \omega^a_b\rangle = \langle\nabla_V U_b, \theta^a\rangle$, for every vector field V. The torsion 1-forms Θ^a are defined by $\Theta^a(V,W) = \langle T(V,W), \theta^a\rangle$. The curvature

2-forms Ω^a_b are defined by $\Omega^a_b(V,W) = \langle R(V,W)U_b, \theta^a \rangle$. The forms θ^a and Θ^a are identified as entries in (column) vectors θ and Θ, the forms ω^a_b and Ω^a_b as entries in (square) matrices ω and Ω. In this notation Cartan's structure equations are

$$d\theta + \omega \wedge \theta = \Theta \qquad d\omega + \tfrac{1}{2}[\omega \wedge \omega] = \Omega,$$

and their exterior derivatives, which are equivalent to the Bianchi identities, are

$$d\Theta + \omega \wedge \Theta = \Omega \wedge \theta \qquad d\Omega + [\omega \wedge \Omega] = 0.$$

If A is a matrix-valued function specifying a change of basis $\hat{\theta} = A^{-1}\theta$, then $\hat{\omega} = A^{-1}dA + A^{-1}\omega A$, $\hat{\Theta} = A^{-1}\Theta$ and $\hat{\Omega} = A^{-1}\Omega A$. Exploiting this formalism one can show from Frobenius's theorem that if a manifold has a connection with vanishing curvature, then the connection is a complete parallelism, which means that there exists about each point a local basis of parallel vector fields. If, further, the connection is symmetric then there is a local coordinate system in which the coordinate fields are parallel; the connection is then flat.

A geodesic is a path whose tangent spaces are parallel along it. An affinely parametrised geodesic is a curve with geodesic image whose tangent vector is parallely transported along it. The tangent vector field to an affinely parametrised geodesic γ satisfies $\nabla_{\dot\gamma}\dot\gamma = 0$. Any geodesic may be affinely parametrised. In local coordinates the equations of an affinely parametrised geodesic are

$$\frac{d^2\gamma^a}{dt^2} + \Gamma^a_{bc}\frac{d\gamma^b}{dt}\frac{d\gamma^c}{dt} = 0.$$

The exponential map at x is defined by $\exp(v) = \gamma_v(1)$, where γ_v is the geodesic with $\gamma_v(0) = x$ and $\dot\gamma_v(0) = v$. It is a diffeomorphism on some neighbourhood of $0 \in T_x\mathcal{M}$. If exp is globally defined on $T_x\mathcal{M}$ whatever the choice of x then $\mathcal{M}$ is called geodesically complete. Exp maps affine coordinates in $T_x\mathcal{M}$, with origin 0, into coordinates around x, called normal coordinates, in which each geodesic through x has the form $t \mapsto (tx^a)$; in these coordinates $\Gamma^a_{bc}(x) + \Gamma^a_{cb}(x) = 0$.

A map of manifolds with connection, which preserves parallelism, is called an affine map. If X generates a one-parameter group of affine diffeomorphisms of $\mathcal{M}$ then $\mathcal{L}_X\nabla_V - \nabla_V\mathcal{L}_X = \nabla_{[X,V]}$ for any vector field V and any argument, be it function, form, vector or tensor field. Moreover, $\mathcal{L}_X T = \mathcal{L}_X R = 0$. The set of infinitesimal affine transformations is a Lie algebra, of dimension at most $m^2 + m$.

On a (pseudo-)Riemannian manifold there is a unique symmetric metric-preserving connection, that is, one with the properties $\nabla_V W - \nabla_W V = [V,W]$; $\nabla_V g = 0$. This connection is called the Levi-Civita connection. It is given by

$$\begin{aligned} g(\nabla_U V, W) = \tfrac{1}{2}\{&U(g(V,W)) + V(g(U,W)) - W(g(U,V)) \\ &+ g([U,V],W) - g([U,W],V) - g(U,[V,W])\} \end{aligned}$$

and in local coordinates the connection coefficients are

$$\Gamma^a_{bc} = \tfrac{1}{2}g^{ad}\left(\partial_b g_{cd} + \partial_c g_{bd} - \partial_d g_{bc}\right).$$

The curvature tensor of a Levi-Civita connection satisfies the identities

$$g(R(U_1,U_2)V_2,V_1) = -g(R(U_1,U_2)V_1,V_2)$$

and

$$g(R(V_1, V_2)U_1, U_2) = g(R(U_1, U_2)V_1, V_2).$$

A conformal change of metric does not preserve the connection or geodesics unless it is homothetic, except for null geodesics (in a pseudo-Riemannian manifold), though even then the affine parametrisation is altered. By subtracting suitable combinations of the Ricci tensor and the curvature scalar one may construct from the curvature tensor a conformally invariant tensor called the Weyl tensor.

Notes to Chapter 11

1. Convex normal neighbourhoods. Let $\mathcal{M}$ be a manifold with connection. That each point of $\mathcal{M}$ lies in a convex normal neighbourhood is proved in, for example, Helgason [1978].

2. Torsion in physical theory. A great many authors have written about the physical significance of torsion. There is a survey article by Hehl and others in Held [1980]. The little book by Schrodinger [1954] is less comprehensive but very readable.

12. LIE GROUPS

A group whose elements are labelled by one or more continuously variable parameters may be considered also to be a manifold; one has merely to take the parameters as coordinates. This is the basic idea of the theory of Lie groups. The groups in question might well have been called differentiable groups, but the conventional association with the name of Sophus Lie, who revolutionised the theory of differentiable groups in the last decades of the nineteenth century, is too deeply ingrained in the literature to admit any change.

Many examples of Lie groups have already arisen in this book. The affine group introduced in Chapter 1 is a Lie group. So also are the rotation, Euclidean, Lorentz and Poincaré groups of Chapter 8. The one-parameter groups of transformations introduced in Chapter 3 are (1-dimensional) Lie groups.

The discussion of these groups in this chapter differs in emphasis from that of the preceding chapters. The groups just mentioned arose as groups of transformations of other manifolds. We have hinted already that one can abstract the group structure from the idea of a transformation group and consider the group in its own right without regard to the manifold on which it acts. One can go further than this, and define a Lie group abstractly in the first place, as a manifold endowed with maps defining group multiplication and formation of inverses. This is how the definition is usually presented nowadays. We prefer to begin with some examples, showing how a group may be thought of as a manifold. After giving the formal definitions we go on to consider a certain collection of vector fields on any Lie group, called the Lie algebra of the group. It turns out that a Lie group may be reconstructed "almost uniquely" from its Lie algebra. We discuss the extent to which this is the case, and describe the exponential map, by which one can move from the Lie algebra to the group. The chapter concludes with a re-examination of some aspects of groups of transformations.

1. Introductory Examples

We begin with a fairly detailed treatment of two examples of Lie groups, in order to motivate the definition.

Orientation-preserving isometries of the plane. Our first example is the group E of orientation-preserving isometries of the Euclidean plane $\mathcal{E}^2$. With (x^1, x^2) as Cartesian coordinates, let $\tau_{(\xi^1,\xi^2)}$ denote the translation $(x^1, x^2) \mapsto (x^1 + \xi^1, x^2 + \xi^2)$, and let R_ϑ denote the counter-clockwise rotation $(x^1, x^2) \mapsto (x^1 \cos\vartheta - x^2 \sin\vartheta, x^1 \sin\vartheta + x^2 \cos\vartheta)$. Then

$$\tau_{(\xi^1,\xi^2)} \circ \tau_{(\eta^1,\eta^2)} = \tau_{(\xi^1+\eta^1,\xi^2+\eta^2)}$$
$$R_\vartheta \circ R_\varphi = R_{\vartheta+\varphi}$$
$$R_\vartheta \circ \tau_{(\xi^1,\xi^2)} = \tau_{(\xi^1 \cos\vartheta - \xi^2 \sin\vartheta, \xi^1 \sin\vartheta + \xi^2 \cos\vartheta)} \circ R_\vartheta.$$

Also, $R_{\vartheta+2\pi} = R_\vartheta$. Here it is conventional that the operation on the right is carried out first. It follows that any succession of translations and rotations may by suitable choice of ξ^1, ξ^2 and ϑ be written in the form $\tau_{(\xi^1,\xi^2)} \circ R_\vartheta$, with ξ^1 and ξ^2 uniquely determined and ϑ determined modulo 2π. On the other hand, every triple $(\xi^1, \xi^2, \vartheta)$ determines a unique isometry $\tau_{(\xi^1,\xi^2)} \circ R_\vartheta$. Thus ξ^1, ξ^2 and ϑ label the elements of the group E, and group elements labelled by neighbouring values of these variables produce nearby effects on $\mathcal{E}^2$. Thus ξ^1, ξ^2 and ϑ may be regarded as coordinates of the elements of E, and E, with these coordinates, and the usual provisos about the angular coordinate ϑ, may be considered to be a 3-dimensional manifold. Taking into account the ranges of ξ^1, ξ^2 and ϑ one should recognise that, as a manifold, E is diffeomorphic to $\mathbf{R}^2 \times S^1$ (here S^1 represents the circle).

The group E thus arises in the first place as a group of transformations of the plane, but on further consideration may be recognised as a differentiable manifold in its own right. From this point of view the group-theoretical processes of multiplication and formation of inverses are to be thought of as maps of manifolds.

Exercise 1. Write $(\xi^1, \xi^2, \vartheta)$ for $\tau_{(\xi^1,\xi^2)} \circ R_\vartheta$. Verify that

$$(\xi^1, \xi^2, \vartheta) \circ (\eta^1, \eta^2, \varphi) = (\xi^1 + \eta^1 \cos\vartheta - \eta^2 \sin\vartheta, \xi^2 + \eta^1 \sin\vartheta + \eta^2 \cos\vartheta, \vartheta + \varphi)$$

and that

$$(\xi^1, \xi^2, \vartheta)^{-1} = (-\xi^1 \cos\vartheta + \xi^2 \sin\vartheta, -\xi^1 \sin\vartheta - \xi^2 \cos\vartheta, -\vartheta).$$

□

From this exercise it is seen that multiplication and formation of inverses are given by smooth, indeed analytic, functions. Thus multiplication may be described as a smooth map $E \times E \to E$ and formation of the inverse as a smooth map $E \to E$.

We have chosen to deal with this example here because it is simple enough to allow explicit computations, but complicated enough to exhibit the most important features of Lie groups. It has the advantage that the group is of dimension 3, while the manifold on which it acts is of dimension 2, so that from the outset there is a clear distinction between them. In this respect it is more useful, as an example, than the rotation group, which arises as a group of transformations of $\mathcal{E}^3$ and is itself 3-dimensional. In that case some confusion can arise between the group and the space on which it acts, and, in particular, between the tangent spaces to the group, on the one hand, and to $\mathcal{E}^3$, on the other.

Matrix groups. The group $GL(\mathcal{V})$ of non-singular linear transformations of a real vector space $\mathcal{V}$ attains concrete form, when a basis is chosen for $\mathcal{V}$, as the group of non-singular matrices $GL(n, \mathbf{R})$ acting on $\mathbf{R}^n$. As was pointed out in Section 1 of Chapter 10, the $n \times n$ non-singular matrices constitute an n^2-dimensional manifold with coordinate functions (x^a_b), where $x^a_b(g)$ is the entry in the ath row and bth column of $g \in GL(n, \mathbf{R})$. The group multiplication is bilinear in these coordinates, and formation of the inverse yields for each entry a quotient of polynomials, with non-singular denominator, so that both operations may be expressed as smooth maps. Thus $GL(n, \mathbf{R})$ is a group which may at the same time be considered to be a manifold, and whose group-theoretic operations of multiplication and formation of inverses are given by smooth maps. The vector space $\mathbf{R}^n$ on which the group acts may be left out of consideration and the group considered as an object in its own right.

2. Definition and Further Examples

A *Lie group* is a group which is at the same time a manifold, in such a way that the group operations are smooth maps. These operations are (with G as the group and g, g_1, g_2 elements of it)

multiplication : $G \times G \to G$ by $(g_1, g_2) \mapsto g_1 g_2$.

formation of inverses:

$G \to G$ by $g \mapsto g^{-1}$.

It turns out that the group structure restricts the manifold structure so much that one may even assume the group operations to be analytic without losing any generality.

Examples. We next describe some further examples of Lie groups.

(1) The real line $\mathbf{R}$, with addition as the group multiplication, is a 1-dimensional Lie group.

(2) The unit circle S^1 in the complex plane, with multiplication of complex numbers $(e^{i\vartheta_1}, e^{i\vartheta_2}) \mapsto e^{i(\vartheta_1 + \vartheta_2)}$ as the group multiplication, is a 1-dimensional Lie group.

(3) Two Lie groups are said to be *isomorphic* if they are isomorphic as groups and if the isomorphism is a diffeomorphism of manifolds. The group of real numbers $\{ x \mid 0 \leq x < 1 \}$ under addition modulo 1 is a 1-dimensional Lie group isomorphic to S^1.

(4) If G_1 and G_2 are Lie groups, then the *product group* $G_1 \times G_2$, with multiplication given by $(g_1, g_2)(g_1', g_2') = (g_1 g_1', g_2 g_2')$, endowed with the structure of product manifold, is a Lie group. The torus $T^2 = S^1 \times S^1$ is an example.

(5) As we have already pointed out, the group $GL(n, \mathbf{R})$ of $n \times n$ non-singular matrices is a Lie group. Many important Lie groups are groups of matrices, that is, subgroups of $GL(n, \mathbf{R})$ for some n. For example: the special linear group $SL(n, \mathbf{R})$ of $n \times n$ matrices of determinant 1; the orthogonal group $O(n)$ of $n \times n$ matrices g satisfying $gg^T = I_n$; the special orthogonal group $SO(n)$ of $n \times n$ orthogonal matrices with determinant 1; orthogonal and special orthogonal groups corresponding to scalar products of other signatures.

(6) The group $GL(n, \mathbf{C})$ of non-singular $n \times n$ matrices with complex entries is also a Lie group, where we take for coordinate functions the real and imaginary parts of the entries. Subgroups of $GL(n, \mathbf{C})$ also furnish important examples of Lie groups. For example: $SL(n, \mathbf{C})$, the group of $n \times n$ complex matrices of determinant 1; $U(n)$, the unitary group, consisting of $n \times n$ complex matrices g satisfying $gg^\dagger = I_n$, where $\dagger$ denotes the complex conjugate transpose; $SU(n)$, the special unitary group, which is the subgroup of $U(n)$ of elements with determinant 1.

(7) The affine group of an n-dimensional affine space is a Lie group, and so also are subgroups of it which preserve additional structure, such as the Euclidean and the Poincaré groups.

Subgroups. Many of these examples are subgroups of larger Lie groups, and at the same time submanifolds of them. It is natural to use the term "Lie subgroup" in this context. Unfortunately, one has to be rather careful about the meaning of the word "submanifold", as the following example shows. Let T^2 be the torus, defined now as the set $\{ (x^1, x^2) \mid 0 \leq x^1, x^2 < 1 \}$, under addition defined componentwise mod 1. Then for each real α the subset $R = \{ (x^1, x^2) \mid x^2 = \alpha x^1 \}$ of T^2 is a subgroup; R is a homomorphic image of $\mathbf{R}$ by a $1:1$ immersion; but how the subset sits in T^2 depends on whether α is rational or irrational. When α is rational R is a closed imbedded submanifold of T^2 which is actually diffeomorphic to a circle, since the line $x^2 = \alpha x^1$ in $\mathbf{R}^2$ passes through points, other than $(0,0)$, both of whose coordinates are integers, and the subset R therefore eventually returns to its starting point. When α is irrational, on the other hand, R is not closed—in fact it is a dense subset of T^2—and is an immersed but not an imbedded submanifold. In this case the subgroup is isomorphic to $\mathbf{R}$, algebraically; but topologically, it is not a nice subset of T^2 and in particular is not homeomorphic to $\mathbf{R}$ in its induced topology. The topological difficulties associated with this example have already been pointed out in the context of the theory of integrable distributions in Chapter 10, Section 7.

We define a *Lie subgroup* of a Lie group G to be the image in G of a Lie group H by a $1:1$ immersion. It should be noted that the topology of a Lie subgroup is not necessarily the same as the topology induced on it as a subset of the larger group. The case α irrational described above is a case in point. This example is actually a paradigm, in that being a closed subset of G is a necessary and sufficient condition for a Lie subgroup to be an imbedded submanifold, and thus to have the same topology as is induced from the topology of G. The details of this take one too deep into the realms of topology to be worth repeating here; however, it is as well to be aware of the reason for the definition, and the significance of closure for a subgroup.

3. Computations in Coordinates

Many of the constructions in the theory of Lie groups can be worked out explicitly, using coordinates, in terms of the functions which represent the group operations, and most of them were first discovered in this way. It is useful to be able to carry out explicit calculations, as well as to be able to employ the more abstract methods to be explained later in this chapter. We begin the description of computational methods here.

Let G be a Lie group of dimension n and let (x^a) be coordinate functions defined on a neighbourhood of the identity e of G. Suppose that $x^a(e) = 0$: this mild restriction, which can always be complied with by a translation of coordinates, saves a good deal of writing. Multiplication functions Ψ^a and inversion functions I^a, the former from some neighbourhood of $\mathbf{0}$ in $\mathbf{R}^n \times \mathbf{R}^n$ to $\mathbf{R}^n$, the latter from some neighbourhood of $\mathbf{0}$ in $\mathbf{R}^n$ to $\mathbf{R}^n$, are defined as follows:

$$x^a(gh) = \Psi^a\left(x^b(g), x^c(h)\right)$$
$$x^a(g^{-1}) = I^a\left(x^b(g)\right)$$

(when both sides make sense).

Exercise 2. Show that $\Psi^a(\xi^b, 0) = \xi^a$ and $\Psi^a(0, \eta^b) = \eta^a$; that $\Psi^a(\xi^b, I^c(\xi^d)) = \Psi^a(I^b(\xi^d), \xi^c) = 0$; and that $\Psi^a(\xi^b, \Psi^c(\eta^d, \varsigma^e)) = \Psi^a(\Psi^b(\xi^d, \eta^e), \varsigma^c)$. □

Since the group operations may be assumed analytic, the functions Ψ^a and I^a may be expanded in Taylor series. From the results of the exercise it follows that

$$\Psi^a(\xi^d, \eta^e) = \xi^a + \eta^a + O_2,$$

where O_k will denote terms of order k and higher. Applying this with $\eta^c = I^c(\xi^d)$ one obtains

$$I^a(\xi^b) = -\xi^a + O_2.$$

Exercise 3. Show that there are no terms in Ψ^a of order 2 or higher which contain only ξs or only ηs. □

From the result of this exercise one may suppose that

$$\Psi^a(\xi^b, \eta^c) = \xi^a + \eta^a + \alpha^a_{bc}\xi^b\eta^c + O_3$$

for some numbers α^a_{bc}.

Exercise 4. Show that $I^a(\xi^d) = -\xi^a + \alpha^a_{bc}\xi^b\xi^c + O_3$. □

An analytic coordinate transformation, which leaves the coordinates of the identity unchanged, may be expanded in a Taylor series: if $(\acute{x}^a)$ are the new coordinate functions, then for any g, with $x^a(g) = \xi^a$,

$$\acute{x}^a(g) = \acute{\xi}^a = A^a_b\xi^b + B^a_{bc}\xi^b\xi^c + O_3$$

say, where the matrix of coefficients (A^a_b) must be invertible so that the coordinate transformation is invertible in a neighbourhood of 0.

Exercise 5. Compute the multiplication functions in the new coordinates, and show that by a suitable choice of B^a_{bc}, which may be assumed to be symmetric in its lower indices, the symmetric part $\frac{1}{2}(\alpha^a_{bc} + \alpha^a_{cb})$ of α^a_{bc} may be eliminated, while the skew-symmetric part $\frac{1}{2}(\alpha^a_{bc} - \alpha^a_{cb})$ transforms tensorially by A^a_b. □

The special choice of coordinates described in this exercise can be specified neatly in terms of the inversion functions I^a: it is the choice for which $I^a(\xi^a) = -\xi^a + O_3$. By considering higher order terms one can easily convince oneself that in a formal sense, without regard to questions of convergence, one can choose coordinates so that $I^a(\xi^a) = -\xi^a$ exactly. It follows that the symmetric part of α^a_{bc} has no invariant significance. However, the skew-symmetric part does have such significance. We write $C^a_{bc} = \frac{1}{2}(\alpha^a_{bc} - \alpha^a_{cb})$.

Exercise 6. Show that $2C^a_{bc}$ is the coefficient of the leading term in the expansion of the commutator $ghg^{-1}h^{-1}$ of group elements g and h: $x^a(ghg^{-1}h^{-1}) = 2C^a_{bc}x^b(g)x^c(h) + O_3$. □

Exercise 7. Show that the associativity condition $g(hk) = (gh)k$ leads at order 3 to the "Jacobi identity"

$$C^a_{eb}C^e_{cd} + C^a_{ec}C^e_{db} + C^a_{ed}C^e_{bc} = 0.$$ □

4. Transformation Groups

As we have already pointed out, Lie groups appeared in the first place as groups of transformations of other manifolds, and still play this role in many applications. In this section we describe some general features of transformation groups which do not depend primarily on their differentiability properties.

An *action* of a group G on a set $\mathcal{M}$ is a homomorphism $G \to S(\mathcal{M})$ where $S(\mathcal{M})$ is the group of "permutations" of $\mathcal{M}$, in other words, bijective maps of $\mathcal{M}$ onto itself. If ϕ is an action of G on $\mathcal{M}$ we shall here write ϕ_g to denote the transformation of $\mathcal{M}$ determined by $g \in G$; in this notation the conditions that ϕ be an action are

$$\phi_g \circ \phi_h = \phi_{gh} \qquad \phi_e = \mathrm{id}_{\mathcal{M}}.$$

Exercise 8. Show that an action of the real line, considered as a group under addition, on a manifold $\mathcal{M}$ is a one-parameter group of transformations of $\mathcal{M}$. □

Left and right actions. The content of this definition of an action depends on the convention that if transformations are written composed, then the rightmost transformation is to be carried out first. This is to say that if x is any element of $\mathcal{M}$, $(\phi_g \circ \phi_h)(x)$ means $\phi_g(\phi_h(x))$, and the condition for ϕ to be an action is $\phi_g(\phi_h(x)) = \phi_{gh}(x)$. This is reflected in the custom of abbreviating $\phi_g(x)$ to gx, leaving ϕ understood, for then $g(hx) = (gh)x$. However, the stated condition can be too restrictive. Suppose that $\mathcal{M}$ is $\mathbf{R}^n$ and that G is a group of $n \times n$ matrices. If ϕ denotes matrix multiplication, with each matrix represented by itself, and x is a column vector then indeed $\phi_g(x) = gx$ and $g(hx) = (gh)x$. However, if x is a row vector then $\phi_g(x)$ must mean xg, and then $(xg)h = x(gh)$; but written in terms of ϕ this becomes $\phi_h(\phi_g(x)) = \phi_{gh}(x)$, so that $\phi_h \circ \phi_g = \phi_{gh}$ with the order of the factors on the left opposite to what it was before.

Such situations occur sufficiently often that they must be allowed for, and accordingly one distinguishes two kinds of group actions, *left actions*, where $\phi_{gh} = \phi_g \circ \phi_h$ as in the definition at the beginning of this section, and as is the case for matrix multiplication of column vectors; and *right actions*, where $\phi_{gh} = \phi_h \circ \phi_g$, as is the case for matrix multiplication of row vectors. The simplified notations are gx for $\phi_g(x)$ when ϕ is a left action; xg for $\phi_g(x)$ when ϕ is a right action.

Exercise 9. Verify that if ϕ is a left action then $g \mapsto \phi_{g^{-1}}$ is a right action, and *vice versa*. □

If a left action of G on $\mathcal{M}$ is given, then $\mathcal{M}$ is sometimes called a *left G-space*, while if a right action of G on $\mathcal{M}$ is given, then $\mathcal{M}$ is sometimes called a *right G-space*.

Orbits and homogeneous spaces. An action ϕ of G on $\mathcal{M}$ distinguishes certain subgroups of G and subsets of $\mathcal{M}$. The subgroups are those which leave points of $\mathcal{M}$ fixed, and the subsets are those which are preserved by G. For example, if G is the rotation group and $\mathcal{M}$ is $\mathcal{E}^3$, then the subgroup leaving the origin fixed is the whole of G, while the subgroup leaving any other point fixed is the one-parameter group of rotations about the radius vector to that point. The subsets of $\mathcal{M}$ preserved by G are the origin and each sphere with centre at the origin.

Let ϕ be an action of G on $\mathcal{M}$. The set of points which may be reached from $x \in \mathcal{M}$ by the action of G is called the *orbit*, or in case of ambiguity the G-orbit, of x, and denoted Gx (or xG if the action is a right action). This concept is an obvious generalisation of the corresponding one for one-parameter groups of transformations which we have used so frequently in previous chapters.

If a subset $\mathcal{N}$ of $\mathcal{M}$ lies on a single orbit, then G is said to *act transitively* on $\mathcal{N}$, and $\mathcal{N}$ is called a *homogeneous space* of G. Every G-space is partitioned by the action of G into G-orbits, each of which is a homogeneous space of G. The set of G-orbits for a left G-action is denoted $\mathcal{M}/G$, and $\pi: \mathcal{M} \to \mathcal{M}/G$ by $x \mapsto Gx$ is the projection taking each point of $\mathcal{M}$ into the G-orbit on which it lies. The set of orbits for a right G-action is denoted $G\backslash\mathcal{M}$. In case $\mathcal{M}$ is a smooth manifold and G a Lie transformation group of $\mathcal{M}$, the set $\mathcal{M}/G$ may be endowed with the quotient topology induced by the projection: a set $\mathcal{U}$ is open in $\mathcal{M}/G$ if and only if its pre-image $\pi^{-1}(\mathcal{U})$ is open in $\mathcal{M}$. However $\mathcal{M}/G$ is not in general a manifold and need not even be a Hausdorff space. Similar comments apply in the case of a right action.

Let ϕ be an action of a group G on a set $\mathcal{M}$. The set of elements of G leaving fixed a chosen element x of $\mathcal{M}$ is a subgroup of G called the *isotropy group* of x and denoted G_x. Thus $G_x = \{ g \in G \mid \phi_g(x) = x \}$.

Exercise 10. Show that points on the same orbit have conjugate isotropy groups. □

If the isotropy group of every point is the identity, then G is said to act *freely* on $\mathcal{M}$; in this case no element except the identity leaves any point fixed. If the intersection of the isotropy groups of all points of $\mathcal{M}$ is the identity, then G is said to act *effectively* on $\mathcal{M}$; in this case no element except the identity leaves every point fixed. For example, the group E of translations and rotations of the plane acts effectively on the plane; the subgroup of translations acts freely. In fact our definition of an affine space in Chapter 1 may be paraphrased thus: an affine space is a space on which a vector space (considered as a group) acts freely and transitively.

Exercise 11. Show that every element of E, acting on the plane, which is not a translation or the identity has a unique fixed point. □

Let K be a subgroup of a group G. By restriction of the multiplication in G one obtains a map $K \times G \to G$ which is a left action of K on G. The orbits of G under this action are called *right cosets* of K in G; they are the sets $Kg = \{ kg \mid k \in K \}$ (unfortunately, some authors call them left cosets, so one should always check the definition). Since G is partitioned by the action of K, each element of G belongs to exactly one right coset of the subgroup K. The subgroup K also acts on G by restriction of the multiplication $G \times K \to G$, which is a right action of K on G; its orbits gK are called *left cosets* of K in G.

Exercise 12. Let $\mathcal{M}$ be a left or right G-space. Show that the set of elements of G taking a chosen point x to a chosen point on its orbit is a left or right coset of G_x. □

If $\mathcal{N}$ is a homogeneous space of G, and x a chosen point of $\mathcal{N}$, then every other point of $\mathcal{N}$ is on the orbit of x, and so to every other point there is, by the exercise, a corresponding coset of G_x. Thus there is a $1:1$ correspondence between points of $\mathcal{N}$ and cosets of G_x, and thus any homogeneous G-space may be identified with a space of cosets of G_x—a left G-space with the space of left cosets G/G_x, a right G-space with the space of right cosets $G_x\backslash G$. For example, if G is the rotation group $SO(3)$ and $\mathcal{N}$ is a sphere S^2 with centre at the origin then, for any $x \in \mathcal{N}$, G_x is the group of rotations about the radius vector through x, which is the group S^1. Thus $S^2 = SO(3)/S^1$.

Actions of a group on itself. The actions of a group on itself are of great importance. Three of them will be defined here; they are used repeatedly in the rest of this chapter. Taking $\mathcal{M}$ to be G itself in the above construction one obtains: *left translation* $L_g: G \to G$ by $h \mapsto gh$ for all $h \in G$; and *right translation* $R_g: G \to G$ by $h \mapsto hg$ for all $h \in G$. Besides these, *inner automorphism* or *conjugation* is the action of G on itself defined by $I_g = L_g \circ R_{g^{-1}}$, so that $I_g(h) = ghg^{-1}$.

Exercise 13. Show that left (right) translation is a free transitive left (right) action of G on itself. □

Exercise 14. Show that $g \mapsto R_{g^{-1}}$ is a free transitive left action of G on itself. □

Exercise 15. Show that inner automorphism is a left action of G on itself. Describe the isotropy group of any element of G under this action, and in particular the isotropy group of the identity. Show that this action is never transitive or free, for non-trivial G. □

An *automorphism* of a group G is an isomorphism $G \to G$.

Exercise 16. Show that the automorphisms of G form a group under composition. □

This group is denoted $\operatorname{aut} G$. An automorphism of a Lie group must be a diffeomorphism. It is known that if G is a connected Lie group then $\operatorname{aut} G$ is itself a Lie group.

Exercise 17. Show that for any $g \in G$ the conjugation I_g is an automorphism of G (thus justifying the alternative name "inner automorphism"). Show that the map $G \to \operatorname{aut} G$ by $g \mapsto I_g$ is a homomorphism. Show that the inner automorphisms form a normal subgroup of $\operatorname{aut} G$. □

Exercise 18. Let $\psi: G \to H$ be a group homomorphism. Show that for each $g \in G$, $\psi \circ L_g = L_{\psi(g)} \circ \psi$, $\psi \circ R_g = R_{\psi(g)} \circ \psi$ and $\psi \circ I_g = I_{\psi(g)} \circ \psi$. □

Exercise 19. Let Ψ^a be the multiplication functions for a Lie group G. Show that the coordinate presentation of left translation by a fixed element g of G is $\eta^a \mapsto \Psi^a(\xi^b, \eta^c)$, where $\xi^a = x^a(g)$. □

5. The Lie Algebra of a Lie Group

The elegant structure of Lie groups does not become fully apparent until actions of the group on itself are extended to actions on tangent vectors. This structure was discovered through the study of neighbourhoods of the identity in transformation groups, but the approach to be described here, which has been developed during the last half century, is more direct and perhaps simpler. The general idea is to construct a vector field on the group, from a vector assigned at one point, by some transitive action of the group on itself. It is customary to choose left translation for this action.

Let G be a Lie group, and let L_g denote left translation by the element $g \in G$, as before. Each map L_g is a diffeomorphism of the manifold G to itself; let L_{g*} denote the induced map of tangent spaces in the usual way. Because $g \mapsto L_g$ is an action, L_{g*} has the property $L_{g*} \circ L_{h*} = (L_g \circ L_h)_* = L_{gh*}$. Now assign any tangent vector X_e at the identity element e of G. Define a tangent vector X_g at g by left translation: $X_g = L_{g*}X_e$. A tangent vector may be defined in this way at every point of G, and the result is actually a smooth vector field, because of the smoothness of group multiplication. Moreover, by the property of L_g just stated,

$$L_{g*}X_h = L_{g*}L_{h*}X_e = L_{gh*}X_e = X_{gh}.$$

This says that if you left translate X_e from e first to h, then from there to gh, you get the same result at gh as if you translated there directly. In other words, the vector field X defined by this construction is taken into itself by any left translation: thus $L_{g*}X = X$.

A vector field on a Lie group which is taken into itself by left translations is called *left-invariant*. Conversely, any left-invariant vector field X may be reconstructed by the method above from X_e, its value at e, since $X_g = L_{g*}X_e$ by virtue of its left invariance.

If X and Y are left-invariant vector fields on a Lie group G, so is $kX + lY$ where $k, l \in \mathbf{R}$, because of the linearity of L_{g*}. The left-invariant vector fields on G therefore constitute a vector space, which is denoted $\mathcal{G}$. Since a left-invariant vector field is determined completely by its value at e, or indeed at any point of G, the dimension of $\mathcal{G}$ is the same as the dimension of any tangent space to G, which is just the dimension of G itself. It is often convenient to identify $\mathcal{G}$ with T_eG, the tangent space at the identity of G, by the correspondence between left-invariant vector fields and their values at e.

Since the brackets of L_g-related vector fields are L_g-related, if X and Y are left-invariant then $L_{g*}[X, Y] = [L_{g*}X, L_{g*}Y] = [X, Y]$, so that the bracket of left-invariant vector fields is left-invariant. It follows that the left-invariant vector fields on a Lie group G form a finite-dimensional vector space equipped with a bilinear skew-symmetric product or bracket operation which satisfies the Jacobi identity. Such a structure is called a *Lie algebra*. The Lie algebra $\mathcal{G}$ is called the Lie algebra of G. We have discussed Lie algebras before in the context of groups of transformations: for example, Lie algebras of infinitesimal isometries in Chapter 8. The present construction shows how the Lie algebra of a Lie group may be defined in terms of the group itself.

The bracket may also be defined on T_eG, by $[X_e, Y_e] = [X, Y]_e$, where X, Y are the left-invariant vector fields determined by X_e, Y_e. If $\{X_a\}$ is a basis for $\mathcal{G}$ then $[X_b, X_c]$ must be a linear combination of basis vectors, say $[X_b, X_c] = C^a_{bc}X_a$, where the C^a_{bc} are numbers. These numbers are called the ***structure constants*** of $\mathcal{G}$, in the chosen basis. We shall show eventually that these numbers are the same as those defined in Section 3, immediately before Exercise 6.

Exercise 20. Show that under change of basis the structure constants transform as a tensor of the indicated type. □

Exercise 21. Show from the skew-symmetry of the bracket that
$$C^a_{cb} = -C^a_{bc}$$
and from the Jacobi identity that
$$C^a_{eb}C^e_{cd} + C^a_{ec}C^e_{db} + C^a_{ed}C^e_{bc} = 0. \quad □$$

Exercise 22. Let E be the group of orientation-preserving isometries of the plane, described in Section 1, and let $(\xi^1, \xi^2, \vartheta)$ be the coordinates for E introduced there. Construct the left-invariant vector fields whose values at the identity are $\partial/\partial\xi^1$, $\partial/\partial\xi^2$ and $\partial/\partial\vartheta$ respectively, and calculate the structure constants with these vector fields as basis for the Lie algebra $\mathcal{E}$. □

The Lie algebra of a matrix group. It is useful to be able to compute the Lie algebra explicitly for a matrix group. The key to such computations is the identification of the tangent space at any point of a matrix group with a vector space of matrices. First let G be $GL(n,\mathbf{R})$ and let (x^a_b) be the coordinate functions on G, described earlier. Let γ be any curve in G, with coordinate expression $\gamma^a_b(t) = x^a_b(\gamma(t))$, and set $\gamma(0) = g$. If f is any function on G, then

$$\frac{d}{dt}(f \circ \gamma)(0) = \frac{\partial f^x}{\partial x^a_b}\dot{\gamma}^a_b(0),$$

the partial derivatives being evaluated at $(x^a_b(g))$. The $\dot{\gamma}^a_b(0)$ are merely numbers, which form an $n \times n$ array: in other words, there is an $n \times n$ matrix, A, an element of $M_n(\mathbf{R})$, such that $A^a_b = \dot{\gamma}^a_b(0)$. Thus every tangent vector to $GL(n,\mathbf{R})$ at g may be written in coordinates $A^a_b\partial/\partial x^a_b$ for some $A \in M_n(\mathbf{R})$. To check that the whole of $M_n(\mathbf{R})$ may be reached in this way, one has merely to observe that if g is a non-singular matrix and A any matrix then $g + tA$ is non-singular for small enough $|t|$, and that the curve $t \mapsto g + tA$ in $GL(n,\mathbf{R})$ has tangent vector $A^a_b\partial/\partial x^a_b$ at g. This argument shows that the tangent space to $GL(n,\mathbf{R})$ at any point may be identified with $M_n(\mathbf{R})$. We shall use this result most often when the point in question is the unit matrix, the identity of the group.

Now let G be a Lie group of matrices, in other words a Lie subgroup of $GL(n,\mathbf{R})$ for some n. If the curve γ lies in G, then its tangent vectors must be tangent to G. We may regard G as an (immersed) submanifold of $GL(n,\mathbf{R})$, so that its tangent space at any one point is a subspace of the tangent space to $GL(n,\mathbf{R})$. Thus the tangent spaces to G may be identified as vector subspaces of $M_n(\mathbf{R})$. It is not difficult to find conditions determining these subspaces from the conditions satisfied by elements of G. For example, if G is $O(n)$ or $SO(n)$ then $\gamma(t)\gamma(t)^T = I_n$ for each t. Differentiation with respect to t yields $\dot{\gamma}(t)\gamma(t)^T + \gamma(t)\dot{\gamma}(t)^T = 0$, so that if γ is a curve at the identity, with $\gamma(0) = I_n$, then $\dot{\gamma}(0) + \dot{\gamma}(0)^T = 0$; the tangent space to $O(n)$ or $SO(n)$ at the identity may therefore be identified with the vector space of antisymmetric $n \times n$ matrices.

Exercise 23. Show that the tangent space to $U(n)$ at the identity comprises all skew-Hermitian matrices, that is, all complex $n \times n$ matrices A satisfying $A + A^\dagger = 0$. □

Exercise 24. If γ is a curve in $SL(n,\mathbf{R})$ then $\det\gamma(t) = 1$ for each t. Using the formula for the derivative of a determinant show that $0 = d/dt(\det\gamma(t))(0) = \operatorname{tr}(\dot{\gamma}(0)\gamma(0)^{-1})$, and infer that the tangent space to $SL(n,\mathbf{R})$ at the identity comprises all trace-free $n \times n$ matrices. □

The identification of the Lie algebra $\mathcal{G}$ with T_eG bears fruit if one computes the commutator of Lie algebra elements. To do this it is first necessary to compute the left-invariant vector fields.

Let h be any point of $GL(n,\mathbf{R})$, γ a curve at h, $\dot{\gamma}(0)$ the tangent vector to γ at h, and f any function on the group. For any $g \in GL(n,\mathbf{R})$,

$$\frac{d}{dt}f(g\gamma(t))_{t=0} = \frac{\partial f^x}{\partial x^a_b}\frac{d}{dt}(x^a_c(g)\gamma^c_b(t))_{t=0} = x^a_c(g)\dot{\gamma}^c_b(0)\frac{\partial f^x}{\partial x^a_b},$$

the partial derivatives being evaluated at $x^a_b(gh)$. The first expression gives precisely the directional derivative of f along the vector $L_{g*}\dot{\gamma}(0)$, and so from the last expression

$$L_{g*}\left(\dot{\gamma}^a_b(0)\frac{\partial}{\partial x^a_b}\right) = x^a_c(g)\dot{\gamma}^c_b(0)\frac{\partial}{\partial x^a_b}.$$

It follows that if the tangent spaces at h and gh are both identified with $M_n(\mathbf{R})$ then the linear map L_{g*} is represented by left matrix multiplication by g. Now to any matrix (X^a_b) there corresponds a tangent vector $X^a_b\partial/\partial x^a_b$ to $GL(n,\mathbf{R})$ at the identity. This tangent vector determines a left-invariant vector field X given by

$$X_g = L_{g*}\left(X^a_b\frac{\partial}{\partial x^a_b}\right) = x^a_c(g)X^c_b\frac{\partial}{\partial x^a_b}$$

or equivalently

$$X = x^a_c X^c_b \frac{\partial}{\partial x^a_b}.$$

This is the form which any left-invariant vector field takes when it is expressed in terms of standard coordinates on $GL(n,\mathbf{R})$.

Exercise 25. Show that if $X = x^a_c X^c_b \partial/\partial x^a_b$ and $Y = x^a_c Y^c_b \partial/\partial x^a_b$ are two left-invariant vector fields on $GL(n,\mathbf{R})$ then

$$[X,Y] = x^a_c(X^c_d Y^d_b - Y^c_d X^d_b)\partial/\partial x^a_b. \qquad \square$$

Thus the matrix corresponding to $[X,Y]$ is just the commutator of the matrices corresponding to X and Y. In this way the Lie algebra of $GL(n,\mathbf{R})$ may be identified with $M_n(\mathbf{R})$ equipped with the bracket operation of the commutator of matrices.

Exercise 26. A basis for the Lie algebra of $GL(n,\mathbf{R})$ comprises matrices E^q_p with entry unity in the pth row and qth column and all other entries zero. Show that $[E^q_p, E^s_r] = \delta^q_r E^s_p - \delta^s_p E^q_r$. $\square$

Consider now a Lie group G of matrices, which is a Lie subgroup of $GL(n,\mathbf{R})$. If X is a left-invariant vector field on $GL(n,\mathbf{R})$ such that $X_e \in T_eG$, then the restriction of X to G, regarded as a submanifold of $GL(n,\mathbf{R})$, is everywhere tangent to G, and is left-invariant under the action of G. Every left-invariant vector field on G may be regarded in this way. Since the bracket of two vector fields tangent to a submanifold is also tangent to the same submanifold it follows that the bracket operation in the Lie algebra $\mathcal{G}$ of G responds to the commutator of matrices exactly as before. Thus $\mathcal{G}$, when identified with a subspace of $M_n(\mathbf{R})$ *via* T_eG, must be closed under formation of commutators.

Exercise 27. Confirm that the space of antisymmetric $n \times n$ matrices (the Lie algebra of $O(n)$), the space of skew-Hermitian $n \times n$ complex matrices (the Lie algebra of $U(n)$) and the space of trace-free $n \times n$ matrices (the Lie algebra of $SL(n,\mathbf{R})$) are all closed under formation of commutators. $\square$

Exercise 28. Show that, just as the map of $M_n(\mathbf{R})$ induced by left translation in $GL(n,\mathbf{R})$ corresponds to left matrix multiplication, so the map induced by right translation corresponds to right matrix multiplication and the map induced by inner automorphism corresponds to matrix conjugation, for any matrix Lie group. $\square$

6. Left-invariant Forms

Not surprisingly, the left-invariant forms on a Lie group constitute the vector space dual to the vector space of left-invariant vector fields. The formulae have only to be adapted to the fact that maps of cotangent spaces are contragredient to the maps of manifolds which induce them.

Let G be any Lie group, and let L_g denote left translation by $g \in G$, as before. Then $L_g{}^*$ pulls covectors back from gh to h, say, and to map covectors from h to gh one must use $L_{g^{-1}}{}^*$. Let α_e denote a covector at e, and define $\alpha_g = L_{g^{-1}}{}^*\alpha_e$. A cotangent vector is defined in this way at every point of G, and the result is a smooth 1-form.

Exercise 29. Show that, with α thus defined, $\alpha_{gh} = L_{g^{-1}}{}^*\alpha_h$ and that therefore $L_g{}^*\alpha = \alpha$ for all $g, h \in G$. □

The 1-form α is a *left-invariant 1-form* on G.

Exercise 30. Show that if X is a left-invariant vector field and α a left-invariant 1-form on G then $\langle X_g, \alpha_g \rangle = \langle X_e, \alpha_e \rangle$ for all $g \in G$. □

Since $L_g{}^*$ is linear, left-invariant 1-forms on G constitute a vector space. So in fact do *left-invariant p-forms*, which satisfy $L_g{}^*\omega = \omega$ for all $g \in G$. It follows from the last exercise that if the vector field X and the 1-form α are left-invariant then $\langle X, \alpha \rangle$ is constant on G. Consequently the vector space of left-invariant 1-forms may be identified with the dual $\mathcal{G}^*$ of the Lie algebra $\mathcal{G}$ (considered as a vector space).

If $X, Y \in \mathcal{G}$ and $\alpha \in \mathcal{G}^*$ then the formula for the exterior derivative of a 1-form yields

$$d\alpha(X, Y) = -\langle [X, Y], \alpha \rangle,$$

when one takes account of the constancy of $\langle X, \alpha \rangle$ and $\langle Y, \alpha \rangle$. In particular, if $\{X_a\}$ is a basis for $\mathcal{G}$ and $\{\alpha^a\}$ the dual basis for $\mathcal{G}^*$ then

$$d\alpha^a(X_b, X_c) = -C_{bc}^d \langle X_d, \alpha^a \rangle = -C_{bc}^a,$$

where the C_{bc}^a are the structure constants for $\mathcal{G}$.

Exercise 31. Show that if α is a left-invariant 1-form on G, then $d\alpha$ is a left-invariant 2-form. Show that if $\{\alpha^a\}$ is a basis for $\mathcal{G}^*$ then

$$d\alpha^a = -\tfrac{1}{2} C_{bc}^a \alpha^b \wedge \alpha^c.$$ □

These equations are called the *Maurer-Cartan equations* for G.

Exercise 32. Show that the 1-forms α_b^a on $GL(n, \mathbf{R})$ given by $\alpha_b^a|_g = x_c^a(g^{-1}) dx_b^c|_g$ constitute a basis for left-invariant 1-forms on this group. □

These 1-forms are often combined into a single matrix-valued 1-form, written $g^{-1}dg$.

7. One-parameter Subgroups

A *one-parameter subgroup* of a Lie group G is a smooth homomorphism of Lie groups $\phi: \mathbf{R} \to G$ where $\mathbf{R}$ is the real line with its additive Lie group structure (the nomenclature survives from the time when coordinates on Lie groups were called parameters—"one-dimensional subgroup" would serve just as well). Thus ϕ is a smooth curve in G such that

$$\phi(s + t) = \phi(s)\phi(t) \qquad \text{for all } s, t \in \mathbf{R}$$
$$\phi(0) = e.$$

Exercise 33. Show that rotations about a fixed axis, parametrised by angle of rotation, constitute a one-parameter subgroup of $SO(3)$. □

Exercise 34. Show that if $\mathcal{V}$ is a vector space, regarded as a (commutative) Lie group, then for any $v \in \mathcal{V}$ the map $t \mapsto tv$ is a one-parameter subgroup of $\mathcal{V}$, and that every one-parameter subgroup of $\mathcal{V}$ is of this form. □

If $\mathcal{M}$ is a manifold on which G acts to the left, say, and if ϕ is a one-parameter subgroup of G, then ϕ defines a one-parameter group of diffeomorphisms of $\mathcal{M}$ by $\phi_t(x) = \phi(t)x$, and the ideas of Chapters 3 and 10 may again be put to use. In particular the actions of G on itself may be combined with one-parameter subgroups of G: if ϕ is a one-parameter subgroup of G, then $R_{\phi(t)}$, $L_{\phi(t)}$ and $I_{\phi(t)}$ are all one-parameter groups of diffeomorphisms of G. Moreover, left and right translations commute, so that for any $g \in G$, $L_g \circ R_{\phi(t)} = R_{\phi(t)} \circ L_g$. It follows that the vector field X which is the infinitesimal generator of the one-parameter group of diffeomorphisms $R_{\phi(t)}$ satisfies $L_{g*}X = X$ for every g: in other words, it is a left-invariant vector field.

Exercise 35. Show that the generators of left translations are right-invariant. □

Conversely, any left-invariant vector field on G generates a one-parameter group of right translations. Let X be a left-invariant vector field and ϕ its integral curve at e, so that $\phi(0) = e$. The smoothness of X ensures that ϕ is defined on some interval of $\mathbf{R}$ containing 0, say $|t| < \varepsilon$. The group structure of G then makes it possible to extend ϕ indefinitely, as follows. Apply the diffeomorphism L_g to everything in sight: $t \mapsto g\phi(t)$ is the integral curve of $L_{g*}X$ at g. But X is left-invariant, so this actually gives the integral curve of X at g. Thus left translation of ϕ yields bits of integral curves of X all over G. By piecing these bits together one may extend each integral curve indefinitely: choose $g = \phi(s)$, $|s| < \varepsilon$, and then the integral curve through e may be extended to all $|t| < 2\varepsilon$, and so on.

Now $t \mapsto \phi(s+t)$ is the integral curve of X at $\phi(s)$, by the congruence property of integral curves; but this integral curve is $t \mapsto \phi(s)\phi(t)$ by left-invariance, so $\phi(s+t) = \phi(s)\phi(t)$ and ϕ is a one-parameter subgroup of G. Moreover, since the integral curve of X through g is $t \mapsto g\phi(t) = R_{\phi(t)}g$, X generates the one-parameter group of right translations $R_{\phi(t)}$. Since X_e determines the vector field X, it also determines ϕ, and conversely: $\dot{\phi}(0) = X_e$.

If X is a left-invariant vector field so is rX, for any constant r. Multiplication of X by a constant factor does not change its integral curves as point sets, it only reparametrises them. The relation between the one-parameter subgroups corresponding to rX and X is given by $\phi_{rX}(t) = \phi_X(rt)$, where the subscript indicates the generator (of the one-parameter group of right translations).

Exercise 36. Show that if X is a left-invariant vector field and Y a right-invariant vector field then $[X,Y] = 0$. □

If G is a matrix group and X a left-invariant vector field on G, and if ϕ is an integral curve of X, then ϕ must satisfy $d\phi/dt = \phi A$ where A is the matrix corresponding to X_e under the identification of T_eG with a space of matrices. Thus, differentiating repeatedly, $d^k\phi/dt^k = \phi A^k$. If ϕ is the one-parameter subgroup corresponding to X then $\phi(0) = I_n$, so that

$$\phi(t) = I_n + tA + \frac{1}{2!}t^2A^2 + \cdots$$

which is just the matrix exponential $\exp(tA)$.

Having got this far one is not hard put to it to recognise that the exponential can be defined whether G is a matrix group or not; one has only to get the domain right.

8. The Exponential Map

Let G be a Lie group and $\mathcal{G}$ its Lie algebra. The *exponential map* $\exp: \mathcal{G} \to G$ is given by

$$\exp X = \phi_X(1)$$

where X is any element of $\mathcal{G}$ and ϕ_X is the one-parameter subgroup of G with tangent vector X_e at e.

Interchanging r and t in the formula for change of parametrisation and then setting $r = 1$ one finds that

$$\exp tX = \phi_{tX}(1) = \phi_X(t),$$

and therefore $\exp tX$ is the one-parameter subgroup of G whose action by right translation is generated by the left invariant vector field X. Moreover, $\exp(s+t)X = \exp sX \exp tX$; in particular, $\exp 0 = e$ and $\exp(-X) = (\exp X)^{-1}$.

The tangent space at 0 to the vector space $\mathcal{G}$ may be identified with $\mathcal{G}$ itself, and the tangent space at e to the group G may also be identified with $\mathcal{G}$, so that the map $\exp_*$ of vectors induced by exp at 0 is a map $\mathcal{G} \to \mathcal{G}$. To compute $\exp_* X$ one may take any curve in $\mathcal{G}$ whose tangent vector at 0 is X and find the tangent vector at e to the image of this curve in G. The ray $t \mapsto tX$ is a suitable curve in $\mathcal{G}$; its image in G is $t \mapsto \exp tX$, whose tangent vector at e is again X. Thus at 0 in $\mathcal{G}$, $\exp_*$ is the identity map. It follows from the inverse function theorem that exp is a diffeomorphism of some open neighbourhood of 0 in $\mathcal{G}$ onto some open neighbourhood of e in G.

This is reminiscent of the exponential map associated with a connection, as defined in Chapter 11. In fact one may define a connection on a Lie group with the aid of left translation, by using $L_{hg^{-1}*}: T_gG \to T_hG$ to define parallel translation. This parallel translation is path independent, so one has in fact defined a complete parallelism; the corresponding connection therefore has zero curvature, but in general it is not a symmetric connection: the components of its torsion tensor with respect to a basis of left-invariant vector fields are the structure constants of the Lie algebra. Now any integral curve of a left-invariant vector field has parallel tangent vectors, according to this connection, and is therefore a geodesic. Thus the exponential map defined above is just the exponential map of the connection based at the identity.

We may define normal coordinates on the group, in a neighbourhood of the identity, by first choosing a basis, say $\{X_a\}$, for $\mathcal{G}$. Then for any g close enough to the identity we may write $g = \exp X$ for some $X \in \mathcal{G}$, and the normal coordinates $x^a(g)$ are given by $x^a(g) = \xi^a$ where $X = \xi^a X_a$. Note that with respect to normal coordinates $x^a(g^{-1}) = -x^a(g)$, since $(\exp X)^{-1} = \exp(-X)$. Thus normal coordinates have the special property described in the paragraph following Exercise 5. Note also that if one expands the multiplication functions Ψ^a for normal coordinates in series, in the form (see Exercise 3 and the following remarks)

$$\Psi^a(\xi^d, \eta^e) = \xi^a + \eta^a + \alpha^a_{bc}\xi^b\eta^c + \cdots$$

then the array of coefficients α^a_{bc} must be skew-symmetric in b and c. This follows from the one-parameter subgroup property $\exp sX \exp tX = \exp(s+t)X$: for if $X = \xi^a X_a$ then

$$\begin{aligned} x^a(\exp sX \exp tX) &= \Psi^a(s\xi^b, t\xi^c) \\ &= s\xi^a + t\xi^a + st\alpha^a_{bc}\xi^b\xi^c \cdots \\ &= x^a\big(\exp(s+t)X\big) = (s+t)\xi^a, \end{aligned}$$

whence $\alpha^a_{bc}\xi^b\xi^c = 0$ for every (ξ^a). That the α^a_{bc} are actually the structure constants is the content of Exercise 38 below.

Exercise 37. Let $\mathcal{V}$ be a vector space, considered as additive group G, so that $\mathcal{G}$, as tangent space at 0, may be identified with $\mathcal{V}$. Verify that exp is a diffeomorphism of $\mathcal{V}$ onto $\mathcal{V}$. □

The result of this exercise holds for some other groups, but the exponential map for the circle group $t \mapsto e^{it}$ shows that the exponential map need not be 1 : 1, and it may not even be surjective, as may be seen from the following example.

Exponentiation in SL(2,R). The group $SL(2,\mathbf{R})$ of 2×2 matrices with real entries and unit determinant is a 3-dimensional Lie group. Its Lie algebra $\mathcal{SL}(2,\mathbf{R})$ comprises 2×2 matrices with trace zero. If $X \in \mathcal{SL}(2,\mathbf{R})$ then $\exp tX$ is the solution to the equation $d/dt(\exp tX) = (\exp tX)X$ with $\exp 0 = I_2$. Differentiating again, one sees that $\exp tX$ is the solution to $d^2/dt^2(\exp tX) = (\exp tX)X^2$ with $\exp 0 = I_2$ and $d/dt\big(\exp tX\big)(0) = X$. This second order equation is easy to solve because if

$$X = \begin{pmatrix} u & v \\ w & -u \end{pmatrix}$$

then

$$X^2 = \begin{pmatrix} u^2 + vw & 0 \\ 0 & u^2 + vw \end{pmatrix} = (u^2 + vw)I_2.$$

There are three cases, depending on the value of $(u^2 + vw)$.

(1) $u^2 + vw > 0$: then

$$\exp tX = \begin{pmatrix} \cosh \rho t + \frac{u}{\rho}\sinh \rho t & \frac{v}{\rho}\sinh \rho t \\ \frac{w}{\rho}\sinh \rho t & \cosh \rho t - \frac{u}{\rho}\sinh \rho t \end{pmatrix}$$

where $\rho = \sqrt{u^2 + vw}$. Call this the hyperbolic case.

(2) $u^2 + vw = 0$: then

$$\exp tX = \begin{pmatrix} 1 + tu & tv \\ tw & 1 - tu \end{pmatrix}.$$

Call this the parabolic case.

(3) $u^2 + vw < 0$: then

$$\exp tX = \begin{pmatrix} \cos\rho t + \frac{u}{\rho}\sin\rho t & \frac{v}{\rho}\sin\rho t \\ \frac{w}{\rho}\sin\rho t & \cos\rho t - \frac{u}{\rho}\sin\rho t \end{pmatrix}$$

where $\rho = \sqrt{-(u^2 + vw)}$. Call this the elliptic case.

Now observe that

$$\begin{array}{ll} \text{in the hyperbolic case,} & \operatorname{tr}(\exp tX) \geq 2 \\ \text{in the parabolic case,} & \operatorname{tr}(\exp tX) = 2 \\ \text{in the elliptic case,} & -2 \leq \operatorname{tr}(\exp tX) \leq 2. \end{array}$$

Thus any element of $SL(2,\mathbf{R})$ which is an exponential must have trace at least -2. However, there are elements of $SL(2,\mathbf{R})$ with trace less than -2, for example

$$\begin{pmatrix} -2 & 0 \\ 0 & -\frac{1}{2} \end{pmatrix}.$$

The exponential map is therefore not surjective in this group.

9. Homomorphisms of Lie Groups and their Lie Algebras

The idea of the exponential map may be used to simplify results about homomorphisms. In the first place, if $\psi: G \to H$ is a homomorphism of Lie groups and $\exp tX$ is a one-parameter subgroup of G then $\psi(\exp tX)$ is a one-parameter subgroup of H. Moreover,

$$\psi(R_{\exp tX}g) = \psi(g \exp tX) = \psi(g)\psi(\exp tX) = R_{\psi(\exp tX)}\psi(g)$$

so that

$$\psi \circ R_{\exp tX} = R_{\psi(\exp tX)} \circ \psi.$$

Consequently the generator of right translations by $\psi(\exp tX)$ is ψ-related to X. We shall denote by $\psi_* X$ the generator of right translations by $\psi(\exp tX)$. Then ψ_* is a map of left-invariant vector fields on G to left-invariant vector fields on H. The vector field $\psi_* X$ could equally well be defined as the left-invariant vector field on H whose value at the identity is $\psi_{*e} X_e$. Either way, we have defined a linear map $\psi_*: \mathcal{G} \to \mathcal{H}$ (as vector spaces).

Now ψ- relatedness preserves brackets, so that $[\psi_* X, \psi_* Y] = \psi_*[X,Y]$. Thus $\psi_*: \mathcal{G} \to \mathcal{H}$ is a Lie algebra homomorphism. Furthermore, $\psi(\exp tX) = \exp(t\psi_* X)$, and this for any $X \in \mathcal{G}$, so that

$$\psi \circ \exp = \exp \circ \psi_*.$$

Of course the exp on the left acts in $\mathcal{G}$, that on the right in $\mathcal{H}$. This formula may be remembered wth the help of the diagram

$$\begin{array}{ccc} \mathcal{G} & \xrightarrow{\psi_*} & \mathcal{H} \\ \downarrow \exp & & \downarrow \exp \\ G & \xrightarrow{\psi} & H \end{array}$$

The adjoint representation. If $\psi: G \to H$ is an isomorphism of Lie groups then the induced map $\psi_*: \mathcal{G} \to \mathcal{H}$ is an isomorphism of Lie algebras; and in particular if ψ is an automorphism of a Lie group G then ψ_* is an automorphism of its Lie algebra. An important application of this result is obtained by specialising to the case of the inner automorphisms of G, that is, to the conjugation maps $I_g: g' \mapsto gg'g^{-1}$. The automorphism of $\mathcal{G}$ induced by I_g is denoted $\operatorname{ad} g$ so that for $X \in \mathcal{G}$

$$\operatorname{ad} g(X) = I_{g*}X.$$

It follows from the general theory that

$$\exp(\operatorname{ad} g(X)) = I_g(\exp X) = g \exp X g^{-1}.$$

Since $g \mapsto I_g$ is a homomorphism $G \to \operatorname{aut} G$ (Exercise 17), $I_{g*}I_{h*} = I_{gh*}$, or

$$\operatorname{ad} g \operatorname{ad} h = \operatorname{ad} gh.$$

Thus ad is a left action of G on $\mathcal{G}$. Moreover, for each $g \in G$, $\operatorname{ad} g$ is an automorphism of $\mathcal{G}$; that is to say, it is a non-singular linear transformation of $\mathcal{G}$ and satisfies

$$[\operatorname{ad} g(X), \operatorname{ad} g(Y)] = \operatorname{ad} g([X, Y]).$$

We summarise these properties by saying that ad is a representation of G on $\mathcal{G}$: it is called the *adjoint representation*.

In the case of a Lie group G of matrices, when $\mathcal{G}$ is identified with a space of matrices, ad corresponds simply to matrix conjugation: $\operatorname{ad} g(X) = gXg^{-1}$.

Using the expression of I_g as $L_g \circ R_{g^{-1}}$, or equivalently $R_{g^{-1}} \circ L_g$ (since right and left translations commute), we deduce that for a left-invariant vector field Y

$$\operatorname{ad} g(Y) = I_{g*}Y = R_{g^{-1}*}L_{g*}Y = R_{g^{-1}*}Y.$$

We shall use this important formula in a later chapter; we also draw an interesting deduction from it here. Suppose that X is another left-invariant vector field. Then X is the generator of the one-parameter group $R_{\exp tX}$. It follows from the result above that

$$\operatorname{ad}(\exp tX)(Y) = R_{(\exp tX)^{-1}*}Y = R_{\exp(-tX)*}Y.$$

On differentiating the right hand expression with respect to t and setting $t = 0$ one obtains the Lie derivative $\mathcal{L}_X Y$, or $[X, Y]$: thus

$$\frac{d}{dt}\left(\operatorname{ad}(\exp tX)(Y)\right)_{t=0} = [X, Y].$$

Exercise 38. Show that with respect to normal coordinates,

$$\left(\operatorname{ad}(\exp tX)(Y)\right)^a = \eta^a + t\alpha^a_{bc}\xi^b\eta^c + \cdots$$

where $X = \xi^a X_a$ and $Y = \eta^a X_a$, and the α^a_{bc} are the coefficients in the multiplication functions; deduce that $\alpha^a_{bc} = C^a_{bc}$, the structure constants. □

This result has a nice interpretation in terms of homomorphisms of Lie groups and their Lie algebras. The group of automorphisms of a Lie algebra $\mathcal{G}$ is itself a Lie group, which we write $\operatorname{aut} \mathcal{G}$. Then $\operatorname{ad}: G \to \operatorname{aut} \mathcal{G}$ is a homomorphism of Lie groups. The Lie algebra $\mathcal{A}(\mathcal{G})$ of aut$\mathcal{G}$ is the space of *derivations* of $\mathcal{G}$, that is of linear maps $D: \mathcal{G} \to \mathcal{G}$ such that

$$D[Y,Z] = [DY,Z] + [Y,DZ].$$

Now for any element X of $\mathcal{G}$ the map $\mathcal{G} \to \mathcal{G}$ by $Y \mapsto [X,Y]$ is a derivation of $\mathcal{G}$, by Jacobi's identity:

$$[X,[Y,Z]] = [[X,Y],Z] + [Y,[X,Z]].$$

From the general theory it follows that the homomorphism ad: $G \to$ aut $\mathcal{G}$ induces a homomorphism $\mathrm{ad}_*: \mathcal{G} \to \mathcal{A}(\mathcal{G})$, and from the result above it is clear what this homomorphism is:

$$\mathrm{ad}_*(X)(Y) = \frac{d}{dt}\big(\mathrm{ad}(\exp tX)(Y)\big)(0) = [X,Y].$$

Thus ad_* is the map which associates with each element of $\mathcal{G}$ the derivation of $\mathcal{G}$ which it defines.

The adjoint representation for SO(3). The Lie algebra $\mathcal{SO}(3)$ of the rotation group consists of 3×3 skew-symmetric matrices. A convenient basis is given by the matrices

$$X_1 = \begin{pmatrix} 0 & 0 & 0 \\ 0 & 0 & -1 \\ 0 & 1 & 0 \end{pmatrix} \quad X_2 = \begin{pmatrix} 0 & 0 & 1 \\ 0 & 0 & 0 \\ -1 & 0 & 0 \end{pmatrix} \quad X_3 = \begin{pmatrix} 0 & -1 & 0 \\ 1 & 0 & 0 \\ 0 & 0 & 0 \end{pmatrix}$$

so that X_a is the matrix whose (b,c) entry is $-\epsilon_{abc}$. The exponentials of these Lie algebra elements are the one-parameter subgroups of rotations about the coordinate axes. Their brackets are given by

$$[X_1,X_2] = X_3 \qquad [X_2,X_3] = X_1 \qquad [X_3,X_1] = X_2.$$

Now $\mathcal{SO}(3)$ is 3-dimensional, and in many constructions it is commonly identified (or confused) with the space $\mathcal{E}^3$ on which $SO(3)$ normally acts. We shall explain the extent to which this is justified in the case of the adjoint representation.

With each element $X = \xi^a X_a \in \mathcal{SO}(3)$ one associates (as in Chapter 8, Section 5) the point $\xi = (\xi^a) \in \mathcal{E}^3$. Thus the correspondence associates (ξ^a) with the matrix $(-\xi^a \epsilon_{abc})$. (It is convenient here to keep all matrix indices in the lower position, which avoids the necessity of inserting numerous Kronecker deltas.) Then

$$[\xi^a X_a, \eta^b X_b] = (\xi^2\eta^3 - \xi^3\eta^2)X_1 + (\xi^3\eta^1 - \xi^1\eta^3)X_2 + (\xi^1\eta^2 - \xi^2\eta^1)X_3.$$

Thus the bracket of Lie algebra elements corresponding to (ξ^a) and (η^a) is the element corresponding to their cross product.

In order to work out the adjoint representation we need the result of the following exercise. (The summation convention still applies for repeated indices even though both may be in lower position.)

Exercise 39. Show that if $g = (g_{ab}) \in O(3)$ then $\epsilon_{ade} g_{bd} g_{ce} = \epsilon_{fbc} g_{fa} \det g$. □

Now for any $g \in SO(3)$ (so that $\det g = 1$) we have with the aid of the exercise

$$\begin{aligned}\left(\operatorname{ad} g(X_a)\right)_{bc} &= (g X_a g^{-1})_{bc} = (g X_a g^T)_{bc} \\ &= -\epsilon_{ade} g_{bd} g_{ce} = -\epsilon_{fbc} g_{fa}.\end{aligned}$$

Thus $\operatorname{ad} g(X_a)$ is the skew-symmetric matrix corresponding to the ath column of g. But the ath column of g is just the image of the ath basis vector under the usual matrix action of g on vectors. Thus when $\mathcal{SO}(3)$ is identified with $\mathcal{E}^3$ the adjoint action simply reproduces the usual action of matrices on $\mathcal{E}^3$.

Notice, however, from the exercise, that this is not so if one considers instead the group $O(3)$, that is, if one allows reflections as well as rotations, for then the factor $\det g$ intrudes. In fact the action of $O(3)$ on vectors induced from its adjoint action on its Lie algebra (which is just $\mathcal{SO}(3)$ again) is $\xi \mapsto (\det g) g\xi$. Objects which transform in this way, acquiring an extra minus sign under reflection, are often called "axial vectors". Thus elements of $\mathcal{SO}(3)$ behave as axial vectors under the adjoint action of $O(3)$.

10. Coverings and Connectedness

A great deal is known about a Lie group if its Lie algebra is given: the algebra is in a sense an infinitesimal version of the group, and a neighbourhood of the identity in the group, at least, can be reconstructed from the algebra by means of the exponential map. However, the algebra alone does not completely determine the group, as is seen from the simple example of the Lie groups $\mathbf{R}$ and S^1, which have isomorphic Lie algebras but are not isomorphic groups.

The relationship between $\mathbf{R}$ and S^1 is indicative of the general case, which is to be described in this section. There is a homomorphism of Lie groups $\mathbf{R} \to S^1$ by $t \mapsto e^{2\pi i t}$; here S^1 is realised as the multiplicative group of complex numbers of unit modulus. This homomorphism has two notable properties.

(1) Its kernel, comprising the set of elements of $\mathbf{R}$ which are mapped to the identity in S^1, consists of all the integers. It is therefore a discrete subgroup of $\mathbf{R}$, which means that each of its elements lies in a neighbourhood containing no others.

(2) The homomorphism is locally an isomorphism, which means that there are neighbourhoods of the identities 0 in $\mathbf{R}$ and 1 in S^1 which are diffeomorphic and within which the group multiplications correspond bijectively, so long as the products remain in these neighbourhoods. By choosing the neighbourhood of 0 in $\mathbf{R}$ to be an open interval of length less than 1, one ensures that the integer translates of this neighbourhood are pairwise disjoint, so that each element of the kernel has a neighbourhood diffeomorphic to a given neighbourhood of the identity in S^1. The same will hold for the set of inverse images of any chosen element of S^1.

One may think of S^1 as obtained from $\mathbf{R}$ by identifying with each other points which differ from one another by an integer. The process of identification does not affect local properties but has dramatic global effects. It is an example of a covering map, an idea which we now proceed to make precise.

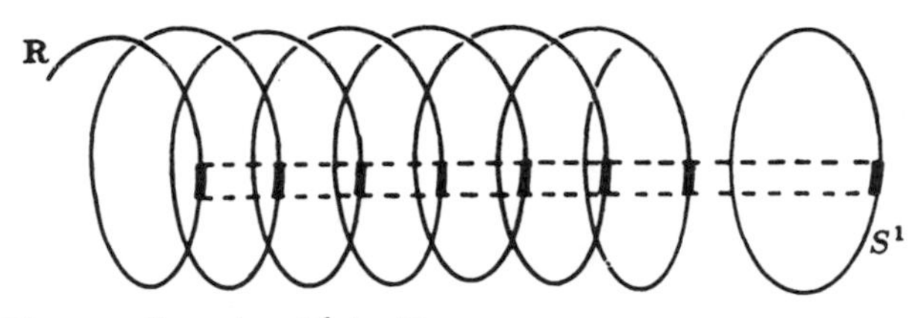

Fig. 1 Covering S^1 by $\mathbf{R}$.

Covering maps. A smooth surjective map of manifolds $\pi: \mathcal{M} \to \mathcal{N}$ is a *covering* of $\mathcal{N}$ by $\mathcal{M}$ if each point $y \in \mathcal{N}$ has a neighbourhood $\mathcal{P}$ such that $\pi^{-1}(\mathcal{P})$ is the union of disjoint open sets $\mathcal{O}$ in $\mathcal{M}$ on each of which the restriction $\pi|_{\mathcal{O}}: \mathcal{O} \to \mathcal{P}$ is a diffeomorphism. The neighbourhood $\mathcal{P}$ is said to be *evenly covered*. The manifold $\mathcal{M}$ is called a *covering space* of the manifold $\mathcal{N}$.

Exercise 40. Show that $\mathbf{R} \to S^1$ by $t \mapsto e^{it}$ is a covering. □

Exercise 41. Show that $\mathbf{R} \to S^1$ by $t \mapsto e^{4i \arctan t}$ is not a covering. □

Exercise 42. Show that $\mathbf{R}^2 \to T^2$ by

$$(x^1, x^2) \mapsto (x^1 - [x^1], x^2 - [x^2])$$

is a covering (where $[x]$ means the integer part of x). □

A covering as here defined is a differentiable idea, invoking no group structure, and in fact this differentiable idea can be weakened to a topological one by requiring only that π be continuous and that each restriction $\pi|_{\mathcal{O}}$ be a homeomorphism. However, the additional structure introduced if it is assumed that π is a homomorphism of Lie groups restricts the possibilities very much; in a Lie group any point is as good as any other, and all considerations may be reduced to the question of what happens in a neighbourhood of the identity.

To state the general position regarding homomorphic Lie groups with isomorphic Lie algebras we first define a discrete subgroup. A subgroup K of a Lie group G is *discrete* if each $k \in K$ has a neighbourhood (in G) which contains no element of K other than k. The general position is as follows. If $\psi: G \to H$ is a homomorphism of Lie groups, if G is connected, and if $\psi_*: \mathcal{G} \to \mathcal{H}$ is an isomorphism, then the kernel of ψ is a discrete normal subgroup of G, and ψ is a covering map. The key to the proof is that, by the inverse function theorem, since ψ_* is an isomorphism, ψ maps a neighbourhood of the identity in G diffeomorphically onto a neighbourhood of the identity in H. By the homogeneity of Lie groups and the homomorphism property of ψ this is, as it turns out, enough to generate the requisite neighbourhoods of other points of G, to prove discreteness of the kernel, and to provide even coverings. An example of this situation is the two-to-one covering of $SO(3)$ by $SU(2)$ described in Chapter 8.

It is natural to ask if there are Lie groups H which cannot be multiply covered, in the way still allowed by this result: if the stated conditions are satisfied, what further condition must be imposed to ensure that ψ is an isomorphism? The answer is that H must be simply connected, a term we now explain.

Simple connectedness. A manifold $\mathcal{M}$ is called *simply connected* if every closed curve γ in the manifold can be smoothly shrunk to a point, that is, if there is a smooth map $\hat{\gamma}: [0,1] \times [0,1] \to \mathcal{M}$ such that $\hat{\gamma}(1,t) = \gamma(t)$ and $t \mapsto \hat{\gamma}(0,t)$ is a constant curve. The real line $\mathbf{R}$ is simply connected, but the circle S^1 is not, because it is itself the image of a closed curve which cannot, within S^1, be smoothly shrunk to a point. The real number space $\mathbf{R}^n$ is simply connected, and so is S^n for $n \geq 2$. The torus, the cylinder, and the projective spaces, all of which are derived from simply connected spaces by making identifications, are not simply connected.

Suppose now that $\pi: \mathcal{M} \to \mathcal{N}$ is a covering of $\mathcal{N}$, and that $\mathcal{N}$ is simply connected. It is not difficult to show that π must be a diffeomorphism. Suppose that there are distinct points $x, y \in \mathcal{M}$ such that $\pi(x) = \pi(y)$. Any curve in $\mathcal{M}$ joining x to y must project into a closed curve in $\mathcal{N}$, which by assumption can be shrunk smoothly to $\pi(x)$; each intermediate curve in the shrinking process can be lifted to a unique curve in $\mathcal{M}$, by the local diffeomorphism property of coverings, and the curves in $\mathcal{M}$ thus obtained constitute a smooth deformation of the original curve, and must all pass through y. But the smoothly shrinking curve in $\mathcal{N}$ must eventually lie within a neighbourhood of $\pi(x)$ which is evenly covered, and the lifted curve must lie in the corresponding diffeomorphic neighbourhood of x. It can therefore no longer reach y, which projects onto the same point as x and so cannot lie in this neighbourhood of x. This contradicts the assumption of distinct points of $\mathcal{M}$ mapping to the same point of $\mathcal{N}$, and shows that π must after all be a diffeomorphism. In the case where $\mathcal{M} = G$ and $\mathcal{N} = H$ are Lie groups and $\pi = \psi$ is a homomorphism as before, it follows that ψ must be an isomorphism.

On the other hand, it is possible to construct, for any connected manifold $\mathcal{N}$, a simply connected covering space, unique up to diffeomorphism, in effect by unwinding those closed curves in $\mathcal{N}$ which cannot be smoothly shrunk to points. But if $\mathcal{N}$ has the property that each covering of it is a diffeomorphism, then it is diffeomorphic to its simply connected covering, and hence is itself simply connected.

The conclusion of these arguments, so far as Lie groups are concerned, is that if $\psi: G \to H$ is a homomorphism of Lie groups, if G is connected, if H is simply connected, and if $\psi_*: \mathcal{G} \to \mathcal{H}$ is an isomorphism, then ψ is an isomorphism. However, a situation which arises often in practice, which is not covered by this result, is that two Lie groups are given which are known to have isomorphic Lie algebras but are not known *ab initio* to be homomorphic groups. In this context, by "isomorphic" we mean isomorphic as Lie algebras, not just as vector spaces. This is in contrast to the situation so far discussed, for the fact that a group homomorphism induces a homomorphism of Lie algebras and not just of vector spaces, though it has been mentioned, has not been made use of. It is in fact possible, if G is simply connected and $\Psi: \mathcal{G} \to \mathcal{H}$ is an isomorphism of Lie algebras, to construct a homomorphism of Lie groups $\psi: G \to H$ such that $\psi_* = \Psi$. Consequently if G and H are Lie groups with isomorphic Lie algebras and G is simply connected then G covers H and H is the quotient of G by a discrete normal subgroup.

As we have already mentioned, the group $SU(2)$ is a covering group of $SO(3)$; it is in fact the simply connected cover, since as a topological space it is just S^3. Thus $SO(3)$ is not simply connected. This topological fact about $SO(3)$ and its relation to

$SU(2)$ is at the bottom of the explanation of intrinsic spin in quantum mechanics. The group $SL(2,\mathbf{C})$ plays the same role in relation to the proper orthochronous Lorentz group as $SU(2)$ does for $SO(3)$ and this has important consequences in relativity.

Finally, we mention Ado's theorem, which asserts that any finite-dimensional Lie algebra is isomorphic to a subalgebra of $M_n(\mathbf{R})$ for some n, whence it can be shown that any Lie algebra is isomorphic to the Lie algebra of a Lie group. Consequently, any Lie algebra is isomorphic to the Lie algebra of a unique simply connected Lie group, and any other connected Lie group with Lie algebra isomorphic to this one is a quotient of the first group by a discrete normal subgroup.

11. Lie Algebras of Transformation Groups

In this section, continuing the development of the theory of transformation groups, we show that given an action of a Lie group G on a manifold $\mathcal{M}$, there is, as one might expect, a homomorphism from the Lie algebra of G to the algebra of generators of one-parameter groups of transformations of $\mathcal{M}$, in other words, vector fields on $\mathcal{M}$.

We suppose for definiteness that G acts to the right on $\mathcal{M}$, and denote the action by ϕ, so that for each $g \in G$, ϕ_g is a diffeomorphism of $\mathcal{M}$. For any $X \in \mathcal{G}$, the Lie algebra of G, $\phi_{\exp tX}$ is a one-parameter group of diffeomorphisms of $\mathcal{M}$; we call the infinitesimal generator of this one-parameter group of transformations the *fundamental vector field* corresponding to $X \in \mathcal{G}$, and denote it $\tilde{X}$.

For purposes of calculation it is sometimes convenient to fix $x \in \mathcal{M}$ and vary $g \in G$ rather than the other way about; we write ϕ_x for the map $G \to \mathcal{M}$ by $g \mapsto \phi_g(x)$, so that $\phi_x(g) = \phi_g(x)$. This is analogous to the switch from transformation to orbit which we have frequently employed when discussing one-parameter groups. Then for any $X \in \mathcal{G}$ the fundamental vector $\tilde{X}_x$ is the tangent at $t = 0$ to the curve $t \mapsto \phi_{\exp tX}(x) = \phi_x(\exp tX)$. It follows that $\tilde{X}_x = \phi_{x*}X_e$, and therefore that the map $X \mapsto \tilde{X}$ is linear.

We shall next evaluate $\phi_{g*}\tilde{X}$ where g is any element of G and $\tilde{X}$ is any fundamental vector field arising from the action. Now $\phi_{g*}\tilde{X}$ is the generator of the one-parameter group $\phi_g \circ \phi_{\exp tX} \circ {\phi_g}^{-1}$; using the fact that ϕ is a right action one may express this as follows:

$$\phi_g \circ \phi_{\exp tX} \circ {\phi_g}^{-1} = \phi_g \circ \phi_{\exp tX} \circ \phi_{g^{-1}} = \phi_{g^{-1}(\exp tX)g}.$$

In other words, $\phi_{g*}\tilde{X}$ is the generator of the one-parameter group of transformations of $\mathcal{M}$ corresponding to the one-parameter subgroup $I_{g^{-1}}(\exp tX)$ of G. But by the results of Section 9

$$I_{g^{-1}} \circ \exp = \exp \circ I_{g^{-1}*} = \exp \circ \operatorname{ad} g^{-1}.$$

Thus $\phi_{g*}\tilde{X}$ is the fundamental vector field on $\mathcal{M}$ corresponding to the element $\operatorname{ad} g^{-1}(X)$ of $\mathcal{G}$:

$$\phi_{g*}\tilde{X} = \widetilde{\operatorname{ad} g^{-1}(X)}.$$

If also $Y \in \mathcal{G}$ and $\tilde{Y}$ is the corresponding fundamental vector field then the bracket $[\tilde{X}, \tilde{Y}]$ may be computed as the Lie derivative $\mathcal{L}_{\tilde{X}}\tilde{Y}$:

$$\mathcal{L}_{\tilde{X}}\tilde{Y} = \frac{d}{dt}\left(\phi_{\exp(-tX)*}\tilde{Y}\right)_{t=0} = \frac{d}{dt}\widetilde{\left(\mathrm{ad}(\exp tX)(Y)\right)}_{t=0}.$$

But as we proved in Section 9

$$\frac{d}{dt}\left(\mathrm{ad}(\exp tX)(Y)\right)_{t=0} = [X, Y].$$

Thus

$$[\tilde{X}, \tilde{Y}] = \widetilde{[X, Y]}.$$

It follows that the set of fundamental vector fields generated by a right action of a Lie group G on a manifold $\mathcal{M}$ is a (finite-dimensional) Lie algebra of vector fields on $\mathcal{M}$, which is a homomorphic image of $\mathcal{G}$.

Exercise 43. Show that in the case of a left action, when $\tilde{X}$ is defined analogously, the corresponding result is $[\tilde{X}, \tilde{Y}] = -\widetilde{[X, Y]}$, and then $X \mapsto -\tilde{X}$ is a Lie algebra homomorphism. □

12. Symmetry Groups and Momentum in Mechanics

It was pointed out in Chapter 6 (Exercise 38) that if ω is a closed 2-form of maximal rank on an even dimensional manifold (actually an affine space in Chapter 6, but it makes little difference), if h is a function, and if V_h is the vector field determined by $V_h \lrcorner \omega = -dh$, then with respect to coordinates in which ω takes the standard form $dp_a \wedge dq^a$ the integral curves of V_h satisfy Hamilton's equations for the Hamiltonian function h. An even dimensional manifold $\mathcal{M}$ with closed maximal rank 2-form ω is thus the natural environment for a generalised form of Hamiltonian mechanics. Such a structure is called a *symplectic manifold*. We shall examine here group actions on a symplectic manifold preserving the symplectic structure.

Let G be a Lie group acting on a symplectic manifold. Its action is said to be a *symplectic action* if for every $g \in G$,

$$\phi_g{}^*\omega = \omega.$$

In this case, for every $X \in \mathcal{G}$

$$\mathcal{L}_{\tilde{X}}\omega = 0.$$

But then since ω is closed $d(\tilde{X} \lrcorner \omega) = 0$. Thus $\tilde{X} \lrcorner \omega$ is locally exact: we suppose that it is actually exact, so that there is some function J_X such that

$$\tilde{X} \lrcorner \omega = -dJ_X.$$

(The minus sign is conventional, like that in the definition of the Hamiltonian vector field.) Note that J_X is determined only up to an additive constant. Since $X \mapsto \tilde{X}$ is a linear map, it follows that J_X depends linearly on X (this requires some fixing of constants, which may be achieved by specifying that $J_X(x_0) = 0$ for all X, where x_0 is some chosen point of $\mathcal{M}$, for example). Then for each $x \in \mathcal{M}$ the map $X \mapsto J_X(x)$

may be regarded as a linear functional on $\mathcal{G}$ and therefore as an element of its dual space $\mathcal{G}^*$. So this symplectic action defines a map $\mathcal{M} \to \mathcal{G}^*$ such that

$$\tilde{X} \lrcorner \omega = -d\langle X, J\rangle$$

where $\langle X, J\rangle$ denotes the function $x \mapsto \langle X, J(x)\rangle = J_X(x)$ on $\mathcal{M}$. The map J, which is determined up to the addition of a constant element of $\mathcal{G}^*$, is called the *momentum map* associated with the action: this is because it is a generalisation of linear and angular momentum in elementary mechanics, as we shall shortly show.

Suppose first, however, that h is a Hamiltonian function and that $\phi_g{}^*h = h$ for all $g \in G$. Then $\tilde{X}h = 0$ for all $X \in \mathcal{G}$ and as a consequence

$$\langle X, V_h J\rangle = V_h\langle X, J\rangle = -V_h \lrcorner(\tilde{X} \lrcorner \omega) = \tilde{X} \lrcorner(V_h \lrcorner \omega) = -\tilde{X}h = 0.$$

Thus $V_h J = 0$: in other words, if a group acts symplectically and defines a momentum map, and if it also acts so as to leave a Hamiltonian function invariant, then the momentum is conserved by the corresponding Hamiltonian vector field. In this case one says that G is a *group of symmetries* of the Hamiltonian system; this result is a general form of the correlation between symmetries and conserved quantities in Hamiltonian mechanics.

The momentum map obeys a transformation law under the action of G, which we shall now derive. For any $g \in G$

$$\begin{aligned}\langle X, \phi_g{}^*J\rangle &= \phi_g{}^*\langle X, J\rangle = -\phi_g{}^*(\tilde{X} \lrcorner \omega)\\ &= -\phi_{g^{-1}*}\tilde{X} \lrcorner \omega = -\widetilde{\operatorname{ad} g(X)} \lrcorner \omega = \langle \operatorname{ad} g(X), J\rangle\end{aligned}$$

where we have used the fact that $\phi_g{}^*\omega = \omega$, and a formula from the previous subsection which holds under the assumption that G acts to the right. We express this conclusion in terms of J and g alone by utilising the action of G on $\mathcal{G}^*$ generated from the adjoint representation by duality. Thus we define $\operatorname{ad} g^*(p)$ for $p \in \mathcal{G}^*$ by $\langle X, \operatorname{ad} g^*(p)\rangle = \langle \operatorname{ad} g(X), p\rangle$ for all $X \in \mathcal{G}$; then

$$\phi_g{}^*J = \operatorname{ad} g^*(J).$$

To understand this equation it is important to realise that on the left-hand side $\phi_g{}^*$ acts on J *via* its argument (in $\mathcal{M}$), while on the right-hand side $\operatorname{ad} g^*$ acts on J *via* its value (in $\mathcal{G}^*$).

Exercise 44. Deduce from the transformation law for the momentum map that for any $X, Y \in \mathcal{G}$, $\tilde{X}\langle Y, J\rangle = \langle [X, Y], J\rangle$. □

It may be the case that ω is exact, say $\omega = d\theta$; we then speak of an *exact symplectic manifold*. If the action then preserves θ we call it an *exact symplectic action*. In this case the momentum map has a straightforward expression in terms of θ and there is no problem of indeterminacy up to additive constants. For if $\mathcal{L}_{\tilde{X}}\theta = 0$ then

$$\tilde{X} \lrcorner \omega = \tilde{X} \lrcorner d\theta = -d\langle \tilde{X}, \theta\rangle$$

and so we may take

$$\langle X, J\rangle = \langle \tilde{X}, \theta\rangle.$$

We now give the examples which explain why the momentum map has its name.

First, let $\mathcal{M} = \mathcal{E}^{2n} = \mathcal{E}^n \times \mathcal{E}^n$, with coordinates (q^a, p_a), and let $\theta = p_a dq^a$. Then $\mathcal{M}$, equipped with $d\theta$, is an exact symplectic manifold. Think of the first factor of $\mathcal{M}$, with coordinates (q^a), as representing configuration space, and the second factor, with coordinates (p_a), as representing momentum space. Let $G = \mathbf{R}^n$ acting on $\mathcal{M}$ by $(q^a, p_a) \mapsto (q^a + v^a, p_a)$. This is an exact symplectic action. If $X = (\xi^a) \in \mathcal{G} = \mathbf{R}^n$ then $\tilde{X} = \xi^a \partial/\partial q^a$ and so $J_X(q^a, p_a) = p_a \xi^a$. Thus in particular if $\xi = (\xi^a)$ is a unit vector then J_X is the component of linear momentum in the ξ-direction. Thus J corresponds to linear momentum in this case. Since G is commutative, $\operatorname{ad} g$ is the identity, and the transformation rule for J says that momentum is translation invariant. If $\tilde{X}h = 0$ then h is invariant under translations in the direction corresponding to X, and the formula $V_h J_X = 0$ shows that the component of momentum in the direction of the invariance translation is constant.

Second, let $\mathcal{M} = \mathcal{E}^6 = \mathcal{E}^3 \times \mathcal{E}^3$ with $\theta = p_a dq^a$ as before (but now $n = 3$). Let $G = SO(3)$ acting by $\phi_g(\mathbf{q}, \mathbf{p}) = (g\mathbf{q}, g\mathbf{p})$ (since this is a left action the transformation law for J will have to be modified in this case). Then representing θ as $\mathbf{p}^T d\mathbf{q}$ it follows that this is an exact symplectic action. Using the basis for $\mathcal{SO}(3)$ given at the end of Section 9, one obtains

$$\tilde{X}_1 = q^2 \frac{\partial}{\partial q^3} - q^3 \frac{\partial}{\partial q^2} + p_2 \frac{\partial}{\partial p_3} - p_3 \frac{\partial}{\partial p_2}$$

with corresponding expressions for $\tilde{X}_2$ and $\tilde{X}_3$. Thus

$$\langle \tilde{X}_1, \theta \rangle = q^2 p_3 - q^3 p_2,$$

the first component of angular momentum; likewise, $\langle \tilde{X}_2, \theta \rangle$ and $\langle \tilde{X}_3, \theta \rangle$ are the second and third components of angular momentum. More generally, if $X \in \mathcal{SO}(3)$ corresponds to a unit vector in $\mathcal{E}^3$ (in the manner explained in Section 9), then J_X is the component of angular momentum in that direction. Thus J corresponds to angular momentum. Since we are dealing with a left action the transformation law for J reads

$$\phi_g{}^* J = \operatorname{ad} g^{-1*} J.$$

As we pointed out in Section 9, when $\mathcal{SO}(3)$ is identified with $\mathcal{E}^3$ the adjoint representation corresponds to the ordinary action of $SO(3)$; and since $(g^{-1})^T = g$ the action $\operatorname{ad} g^{-1*}$ on the dual is again the ordinary action of $SO(3)$. So the transformation law says that angular momentum transforms as a vector under rotations (but again, if reflections are allowed, one has to modify this to say that angular momentum transforms as an axial vector). Finally, a Hamiltonian h such that $\tilde{X}h = 0$ is invariant under rotations about the axis corresponding to X, and the angular momentum component in that direction is conserved.

The coadjoint action. The *coadjoint action* of a Lie group G on $\mathcal{G}^*$, the vector space dual to its Lie algebra, is the right action defined by $g \mapsto \operatorname{ad} g^*$. As we have shown above, the study of symmetry groups of Hamiltonian systems leads to a consideration of this action, *via* the momentum map. It follows from the transformation formula for J that each orbit of the action of G in $\mathcal{M}$ is mapped by J into an orbit of the coadjoint action of G in $\mathcal{G}^*$. It is therefore of interest to study the orbits of the coadjoint action. We shall show that each orbit has its own naturally defined symplectic structure.

For any $X \in \mathcal{G}$ we define a vector field on $\mathcal{G}^*$ as the generator of the one-parameter group $\operatorname{ad}(\exp tX)^*$ in the usual way; we shall denote this vector field X^* to avoid confusion. The linear structure of $\mathcal{G}^*$ allows one to give a simple description of X^* by evaluating it on linear functions. Every linear function on $\mathcal{G}^*$ is determined by an element of $\mathcal{G}$ and is given by $p \mapsto \langle Z, p\rangle$, $p \in \mathcal{G}^*$, $Z \in \mathcal{G}$. We denote by f_Z the linear function corresponding to Z. Then

$$\begin{aligned} X_p^* f_Z &= \frac{d}{dt}\left(\langle Z, \operatorname{ad}(\exp tX)^* p\rangle\right)_{t=0} \\ &= \frac{d}{dt}\langle \operatorname{ad}(\exp tX) Z, p\rangle_{t=0} \\ &= \langle [X, Z], p\rangle = f_{[X,Z]}(p). \end{aligned}$$

The orbit of $p \in \mathcal{G}^*$ under the coadjoint action is a submanifold of $\mathcal{G}^*$ whose tangent space at p consists of all vectors of the form X_p^* where $X \in \mathcal{G}$. In general, the map $X \mapsto X_p^*$ is not injective. Its kernel, that is, the set of elements $X \in \mathcal{G}$ for which $X_p^* = 0$, may be found from the formula just proved: for $X_p^* = 0$ if and only if X_p^* vanishes on all linear functions. Thus the kernel of $X \mapsto X_p^*$ is the subspace $\mathcal{G}_p$ of $\mathcal{G}$ consisting of those elements X such that $\langle [X, Z], p\rangle = 0$ for all $Z \in \mathcal{G}$. In fact $\mathcal{G}_p$ is a subalgebra of $\mathcal{G}$: if $X, Y \in \mathcal{G}_p$ and $Z \in \mathcal{G}$ then $\langle [[X, Y], Z], p\rangle = 0$ by the Jacobi identity. It is the Lie algebra of the isotropy group of p under the coadjoint action. The tangent space to the orbit of p is isomorphic to $\mathcal{G}/\mathcal{G}_p$.

We next define a skew bilinear form Ω_p on the tangent space at p to the orbit of p by setting

$$\Omega_p(X_p^*, Y_p^*) = \langle [X, Y], p\rangle.$$

This form has maximal rank, for if $\Omega_p(X_p^*, Y_p^*) = 0$ for all Y_p^*, then $\langle [X, Y], p\rangle = 0$ for all Y, whence $X \in \mathcal{G}_p$ and $X_p^* = 0$. Using the same definition pointwise over the whole orbit we obtain a 2-form on the orbit. This 2-form, Ω, has the property that

$$\Omega(X^*, Y^*) = f_{[X,Y]}.$$

We show that it is closed. In fact

$$\begin{aligned} d\Omega(X^*, Y^*, Z^*) &= X^*\Omega(Y^*, Z^*) + Y^*\Omega(Z^*, X^*) + Z^*\Omega(X^*, Y^*) \\ &\quad - \Omega([X^*, Y^*], Z^*) - \Omega([Y^*, Z^*], X^*) - \Omega([Z^*, X^*], Y^*) \\ &= X^* f_{[Y,Z]} + Y^* f_{[Z,X]} + Z^* f_{[X,Y]} \\ &\quad - f_{[[X,Y],Z]} - f_{[[Y,Z],X]} - f_{[[Z,X],Y]} \\ &= 2\left\{ f_{[X,[Y,Z]]} + f_{[Y,[Z,X]]} + f_{[Z,[X,Y]]} \right\} = 0 \end{aligned}$$

by the Jacobi identity. Since the vector fields X^* span the tangent space to the orbit at each point of it, this is enough to show that Ω is closed.

Thus each orbit of the coadjoint action in $\mathcal{G}^*$, equipped with the 2-form Ω defined above, is a symplectic manifold. The coadjoint action is a symplectic action:

for any $g \in G$, with ϕ denoting the action, we have

$$\begin{aligned}(\phi_g{}^*\Omega)_p(X_p^*, Y_p^*) &= \Omega(\phi_{g*}X^*, \phi_{g*}Y^*)_{\phi_g(p)} \\ &= \Omega\big(\operatorname{ad} g^{-1}(X)^*, \operatorname{ad} g^{-1}(Y)^*\big)_{\operatorname{ad} g^*(p)} \\ &= \big\langle [\operatorname{ad} g^{-1}(X), \operatorname{ad} g^{-1}(Y)], \operatorname{ad} g^*(p)\big\rangle \\ &= \big\langle \operatorname{ad} g^{-1}[X,Y], \operatorname{ad} g^*(p)\big\rangle \\ &= \big\langle [X,Y], p\big\rangle = \Omega_p(X_p^*, Y_p^*).\end{aligned}$$

Finally, we show how this structure is related to a given symplectic action of G *via* the momentum map. From the transformation law for the momentum map it follows that for any $X \in \mathcal{G}$ and $x \in \mathcal{M}$

$$J_*\tilde{X}_x = \frac{d}{dt}\big(J(\phi_{\exp tX}x)\big)_{t=0} = \frac{d}{dt}\big(\operatorname{ad}(\exp tX)^*J(x)\big)_{t=0} = X^*_{J(x)}$$

(where $\tilde{X}$ is the fundamental vector field on $\mathcal{M}$ generated from X by the group action). Moreover, from Exercise 44,

$$\tilde{X}\langle Y, J\rangle = \big\langle [X,Y], J\big\rangle.$$

Thus regarding J as a map from the orbit of $x \in \mathcal{M}$ to the orbit of $J(x) \in \mathcal{G}^*$ we have

$$\begin{aligned}(J^*\Omega)_x(\tilde{X}_x, \tilde{Y}_x) &= \Omega_{J(x)}(J_*\tilde{X}_x, J_*\tilde{Y}_x) = \Omega_{J(x)}(X^*_{J(x)}, Y^*_{J(x)}) \\ &= \big\langle [X,Y], J(x)\big\rangle = \tilde{X}_x\langle Y, J\rangle \\ &= -\tilde{X}_x \lrcorner (\tilde{Y}_x \lrcorner \omega_x) = \omega_x(\tilde{X}_x, \tilde{Y}_x).\end{aligned}$$

The conclusion is that $J^*\Omega$ coincides with the restriction of ω to the orbit. In particular, if the symplectic action of G on $\mathcal{M}$ happens to be transitive then J maps $\mathcal{M}$ symplectically into a coadjoint orbit in $\mathcal{G}^*$.

Summary of Chapter 12

A Lie group is a group which is at the same time a differentiable manifold, such that multiplication and the formation of inverses are smooth operations. Examples include $\mathbf{R}$, S^1, T^2, $GL(n,\mathbf{R})$, $O(n)$, $GL(n,\mathbf{C})$, $U(n)$ and many other familiar groups.

Lie groups are often encountered in the role of transformation groups. An action of a group G on a set $\mathcal{M}$ is an assignment of a transformation ϕ_g of $\mathcal{M}$ to each $g \in G$ such that either $\phi_{gh} = \phi_g \circ \phi_h$ (left action) or $\phi_{gh} = \phi_h \circ \phi_g$ (right action). The orbit of $x \in \mathcal{M}$ is the set $\{\,\phi_g(x) \mid g \in G\,\}$; its isotropy group the set $\{\,g \in G \mid \phi_g(x) = x\,\}$. If the whole of $\mathcal{M}$ is one single orbit then G acts transitively and $\mathcal{M}$ is a homogeneous space. If the isotropy group of every point is the identity then G acts freely; if the intersection of all isotropy groups is the identity then G acts effectively. Any homogeneous space of G may be identified with the space of cosets of the isotropy group of any one of its points in G.

A Lie group acts smoothly on itself by left translation $L_g: h \mapsto gh$ and by right translation $R_g: h \mapsto hg$. Left (right) translation is a free transitive left (right) action. Conjugation is the left action $L_g \circ R_{g^{-1}}$; it is an automorphism of the group, also called inner automorphism.

A vector field X on a Lie group is left-invariant if $L_{g*}X = X$ for all g. A left-invariant vector field is determined by its value at any one point, usually the identity e. The space of left-invariant vector fields on G is linear and finite dimensional (equal in dimension to the dimension of G), and is closed under bracket. It is the Lie algebra of G, denoted $\mathcal{G}$. If $\{X_a\}$ is a basis for $\mathcal{G}$ then $[X_b, X_c] = C^a_{bc}X_a$ where the numbers C^a_{bc} are the structure constants of the Lie algebra. When G is a group of matrices its Lie algebra may be identified as a space of matrices closed under matrix commutator; for example, the Lie algebra of $O(n)$ or $SO(n)$ is the space of $n \times n$ skew-symmetric matrices. Dually one defines left-invariant forms by $L_g{}^*\alpha = \alpha$; the space of left-invariant 1-forms is dual to the Lie algebra. If $\{\alpha^a\}$ is a basis of left-invariant 1-forms then $d\alpha^a = -\frac{1}{2}C^a_{bc}\alpha^b \wedge \alpha^c$, the Maurer-Cartan equations.

Each left-invariant vector field X determines a one-parameter subgroup of G, denoted $\exp tX$ since in the case of a matrix group it is the matrix exponential. The integral curve of X through g is then $t \mapsto g\exp tX$. The exponential map $\mathcal{G} \to G$ by $X \mapsto \exp X$ is a diffeomorphism of an open neighbourhood of 0 in $\mathcal{G}$ onto an open neighbourhood of e in G. It may be used to introduce normal coordinates (x^a) into the group, with respect to which $x^a(g^{-1}) = -x^a(g)$ and $x^a(gh) = x^a(g) + x^a(h) + C^a_{bc}x^b(g)x^c(h) + \cdots$ for g, h sufficiently close to e. If $\psi: G \to H$ is a homomorphism of Lie groups then $\psi \circ \exp = \exp \circ \psi_*$. Using the inner automorphism $I_g = L_g \circ R_{g^{-1}}$ one defines the adjoint representation of G on $\mathcal{G}$ by $\operatorname{ad} g(X) = I_{g*}X$. Then $\exp(\operatorname{ad} g(X)) = I_g \exp X$. Further, $\operatorname{ad} g$ is an automorphism of $\mathcal{G}$, and $g \mapsto \operatorname{ad} g$ a homomorphism of G into the group of automorphisms of $\mathcal{G}$ (this is the significance of the term representation). From the left-invariance of Y it follows that $\operatorname{ad} g(Y) = R_{g^{-1}*}Y$ whence $d/dt\big(\operatorname{ad}(\exp tX)(Y)\big)_{t=0} = [X, Y]$. In the case of the group $SO(3)$ the Lie algebra, being 3-dimensional, may be identified with $\mathcal{E}^3$, and the adjoint representation corresponds to the usual matrix action; but under $O(3)$ the elements of the algebra transform as axial vectors.

The Lie algebra of a Lie group determines the group in a neighbourhood of the identity, but not globally. There is a unique simply connected Lie group with the given algebra (unique, that is, up to isomorphism); every other Lie group with isomorphic algebra is a quotient of this one by a discrete normal subgroup.

When a group G acts on a manifold $\mathcal{M}$, to the right say, each element X of its Lie algebra determines a vector field $\tilde{X}$ on $\mathcal{M}$ which is the generator of the one-parameter group of transformations of $\mathcal{M}$ induced by $\exp tX$. The map $X \mapsto \tilde{X}$ is linear and preserves brackets, and $\phi_{g*} = \widetilde{\operatorname{ad} g^{-1}(X)}$.

One situation in which these results are repeatedly used is the consideration of symmetry in Hamiltonian mechanics. The arena is an even-dimensional manifold with maximal rank closed 2-form ω—a symplectic manifold. The vector field V_h determined by $V_h \lrcorner \omega = -dh$ corresponds to Hamilton's equations for the Hamiltonian h. If G acts on $\mathcal{M}$ symplectically, so that $\mathcal{L}_{\tilde{X}}\omega = 0$ for all $X \in \mathcal{G}$ and if further $\tilde{X} \lrcorner \omega$ is exact for all $X \in \mathcal{G}$ (not just closed), then one may define a momentum map J by

$\tilde{X} \lrcorner \omega = -d\langle X, J\rangle$. This map generalises linear momentum ($G = \mathbf{R}^n$) and angular momentum ($G = SO(3)$). It satisfies the transformation law $\phi_g{}^*J = \operatorname{ad} g^*(J)$ (for a right action) where $g \mapsto \operatorname{ad} g^*$ is the coadjoint action of G on $\mathcal{G}^*$. Each orbit of the coadjoint action has its own naturally defined symplectic structure, and when G acts transitively (and symplectically) on $\mathcal{M}$, its momentum map (assuming it has one) is a symplectic map of $\mathcal{M}$ into an orbit in $\mathcal{G}^*$.

Notes to Chapter 12

1. Analyticity of group operations. The result that the operations of multiplication and the formation of inverses in any Lie group are analytic is proved in Montgomery and Zippin [1955].

2. The automorphism group of a Lie group. The assertion made near the end of Section 4, that the automorphism group of a simply connected Lie group is in turn a Lie group, is proved by, for example, Warner [1971].

3. Lie algebras and Lie groups. A proof of the theorem that every real Lie algebra is the Lie algebra of some Lie group, referred to in Section 10, is given in Cartan [1936].

13. THE TANGENT AND COTANGENT BUNDLES

The collection of all the tangent vectors at all the points of a differentiable manifold, in other words the union of all its tangent spaces, may in turn be given the structure of a differentiable manifold. It is often convenient to regard the tangent vectors to a manifold as points of a "larger" manifold in this manner: for example, if the original manifold is the space of configurations of a time-independent mechanical system with finitely many degrees of freedom then the manifold of tangent vectors is the space of configurations and generalised velocities, that is, the space on which the Lagrangian function of the system is defined.

1. The Tangent Bundle

For a given differentiable manifold $\mathcal{M}$ we denote by $T\mathcal{M}$ the set of all tangent vectors v at all points $x \in \mathcal{M}$, in other words the union $\bigcup_{x\in\mathcal{M}} T_x\mathcal{M}$ of the tangent spaces to $\mathcal{M}$. This space, together with an appropriate differentiable structure which will be explained in greater detail in this section, is called the *tangent bundle* of $\mathcal{M}$: or to be more precise, as a space it is the bundle space of the tangent bundle of $\mathcal{M}$.

We show first that $T\mathcal{M}$ is itself a differentiable manifold. An atlas for $T\mathcal{M}$ may be constructed out of an atlas for $\mathcal{M}$ as follows. For a chart $(\mathcal{P},\psi)$ on $\mathcal{M}$, let $\hat{\mathcal{P}}$ be the subset of $T\mathcal{M}$ consisting of those tangent vectors whose points of tangency lie in $\mathcal{P}$: thus $\hat{\mathcal{P}} = \bigcup_{x\in\mathcal{P}} T_x\mathcal{M}$. Then if $v \in \hat{\mathcal{P}}$ it may be expressed in the form $v^a\partial_a$ where the ∂_a are the coordinate vector fields associated with the coordinates on $\mathcal{P}$. The coordinates of the tangent vector v are taken to be the coordinates (x^a) of its point of tangency, as given by the chart $(\mathcal{P},\psi)$, and the components (v^a) of v; we write these coordinates collectively (x^a, v^a); they are $2m$ in number, where $m = \dim\mathcal{M}$.

Exercise 1. Write down explicitly the map $\hat{\psi}: \hat{\mathcal{P}} \to \mathbf{R}^{2m}$ thus defined. Show that $\hat{\psi}(\hat{\mathcal{P}})$ is an open subset of $\mathbf{R}^{2m}$ and that $\hat{\psi}$ is a bijective map of $\hat{\mathcal{P}}$ onto $\hat{\psi}(\hat{\mathcal{P}})$. Show that if $\{(\mathcal{P}_\alpha,\psi_\alpha)\}$ is an atlas for $\mathcal{M}$ then the sets $\hat{\mathcal{P}}_\alpha$ cover $T\mathcal{M}$. □

To confirm that $\{(\hat{\mathcal{P}}_\alpha,\hat{\psi}_\alpha)\}$ is an atlas for $T\mathcal{M}$ one has to check that its coordinate transformations are smooth. Suppose that $(\mathcal{P},\psi)$ and $(\mathcal{Q},\phi)$ are charts of the atlas for $\mathcal{M}$, with $\mathcal{P}\cap\mathcal{Q}$ non-empty. Set $\chi = \phi\circ\psi^{-1}$. The appropriate coordinate transformation $\hat{\phi}\circ\hat{\psi}^{-1}$ on $T\mathcal{M}$ over $\mathcal{P}\cap\mathcal{Q}$ will consist of the transformation of coordinates of points of tangency together with the corresponding transformation of tangent vector components. For any denote smooth map $\Phi: \mathbf{R}^m \to \mathbf{R}^m$ we shall denote by Φ' the Jacobian matrix of Φ, as in Chapter 2. Then $\hat{\phi}\circ\hat{\psi}^{-1}$ is given by

$$\hat{\phi}\circ\hat{\psi}^{-1}(x^d, v^d) = \left(\chi^a(x^c), (\chi'(x^c))^a_b v^b\right)$$

for $(x^a, v^a) \in \hat{\psi}(\mathcal{P}\cap\mathcal{Q}) \subset \mathbf{R}^{2m}$. This is evidently smooth, since χ is. Thus $\{(\hat{\mathcal{P}}_\alpha,\hat{\psi}_\alpha)\}$ is indeed an atlas for $T\mathcal{M}$ whenever $\{(\mathcal{P}_\alpha,\psi_\alpha)\}$ is an atlas for $\mathcal{M}$; so that

$T\mathcal{M}$, equipped with the completion of this atlas, is a differentiable manifold. Of course, an arbitrary chart from the complete atlas will not come from a chart on $\mathcal{M}$ in the convenient way that $(\hat{\mathcal{P}}_\alpha, \hat{\psi}_\alpha)$ comes from $(\mathcal{P}_\alpha, \psi_\alpha)$: it is usually sufficient, however, and most convenient, to deal only with those charts on $T\mathcal{M}$ which do.

Exercise 2. Show that the map $\pi: T\mathcal{M} \to \mathcal{M}$ by $\pi(v) = x$ if $v \in T_x\mathcal{M}$ is smooth and surjective, and that for each $x \in \mathcal{M}$, $\pi^{-1}(x) = T_x\mathcal{M} \subset T\mathcal{M}$ is an imbedded submanifold of $T\mathcal{M}$. □

This map π is called the tangent bundle *projection*.

The tangent spaces at different points of $\mathcal{M}$ are identical, in the sense that each is isomorphic to $\mathbf{R}^m$ and hence to every other. On the other hand, the realisation of such an isomorphism between two tangent spaces depends on the choice of a basis for each space, and in general there will be no obvious candidates to choose: in this sense, the tangent spaces are distinct. Were it not for this complication one might imagine that $T\mathcal{M}$ was simply $\mathcal{M} \times \mathbf{R}^m$, the product of the two manifolds, with π projection onto the first factor. But if this were the case then one could find, for any manifold $\mathcal{M}$, smooth nowhere-vanishing vector fields defined on the whole of $\mathcal{M}$ (by fixing a nowhere-zero vector $\xi \in \mathbf{R}^m$ and choosing at each $x \in \mathcal{M}$ the vector in $T_x\mathcal{M}$ corresponding to ξ). However, the two-sphere, to give one example, supports no smooth nowhere-vanishing globally defined vector field, as one may easily convince oneself by trying to construct one (though a proof is not so straightforward).

Thus the tangent bundle of a manifold need not be a product, as a whole. The most that one can say in general is that $T\mathcal{M}$ is locally a product, as follows. Given any point x in $\mathcal{M}$ there is a neighbourhood $\mathcal{P}$ of x such that $\pi^{-1}(\mathcal{P})$ is diffeomorphic to the product $\mathcal{P} \times \mathbf{R}^m$. In fact if $\mathcal{P}$ is a coordinate patch then the map $\pi^{-1}(\mathcal{P}) \to \mathcal{P} \times \mathbf{R}^m$ by $v \mapsto \big(\pi(v), (v^a)\big)$, where v^a are the components of v with respect to the coordinate fields on $\mathcal{P}$, defines such a *local product structure*. Of course, two coordinate neighbourhoods $\mathcal{P}$ and $\mathcal{Q}$ of x produce different product decompositions of $\pi^{-1}(\mathcal{P} \cap \mathcal{Q})$.

Exercise 3. Let $\{\mathcal{P}_\alpha\}$ be coordinate patches on $\mathcal{M}$ and define maps $\Psi_\alpha: \psi^{-1}(\mathcal{P}_\alpha) \to \mathbf{R}^m$ by $\Psi_\alpha(v) = (v^a_\alpha)$, the vector of components of v with respect to the coordinate vector fields on $\mathcal{P}_\alpha$. Thus the local product decomposition $\pi^{-1}(\mathcal{P}_\alpha) \to \mathcal{P}_\alpha \times \mathbf{R}^m$ is given by (π, Ψ_α). Show that if $\alpha \neq \beta$ and $\mathcal{P}_\alpha \cap \mathcal{P}_\beta$ is non-empty then for $x \in \mathcal{P}_\alpha \cap \mathcal{P}_\beta$, $\Psi_\beta(v) = \chi'_{\alpha\beta}(x)\Psi_\alpha(v)$, where $\chi_{\alpha\beta}$ is the coordinate transformation function on $\mathcal{P}_\alpha \cap \mathcal{P}_\beta$, and matrix multiplication by the Jacobian matrix is implied on the right-hand side but not summation over α. Show that the "transition functions" $\chi'_{\alpha\beta}: \mathcal{P}_\alpha \cap \mathcal{P}_\beta \to GL(m, \mathbf{R})$ satisfy $\chi'_{\beta\alpha} = (\chi'_{\alpha\beta})^{-1}$ and $\chi'_{\alpha\beta}\chi'_{\beta\gamma} = \chi'_{\alpha\gamma}$ on $\mathcal{P}_\alpha \cap \mathcal{P}_\beta \cap \mathcal{P}_\gamma$ when α, β, γ are distinct and the triple intersection is non-empty. □

The tangent bundle $T\mathcal{M}$ is thus a differentiable manifold of a rather special kind: it has a projection map $\pi: T\mathcal{M} \to \mathcal{M}$, the "fibres" $\pi^{-1}(x)$ are all diffeomorphic (to $\mathbf{R}^m$) and $T\mathcal{M}$ is a local product manifold with transition functions (which relate different local product decompositions) taking their values in a Lie group ($GL(m, \mathbf{R})$). These are the essential features of what is known as a fibre bundle, of which more in Chapter 14. When it is desirable to emphasise that all these features are essential parts of the fibre bundle structure of the tangent bundle, one calls $T\mathcal{M}$ the *bundle space* and $\mathcal{M}$ the *base*. For each $x \in \mathcal{M}$, $\pi^{-1}(x)$ is called the *fibre* over x.

The set of cotangent vectors on $\mathcal{M}$, or the union of its cotangent spaces, may also be made into a differentiable manifold, which is also a fibre bundle, the *cotangent bundle* $T^*\mathcal{M}$ of $\mathcal{M}$.

Exercise 4. By adapting the argument given for $T\mathcal{M}$ show that $T^*\mathcal{M}$ is a differentiable manifold of dimension $2\dim\mathcal{M} = 2m$. Show that the map τ which sends each cotangent vector to the point of $\mathcal{M}$ at which it acts is a smooth, surjective map $T^*\mathcal{M} \to \mathcal{M}$, the projection. Show that the fibres $\tau^{-1}(x)$ are all diffeomorphic to $\mathbf{R}^m$ and that $T^*\mathcal{M}$ is a local product manifold whose transition functions are $(\chi'_{\alpha\beta})^\times$, where, for $g \in GL(m,\mathbf{R})$, $g^\times$ is the inverse of the transpose of g. Show that $(\chi'_{\beta\alpha})^\times = \left((\chi'_{\alpha\beta})^\times\right)^{-1}$ and that $(\chi'_{\alpha\beta})^\times(\chi'_{\beta\gamma})^\times = (\chi'_{\alpha\gamma})^\times$ on appropriate domains. □

The differential geometry of $T\mathcal{M}$, and of $T^*\mathcal{M}$, is affected by the fact that each has this structure of a fibre bundle. It is also flavoured by the option one always has of regarding a point of $T\mathcal{M}$ ($T^*\mathcal{M}$) as a tangent (cotangent) vector to $\mathcal{M}$, and *vice versa*. Thus geometrical constructions on $T\mathcal{M}$ or $T^*\mathcal{M}$ may often be interpreted on two levels, in terms of the bundle itself, or in terms of structures on $\mathcal{M}$. Again, many geometric objects on $\mathcal{M}$ have an alternative form of existence in which they are interpreted as geometric objects, perhaps of a superficially quite different kind, on $T\mathcal{M}$ or $T^*\mathcal{M}$. We now begin to give examples of these ideas, first in $T\mathcal{M}$.

In the sequel it will be necessary from time to time to show explicitly the point of attachment of a tangent vector v to $\mathcal{M}$ when considering it as a point of $T\mathcal{M}$. We shall therefore sometimes write (x,v) for a generic point of $T\mathcal{M}$, where $v \in T_x\mathcal{M}$, or in other words $x = \pi(v)$.

2. Lifts

A curve σ in $\mathcal{M}$ with a vector field V defined along it generates a curve $t \mapsto (\sigma(t), V(t))$ in $T\mathcal{M}$, whose projection is just σ. In particular, since every curve has a vector field naturally defined along it, namely its field of tangent vectors, every curve σ in $\mathcal{M}$ determines a curve $(\sigma,\dot\sigma)$ in $T\mathcal{M}$, which we call its *natural lift*. The projection map $\pi: T\mathcal{M} \to \mathcal{M}$, being smooth, induces linear maps $\pi_*: T_vT\mathcal{M} \to T_{\pi(v)}\mathcal{M}$. Since every curve in $\mathcal{M}$ through $\pi(v)$ may be obtained as the projection of a curve in $T\mathcal{M}$ through v, π_* is surjective. On the other hand, since curves in the fibre over $\pi(v)$ project onto the constant curve $\pi(v)$, π_* maps vectors tangent to the fibre to zero. Such vectors are said to be *vertical* and the subspace of $T_vT\mathcal{M}$ consisting of vectors tangent to the fibre, which by dimension is precisely the kernel of π_*, is called its *vertical subspace*.

Exercise 5. This exercise is concerned with the coordinate vector basis $\{\partial/\partial x^a, \partial/\partial v^a\}$ for coordinates (x^a, v^a) adapted to the local product structure of $T\mathcal{M}$. Show that the vertical subspace at a point is spanned by $\{\partial/\partial v^a\}$. Show that a coordinate transformation induced by a change of coordinates in $\mathcal{M}$ does not affect this conclusion. Show that, on the other hand, the subspace of $T_vT\mathcal{M}$ spanned by $\{\partial/\partial x^a\}$ is not invariant under changes of coordinates of this kind. □

The point of Exercise 5 is that, while the bundle structure of $T\mathcal{M}$ picks out the vertical subspaces in an invariant way, it does not invariantly distinguish a complementary subspace at each point. This is another manifestation of the fact that $T\mathcal{M}$ is not in general a product manifold.

The fibres of $T\mathcal{M}$ are vector spaces and therefore have the usual affine structure of vector spaces so far as vectors tangent to them are concerned. Thus given a vertical vector $w \in T_vT\mathcal{M}$ we can define a vertical vector field all over the fibre through v by affine parallelism. Moreover, a point of a vector space may equally well be thought of as a vector tangent to it at the origin, or indeed at any other point. Thus given any $u \in T_{\pi(v)}\mathcal{M}$ one may define a vertical vector at v (or at any other point in the fibre) by regarding u as tangent to $T_{\pi(v)}\mathcal{M}$ at the origin and applying affine parallel translation to bring it to v. The vector field thus obtained is called the *vertical lift* of u from $\pi(v)$ to v and is denoted $u^\uparrow$.

Exercise 6. Show that $u^\uparrow$ is the tangent at $t = 0$ to the curve $t \mapsto v + tu$ in the fibre through v. Show that if $u = u^a\partial/\partial x^a$ with respect to coordinates (x^a) on $\mathcal{M}$ then $u^\uparrow = u^a\partial/\partial v^a$. □

Exercise 7. Show that for any vector field V on $\mathcal{M}$, the transformations $\phi_t: T\mathcal{M} \to T\mathcal{M}$ by $\phi_t(v) = v + tV_{\pi(v)}$ form a one-parameter group mapping each fibre to itself, whose infinitesimal generator is the vector field $V^\uparrow$ whose value at each point v is just the vertical lift of $V_{\pi(v)}$ to v. □

We define next another way of lifting a vector field, say W, from $\mathcal{M}$ to $T\mathcal{M}$, which leads not to a vertical vector field but to one which projects onto W. Given a point $v \in T\mathcal{M}$, let σ be the integral curve of W with $\sigma(0) = \pi(v)$ and let V be the vector field along σ obtained by Lie transport of v by W. Then (σ, V) defines a curve in $T\mathcal{M}$ through v; its tangent vector there evidently projects onto $\dot\sigma(0) = W_{\pi(v)}$. We therefore define $\tilde{W}$, the *complete lift* of W to $T\mathcal{M}$, as follows: $\tilde{W}_v$ is the tangent vector at $t = 0$ to the curve (σ, V).

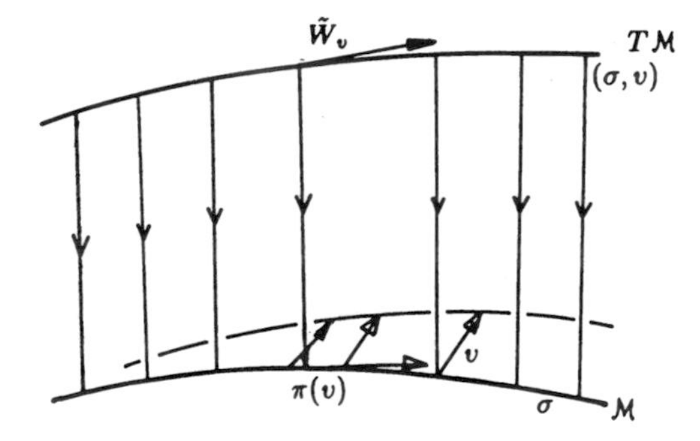

Fig. 1 The complete lift construction.

Exercise 8. Show that the flow $\tilde\phi$ on $T\mathcal{M}$ generated by $\tilde{W}$ is given by $\tilde\phi_t(v) = \phi_{t*}v$, $\pi \circ \tilde\phi_t = \phi_t \circ \pi$, where ϕ is the flow generated by W. □

Exercise 9. Show that if $W = W^a\partial/\partial x^a$ then

$$\tilde{W} = W^a\frac{\partial}{\partial x^a} + v^b\frac{\partial W^a}{\partial x^b}\frac{\partial}{\partial v^a}.$$ □

This is an appropriate point at which to make some comments about maps of manifolds and of tangent bundles. One can build out of a map $\phi\colon \mathcal{M} \to \mathcal{N}$ of manifolds an induced map $T\phi\colon T\mathcal{M} \to T\mathcal{N}$ of tangent bundles by combining into one object all the maps of tangent spaces induced by ϕ: that is, for each $v \in T\mathcal{M}$,

$$T\phi(v) = \phi_{*\pi(v)}v.$$

Exercise 10. Show that $T\phi$ is a smooth map of manifolds and satisfies $\pi_{\mathcal{N}} \circ T\phi = \phi \circ \pi_{\mathcal{M}}$, where $\pi_{\mathcal{M}}$ and $\pi_{\mathcal{N}}$ are the projections on $\mathcal{M}$ and $\mathcal{N}$. Show that if $\phi\colon \mathcal{M} \to \mathcal{N}$ and $\psi\colon \mathcal{M} \to \mathcal{P}$ then $T(\phi \circ \psi) = T\phi \circ T\psi$. □

The map $T\phi$ respects the bundle structure of $T\mathcal{M}$ and $T\mathcal{N}$ in the sense that if v and v' belong to the same fibre of $T\mathcal{M}$ then their images $T\phi(v)$ and $T\phi(v')$ belong to the same fibre of $T\mathcal{N}$; this is the content of the property $\pi_{\mathcal{N}} \circ T\phi = \phi \circ \pi_{\mathcal{M}}$. More generally, a map $\Phi\colon T\mathcal{M} \to T\mathcal{N}$ which preserves fibres in this way is called a *bundle map*: Φ is a bundle map if and only if there is a smooth map $\phi\colon \mathcal{M} \to \mathcal{N}$ such that $\pi_{\mathcal{N}} \circ \Phi = \phi \circ \pi_{\mathcal{M}}$.

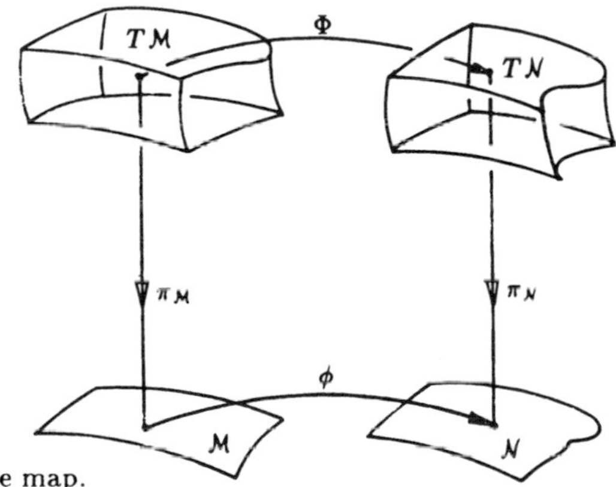

Fig. 2 A bundle map.

Exercise 11. Show that a map $\Phi\colon T\mathcal{M} \to T\mathcal{M}$ which maps each fibre to itself is a particular case of a bundle map in which the corresponding map of the base is just the identity map of $\mathcal{M}$. □

The transformations of the one-parameter group generated by $V^{\uparrow}$ are examples.

We say that a vector field V on $T\mathcal{M}$ is *projectable* if there is a vector field on $\mathcal{M}$ which is π-related to V. When V is projectable we write $\pi_* V$ for the π-related field on $\mathcal{M}$.

Exercise 12. Show that a vector field on $T\mathcal{M}$ (assumed complete, for simplicity) is projectable if and only if its one-parameter group consists of bundle maps. □

The one-parameter group of a complete lift consists of bundle maps.

Exercise 13. Show that, for vector fields V, W on $\mathcal{M}$,

$$[V^{\uparrow}, W^{\uparrow}] = 0 \qquad [\tilde{V}, W^{\uparrow}] = [V, W]^{\uparrow} \qquad [\tilde{V}, \tilde{W}] = \widetilde{[V, W]}.$$ □

Exercise 14. The corresponding lifts from $\mathcal{M}$ to $T^*\mathcal{M}$ are defined as follows. Let α be a 1-form on $\mathcal{M}$: its vertical lift $\alpha^\uparrow$ is the vector field which generates the one-parameter group $p \mapsto p + t\alpha_{\tau(p)}$ ($p \in T^*\mathcal{M}$). Let W be a vector field on $\mathcal{M}$: its complete lift $\tilde{W}$ to $T^*\mathcal{M}$ is the vector field which generates the flow $\tilde{\phi}_t\colon p \mapsto \phi_{-t}{}^*{}_{\tau(p)}(p)$. Show that if $\alpha_a dx^a$ and $W = W^a\partial_a$ then

$$\alpha^\uparrow = \alpha_a\partial/\partial p_a \quad \text{and} \quad \tilde{W} = W^a\partial/\partial x^a - p_b(\partial W^b/\partial x^a)\partial/\partial p_a.$$

(Here the coordinates of a generic point $p \in T^*\mathcal{M}$ are (x^a, p_a) where $p = p_a dx^a$.) Show that $\tau_*\alpha^\uparrow = 0$ and $\tau_*\tilde{W} = W$. □

Exercise 15. Show that on $T^*\mathcal{M}$

$$[\alpha^\uparrow, \beta^\uparrow] = 0 \qquad [\tilde{V}, \alpha^\uparrow] = (\mathcal{L}_V\alpha)^\uparrow \qquad [\tilde{V}, \tilde{W}] = \widetilde{[V, W]}.$$ □

Another construction on $T\mathcal{M}$, which exploits the linear structure of its fibres, leads to the definition of the *dilation field* Δ. A one-parameter group of dilation maps δ_t is defined by $\delta_t v = e^t v$ for all $t \in \mathbf{R}$; Δ is its infinitesimal generator. The dilation field is of considerable use when one wants to deal with objects on $T\mathcal{M}$ which are homogeneous in the fibre coordinates v^a.

Exercise 16. Show that $\Delta = v^a\partial/\partial v^a$. □

Exercise 17. Show that the vector field $p_a\partial/\partial p_a$ on $T^*\mathcal{M}$ may be similarly described, globally, in terms of a one-parameter group of dilations of the fibres. □

Exercise 18. Show that, on $T\mathcal{M}$, $[\Delta, V^\uparrow] = -V^\uparrow$ and $[\Delta, \tilde{V}] = 0$. □

In effect, $V^\uparrow$ is homogeneous of degree -1 and $\tilde{V}$ homogeneous of degree 0 in the fibre coordinates; these results are related to Euler's theorem on homogeneous functions. The dilation field is a convenient example of a vertical vector field which is not a vertical lift.

The reader may have noticed that, although we defined the vertical lift for a vector at a point of $\mathcal{M}$, we defined the complete lift only for vector fields. A comparison of the definitions will make it clear why: the complete lift depends on the action of the flow generated by a vector field on $\mathcal{M}$ and this in turn depends, not just on the value of the vector field at a point, but on its values in a neighbourhood of the point. The vertical lift, on the other hand, is a truly pointwise construction.

In fact, for each $v \in T\mathcal{M}$ the vertical lift gives an isomorphism of $T_{\pi(v)}\mathcal{M}$ with the vertical subspace of $T_vT\mathcal{M}$. With this in mind, one may construct a type (1,1) tensor field on $T\mathcal{M}$ which represents the projection of vectors, namely the tensor field S defined by

$$S_v(w) = (\pi_* w)^\uparrow \qquad \text{for all } w \in T_vT\mathcal{M},$$

the lift being to v also. Notice that the vertical subspace of $T_vT\mathcal{M}$ is distinguished by being simultaneously the kernel and the image of the linear map $S_v\colon T_vT\mathcal{M} \to T_vT\mathcal{M}$. Thus $S^2 = S \circ S = 0$. In terms of local coordinates, $S = \partial/\partial v^a \otimes dx^a$. The tensor field S, considered as defining linear maps of tangent spaces, is sometimes called the *vertical endomorphism*.

Exercise 19. Any type (1,1) tensor field A on $\mathcal{M}$ may be lifted to give a type (1,1) tensor field $A^\uparrow$ on $T\mathcal{M}$ by setting $A^\uparrow_v(w) = \left(A_{\pi(v)}(\pi_* w)\right)^\uparrow$; show that $S = I^\uparrow$ where I is the identity tensor field on $\mathcal{M}$. □

Exercise 20. It follows from Chapter 10, Exercise 17 that for any tensor field A of type $(1,1)$ on any manifold the rule

$$N_A(V,W) = A^2([V,W]) + [A(V),A(W)] - A([A(V),W]) - A([V,A(W)]),$$

where V, W are any vector fields, defines a tensor field N_A of type $(1,2)$. Show that $N_S = 0$ (as a tensor field of type $(1,2)$ on $T\mathcal{M}$). (Since for any local basis of vector fields $\{V_a\}$ on $\mathcal{M}$ the vector fields $\{V_a^\uparrow, \tilde{V}_a\}$ form a local basis of vector fields on $T\mathcal{M}$, it is sufficient to show that N_S vanishes when its arguments are vertical or complete lifts.) Show that, conversely, if a manifold admits a type $(1,1)$ tensor field S whose kernel and image (as a linear map of tangent spaces) coincide everywhere, and which satisfies $N_S = 0$, then the image subspaces form a distribution which is integrable in the sense of Frobenius's theorem. Show further that local coordinates (x^a, v^a) may be found such that the integral manifolds are given by $x^a =$ constant and S takes the form $\partial/\partial v^a \otimes dx^a$. □

Such a construction is not available on the cotangent bundle, but the cotangent bundle does nevertheless carry a canonical object which is closely related to its projection map; it happens that in the case of the cotangent bundle this object is a 1-form. The *canonical 1-form* θ is defined as follows: for $w \in T_pT^*\mathcal{M}$

$$\langle w, \theta_p \rangle = \langle \tau_* w, p \rangle.$$

The definition simply takes advantage of the fact that a point p of $T^*\mathcal{M}$ is a cotangent vector to $\mathcal{M}$. In terms of local coordinates (x^a, p_a),

$$\left\langle \frac{\partial}{\partial x^a}, \theta \right\rangle = p_a \qquad \left\langle \frac{\partial}{\partial p_a}, \theta \right\rangle = 0,$$

and therefore
$$\theta = p_a dx^a.$$

Thus $d\theta = dp_a \wedge dx^a$. The 2-form $d\theta$ is therefore already in the Darboux form in terms of standard coordinates on $T^*\mathcal{M}$ (see Section 6 of Chapter 6).

Exercise 21. (On $T\mathcal{M}$.) Show that $\mathcal{L}_{W^\uparrow} S = 0$ and $\mathcal{L}_\Delta S = -S$. Show that if V is vertical and $\mathcal{L}_V S = 0$ then V is a vertical lift. □

Exercise 22. (On $T^*\mathcal{M}$.) Show that $\mathcal{L}_{\alpha^\uparrow} \theta = \tau^*\alpha$ and $\mathcal{L}_\Delta \theta = \theta$. □

Exercise 23. Show that for any complete lift $\tilde{W}$ on $T\mathcal{M}$, $\mathcal{L}_{\tilde{W}} S = 0$, and for any complete lift $\tilde{W}$ on $T^*\mathcal{M}$, $\mathcal{L}_{\tilde{W}} \theta = 0$. □

3. Connections and the Tangent Bundle

We show how a connection on $\mathcal{M}$ may be described by structures on $T\mathcal{M}$.

We may use the connection, together with the observation that a curve in $\mathcal{M}$ with a vector field along it defines a curve in $T\mathcal{M}$, to define a new process for lifting curves from $\mathcal{M}$ to $T\mathcal{M}$. Let σ be a curve in $\mathcal{M}$, v a point of $T\mathcal{M}$ with $\pi(v) = \sigma(0)$. Let V be the vector field along σ obtained by parallel translation of v. Then we call the curve σ^h in $T\mathcal{M}$ defined by $\sigma^h(t) = (\sigma(t), V(t))$ the *horizontal lift* of σ through v. Evidently $\pi \circ \sigma^h = \sigma$. By means of this construction for curves we may also define a horizontal lift of vectors tangent to $\mathcal{M}$, that is, a map $T_x\mathcal{M} \to T_vT\mathcal{M}$, as follows. Given $u \in T_x\mathcal{M}$ let σ be a curve through x such that $\dot{\sigma}(0) = u$. Now if V is parallel along σ and $V(0) = v$ then at $\sigma(0) = x$

$$\dot{V}^a(0) + \Gamma^a_{bc}(x) v^b u^c = 0$$

and so the tangent vector to σ^h at $t = 0$ is

$$\dot{\sigma}^a(0)\frac{\partial}{\partial x^a} + \dot{V}^a(0)\frac{\partial}{\partial v^a} = u^c\left(\frac{\partial}{\partial x^c} - \Gamma^a_{bc}(x)v^b\frac{\partial}{\partial v^a}\right).$$

Note first of all that this is independent of the choice of σ (subject to the condition that its initial tangent vector be u), and so the rule is adequate to define a map from $T_x\mathcal{M} \to T_vT\mathcal{M}$. We denote by u^h the image of u, which we call the *horizontal lift* of u to $T_vT\mathcal{M}$. The following conditions are satisfied by the horizontal lift:

(1) $\pi_* u^h = u$

(2) u^h depends linearly on u

(3) $u^h = 0$ if and only if $u = 0$.

The map $u \mapsto u^h$ is thus a linear map $T_{\pi(v)}\mathcal{M} \to T_vT\mathcal{M}$ which is injective, having π_* as a left inverse. Its image is a subspace of $T_vT\mathcal{M}$ isomorphic to $T_{\pi(v)}\mathcal{M}$ and complementary to the vertical subspace; we call this the *horizontal subspace* defined by the connection (the reason for the use of the word "horizontal" to describe this general construction should now be clear).

Exercise 24. Show that the horizontal subspaces have the vector fields $H_a = \partial/\partial x^a - \Gamma^c_{ba}v^b\partial/\partial v^c$ as local bases, so that these and the vertical vector fields $V_a = \partial/\partial v^a$ form a local basis of vector fields on $T\mathcal{M}$ adapted to the connection. Show that the dual basis of 1-forms is $\{\theta^a, \phi^a\}$ where $\theta^a = dx^a$ and $\phi^a = dv^a + \Gamma^a_{bc}v^b dx^c$. □

Exercise 25. Show that if W^h is the horizontal lift of any vector field W on $\mathcal{M}$ then $[\Delta, W^h] = 0$. □

Thus a connection on $\mathcal{M}$ defines a distribution of subspaces on $T\mathcal{M}$ which is horizontal, in the sense of being complementary to the vertical subspace at each point, and which satisfies a homogeneity condition expressed by the result of Exercise 25. In fact this structure is equivalent to the existence of a connection. For suppose there is given on $T\mathcal{M}$ a distribution of horizontal subspaces which is smooth, in the following sense. Since these subspaces are horizontal, that is, complementary to the vertical, the map $\pi_*: T_vT\mathcal{M} \to T_{\pi(v)}\mathcal{M}$ when restricted to the horizontal subspace of $T_vT\mathcal{M}$ is an isomorphism. The horizontal lift is therefore well-defined, so that given any vector field W on $\mathcal{M}$ there is a unique horizontal vector field W^h such that $\pi_*W^h = W$. The smoothness condition we require is that the horizontal lifts of smooth (local) vector fields be smooth. Now given a local coordinate system on $\mathcal{M}$ with corresponding coordinates (x^a, v^a) on $T\mathcal{M}$, we may write

$$\left(\frac{\partial}{\partial x^a}\right)^h = \frac{\partial}{\partial x^a} - \Gamma^b_a\frac{\partial}{\partial v^b}$$

where Γ^b_a are local functions on $T\mathcal{M}$. But by assumption $[\Delta, W^h] = 0$ for all horizontal lifts and therefore

$$\left[v^c\frac{\partial}{\partial v^c}, \frac{\partial}{\partial x^a} - \Gamma^b_a\frac{\partial}{\partial v^b}\right] = \left(-v^c\frac{\partial}{\partial v^c}(\Gamma^b_a) + \Gamma^b_a\right) = 0$$

so that

$$v^c\frac{\partial}{\partial v^c}(\Gamma^b_a) = \Gamma^b_a.$$

This means that as functions of v^c the Γ_a^b are homogeneous of the first order; but they are also smooth functions, and in particular smooth at $v^c = 0$, and the only first order homogeneous functions on $\mathbf{R}^m$ smooth everywhere including the origin are linear functions. Thus

$$\Gamma_a^b = \Gamma_{ca}^b v^c,$$

where the Γ_{ca}^b are now local functions on $\mathcal{M}$.

Exercise 26. Show that under a coordinate transformation on $\mathcal{M}$, and the corresponding induced coordinate transformation on $T\mathcal{M}$, the functions Γ_{ca}^b transform as the components of a connection. □

Note the importance of the homogeneity condition: it is not sufficient to give a distribution of horizontal subspaces; unless the homogeneity condition is satisfied the horizontal distribution will not define a connection.

We next describe how to represent the covariant derivative as an operation on $T\mathcal{M}$. As a guide, note that the complete lift of a vector field is defined in terms of Lie transport, and that $[\tilde{V}, W^\uparrow] = [V,W]^\uparrow = (\mathcal{L}_V W)^\uparrow$ gives the corresponding derivative operation. Now the horizontal lift is defined analogously in terms of parallel transport. It should seem appropriate, therefore, to consider, for any pair of vector fields V, W on $\mathcal{M}$ the vector field $[V^h, W^\uparrow]$. This is a vertical vector field, since $\pi_* V^h = V$, $\pi_* W^\uparrow = 0$, and therefore $\pi_*[V^h, W^\uparrow] = [V, 0] = 0$. Furthermore,

$$\begin{aligned}[\Delta, [V^h, W^\uparrow]] &= [[\Delta, V^h], W^\uparrow] + [V^h, [\Delta, W^\uparrow]] \\ &= 0 + [V^h, -W^\uparrow] = -[V^h, W^\uparrow]\end{aligned}$$

and therefore (arguing just as we did to show the linearity of Γ_a^b above) it follows that $[V^h, W^\uparrow]$ is actually the vertical lift of a vector field on $\mathcal{M}$. We denote this vector field $\nabla(V,W)$, so as not to prejudge the issue too blatantly. Clearly $\nabla(V,W)$ is linear (over $\mathbf{R}$) in both arguments. To determine whether this procedure defines a covariant derivative we have to examine the effect of multiplying V or W by a function f on $\mathcal{M}$.

Exercise 27. Show that $(fV)^h = (f \circ \pi)V^h$ and $(fV)^\uparrow = (f \circ \pi)V^\uparrow$. Show that for any projectable vector field U on $T\mathcal{M}$, $U(f \circ \pi) = ((\pi_* U)f) \circ \pi$, while for any vertical vector field U, $U(f \circ \pi) = 0$. □

Thus

$$\begin{aligned}\nabla(fV, W)^\uparrow &= [(f \circ \pi)V^h, W^\uparrow] = (f \circ \pi)[V^h, W^\uparrow] - W^\uparrow(f \circ \pi)V^h \\ &= (f \circ \pi)[V^h, W^\uparrow] = (f\nabla(V,W))^\uparrow\end{aligned}$$

and

$$\begin{aligned}\nabla(V, fW)^\uparrow &= [V^h, (f \circ \pi)W^\uparrow] = (f \circ \pi)[V^h, W^\uparrow] + V^h(f \circ \pi)W^\uparrow \\ &= (f \circ \pi)[V^h, W^\uparrow] + (Vf \circ \pi)W^\uparrow \\ &= (f\nabla(V,W) + (Vf)W)^\uparrow.\end{aligned}$$

Thus $\nabla(V,W) = \nabla_V W$ satisfies the conditions of a covariant differentiation operator on $\mathcal{M}$. (It is worth spending a moment realising where a similar argument purporting to show that $[\tilde{V}, W^\uparrow]$ defines a connection—which could not be correct, of course—breaks down.)

Exercise 28. Show that $[H_a, V_b] = \Gamma^c_{ba} V_c$ (H_a and V_a are defined in Exercise 24). □

It follows that the covariant differentiation operator defined by $(\nabla_V W)^\uparrow = [V^h, W^\uparrow]$ is the one appropriate to the connection we started with.

Curvature. The horizontal distribution defining a connection need not be integrable in the sense of Frobenius's theorem: the bracket of two horizontal vector fields need not be horizontal. In fact the departure from integrability is given by the curvature of the connection. Consider the bracket of two horizontal lifts $[U^h, V^h]$, where U, V are vector fields on $\mathcal{M}$: it projects onto $[U, V]$, and so its horizontal component is $[U,V]^h$. Thus $[U,V]^h - [U^h, V^h]$ is vertical. We set

$$R(U,V) = [U,V]^h - [U^h, V^h];$$

then $R(U,V)$ is a vertical vector field, $\mathbf{R}$-linear and skew-symmetric in U and V. Moreover $[\Delta, R(U,V)] = 0$, so that $R(U,V)$ has components which depend linearly on v^a. It is closely related to the curvature of the connection. In fact

$$\begin{aligned}
(\nabla_U \nabla_V W &- \nabla_V \nabla_U W - \nabla_{[U,V]} W)^\uparrow \\
&= [U^h, (\nabla_V W)^\uparrow] - [V^h, (\nabla_U W)^\uparrow] - [[U,V]^h, W^\uparrow] \\
&= [U^h, [V^h, W^\uparrow]] - [V^h, [U^h, W^\uparrow]] - [[U,V]^h, W^\uparrow] \\
&= [[U^h, V^h], W^\uparrow] - [[U,V]^h, W^\uparrow] = [W^\uparrow, R(U,V)]
\end{aligned}$$

(using Jacobi's identity to rearrange the double bracket terms).

Exercise 29. Show that $[H_a, H_b] = -R^d{}_{cab} v^c \partial/\partial v^d$, and thus that

$$[U,V]^h - [U^h, V^h] = R^d{}_{cab} U^a V^b v^c \frac{\partial}{\partial v^d}.$$ □

Exercise 30. Describe how to construct the torsion of the connection in a similar way. □

Note that the horizontal distribution is integrable if and only if the curvature vanishes, in which case the horizontal subspaces are tangent to submanifolds of $T\mathcal{M}$.

4. The Geodesic Spray

We suppose now that the connection is symmetric.

If γ is a geodesic of the connection then, since its tangent vector is parallel along it, its natural lift $(\gamma, \dot\gamma)$ to $T\mathcal{M}$ is a horizontal curve. If $\dot\gamma(0) = v$ then the tangent vector to the natural lift projects onto v so that it is actually the horizontal lift v^h of v to $T_v T\mathcal{M}$. We therefore consider the vector field Γ on $T\mathcal{M}$ given by

$$\Gamma_v = v^h$$

(horizontal lift of v to $T_v T\mathcal{M}$). The integral curves of Γ consist precisely of the natural lifts to $T\mathcal{M}$ of all geodesics of the connection: from the fact that $\pi_* \Gamma_v = v$, it follows that the integral curves of Γ are natural lifts of curves in $\mathcal{M}$; from the fact that Γ is horizontal it follows that these curves are geodesics. The vector field Γ is called the *geodesic spray*.

Exercise 31. Suppose that a vector field Γ on $T\mathcal{M}$ has the property that, for every $v \in T\mathcal{M}$, $\pi_*\Gamma_v = v$. Show that in local coordinates $\Gamma = v^a\partial/\partial x^a + \Gamma^a\partial/\partial v^a$ for some functions Γ^a locally defined on $T\mathcal{M}$, and that the integral curves of Γ are the natural lifts of solutions of the equations $\ddot{x}^a = \Gamma^a(x^b, \dot{x}^c)$. □

Any vector field Γ satisfying the condition $\pi_*\Gamma_v = v$, which may be equivalently written $S(\Gamma) = \Delta$, is called a *second-order differential equation field*. A geodesic spray satisfies an additional homogeneity condition, which derives from the affine reparametrisation property of geodesics. If γ is a geodesic then, for constant k, $t \mapsto \gamma(kt)$ is the geodesic through $\gamma(0)$ with initial tangent vector $k\dot{\gamma}(0)$. We set $k = e^s$ and denote the natural lift of γ by $\hat{\gamma}$, and then this may be written $\hat{\gamma}(e^s t) = \delta_s\hat{\gamma}(t)$, where δ_s is the one-parameter group of dilations of the fibres of $T\mathcal{M}$. But $\hat{\gamma}$ is an integral curve of Γ, and so by differentiation $e^s\Gamma = \delta_{s*}\Gamma$, from which it follows that

$$[\Delta, \Gamma] = \mathcal{L}_\Delta\Gamma = \frac{d}{ds}\left(\delta_{s*}\Gamma\right)_{s=0} = \Gamma.$$

Exercise 32. Check, by writing $\Gamma = v^a H_a$ and computing $[\Delta, \Gamma]$ directly. □

We shall show that, conversely, a second-order differential equation field Γ satisfying $[\Delta, \Gamma] = \Gamma$ is the geodesic spray of a symmetric connection. To do so, we shall have to construct the horizontal distribution defining the connection. A clue as to how this may be done is provided by the following exercise.

Exercise 33. Show that if Γ is the geodesic spray of a symmetric connection, whose basic horizontal vector fields are H_a, then $(\mathcal{L}_\Gamma S)(V_a) = V_a$ but $(\mathcal{L}_\Gamma S)(H_a) = -H_a$. Show that a vector w tangent to $T\mathcal{M}$ is vertical if and only if $(\mathcal{L}_\Gamma S)(w) = w$ and horizontal if and only if $(\mathcal{L}_\Gamma S)(w) = -w$. □

Thus when the connection is known the direct sum decomposition of the tangent spaces to $T\mathcal{M}$ into vertical and horizontal subspaces may be defined in terms of $\mathcal{L}_\Gamma S$: as a linear map of tangent spaces $\mathcal{L}_\Gamma S$ has eigenvalues ± 1, its eigenspace at any point corresponding to the eigenvalue $+1$ being the vertical subspace and its eigenspace corresponding to the eigenvalue -1 being the horizontal subspace.

When only a second-order differential equation field Γ is known, one may use this construction to define the horizontal distribution, once it has been confirmed that $\mathcal{L}_\Gamma S$ still has these properties. We show first that, for any vertical vector field V, $(\mathcal{L}_\Gamma S)(V) = V$. It is sufficient to do this when V is a vertical lift, since the given equation is tensorial and the vertical lifts span the vertical subspace at each point. Now

$$(\mathcal{L}_\Gamma S)(U^\uparrow) = [\Gamma, S(U^\uparrow)] - S([\Gamma, U^\uparrow]) = S([U^\uparrow, \Gamma]).$$

On the other hand, since the Lie derivative of S by a vertical lift is zero,

$$0 = (\mathcal{L}_{U^\uparrow} S)(\Gamma) = [U^\uparrow, S(\Gamma)] - S([U^\uparrow, \Gamma])$$

and so

$$S([U^\uparrow, \Gamma]) = [U^\uparrow, S(\Gamma)] = [U^\uparrow, \Delta] = U^\uparrow.$$

Thus

$$(\mathcal{L}_\Gamma S)(U^\uparrow) = U^\uparrow$$

as required. Conversely, if for some vector field W on $T\mathcal{M}$, $(\mathcal{L}_\Gamma S)(W) = W$, then $S(W) = S\big((\mathcal{L}_\Gamma S)(W)\big) = -(\mathcal{L}_\Gamma S)\big(S(W)\big)$ (using the fact that, since $S^2 = 0$, $S \circ \mathcal{L}_\Gamma S = -\mathcal{L}_\Gamma S \circ S$). But since $S(W)$ is vertical, $(\mathcal{L}_\Gamma S)\big(S(W)\big) = S(W)$, whence $S(W) = -S(W) = 0$ and W is therefore vertical. This establishes that $\mathcal{L}_\Gamma S$ has eigenvalue $+1$ with the vertical subspace as eigenspace.

Consider next $(\mathcal{L}_\Gamma S)(\tilde{U})$ for any complete lift $\tilde{U}$. Using a similar argument to the one for $U^\uparrow$

$$(\mathcal{L}_\Gamma S)(\tilde{U}) = [\Gamma, S(\tilde{U})] - S\big([\Gamma, \tilde{U}]\big) = [\Gamma, U^\uparrow] + S\big([\tilde{U}, \Gamma]\big).$$

On the other hand

$$\begin{aligned} 0 = \mathcal{L}_{\tilde{U}} S(\Gamma) &= [\tilde{U}, S(\Gamma)] - S\big([\tilde{U}, \Gamma]\big) \\ &= [\tilde{U}, \Delta] - S\big([\tilde{U}, \Gamma]\big) = -S\big([\tilde{U}, \Gamma]\big). \end{aligned}$$

(Notice, in passing, that this shows that $[\tilde{U}, \Gamma]$ is vertical.) Thus

$$(\mathcal{L}_\Gamma S)(\tilde{U}) = [\Gamma, U^\uparrow].$$

Now in the earlier argument it was shown that $S\big([U^\uparrow, \Gamma]\big) = U^\uparrow$, or in other words, $S\big(\tilde{U} + [\Gamma, U^\uparrow]\big) = 0$; thus $\tilde{U} + [\Gamma, U^\uparrow]$, or equally $\tilde{U} + (\mathcal{L}_\Gamma S)(\tilde{U})$, is vertical. It follows, by what we have already proved, that

$$(\mathcal{L}_\Gamma S)\big(\tilde{U} + (\mathcal{L}_\Gamma S)(\tilde{U})\big) = \tilde{U} + (\mathcal{L}_\Gamma S)(\tilde{U})$$

whence

$$(\mathcal{L}_\Gamma S)^2(\tilde{U}) = \tilde{U}.$$

Thus $(\mathcal{L}_\Gamma S)^2$ is the identity tensor on $T\mathcal{M}$: it is certainly the identity on vertical vectors, and these together with complete lifts span the tangent space to $T\mathcal{M}$ at each point.

Consider now the tensor fields P and Q on $T\mathcal{M}$ given by

$$P = \tfrac{1}{2}(I - \mathcal{L}_\Gamma S) \qquad\qquad Q = \tfrac{1}{2}(I + \mathcal{L}_\Gamma S).$$

It follows from the fact that $(\mathcal{L}_\Gamma S)^2 = I$ that P and Q have the following properties:

$$\begin{aligned} P^2 &= P & Q^2 &= Q \\ P \circ Q = Q \circ P &= 0 & P + Q &= I. \end{aligned}$$

Such tensor fields are projection operators corresponding to a direct sum decomposition of tangent spaces: at each point of $T\mathcal{M}$, the kernel of P coincides with the image of Q, and *vice versa*; the kernels of P and Q are complementary subspaces. We have already established that the kernel of P is the vertical subspace. We call the kernel of Q the horizontal subspace determined by Γ.

Note that we have not used any homogeneity property of Γ in this argument, only the fact that it is a second-order differential equation field. This construction of a horizontal distribution works for any second-order differential equation field.

We have still to show that if Γ is a spray then the horizontal distribution thus defined is a connection (that is, that it satisfies the homogeneity condition $[U^h, \Delta] = 0$); that it is symmetric; and that Γ is its geodesic spray.

For the first, we may define the horizontal lift of any vector field U on $\mathcal{M}$ as the horizontal projection of any vector field on $T\mathcal{M}$ that projects onto U: $\tilde{U}$ for example. Then

$$U^h = Q(\tilde{U}) = \tfrac{1}{2}(\tilde{U} + (\mathcal{L}_\Gamma S)(\tilde{U})) = \tfrac{1}{2}(\tilde{U} + [\Gamma, U^\uparrow]).$$

Thus when $[\Delta, \Gamma] = \Gamma$

$$\begin{aligned} [\Delta, U^h] &= \tfrac{1}{2}([\Delta, \tilde{U}] + [\Delta, [\Gamma, U^\uparrow]]) \\ &= \tfrac{1}{2}([[\Delta, \Gamma], U^\uparrow] + [\Gamma, [\Delta, U^\uparrow]]) \\ &= \tfrac{1}{2}([\Gamma, U^\uparrow] - [\Gamma, U^\uparrow]) = 0. \end{aligned}$$

Thus homogeneity is established. For symmetry, observe that

$$\begin{aligned} &[U^h, V^\uparrow] - [V^h, U^\uparrow] \\ &\qquad = \tfrac{1}{2}([\tilde{U}, V^\uparrow] + [[\Gamma, U^\uparrow], V^\uparrow] - [\tilde{V}, U^\uparrow] - [[\Gamma, V^\uparrow], U^\uparrow]) \\ &\qquad = \tfrac{1}{2}([\Gamma, [U^\uparrow, V^\uparrow]] + [U, V]^\uparrow - [V, U]^\uparrow) = [U, V]^\uparrow, \end{aligned}$$

and so the connection is symmetric. Finally, since it is already known that Γ is a spray we have only to show that it is horizontal to show that it is the geodesic spray of the connection. Now

$$(\mathcal{L}_\Gamma S)(\Gamma) = [\Gamma, S(\Gamma)] - S([\Gamma, \Gamma]) = [\Gamma, \Delta] = -\Gamma$$

and so Γ is horizontal.

In effect, we have shown that a symmetric connection is uniquely determined by its geodesic spray, and we have shown how to reconstruct the connection from the spray. It is also important to note, for later use, that the construction of a horizontal distribution described above will work for any second-order differential equation field, not only for a spray—though it will be a connection only for a spray.

5. The Exponential Map and Jacobi Fields

The exponential map (Chapter 11, Section 7) is the map $T_x\mathcal{M} \to \mathcal{M}$ defined by

$$\exp(v) = \gamma_v(1) \qquad v \in T_x\mathcal{M}$$

where γ_v is the geodesic satisfying $\gamma_v(0) = x$ and $\dot{\gamma}_v(0) = v$. In terms of the geodesic spray, whose flow is $\hat{\gamma}$, say, one may redefine exp as follows. Regard $T_x\mathcal{M}$ as a fibre of $T\mathcal{M}$. There will be some neighbourhood of 0 in $T_x\mathcal{M}$ on which $\hat{\gamma}_1(v)$, the flow through parameter value 1, is defined, since $\hat{\gamma}_t(0)$ is defined for all t. Combining the map $\hat{\gamma}_1$ with projection gives the exponential:

$$\exp = \pi \circ \hat{\gamma}_1.$$

In Section 7 of Chapter 11 we showed that exp is a diffeomorphism on a neighbourhood of 0 in $T_x\mathcal{M}$ by computing $\exp_*$ at 0 and showing it to be the identity. In the present context, the argument goes like this. For $u \in T_x\mathcal{M}$, $\exp_* u$ is the tangent at $t = 0$ to the curve $t \mapsto \exp(tu)$. Now

$$\exp(tu) = \pi(\hat{\gamma}_1(tu)) = \pi(\hat{\gamma}_t(u))$$

since Γ is a spray, and so

$$\exp_*(u) = \pi_*\Gamma_u = u$$

as required.

The exponential map will not necessarily be a diffeomorphism of the whole of $T_x\mathcal{M}$. We describe how it may fail to be so, at the infinitesimal level, by examining $\exp_{*w}$ for $w \in T_x\mathcal{M}$ different from 0, and showing how it can fail to be an isomorphism. Note that we can describe $\exp_{*w}$ as follows. For any vector u tangent to $T_x\mathcal{M}$ at $w \in T_x\mathcal{M} \subset T\mathcal{M}$ (that is, any vertical vector at w),

$$\exp_{*w}(u) = \pi_* \circ \hat{\gamma}_{1*w}(u).$$

Now $\hat{\gamma}_{1*w}(u)$ is the vector at $\hat{\gamma}_1(w)$ obtained by Lie transport of u by the flow of Γ. Thus $\exp_{*w}$ is given by Lie transport followed by projection, and is applied to vertical vectors. In general, vertical vectors do not stay vertical when Lie transported. However, $\exp_{*w}$ will fail to be an isomorphism if there is a vertical vector u at w for which $\hat{\gamma}_{1*w}(u)$ is also vertical.

Lie transport by the flow of Γ is of interest in its own right. We shall call a vector field defined along an integral curve of Γ by Lie transport a *Jacobi field*. The same term is applied to the projected field along the geodesic obtained by projecting the integral curve. Suppose that J is a Jacobi field along the integral curve $(\gamma, \dot{\gamma})$ of Γ. We write

$$J = \lambda^a H_a + \mu^a V_a.$$

Then the condition $\mathcal{L}_\Gamma J = 0$ is equivalent to

$$\Gamma(\lambda^a)H_a + \lambda^a[\Gamma, H_a] + \Gamma(\mu^a)V_a + \mu^a[\Gamma, V_a] = 0.$$

Now $\Gamma = v^a H_a$, whence

$$[\Gamma, H_a] = v^b[H_b, H_a] - H_a(v^b)H_b = R^d{}_{cab}v^b v^c V_d + \Gamma^b_{ac}v^c H_b$$
$$[\Gamma, V_a] = v^b[H_b, V_a] - H_a = \Gamma^c_{ab}v^b V_c - H_a$$

using the symmetry of the connection; and therefore along the curve, where $v^a = \dot{\gamma}$, we have

$$\dot{\lambda}^a H_a + R^d{}_{cab}\lambda^a\dot{\gamma}^b\dot{\gamma}^c V_d + \Gamma^b_{ac}\lambda^a\dot{\gamma}^c H_b + \dot{\mu}^a V_a + \Gamma^c_{ab}\mu^a\dot{\gamma}^b V_c - \mu^a H_a = 0.$$

Equating horizontal and vertical components to zero separately, we obtain

$$\mu^a = \dot{\lambda}^a + \Gamma^a_{bc}\lambda^b\dot{\gamma}^c$$
$$\dot{\mu}^a + \Gamma^a_{bc}\mu^b\dot{\gamma}^c + R^a{}_{bcd}\dot{\gamma}^b\lambda^c\dot{\gamma}^d = 0.$$

We denote by λ the vector field $\pi_* J$ along γ in $\mathcal{M}$, so that $\lambda = \lambda^a\partial/\partial x^a$; then these equations may be written

$$\mu = \nabla_{\dot{\gamma}}\lambda$$
$$\nabla_{\dot{\gamma}}{}^2\lambda + R(\lambda, \dot{\gamma})\dot{\gamma} = 0.$$

The second-order differential equation is known as *Jacobi's equation*. It is a linear equation, as was to be expected from its construction. Any solution is called a Jacobi field along γ. The solutions constitute a 2(dim $\mathcal{M}$)-dimensional vector space; a Jacobi field λ is determined by its initial value and the initial value of its covariant derivative $\nabla_{\dot{\gamma}}\lambda$ (at $\gamma(0)$ say).

Exercise 34. The Jacobi field λ along γ with initial conditions $\lambda(0) = \dot\gamma(0)$, $(\nabla_{\dot\gamma}\lambda)(0) = 0$ is just $\dot\gamma$, while the one with initial conditions $\lambda(0) = 0$, $(\nabla_{\dot\gamma}\lambda)(0) = \dot\gamma(0)$ is given by $\lambda(t) = t\dot\gamma(t)$. □

This discussion may be summarised by saying that $\exp_{*w}$ fails to be an isomorphism if there is a non-zero Jacobi field along the geodesic fixed by $\dot\gamma(0) = w$ which vanishes at $\gamma(0)$ and $\gamma(1)$. Clearly, affine reparametrisation makes no significant difference. Points x and y on a geodesic in $\mathcal{M}$ are said to be *conjugate* if there is a Jacobi field, not identically zero, which vanishes at x and at y.

A Lie transported vector field may be thought of as defining connecting vectors between neighbouring integral curves of the vector field with respect to which it is transported. Consequently a Jacobi field (along a geodesic in $\mathcal{M}$) may be thought of as defining connecting vectors between neighbouring geodesics.

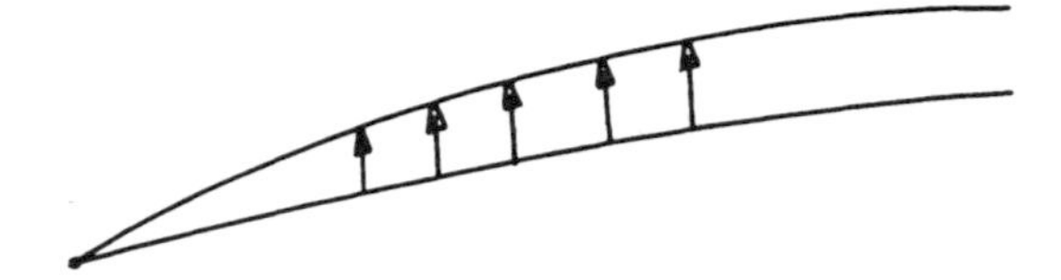

Fig. 3 Two geodesics and a field of connecting vectors.

Let us be more explicit. Suppose that J is a Jacobi field in $T\mathcal{M}$ along the integral curve $(\gamma, \dot\gamma)$ of Γ. Let σ be a curve in $T\mathcal{M}$ through $(\gamma(0), \dot\gamma(0))$ with tangent vector $J(0)$ there. For each fixed s in a neighbourhood of 0 the curve $t \mapsto \hat\gamma_t(\sigma(s))$ is an integral curve of Γ, and so $t \mapsto \pi(\hat\gamma_t(\sigma(s)))$ is a geodesic. One therefore obtains, by varying s, a one-parameter family of geodesics, with γ corresponding to $s = 0$. For each fixed t, on the other hand, the tangent vector to the curve $s \mapsto \hat\gamma(\sigma(s))$ at $s = 0$ is $\hat\gamma_{t*}(\dot\sigma(0)) = \hat\gamma_{t*}(J(0)) = J(t)$. Thus the Jacobi field $\pi_* J$ in $\mathcal{M}$ is the field of tangent vectors to the transverse curves, joining points of the same affine parameter value on the one-parameter family of geodesics, where they cross the central geodesic γ. Two points on a geodesic are conjugate, therefore, if (roughly speaking) there is a one-parameter family of geodesics emanating from the first which focuses at the second.

Cross-sections and geodesic fields. A vector field on $\mathcal{M}$ has been described up to now as a "choice of tangent vector at each point of $\mathcal{M}$". The terminology suggests that there is potentially a map involved in this definition, though it has not been clear what the image space should be. This question can be cleared up with the help of the tangent bundle. A vector field V on $\mathcal{M}$ defines a map $\mathcal{M} \to T\mathcal{M}$ by $x \mapsto V_x$. This map has the special property that the image of x lies in the fibre over x. Such a map is called a *cross-section* of the tangent bundle or, more correctly, a cross-section of $\pi: T\mathcal{M} \to \mathcal{M}$: if, as may happen, a manifold is a fibre bundle in more than one way, it is necessary to make it clear which fibre the image belongs to. The property which defines a cross-section may be most succinctly expressed as follows: a map $\sigma: \mathcal{M} \to T\mathcal{M}$ is a cross-section if $\pi \circ \sigma$ is the identity map of $\mathcal{M}$.

Exercise 35. Show that every cross-section of $T\mathcal{M}$ determines a vector field on $\mathcal{M}$, which is smooth provided that the cross-section is smooth as a map of differentiable manifolds. Show likewise that 1-forms correspond to cross-sections of $T^*\mathcal{M}$. □

A vector field V on $\mathcal{M}$ is geodesic (that is, has geodesics for all of its integral curves) if and only if Γ is tangent to the cross-section σ which defines V. (Here the section is thought of as defining a submanifold of $T\mathcal{M}$.) For suppose that γ is an integral curve of V. Then $\sigma(\gamma(t)) = (\gamma(t), V_{\gamma(t)}) = (\gamma(t), \dot{\gamma}(t))$ is the natural lift of γ. Therefore the natural lift of γ lies in the section, and V is geodesic if and only if the natural lifts of its integral curves are integral curves of Γ.

Exercise 36. Show that if V is geodesic then the vector fields $\tilde{V}$, V^h and Γ all coincide with $\sigma_* V$ on the image of the section σ. □

One may use this observation as a guide to constructing geodesic fields, at least locally. For suppose that $\mathcal{S}$ is a submanifold of $T\mathcal{M}$, of dimension $\dim \mathcal{M} - 1$, which is transverse to Γ. Define a new $\dim \mathcal{M}$-dimensional submanifold $\hat{\mathcal{S}}$ by operating on points of $\mathcal{S}$ with $\hat{\gamma}$: thus $\hat{\mathcal{S}} = \{\, \hat{\gamma}_t(v) \mid t \in \mathbf{R},\ v \in \mathcal{S} \,\}$, for those v for which $\hat{\gamma}_t(v)$ is defined. Then provided that $\hat{\mathcal{S}}$ intersects each fibre of $T\mathcal{M}$ in one and only one point it will define a section of $T\mathcal{M}$, which will be geodesic. Thus $\hat{\mathcal{S}}$ will define a geodesic section near enough to $\mathcal{S}$. However, the tangent space to $\hat{\mathcal{S}}$ may contain vertical vectors if one goes sufficiently far from $\mathcal{S}$, in which case the geodesic field develops "singularities".

6. Symmetries of the Geodesic Spray

A diffeomorphism of $T\mathcal{M}$ which maps a geodesic spray to itself is called a symmetry of the spray. As usual the symmetries of greatest interest are the infinitesimal ones, that is, the vector fields which generate flows of symmetries. A vector field W is an infinitesimal symmetry of the geodesic spray Γ if and only if

$$\mathcal{L}_W \Gamma = [W, \Gamma] = 0.$$

We shall deal first with the case in which the generator is a complete lift. It turns out, perhaps not very surprisingly, that a complete lift $\tilde{X}$ of a vector field X on $\mathcal{M}$ is an infinitesimal symmetry of the geodesic spray Γ if and only if X is an infinitesimal affine transformation of the corresponding affine connection. We embody the necessary computations in a series of exercises. The first four exercises characterise an infinitesimal affine transformation in terms of properties of lifts. Recall that X is an infinitesimal affine transformation of $\mathcal{M}$ if and only if $\mathcal{L}_X(\nabla_V W) - \nabla_V(\mathcal{L}_X W) - \nabla_{[X,V]} W = 0$ for all vector fields V and W on $\mathcal{M}$ (Chapter 11, Section 8).

Exercise 37. With the help of Jacobi's identity, and the relations $[\tilde{V}, W^\uparrow] = [V, W]^\uparrow$ (Exercise 13) and $[V^h, W^\uparrow] = (\nabla_V W)^\uparrow$ (the argument following Exercise 26), show that X is an infinitesimal affine transformation of $\mathcal{M}$ if and only if $[[\tilde{X}, V^h] - [X, V]^h, W^\uparrow] = 0$ (on $T\mathcal{M}$) for all vector fields V and W on $\mathcal{M}$. □

Exercise 38. Show that for any vector fields V and W on $\mathcal{M}$, $\pi_*([\tilde{V}, W^h]) = [V, W]$. Infer that $[\tilde{V}, W^h] - [V, W]^h$ is vertical. □

Exercise 39. Show that a vertical vector field on $T\mathcal{M}$ which commutes with every vertical lift must itself be a vertical lift. Deduce from the results of the previous two exercises that X is an infinitesimal affine transformation of $\mathcal{M}$ if and only if, for all vector fields V on $\mathcal{M}$, $[\tilde{X}, V^h] - [X, V]^h$ is a vertical lift. □

Exercise 40. From the results of Exercises 18 and 25, deduce that for any vector fields X and V on $\mathcal{M}$, $[\Delta, [\tilde{X}, V^h] - [X, V]^h] = 0$, where Δ is the dilation vector field. Infer that X is an infinitesimal affine transformation of $\mathcal{M}$ if and only if $[\tilde{X}, V^h] = [X, V]^h$ for every V. □

In the next exercise, $\mathcal{L}_\Gamma S$ is evaluated on vertical and horizontal vectors: for this recall that, because the Lie derivative is a derivation, $\big(\mathcal{L}_V(\mathcal{L}_\Gamma S)\big)(W) = \mathcal{L}_V\big((\mathcal{L}_\Gamma S)(W)\big) - (\mathcal{L}_\Gamma S)(\mathcal{L}_V W)$ for any vector fields V and W on $T\mathcal{M}$.

Exercise 41. With the help of Exercise 35 show that, for any vector fields V and W on $\mathcal{M}$, $\big(\mathcal{L}_{\tilde{V}}(\mathcal{L}_\Gamma S)\big)(W^\uparrow) = 0$. Show similarly that $\big(\mathcal{L}_{\tilde{V}}(\mathcal{L}_\Gamma S)\big)(W^h) = -2Q([\tilde{V}, W^h])$ (Q is one of the projection operators defined in Section 4). Infer that $\mathcal{L}_{\tilde{X}}(\mathcal{L}_\Gamma S) = 0$ if and only if $[\tilde{X}, V^h]$ is horizontal for every V. □

From Exercise 40 it now follows that $\mathcal{L}_{\tilde{X}}(\mathcal{L}_\Gamma S) = 0$ if and only if $[\tilde{X}, V^h] = [X, V]^h$ for every V. Combining this with the result of Exercise 41 one may infer immediately that $\mathcal{L}_{\tilde{X}}(\mathcal{L}_\Gamma S) = 0$ if and only if X is an infinitesimal affine transformation of $\mathcal{M}$. The next two exercises translate this into the desired condition for a symmetry.

Exercise 42. From the result of Exercise 23, show that for any vector field V on $\mathcal{M}$, $\mathcal{L}_{\tilde{V}}(\mathcal{L}_\Gamma S) = \mathcal{L}_{[\tilde{V}, \Gamma]} S$. Deduce, with the help of Exercise 21, that X is an infinitesimal affine transformation of $\mathcal{M}$ if and only if $[\tilde{X}, \Gamma]$ is a vertical lift. □

Exercise 43. Recall that $[\Delta, \Gamma] = \Gamma$ (Section 4). With the help of Exercise 18 and Jacobi's identity deduce that, for any vector field V on $\mathcal{M}$, $[\Delta, [\tilde{V}, \Gamma]] = [\tilde{V}, \Gamma]$. Again appealing to Exercise 18, which shows for a vertical lift $U^\uparrow$ that $[\Delta, U^\uparrow] = -U^\uparrow$, conclude from the result of the previous exercise that X is an infinitesimal affine transformation of $\mathcal{M}$ if and only if $[\tilde{X}, \Gamma] = 0$. □

Our definition of a symmetry of a spray allows for the possibility of symmetries which are not bundle maps, and which therefore (in the terminology of mechanics) thoroughly mix up positions and velocities. There are infinitely many such maps, not often very interesting. One point is worth making, however. We have defined a Jacobi field as a vector field along an integral curve of Γ which is Lie transported along it by the flow of Γ. Now any infinitesimal symmetry W of Γ satisfies $\mathcal{L}_\Gamma W = 0$ and therefore defines a Jacobi field on every geodesic.

Exercise 44. Let $W = \lambda^a H_a + \mu^a V_a$ be a vector field on $T\mathcal{M}$. Show, using the symmetry of the connection, that the necessary and sufficient conditions for W to be a symmetry of Γ are

$$\mu^a = \Gamma(\lambda^a) + \Gamma^a_{bc}\lambda^b v^c$$

$$\Gamma(\mu^a) + \Gamma^a_{bc}\mu^b v^c + R^a{}_{bcd}\lambda^b v^c \lambda^d = 0.$$ □

Exercise 45. Show that an infinitesimal affine transformation is a Jacobi field along any geodesic. □

Exercise 46. Show that Γ has no non-zero vertical infinitesimal symmetries. Show that, more generally, if W is vertical and $[W, \Gamma]$ is vertical also, then $W = 0$. □

7. Projective Transformations

Until now we have been dealing entirely with affinely parametrised geodesics. We now relax this definition a little and consider geodesic paths: that is to say, we allow more general parametrisations. A representative curve γ of a geodesic path satisfies $\nabla_{\dot\gamma}\dot\gamma = k\dot\gamma$, where k is some function along γ; it follows that there is a reparametrisation of γ which makes it into an affinely parametrised geodesic (Chapter 11, Section 7). The collection of geodesic paths on $\mathcal{M}$ defines a structure on $T\mathcal{M}$ a little more complicated than the geodesic spray; we now describe this structure and some of the transformations of $T\mathcal{M}$ which leave it unchanged.

Consider, first of all, the natural lift of a representative curve γ of a geodesic path to $T\mathcal{M}$: it is a curve (γ, V) where $V = \dot\gamma$ and

$$\dot V^a(t) + \Gamma^a_{bc}(\gamma(t))V^b(t)V^c(t) = k(t)V^a(t).$$

Its tangent vector at $v = V(t)$ is

$$v^a\frac{\partial}{\partial x^a} + \left(kv^a - \Gamma^a_{bc}v^bv^c\right)\frac{\partial}{\partial v^a} = v^aH_a + kv^aV_a$$

and therefore belongs to the 2-dimensional subspace of $T_vT\mathcal{M}$ spanned by Γ and Δ. Now Γ and Δ span a 2-dimensional distribution on $T\mathcal{M}$, less the zero section, which is integrable, by Frobenius's Theorem, because $[\Delta, \Gamma] = \Gamma$. Suppose that (γ, V) is any curve lying in an integral submanifold of this distribution. Its tangent vector at each point is a linear combination of Γ and Δ, say

$$\dot\gamma^a\frac{\partial}{\partial x^a} + \dot V^a\frac{\partial}{\partial v^a} = r\Gamma + s\Delta = rV^a\frac{\partial}{\partial x^a} + \left(sV^a - r\Gamma^a_{bc}V^bV^c\right)\frac{\partial}{\partial v^a}$$

where r and s are functions along the curve. Then

$$\dot\gamma^a = rV^a \qquad \dot V^a = sV^a - r\Gamma^a_{bc}V^bV^c,$$

whence

$$\ddot\gamma^a = \dot rV^a + r\left(sV^a - r\Gamma^a_{bc}V^bV^c\right)$$

and so

$$\ddot\gamma^a + \Gamma^a_{bc}\dot\gamma^b\dot\gamma^c = (\dot r/r + s)\dot\gamma^a.$$

Thus provided that r is nowhere zero the projected curve is a non-affinely parametrised geodesic; the original curve is not its natural lift in this case (unless $r = 1$) but a dilated version of it. We take the distribution $\mathcal{D}$ spanned by Γ and Δ as the object in $T\mathcal{M}$ corresponding to the geodesic paths.

Notice, first of all, that the distribution $\mathcal{D}$ may contain geodesic sprays other than Γ. In fact if $\hat\Gamma = \Gamma + f\Delta$ satisfies $[\Delta, \hat\Gamma] = \hat\Gamma$ then $\hat\Gamma$ will be a spray; this occurs for

$$\Delta f = f.$$

Thus f must be a function on $T\mathcal{M}$ homogeneous of degree 1 in the fibre coordinates, and since it must be smooth at $v^a = 0$ it must be linear in them. Thus $f = \phi_a v^a$ for some locally defined functions ϕ_a on $\mathcal{M}$, which are in fact the coefficients of a 1-form on $\mathcal{M}$.

Exercise 47. Show, by considering coordinate transformations, that this is so. Show also that if ϕ is a 1-form on $\mathcal{M}$, then the function $\hat{\phi}$ on $T\mathcal{M}$ given by $\hat{\phi}(v) = \langle \pi_* v, \phi_{\pi(v)} \rangle$ satisfies $\Delta\hat{\phi} = \hat{\phi}$ and has the coordinate presentation $\phi_a v^a$. □

Thus $\hat{\Gamma} = \Gamma + \hat{\phi}\Delta$ is a geodesic spray which has the same geodesic paths as does Γ, and any geodesic spray with this property must have this form.

Exercise 48. Show that the components $\hat{\Gamma}^a_{bc}$ of the symmetric connection defined by $\hat{\Gamma}$ are given in terms of those of Γ by

$$\hat{\Gamma}^a_{bc} = \Gamma^a_{bc} + \phi_b\delta^a_c + \phi_c\delta^a_b.$$ □

Two symmetric connections (geodesic sprays) which define the same set of geodesic paths are said to be *projectively related*. The expression for the components of projectively related connections given in the exercise is classical.

We consider next the transformations of $\mathcal{M}$ whose induced transformations of $T\mathcal{M}$ preserve $\mathcal{D}$. The conditions for a vector field W on $T\mathcal{M}$ to generate transformations which preserve $\mathcal{D}$ are these:

$$[W, \Gamma] = a\Gamma + b\Delta$$
$$[W, \Delta] = c\Gamma + d\Delta$$

for functions a, b, c, d on $T\mathcal{M}$.

If $W = \tilde{V}$ is a complete lift then the second condition is satisfied automatically (with $c = d = 0$). Since $[\tilde{V}, \Gamma]$ is necessarily vertical, $a = 0$ in the first condition; and since $[\Delta, [\tilde{V}, \Gamma]] = [\tilde{V}, \Gamma]$ it follows that $b = \hat{\phi}$ for some 1-form ϕ on $\mathcal{M}$. The only surviving condition is thus

$$[\tilde{V}, \Gamma] = \hat{\phi}\Delta.$$

If V satisfies this condition then it generates transformations which preserve geodesic paths, that is, map geodesics on $\mathcal{M}$ to geodesics on $\mathcal{M}$, albeit with loss of affine parametrisation. Such transformations are called *projective transformations* of $\mathcal{M}$. Affine transformations are a special case.

Exercise 49. Investigate the projective transformations of an affine space. □

Exercise 50. Show that if V is a projective transformation of $\mathcal{M}$, and $[\tilde{V}, \Gamma] = \hat{\phi}\Delta$, then $(\mathcal{L}_{\tilde{V}}(\mathcal{L}_\Gamma S))(W^\uparrow) = 0$, while $(\mathcal{L}_{\tilde{V}}(\mathcal{L}_\Gamma S))(W^h) = -(\langle W, \phi\rangle \circ \pi)\Delta - \hat{\phi}W^\uparrow$, for any vector field W on $\mathcal{M}$. By adapting the argument concerning affine transformations show that V is a projective transformation if and only if, for every $U, W \in \mathcal{X}(\mathcal{M})$,

$$[\mathcal{L}_V, \nabla_W]U - \nabla_{[V,W]}U = -\tfrac{1}{2}(\langle U, \phi\rangle W + \langle W, \phi\rangle U).$$ □

8. Euler-Lagrange Fields

We have dealt up to now with the geodesic spray of an arbitrary symmetric connection: we want next to consider the Levi-Civita connection of a metric from the present point of view. Now a metric on $\mathcal{M}$ may be used to construct a function on $T\mathcal{M}$ called, by analogy with dynamics, its kinetic energy; the equations for the geodesics of the Levi-Civita connection are the Euler-Lagrange equations obtained when the energy is taken for Lagrangian function. In this section we shall describe how the theory of Euler-Lagrange equations may be set up on the tangent bundle; properties of the Levi-Civita connection follow as a special case.

By the Euler-Lagrange equations we mean the equations of dynamics, derived from a function L, the Lagrangian, which (for an autonomous system) is a function of "generalised coordinates" (conventionally denoted (q^a)) and "generalised velocities" $(\dot{q}^a)$; the equations are usually written

$$\frac{d}{dt}\left(\frac{\partial L}{\partial \dot{q}^a}\right) - \frac{\partial L}{\partial q^a} = 0.$$

When the t-derivative is performed explicitly, provided that the matrix of second partial derivatives $(\partial^2 L/\partial \dot{q}^a \partial \dot{q}^b)$ is everywhere non-singular, the Euler-Lagrange equations may be expressed in the form

$$\ddot{q}^a = \Lambda^a(q^b, \dot{q}^c)$$

for certain functions Λ^a. When L is the kinetic energy minus the potential energy of a dynamical system these are the equations of motion of the system.

We shall interpret generalised coordinates as coordinates on a manifold $\mathcal{M}$, the configuration space of the system, and generalised velocities as the corresponding fibre coordinates on $T\mathcal{M}$. A Lagrangian is then simply a function on $T\mathcal{M}$. We seek a canonical and coordinate-independent way of constructing a second-order differential equation field from a Lagrangian which satisfies a non-degeneracy condition corresponding to the non-singularity of the matrix of second partial derivatives mentioned above. We shall call this second-order differential equation field, whose projected integral curves will satisfy the Euler-Lagrange equations, the *Euler-Lagrange field* of the given Lagrangian.

The construction involves the use of the two canonical geometric objects on $T\mathcal{M}$ which we introduced earlier, the vertical endomorphism S and the dilation field Δ. It will also make use of the construction of a horizontal distribution from a second-order differential equation field.

First, we observe that S may be made to act on cotangent vectors to $T\mathcal{M}$ by duality: for any $v \in T\mathcal{M}$ we define a linear map $S_v^*: T_v^*T\mathcal{M} \to T_v^*T\mathcal{M}$ by

$$\langle w, S_v^*(\alpha)\rangle = \langle S_v(w), \alpha\rangle \qquad w \in T_vT\mathcal{M},\ \alpha \in T_v^*T\mathcal{M}.$$

Exercise 51. Show that $(\mathcal{L}_W S^*) = (\mathcal{L}_W S)^*$ for any vector field W. □

Then given any 1-form θ on $T\mathcal{M}$ there is a corresponding 1-form $S^*(\theta)$; note that $S^*(\theta)$ vanishes on vertical vectors. In this way we can define, for any function L on $T\mathcal{M}$, first a 1-form $S^*(dL)$, and then a 2-form $\omega_L = d(S^*(dL))$; these are called the *Cartan forms* associated with L.

Exercise 52. Compute the Cartan forms in coordinates. □

The function L is said to be a *non-degenerate* Lagrangian if ω_L has maximum rank, that is to say, if the m-fold exterior product $\omega_L \wedge \omega_L \wedge \cdots \wedge \omega_L$ is nowhere vanishing. The Cartan 2-form itself is then also called non-degenerate. In this case the map of vector fields to forms defined by $v \mapsto v \lrcorner \omega_L$ is an isomorphism.

Exercise 53. Show that L is non-degenerate if and only if the matrix$(\partial^2 L/\partial v^a \partial v^b)$ is everywhere non-singular. □

The *energy* associated with a Lagrangian L is the function E_L on $T\mathcal{M}$ defined by

$$E_L = \Delta(L) - L.$$

In the case of a non-degenerate Lagrangian there is a unique vector field Γ such that

$$\Gamma \lrcorner \omega_L = -dE_L.$$

We shall show that Γ is a second-order differential equation field and that its projected integral curves satisfy the Euler-Lagrange equations: it is the Euler-Lagrange field of L.

We shall first show that Γ is a second-order differential equation field by showing that $S(\Gamma) = \Delta$. To do so, we must establish a basic property of the Cartan 2-form: for any vector fields V and W on $T\mathcal{M}$,

$$\omega_L(S(V), W) + \omega_L(V, S(W)) = 0.$$

In fact, from the definition of ω_L and the formula for an exterior derivative it follows that

$$\begin{aligned} &\omega_L(S(V), W) \\ &\quad = S(V)(\langle W, S^*(dL)\rangle) - W(\langle S(V), S^*(dL)\rangle) - \langle [S(V), W], S^*(dL)\rangle \\ &\quad = S(V)S(W)(L) - S([S(V), W])(L) \end{aligned}$$

since $S^2 = 0$. Thus

$$\begin{aligned} &\omega_L(S(V), W) + \omega_L(V, S(W)) \\ &\quad = S(V)S(W)(L) - S(W)S(V)(L) - S([S(V), W])(L) - S([V, S(W)])(L) \\ &\quad = ([S(V), S(W)] - S([S(V), W]) - S([V, S(W)]))(L). \end{aligned}$$

The vector field operating on L in this final expression vanishes, as was shown in Exercise 20, and the result is therefore established. To show that $S(\Gamma) = \Delta$ we substitute Γ for V in this identity: then

$$\begin{aligned} \omega_L(S(\Gamma), W) &= -\omega_L(\Gamma, S(W)) = S(W)(E_L) \\ &= [S(W), \Delta](L) + \Delta(S(W)(L)) - S(W)(L) \\ &= -(\mathcal{L}_\Delta S)(W)(L) - S([\Delta, W])(L) + \Delta(S(W)(L)) - S(W)(L) \\ &= S(W)(L) - S([\Delta, W])(L) + \Delta(S(W)(L)) - S(W)(L) \\ &= \Delta(\langle W, S^*(dL)\rangle) - \langle [\Delta, W], S^*(dL)\rangle = \omega_L(\Delta, W). \end{aligned}$$

This holds for every vector field W, whence $(S(\Gamma) - \Delta) \lrcorner \omega_L = 0$; but since ω_L is non-degenerate this means that

$$S(\Gamma) = \Delta.$$

We show now that the projected integral curves of Γ satisfy the Euler-Lagrange equations for L. We shall derive these equations in a coordinate free form; this requires a further transformation of the equation defining Γ. By substituting $S(\Gamma)$ for Δ in the definition of E_L we obtain

$$E_L = S(\Gamma)(L) - L = \langle \Gamma, S^*(dL)\rangle - L.$$

Thus

$$\Gamma \lrcorner d\big(S^*(dL)\big) + d\big(\Gamma \lrcorner S^*(dL)\big) = dL$$

or

$$\mathcal{L}_\Gamma\big(S^*(dL)\big) = dL.$$

It follows that

$$(\mathcal{L}_\Gamma S^*)(dL) + S^*\big(d(\Gamma(L))\big) = (\mathcal{L}_\Gamma S)^*(dL) + S^*\big(d(\Gamma(L))\big) = dL.$$

This equation is an identity on vertical vector fields since, for any second-order differential equation field Γ, $\mathcal{L}_\Gamma S$ acts as the identity on vertical vector fields, and for any 1-form θ, $S^*(\theta)$ vanishes on vertical vector fields. If, however, the equation is evaluated with argument V^h, the horizontal lift of a vector field V on $\mathcal{M}$ to $T\mathcal{M}$ with respect to the horizontal distribution defined by Γ, one obtains

$$-V^h(L) + V^\uparrow\big(\Gamma(L)\big) = V^h(L).$$

But $2V^h = [V^\uparrow, \Gamma] + \tilde{V}$; thus

$$\Gamma\big(V^\uparrow(L)\big) - \tilde{V}(L) = 0.$$

This is a coordinate free version of the Euler-Lagrange equations: if $V = \partial/\partial q^a$ is a coordinate vector field on $\mathcal{M}$, it becomes

$$\frac{d}{dt}\left(\frac{\partial L}{\partial v^a}\right) - \frac{\partial L}{\partial q^a} = 0,$$

where d/dt means differentiation along any integral curve of Γ.

Exercise 54. Show that the energy E_L is a constant of the motion, that is, that $\Gamma(E_L) = 0$. □

Exercise 55. Let g be a metric on $\mathcal{M}$ and L the function on $T\mathcal{M}$ defined by $L(v) = \frac{1}{2}g_{\pi(v)}(v, v)$. Show that $\Delta(L) = 2L$, that $E_L = L$, and that the corresponding Euler-Lagrange field Γ is a spray. Show that $\Gamma(L) = 0$ and deduce that Γ is the geodesic spray of the Levi-Civita connection of g. □

Exercise 56. Show that the Euler-Lagrange field Γ and Cartan 2-form ω_L of any non-degenerate Lagrangian satisfy $\mathcal{L}_\Gamma \omega_L = 0$. □

Exercise 57. By taking the Lie derivative with respect to Γ of the identity $\omega_L\big(S(V), W\big) + \omega_L\big(V, S(W)\big) = 0$, show that the projection operators P and Q for the horizontal distribution defined by Γ satisfy

$$\omega_L\big(P(V), W\big) + \omega_L\big(V, P(W)\big) = \omega_L(V, W)$$
$$\omega_L\big(Q(V), W\big) + \omega_L\big(V, Q(W)\big) = \omega_L(V, W)$$
$$\omega_L\big(P(V), W\big) = \omega_L\big(V, Q(W)\big).$$

Deduce that ω_L vanishes when both of its arguments are vertical, and when both of its arguments are horizontal. Show that with respect to a local basis of 1-forms $\{\theta^a, \phi^a\}$ adapted to the horizontal distribution (as described in Exercise 24 in the case of a spray) ω_L is given by

$$\omega_L = \frac{\partial^2 L}{\partial v^a \partial v^b}\phi^a \wedge \theta^b.$$ □

Symmetries. A vector field V on $T\mathcal{M}$ which generates transformations preserving ω_L and E_L, in the sense that $\mathcal{L}_V\omega_L = 0$ and $V(E_L) = 0$, is called a *Cartan symmetry* of the Lagrangian system. A Cartan symmetry is necessarily a symmetry of the Euler-Lagrange field Γ, for

$$[V,\Gamma] \lrcorner \omega_L = \mathcal{L}_V(\Gamma \lrcorner \omega_L) - \Gamma \lrcorner \mathcal{L}_V\omega_L = -d(V(E_L)) = 0$$

and therefore by the non-degeneracy of ω_L

$$[V,\Gamma] = 0.$$

The transformations of the flow generated by V therefore permute the integral curves of Γ. If V is a Cartan symmetry then $\mathcal{L}_V(S^*(dL))$ is closed; if it is exact, say $\mathcal{L}_V(S^*(dL)) = df$, we call V an *exact Cartan symmetry*; then

$$\begin{aligned} V \lrcorner \omega_L &= \mathcal{L}_V(S^*(dL)) - d\langle V, S^*(dL)\rangle \\ &= d(f - S(V)(L)) = -dF \end{aligned}$$

say; and

$$\Gamma(F) = \Gamma \lrcorner (V \lrcorner \omega_L) = -V \lrcorner dE_L = 0$$

so that F is a constant of the motion. Thus to every exact Cartan symmetry there corresponds a constant of the motion. Conversely, if F is a constant of the motion then the vector field V defined by $V \lrcorner \omega_L = -dF$ is an exact Cartan symmetry. There is thus a 1 : 1 correspondence between exact Cartan symmetries and constants of the motion. This is an important result, since knowledge of constants of the motion helps one to integrate the Euler-Lagrange equations, and much of classical mechanics is concerned with quantities such as energy, momentum and angular momentum which are constants under suitable hypotheses (see also the related discussion in Chapter 12). The correspondence of symmetries and constants which we have just derived is a general form of a class of results of which the first was found by E. Noether; it may be described as a generalised Noether theorem. The original Noether theorem covered the case in which V is the complete lift of a vector field on $\mathcal{M}$, a so-called point symmetry of the system.

Exercise 58. Show that if, for a vector field W on $\mathcal{M}$, $\tilde{W}(L) = 0$ then $\tilde{W}$ is a Cartan symmetry. □

Exercise 59. Let g be a metric on $\mathcal{M}$ and Γ its Levi-Civita spray, in other words the Euler-Lagrange field of the Lagrangian function L where $L(v) = \frac{1}{2}g(v,v)$. Show that $\tilde{W}$ is a Cartan symmetry if and only if W is an isometry of g. □

Exercise 60. Let g be a metric on $\mathcal{M}$, $T(v) = \frac{1}{2}g(v,v)$ its "kinetic energy", and ϕ a function on $\mathcal{M}$, the "potential energy". Let $L = T - \phi \circ \pi$. Show that if a vector field W on $\mathcal{M}$ is an isometry of g and satisfies $W(\phi) = 0$ then $\tilde{W}$ is a Cartan symmetry, and the corresponding constant of the motion F is given by $F(v) = g(v, W_{\pi(v)})$. □

This is the usual situation encountered in Lagrangian dynamics.

9. The Hamiltonian Formulation

A non-degenerate Cartan 2-form defined on $T\mathcal{M}$ gives it a symplectic structure (as defined in the final section of Chapter 12). When Darboux coordinates (Section 6 of Chapter 6) are chosen, the equations for the integral curves of Γ will take the Hamiltonian rather than the Euler-Lagrange form. However, passage from the Lagrangian to the Hamiltonian formulation is better expressed in a different way which exploits the fact remarked on in Section 2 above that the canonical 2-form $d\theta$ on the cotangent bundle is already in Darboux form when expressed in terms of coordinates induced from coordinates on the base.

Given a non-degenerate Lagrangian L on $T\mathcal{M}$ we define a bundle map $\hat{L}: T\mathcal{M} \to T^*\mathcal{M}$ inducing the identity on $\mathcal{M}$, as follows: for each $v \in T\mathcal{M}$, $\hat{L}(v)$ is the cotangent vector at $\pi(v) = x$ defined by

$$\langle u, \hat{L}(v)\rangle = u^{\uparrow}(L) \qquad \text{for all } u \in T_x\mathcal{M},$$

where the vertical lift is to v. The linearity of the vertical lift ensures that $\hat{L}(v)$ is indeed a cotangent vector. The map $\hat{L}$ is called the *Legendre map* associated with L.

We shall show that the pull-back $\hat{L}^*\theta$ of the canonical 1-form θ on $T^*\mathcal{M}$ is just the Cartan 1-form of L. For $v \in T\mathcal{M}$ and $w \in T_vT\mathcal{M}$,

$$\begin{aligned}\langle w, \hat{L}^*\theta\rangle &= \langle \hat{L}_*w, \theta\rangle = \langle \tau_*\hat{L}_*w, \hat{L}(v)\rangle \\ &= \langle \pi_*w, \hat{L}(v)\rangle = (\pi_*w)^{\uparrow}(L) \\ &= S_v(w)(L) = \langle w, S^*(dL)\rangle,\end{aligned}$$

as required; we have used the fact that $\tau \circ \hat{L} = \pi$ where $\tau: T^*\mathcal{M} \to \mathcal{M}$ is the cotangent bundle projection.

If $\hat{L}$ is a diffeomorphism then $\omega_L = \hat{L}^*(d\theta)$ must have maximum rank, since $d\theta$ has: thus a necessary condition for $\hat{L}$ to be a diffeomorphism is that ω_L should be non-degenerate. When $\hat{L}$ is a diffeomorphism, the vector field $V = \hat{L}_*\Gamma$ on $T^*\mathcal{M}$ satisfies

$$V \lrcorner d\theta = \hat{L}_*\Gamma \lrcorner d\theta = \hat{L}^{-1*}(\Gamma \lrcorner \omega_L) = -dh$$

where h is the function defined by $h = E_L \circ \hat{L}^{-1}$. This is the Hamiltonian function corresponding to the Lagrangian L, and the correspondence between Lagrangian and Hamiltonian formulations is clear.

Summary of Chapter 13

The tangent (cotangent) bundle $T\mathcal{M}$ ($T^*\mathcal{M}$) of a differentiable manifold $\mathcal{M}$ is the collection of all its tangent (cotangent) vectors regarded as a differentiable manifold. Coordinates may be defined on $T\mathcal{M}$ by taking charts $(\mathcal{P}, \psi)$ on $\mathcal{M}$ and ascribing to each $v \in T\mathcal{M}$ whose point of tangency lies in $\mathcal{P}$ the coordinates (x^a, v^a), where (x^a) are the coordinates of the point of tangency and (v^a) the components of v with respect to the coordinate vector fields; a similar construction works for $T^*\mathcal{M}$. Tangent and cotangent bundles are examples of fibre bundles: each has a smooth projection map which maps it surjectively onto $\mathcal{M}$, and each is locally a product manifold, that is, locally diffeomorphic to $\mathcal{P} \times \mathbf{R}^m$ where $\mathcal{P}$ is an open subset of $\mathcal{M}$. The transition functions, which relate different local product decompositions, take their values in $GL(m, \mathbf{R})$.

A vector field V on $\mathcal{M}$ may be lifted to a vector field on $T\mathcal{M}$ in (at least) two ways. First, the vertical lift $V^\uparrow$ is the generator of the one-parameter group $v \mapsto v + tV_{\pi(v)}$ of transformations of $T\mathcal{M}$; $\pi_* V^\uparrow = 0$, whence the name. Second, the complete lift $\tilde{V}$ is the generator of the flow $v \mapsto \phi_{t*}v$ (where ϕ is the flow of V); $\pi_* \tilde{V} = V$. For any V, W on $\mathcal{M}$, $[V^\uparrow, W^\uparrow] = 0$; $[\tilde{V}, W^\uparrow] = [V,W]^\uparrow$; $[\tilde{V}, \tilde{W}] = \widetilde{[V,W]}$. On $T^*\mathcal{M}$ there are vertical lifts of 1-forms and complete lifts of vector fields from $\mathcal{M}$, satisfying $[\alpha^\uparrow, \beta^\uparrow] = 0$; $[\tilde{V}, \alpha^\uparrow] = (\mathcal{L}_V \alpha)^\uparrow$; $[\tilde{V}, \tilde{W}] = \widetilde{[V,W]}$. The dilation field Δ is the generator of the one-parameter group of dilations $v \mapsto e^t v$ of $T\mathcal{M}$. The type $(1,1)$ tensor field S on $T\mathcal{M}$ defined by $S_v(w) = (\pi_* w)^\uparrow$ is called the vertical endomorphism: the subspace of vertical vectors at each point is simultaneously its kernel and its image. The 1-form θ on $T^*\mathcal{M}$ defined by $\langle w, \theta_p \rangle = \langle \tau_* w, p \rangle$ is called the canonical 1-form; its exterior derivative is already in Darboux form in terms of standard coordinates. These constructions and geometric objects are the basic features of the differential geometry of tangent and cotangent bundles.

A connection on $\mathcal{M}$ is equivalent to a distribution on $T\mathcal{M}$ which is horizontal in the sense of being everywhere complementary to the vertical, and homogeneous in the sense that $[\Delta, W^h] = 0$ where W^h is the horizontal lift of W, that is, the unique horizontal vector field which projects onto W. The horizontal curves represent curves in $\mathcal{M}$ with parallel vector fields defined along them. The covariant differentiation operator is defined by $(\nabla_V W)^\uparrow = [V^h, W^\uparrow]$. Curvature R and torsion T are given by $\left(R(U,V)W\right)^\uparrow = [W^\uparrow, R(U,V)]$, where $R(U,V)$ is the vertical vector field $[U,V]^h - [U^h, V^h]$; $T(U,V)^\uparrow = [U^h, V^\uparrow] - [V^h, U^\uparrow] - [U,V]^\uparrow$.

A vector field Γ on $T\mathcal{M}$ with the property $\pi_* \Gamma_v = v$ is called a second-order differential equation field, because the projections of its integral curves are solutions of a system of second-order ordinary differential equations. A particular case is the spray of a symmetric connection, whose integral curves project onto the geodesics. The condition for a vector field to be a second-order differential equation field may be written $S(\Gamma) = \Delta$; a spray must satisfy, in addition, $[\Delta, \Gamma] = \Gamma$. For any second-order differential equation field Γ, $(\mathcal{L}_\Gamma S)^2$ is the identity tensor on $T\mathcal{M}$. The tensor fields $P = \frac{1}{2}(I - \mathcal{L}_\Gamma S)$ and $Q = \frac{1}{2}(I + \mathcal{L}_\Gamma S)$ are complementary projection operators and their kernels give a direct sum decomposition of the tangent space at each point: that of P is the vertical subspace, so that of Q is horizontal. Thus every second-order differential equation field defines a horizontal distribution. When Γ is a spray this distribution defines a connection, which is symmetric, and has Γ for its spray; otherwise the distribution is not homogeneous.

The exponential map is given by $\exp(v) = \pi\left(\hat{\gamma}_1(v)\right)$ where $\hat{\gamma}$ is the flow of the spray Γ. A Jacobi field is a vector field along an integral curve of Γ obtained by Lie transport of a given vector at one point of it. The projected vector field satisfies $\nabla_{\dot{\gamma}}{}^2 \lambda + R(\lambda, \dot{\gamma})\dot{\gamma} = 0$ along the geodesic γ. A Jacobi field may be thought of as defining connecting vectors between neighbouring geodesics.

A vector field V on $\mathcal{M}$ defines a map $\sigma\colon \mathcal{M} \to T\mathcal{M}$ by $x \mapsto V_x$. It is a cross-section of the projection $\pi\colon T\mathcal{M} \to \mathcal{M}$; that is, it satisfies $\pi \circ \sigma = \mathrm{id}_\mathcal{M}$, the identity on $\mathcal{M}$. Every cross-section of $T\mathcal{M}$ defines a vector field; every cross-section of $\tau\colon T^*\mathcal{M} \to \mathcal{M}$ defines a 1-form on $\mathcal{M}$. A vector field is geodesic if and only if the spray Γ is tangent to the cross-section which defines it.

A symmetry of a spray Γ is the infinitesimal generator of a flow of transformations of $T\mathcal{M}$ which map Γ to itself: so a symmetry W satisfies $[W, \Gamma] = 0$. If $\tilde{V}$ is a symmetry then the vector field V on $\mathcal{M}$ is an affine transformation, and conversely. If ϕ is a 1-form on $\mathcal{M}$, $\hat{\phi}$ the fibre-linear function on $T\mathcal{M}$ it defines, and V a vector field on $\mathcal{M}$ which satisfies $[\tilde{V}, \Gamma] = \hat{\phi}\Delta$ then V preserves the 2-dimensional distribution spanned by Γ and Δ; all sprays in this distribution have the same geodesic paths as Γ but with different affine parametrisations; V is an infinitesimal projective transformation of $\mathcal{M}$.

A function L on $T\mathcal{M}$ is often called, in the context of dynamics, a Lagrangian. It defines a 1-form $S^*(dL)$ and a 2-form $\omega_L = d(S^*(dL))$, its Cartan forms. A Lagrangian is non-degenerate if its Cartan 2-form is non-degenerate, that is, has maximum rank. The vector field Γ defined by $\Gamma \lrcorner \omega_L = -dE_L$, where $E_L = \Delta(L) - L$ is the energy, is a second-order differential equation field, whose projected integral curves satisfy the Euler-Lagrange equations of L: it is called the Euler-Lagrange field. A vector field V which satisfies $\mathcal{L}_V \omega_L = 0$ and $V(E_L) = 0$ is a Cartan symmetry: it satisfies $[V, \Gamma] = 0$. To every constant of the motion there corresponds a Cartan symmetry. When $L(v) = \frac{1}{2}g(v, v)$, where g is a metric on $\mathcal{M}$, the Euler-Lagrange field is the spray of the Levi-Civita connection, and $\tilde{W}$ is a Cartan symmetry if and only if W is an isometry.

The Legendre map $\hat{L}: T\mathcal{M} \to T^*\mathcal{M}$ of a Lagrangian L is defined by $\langle u, \hat{L}(v) \rangle = u^\uparrow(L)$. If $\hat{L}$ is a diffeomorphism, then L is non-degenerate, $\hat{L}^*\theta$ is the Cartan 1-form, and $\hat{L}_*\Gamma$ is the Hamiltonian vector field corresponding to the Hamiltonian function $E_L \circ \hat{L}^{-1}$.

14. FIBRE BUNDLES

As we have shown in Chapter 13, the manifold $T\mathcal{M}$ of tangent vectors to a given manifold $\mathcal{M}$ has a special structure which may be conveniently described in terms of the projection map which takes each tangent vector to the point of the original manifold at which it is tangent. The set of points of $T\mathcal{M}$ which are mapped by the projection to a particular point of the original manifold $\mathcal{M}$ is just the tangent space to $\mathcal{M}$ at that point: all tangent spaces are copies of the same standard space ($\mathbf{R}^m$) but not canonically so, though a common identification may be made throughout a suitable open subset of the original manifold, for example a coordinate neighbourhood. These are the basic features of what is known as a fibre bundle: roughly speaking a fibre bundle consists of two manifolds, a "larger" and a "smaller", the larger (the bundle space) being a union of "fibres", one for each point of the smaller manifold (the base space); the fibres are all alike, but not necessarily all the same. A product of two manifolds (base and fibre) is a particular case of a fibre bundle, but in general a fibre bundle will be a product only locally, as is the case for the tangent bundle of a differentiable manifold. The projection map, from bundle space to base space, maps each fibre to the associated point of the base. A final component of the definition of a fibre bundle concerns generalisation of the transformation law for the components of a tangent vector with respect to a local basis of vector fields when that basis is changed.

It will be clear that the definition of a fibre bundle (as distinct from the object itself) is a fairly complex matter. The tangent bundle is an accessible and useful paradigm, and the reader may find it helpful to have this reasonably familiar example in mind when it comes to the general definitions.

1. Fibrations

There are sufficiently many circumstances in which manifolds occur with structures of projection and local product decomposition similar to those enjoyed by $T\mathcal{M}$ and $T^*\mathcal{M}$ to make it profitable to abstract these features into a definition. We shall give the definition in two stages: we deal first with a general situation, which we call a fibration, and specialise afterwards to the case of a fibre bundle.

A *fibration* consists of four things: a differentiable manifold $\mathcal{B}$ called the *bundle space*, a differentiable manifold $\mathcal{M}$ called the *base space*, a differentiable manifold $\mathcal{F}$ called the *standard fibre*, and a smooth map $\pi: \mathcal{B} \to \mathcal{M}$ called the *projection*, satisfying the following conditions:

(1) π is surjective

(2) for each point $x \in \mathcal{M}$, $\pi^{-1}(x)$ is an imbedded submanifold of $\mathcal{B}$ which is diffeomorphic to $\mathcal{F}$; $\pi^{-1}(x)$ is the *fibre over* x

(3) π defines a local product structure on $\mathcal{B}$ in the sense that each point of $\mathcal{M}$ has a neighbourhood $\mathcal{O}$ with a diffeomorphism $\psi: \pi^{-1}(\mathcal{O}) \to \mathcal{O} \times \mathcal{F}$ such that

$\Pi_1 \circ \psi = \pi$, where $\Pi_1 : \mathcal{O} \times \mathcal{F} \to \mathcal{O}$ is projection onto the first factor.

Exercise 1. Show that $\dim \mathcal{B} = \dim \mathcal{M} + \dim \mathcal{F}$. □

To emphasise that all components are strictly necessary in the specification of a fibration, it is usual to talk of

$$\text{“the fibration } \pi : \mathcal{B} \xrightarrow{\mathcal{F}} \mathcal{M}\text{”},$$

though mention of the standard fibre may be omitted since it is usually easily inferred from the other information. Depending on the circumstances, a fibration may be regarded in (at least) two different lights. Either one may consider π as the fundamental object, in which case the fibration is a form of decomposition of $\mathcal{B}$; or one may consider $\mathcal{F}$ as the fundamental object, in which case the fibration gives a new manifold built over $\mathcal{M}$. A quotient space of an affine space is an example of a fibration seen from the first point of view, the product of (say) two affine spaces one from the second.

In the case of the tangent and cotangent bundles the fibres are vector spaces. Note that this will not necessarily be the case for a general fibration, since according to the definition neither the standard fibre nor the individual fibres need have any special structure beyond being manifolds. This step is taken in order to broaden the definition. Nevertheless in most cases of interest the fibres will have some additional structure, though not necessarily that of a vector space. We shall now explain how this idea of additional structure in the fibre may be brought into play.

There is more to the vectorial structure of the tangent and cotangent bundles than the mere fact that their fibres are vector spaces. After all, a coordinate patch in any differentiable manifold is modelled on the vector space $\mathbf{R}^m$: but the vector space structure of $\mathbf{R}^m$ does not play any role in the manifold. The fact that the various maps relating fibres, such as the transition functions, are linear maps is also a key factor in the structure of the tangent and cotangent bundles. We shall therefore describe next how these concepts arise in general.

Let $\pi : \mathcal{B} \to \mathcal{M}$ be a fibration, with standard fibre $\mathcal{F}$, and let $\{\mathcal{O}_\alpha\}$ be a covering of $\mathcal{M}$ by open sets over each of which $\pi^{-1}(\mathcal{O}_\alpha)$ is diffeomorphic to $\mathcal{O}_\alpha \times \mathcal{F}$. (The suffix α serves as an identifier of the open set and is not to be summed over: the summation convention is therefore in abeyance for the present.) For each α there is a diffeomorphism $\psi_\alpha : \pi^{-1}(\mathcal{O}_\alpha) \to \mathcal{O}_\alpha \times \mathcal{F}$ such that $\Pi_1 \circ \psi_\alpha = \pi$. We may therefore write

$$\psi_\alpha(b) = \big(x, \Psi_{\alpha,x}(b)\big)$$

where $b \in \pi^{-1}(\mathcal{O}_\alpha)$ and $x = \pi(b)$; then $\Psi_{\alpha,x} : \pi^{-1}(x) \to \mathcal{F}$ is a diffeomorphism of the fibre over x with the standard fibre. If $\beta \neq \alpha$ and $\mathcal{O}_\alpha \cap \mathcal{O}_\beta$ is non-empty then for any $x \in \mathcal{O}_\alpha \cap \mathcal{O}_\beta$ the fibre over x is identified with the standard fibre in two different ways: by $\Psi_{\alpha,x}$ and by $\Psi_{\beta,x}$. This difference is represented by the diffeomorphism of $\mathcal{F}$ given by $\Psi_{\alpha,x} \circ {\Psi_{\beta,x}}^{-1}$. Thus to each point x of $\mathcal{O}_\alpha \cap \mathcal{O}_\beta$ there corresponds a diffeomorphism of $\mathcal{F}$, and we therefore have a map $\Psi_{\alpha\beta}$ from $\mathcal{O}_\alpha \cap \mathcal{O}_\beta$ into the group of diffeomorphisms of $\mathcal{F}$, called the *transition function* on $\mathcal{O}_\alpha \cap \mathcal{O}_\beta$, and defined by

$$\Psi_{\alpha\beta}(x) = \Psi_{\alpha,x} \circ {\Psi_{\beta,x}}^{-1}.$$

Exercise 2. Show that the transition functions satisfy

$$\Psi_{\beta\alpha}(x) = \Psi_{\alpha\beta}(x)^{-1} \qquad x \in \mathcal{O}_\alpha \cap \mathcal{O}_\beta$$
$$\Psi_{\alpha\beta} \circ \Psi_{\beta\gamma}(x) = \Psi_{\alpha\gamma}(x) \qquad x \in \mathcal{O}_\alpha \cap \mathcal{O}_\beta \cap \mathcal{O}_\gamma$$

provided in the latter case that α, β, and γ are distinct and $\mathcal{O}_\alpha \cap \mathcal{O}_\beta \cap \mathcal{O}_\gamma$ is non-empty. □

(The qualifications at the end of this exercise may be partially avoided by the convention that $\Psi_{\alpha\beta}$ is the identity transformation of $\mathcal{F}$ if $\alpha = \beta$.)

In cases of interest, the standard fibre $\mathcal{F}$ has some additional structure: it may be a vector space, a Euclidean space or a Lie group, for example. The diffeomorphisms of $\mathcal{F}$ which preserve this structure usually form a Lie group G in their turn. The structure of the fibration $\pi: \mathcal{B} \to \mathcal{M}$ is compatible with this structure of the standard fibre if each fibre of $\mathcal{B}$ has the same structure as $\mathcal{F}$ (that is, if it is a vector space when $\mathcal{F}$ is a vector space, and so on), and if $\Psi_{\alpha,x}$ is an isomorphism (in the appropriate sense), for each α and x, for some covering $\{\mathcal{O}_\alpha\}$ of $\mathcal{M}$. The transition functions then take their values in G. When this occurs, for a Lie group G, the fibration is called a *fibre bundle*, and G is called the *group of the bundle*.

Reconstructing a bundle from its transition functions. The transition functions may be thought of as playing something of the role of the coordinate transformations in an atlas for a manifold; this provides an alternative way of thinking of the bundle. In fact, if the transition functions for a given covering $\{\mathcal{O}_\alpha\}$ of $\mathcal{M}$ are known, then the bundle may be reconstructed, using a construction very reminiscent of the definition of a contravariant or covariant vector in classical tensor calculus. That is to say, each element of the bundle space is to be considered as an assignment to each point x of $\mathcal{M}$, and to each $\mathcal{O}_\alpha$ containing x, of an element of $\mathcal{F}$, subject to the appropriate transformation law. To be a little more precise: consider the set of triples $\{(x, \alpha, \xi)\}$, where $x \in \mathcal{M}$, $\mathcal{O}_\alpha$ is an open set of the given covering which contains x, and $\xi \in \mathcal{F}$. The transition functions $\Psi_{\alpha\beta}$ associated with the covering being assumed known, and assumed also to satisfy the conditions given in Exercise 2 above, we define a relation $\sim$ on the set of such triples by setting $(x; \alpha, \xi) \sim (y, \beta, \eta)$ if $x = y$ and if $\eta = \Psi_{\beta\alpha}(x)\xi$. This relation is an equivalence relation by virtue of the conditions assumed for $\Psi_{\alpha\beta}$; each equivalence class is taken to be a point of $\mathcal{B}$. We shall not discuss the question of the differentiable structure of $\mathcal{B}$ from this point of view, but the other factors in the definition of a fibre bundle are clear enough. Denote by $[x, \alpha, \xi]$ the equivalence class of (x, α, ξ); then

(1) $\pi([x, \alpha, \xi]) = x$
(2) $\pi^{-1}(x)$ may be identified with $\mathcal{F}$ by fixing α and mapping $[x, \alpha, \xi]$ to ξ
(3) $\pi^{-1}(\mathcal{O}_\alpha)$ may be identified with $\mathcal{O}_\alpha \times \mathcal{F}$ by mapping $[x, \alpha, \xi]$ to (x, ξ).

Exercise 3. Show that the transition functions for the bundle so reconstructed are just those we started with. □

The additional structure of $\mathcal{F}$ may be transferred to the bundle provided that the transition functions take their values in the group G of structure-preserving diffeomorphisms of $\mathcal{F}$: for then two elements $[x, \alpha, \xi]$ and $[x, \alpha, \xi']$ in the fibre over x may be combined by combining ξ and ξ' in the appropriate way; this will be independent of which two representative elements are chosen in the equivalence class, provided that they both have the same value for the set index α, since the transition functions will respect the law of combination. Suppose, for example, that $\mathcal{F}$ is a Euclidean space, so that for any ξ, ξ' one may form the scalar product $\xi \cdot \xi'$, and suppose that transition functions are given which take their values in the orthogonal group of appropriate dimension. Define a scalar product on $\pi^{-1}(x)$ by

$$[x, \alpha, \xi] \cdot [x, \alpha, \xi'] = \xi \cdot \xi'.$$

This makes sense because if

$$[x, \beta, \eta] \sim [x, \alpha, \xi] \qquad \text{and} \qquad [x, \beta, \eta'] \sim [x, \alpha, \xi']$$

then

$$\eta = \Psi_{\beta\alpha}(x)\xi \qquad \text{and} \qquad \eta' = \Psi_{\beta\alpha}(x)\xi'$$

where $\Psi_{\beta\alpha}(x)$ is an orthogonal transformation of $\mathcal{F}$, and so

$$\eta \cdot \eta' = \xi \cdot \xi'.$$

This discussion may be summarised by saying that a fibre bundle consists of a collection of "trivial" pieces $\mathcal{O}_\alpha \times \mathcal{F}$ which are glued together above the intersections $\mathcal{O}_\alpha \cap \mathcal{O}_\beta$ but with the possible addition of a warp or twist to the $\mathcal{F}$-factors in the glueing process; the transition functions codify this twisting. As an example, apparently rather different from the tangent and cotangent bundles which have been our main examples to date, we now show that the usual recipe for making a Möbius band—take a strip of paper, twist one end through 180° relative to the other end and glue—is an instance of the construction of a fibre bundle from transition functions. We take for $\mathcal{M}$ the circle, for $\mathcal{F}$ the real line, and for G the two-element group $\{+1, -1\}$, acting on $\mathbf{R}$ by multiplication. We cover the circle by two open subsets $\mathcal{O}_1$, $\mathcal{O}_2$, each diffeomorphic to an open interval, so that $\mathcal{O}_1 \cap \mathcal{O}_2$ is the union of two disjoint pieces, say $\mathcal{P}$ and $\mathcal{Q}$ (each again diffeomorphic to an open interval). We set $\Psi_{12}(x) = +1$ if $x \in \mathcal{P}$ and $\Psi_{12}(x) = -1$ if $x \in \mathcal{Q}$. The transition function conditions are satisfied. The resulting fibre bundle is the (infinite) Möbius band.

Trivial bundles. The simplest way of forming a fibre bundle with base manifold $\mathcal{M}$ and fibre $\mathcal{F}$ is to take the product manifold $\mathcal{M} \times \mathcal{F}$ as bundle space, with projection onto the first factor as projection. This is called the *trivial bundle* with base $\mathcal{M}$ and fibre $\mathcal{F}$. Every bundle is locally like a trivial bundle, *via* a *local trivialisation* $\pi^{-1}(\mathcal{O}) \to \mathcal{O} \times \mathcal{F}$. A bundle which is globally, and not just locally, a product is simply said to be trivial: to be precise, the bundle $\pi: \mathcal{B} \to \mathcal{M}$ with standard fibre $\mathcal{F}$ is a *trivial bundle* if there is a diffeomorphism $\psi: \mathcal{B} \to \mathcal{M} \times \mathcal{F}$ such that $\Pi_1 \circ \psi = \pi$. Thus a bundle is a trivial bundle if it is diffeomorphic to the trivial bundle with the same base and fibre, by a diffeomorphism which respects the projections.

Note the small but important (and potentially confusing) distinction between the statements "this is a trivial bundle" and "this is the trivial bundle". The point is that a bundle may be trivial without this being obvious from the way it is presented. Thus a manifold $\mathcal{M}$ covered by a single coordinate patch has trivial tangent bundle, but there is a certain difference between "the set of tangent vectors to $\mathcal{M}$" and "the set of pairs (x, ξ) where $x \in \mathcal{M}$ and $\xi \in \mathbf{R}^m$", though each tangent vector to $\mathcal{M}$ corresponds uniquely to a pair (x, ξ) (namely its point of tangency and its components with respect to universal coordinates). Part of the problem is (as is so often the case) that the correspondence is not canonical but coordinate dependent: a different coordinate system, provided it is still universal, gives rise to a different diffeomorphism of $T\mathcal{M}$ with $\mathcal{M} \times \mathbf{R}^m$. It may happen, in fact, that there is a diffeomorphism of $T\mathcal{M}$ with $\mathcal{M} \times \mathbf{R}^m$ respecting the projections, even though $\mathcal{M}$ does not have a universal coordinate patch. Such a $T\mathcal{M}$ is still trivial.

Exercise 4. Show that a tangent bundle $T\mathcal{M}$ is trivial if and only if there is a set of globally defined smooth vector fields on $\mathcal{M}$ which at each point form a basis for the tangent space there. The circle and the torus have trivial tangent bundles. By considering left invariant vector fields, show that the tangent bundle of any Lie group G is necessarily trivial. Show that this trivialisation is given explicitly by the map $T_gG \to G \times T_eG$ by $v \mapsto (g, L_{g^{-1}*}v)$. □

Exercise 5. Show that $T^*\mathcal{M}$ is trivial if and only if $T\mathcal{M}$ is. □

Cross-sections. A smooth map $\sigma: \mathcal{M} \to \mathcal{B}$ such that $\pi \circ \sigma$ is the identity map of $\mathcal{M}$ is called a *cross-section* (sometimes just *section*) of the bundle $\pi: \mathcal{B} \to \mathcal{M}$. A cross-section of a bundle assigns to each point of the base manifold a quantity which may be identified, though not in general canonically, as an element of the standard fibre, and which obeys a transformation law determined by the transition functions of the bundle. The idea of a cross-section of a bundle with standard fibre $\mathcal{F}$ is thus a generalisation of the idea of a map $\mathcal{M} \to \mathcal{F}$; the possibility that the bundle space may not be the product $\mathcal{M} \times \mathcal{F}$ on a global scale leads to interesting complications of a topological nature when one is dealing with cross-sections rather than straightforward maps.

Exercise 6. Show that a smooth map $\mathcal{M} \to \mathcal{F}$ may be described as a cross-section of the trivial bundle with base $\mathcal{M}$ and fibre $\mathcal{F}$. □

A cross-section defines a field of quantities on $\mathcal{M}$, which may be a field of geometric quantities as in the case of the tangent or cotangent bundle, or a physical field, for a suitable choice of bundle. Note that since there may be many different bundles with the same base and standard fibre, when it comes to specifying the appropriate bundle whose cross-sections will be the fields of some physical theory, one must do more than give just the standard fibre: it is necessary to give a family of transition functions as well, or to do something equivalent to this.

2. Vector Bundles

As we mentioned above, a fibre bundle has structure arising from some structure of its standard fibre. We shall deal in this section with bundles in which this extra structure is that of a vector space.

When the fibres of a fibre bundle are (real, finite dimensional) vector spaces, and the map $\Psi_{\alpha,x}$ of the fibre over x to the standard fibre, which we may take to be $\mathbf{R}^k$ for some k, is a linear isomorphism for each x and for some covering $\{\mathcal{O}_\alpha\}$, then the bundle is called a *vector bundle*.

Exercise 7. Show that if the standard fibre of a vector bundle is $\mathbf{R}^k$, each fibre is a vector space of dimension k, and for a suitable covering of $\mathcal{M}$ the transition functions take their values in $GL(k, \mathbf{R})$. □

The tangent and cotangent bundles of a manifold are vector bundles; we shall describe in a later section the construction out of these of further vector bundles, whose sections are tensor fields, and which are therefore called tensor bundles. Many kinds of physical field are represented by sections of vector bundles. Thus vector bundles form a large and important class of fibre bundles.

Just as one may form linear combinations of vector fields and of 1-forms, so one may form linear combinations of sections of any vector bundle. The coefficients of such linear combinations may be functions on the base manifold. Let $\pi: \mathcal{E} \to \mathcal{M}$ be a vector bundle (we shall usually use $\mathcal{E}$ to stand for the bundle space of a vector bundle, hoping that in this new context there will be no confusion with Euclidean spaces, and assume the standard fibre is $\mathbf{R}^k$, where the dimension of the fibre will be made explicit if necessary). Let σ_1 and σ_2 be two cross-sections of $\pi: \mathcal{E} \to \mathcal{M}$ and let f_1 and f_2 be functions on $\mathcal{M}$. Then $f_1\sigma_1 + f_2\sigma_2$ is the cross-section whose value at $x \in \mathcal{M}$ is

$$f_1(x)\sigma_1(x) + f_2(x)\sigma_2(x);$$

this makes sense because $\pi^{-1}(x)$ is a vector space.

Exercise 8. Verify that $f_1\sigma_1 + f_2\sigma_2$, so defined, is a (smooth) cross-section of $\pi: \mathcal{E} \to \mathcal{M}$. □

The space $\Sigma(\pi)$ of cross-sections of $\pi: \mathcal{E} \to \mathcal{M}$ is thus a linear space, and a module over $\mathcal{F}(\mathcal{M})$, the smooth functions on $\mathcal{M}$.

The local triviality of a vector bundle ensures the existence of local bases of sections. That is to say, given any point of the base, there is a neighbourhood $\mathcal{O}$ of that point, and a set of maps $\{\sigma_1, \sigma_2, \ldots, \sigma_k\}$ of $\mathcal{O}$ into $\pi^{-1}(\mathcal{O})$, each satisfying the conditions for a section on $\mathcal{O}$, and such that for each $x \in \mathcal{O}$, $\{\sigma_1(x), \sigma_2(x), \ldots, \sigma_k(x)\}$ is a basis for the fibre over x (where k is of course the dimension of the fibre). Indeed, the sections corresponding to a fixed basis of the standard fibre under a local product decomposition comprise a local basis of sections. Any section may be expressed uniquely as a linear combination of the sections making up a local basis, over the domain of the local basis, the coefficients being smooth local functions on that domain. Thus the vector space structure of a vector bundle allows one to fix the components of a section relative to a local basis, and therefore to represent the section as a k-tuple of functions: but in general this is possible only locally; different choices of local bases of sections will give different representations.

Algebraic constructions with vector bundles. The cotangent bundle of a manifold may be thought of as being constructed by taking a standard vector space construction—the formation of the dual of a vector space—and applying it fibre by fibre to the tangent bundle. Similarly the formation of the space of linear p-forms on a vector space, when applied fibre by fibre to the tangent bundle, leads to the construction of a new vector bundle whose sections are p-forms on the base manifold. This process, of constructing new vector bundles from a given one by applying vector space constructions fibre by fibre, has quite general application. We shall describe the most important examples, beginning with one which we have not described, even as a special case, before: the direct, or Whitney, sum of two vector bundles.

Let $\pi_1: \mathcal{E}_1 \to \mathcal{M}$ and $\pi_2: \mathcal{E}_2 \to \mathcal{M}$ be two vector bundles with the same base and with fibre dimensions k_1 and k_2. Their *Whitney sum* is a vector bundle with the same base whose fibre dimension is $k_1 + k_2$ and whose fibre over $x \in \mathcal{M}$ is the direct sum of those of the component bundles, $\pi_1^{-1}(x) \oplus \pi_2^{-1}(x)$. In order to specify the Whitney sum completely we must describe its local product structure. It will not necessarily be the case that a covering of $\mathcal{M}$, over each open set of which one of the vector bundles is a product, will serve the same purpose for the other. However, it is possible to construct an open covering of $\mathcal{M}$, $\{\mathcal{O}_\alpha\}$, such that both vector bundles are products over each $\mathcal{O}_\alpha$, by choosing a covering of $\mathcal{M}$ trivialising the first bundle, and another trivialising the second, and taking for the sets $\mathcal{O}_\alpha$ all the non-empty intersections of a set from the first cover with a set from the second. In describing the local product structure of the Whitney sum we may therefore assume the existence of a covering of $\mathcal{M}$ which locally trivialises the two bundles simultaneously. This technical point dealt with, we can proceed with the construction.

The bundle space $\mathcal{E}$ of the Whitney sum is the set of all pairs (v_1, v_2), where $v_i \in \mathcal{E}_i$, $i = 1, 2$, and $\pi_1(v_1) = \pi_2(v_2)$. Thus v_1 and v_2 lie in the fibres of their respective bundles over the same point of $\mathcal{M}$. We define a map $\pi: \mathcal{E} \to \mathcal{M}$ by $\pi(v_1, v_2) = \pi_1(v_1) = \pi_2(v_2)$. An atlas for $\mathcal{E}$ is provided by the charts $(\hat{\mathcal{O}}_\alpha, \hat{\phi}_\alpha)$ where $\hat{\mathcal{O}}_\alpha = \pi^{-1}(\mathcal{O}_\alpha)$, $\{\mathcal{O}_\alpha\}$ is a covering of $\mathcal{M}$, by coordinate patches, of the kind described above, and $\hat{\phi}_\alpha: \hat{\mathcal{O}}_\alpha \to \mathbf{R}^{m+k_1+k_2}$ is given by

$$\hat{\phi}_\alpha(v_1, v_2) = (\phi_\alpha(x), \Psi^1_{\alpha,x}(v_1), \Psi^2_{\alpha,x}(v_2))$$

where $x = \pi(v_1, v_2)$, ϕ_α is the coordinate map on $\mathcal{O}_\alpha$, and $\Psi^i_{\alpha,x}$ is the isomorphism of $\pi_i^{-1}(x)$ with $\mathbf{R}^{k_i}$. Note that $\pi^{-1}(x)$ is the set of pairs (v_1, v_2) with $v_i \in \pi_i^{-1}(x)$ and so, with the usual rules of addition and multiplication by scalars, is just $\pi_1^{-1}(x) \oplus \pi_2^{-1}(x)$. The map $(\Psi^1_{\alpha,x}, \Psi^2_{\alpha,x})$ is a linear isomorphism of $\pi^{-1}(x)$ with $\mathbf{R}^{k_1+k_2}$, and the map $(v_1, v_2) \mapsto (x, \Psi^1_{\alpha,x}(v_1), \Psi^2_{\alpha,x}(v_2))$, where $x = \pi(v_1, v_2) \in \mathcal{O}_\alpha$, is a trivialisation of $\pi^{-1}(\mathcal{O}_\alpha)$. Thus $\pi: \mathcal{E} \to \mathcal{M}$ is a vector bundle with the required properties.

We describe next the construction of the *dual* to a given vector bundle: the dual bundle stands in the same relationship to the original bundle as the cotangent bundle does to the tangent bundle. Let $\pi: \mathcal{E} \to \mathcal{M}$ be a vector bundle of fibre dimension k. The bundle space $\mathcal{E}^*$ of the dual bundle is the set of all the elements of the dual spaces of the fibres $\pi^{-1}(x)$ of $\mathcal{E}$, as x ranges over $\mathcal{M}$. Thus any $\lambda \in \mathcal{E}^*$ is a linear functional on the vector space $\pi^{-1}(x)$ for some $x \in \mathcal{M}$, and $\mathcal{E}^* = \bigcup_{x \in \mathcal{M}} \pi^{-1}(x)^*$, where $\pi^{-1}(x)^*$ is the vector space dual to $\pi^{-1}(x)$. The projection $\tau: \mathcal{E}^* \to \mathcal{M}$ is the map which takes the dual space $\pi^{-1}(x)^*$ to x; the fibres of the dual bundle are just the dual spaces of the fibres of the original bundle. Let $\{\mathcal{O}_\alpha\}$ be a covering of $\mathcal{M}$ such that $\pi^{-1}(\mathcal{O}_\alpha)$ is trivial, and let $\Psi_{\alpha,x}$ be the corresponding isomorphism of the fibre $\pi^{-1}(x)$ with $\mathbf{R}^k$. Then the adjoint map $\Psi_{\alpha,x}{}^*$ is an isomorphism of $\mathbf{R}^{k*}$ with $\pi^{-1}(x)^*$ (recall that adjoints map contragrediently), and therefore $(\Psi_{\alpha,x}{}^*)^{-1}$ is an isomorphism of $\pi^{-1}(x)^*$ with $\mathbf{R}^{k*}$, which may be identified with $\mathbf{R}^k$ (by identifying each row vector with the column vector having the same entries). This gives an isomorphism of each fibre of $\mathcal{E}^*$ with $\mathbf{R}^k$, which extends to a local product decomposition of $\mathcal{E}^*$ based on the same covering $\{\mathcal{O}_\alpha\}$ of $\mathcal{M}$.

Exercise 9. Complete the description of the dual bundle by defining a manifold structure on $\mathcal{E}^*$, and by giving the local product decomposition explicitly. □

Exercise 10. Show that if $\Psi_{\alpha\beta}: \mathcal{O}_\alpha \cap \mathcal{O}_\beta \to GL(k, \mathbf{R})$ are the transition functions for $\pi: \mathcal{E} \to \mathcal{M}$ based on the covering $\{\mathcal{O}_\alpha\}$, then the transition functions for $\tau: \mathcal{E}^* \to \mathcal{M}$ based on the same covering are $\Psi^\times_{\alpha\beta}$, where, for each $x \in \mathcal{O}_\alpha \cap \mathcal{O}_\beta$, $\Psi^\times_{\alpha\beta}(x)$ is the inverse of the transpose of the matrix $\Psi_{\alpha\beta}(x)$. □

Exercise 11. Show that the pairing of the spaces of cross-sections $\Sigma(\pi)$ and $\Sigma(\tau)$ to $\mathcal{F}(\mathcal{M})$ defined by $\langle \sigma, \nu \rangle(x) = \langle \sigma(x), \nu(x) \rangle$ for $\sigma \in \Sigma(\pi)$, $\nu \in \Sigma(\tau)$, satisfies the rules

$$\langle f_1\sigma_1 + f_2\sigma_2, \nu \rangle = f_1\langle \sigma_1, \nu \rangle + f_2\langle \sigma_2, \nu \rangle$$
$$\langle \sigma, f_1\nu_1 + f_2\nu_2 \rangle = f_1\langle \sigma, \nu_1 \rangle + f_2\langle \sigma, \nu_2 \rangle \qquad f_1, f_2 \in \mathcal{F}(\mathcal{M}).$$

Show that if $\{\sigma_1, \sigma_2, \dots, \sigma_k\}$ is a local basis of sections of $\pi: \mathcal{E} \to \mathcal{M}$ then there is a local basis of sections $\{\nu^1, \nu^2, \dots, \nu^k\}$ of $\tau: \mathcal{E}^* \to \mathcal{M}$ such that $\langle \sigma_a, \nu^b \rangle = \delta^b_a$, $a, b = 1, 2, \dots, k$; and that if $\sigma = S^a\sigma_a$ and $\nu = N_a\nu^a$ for local functions S^a and N_a (with the summation convention back in force) then $\langle \sigma, \nu \rangle = S^a N_a$. □

Exercise 12. Show how to construct, for a given vector bundle $\pi: \mathcal{E} \to \mathcal{M}$, a new vector bundle whose fibres are the vector spaces of symmetric bilinear maps $\pi^{-1}(x) \times \pi^{-1}(x) \to \mathbf{R}$. Show that if g is a section of this bundle then for sections σ_1, σ_2 of the original vector bundle, $x \mapsto g(x)(\sigma_1(x), \sigma_2(x))$ defines a smooth function on $\mathcal{M}$; if this smooth function be denoted $g(\sigma_1, \sigma_2)$ then $g(\sigma_2, \sigma_1) = g(\sigma_1, \sigma_2)$; and g therefore defines a map $\Sigma(\pi) \times \Sigma(\pi) \to \mathcal{F}(\mathcal{M})$ which is symmetric and bilinear over $\mathcal{F}(\mathcal{M})$. □

3. Tensor Bundles

The dual of a vector bundle, and the bundle of symmetric bilinear forms which was the subject of the last exercise, both have for their sections quantities which behave in a way analogous to tensor fields on a manifold (cross-sections of the dual behave like 1-forms, or type (0,1) tensor fields, and cross-sections of the bundle of symmetric bilinear forms like symmetric type (0,2) tensor fields). We shall now develop the construction, from a given vector bundle, of a family of new vector bundles whose cross-sections will generalise tensor fields. The construction is based on the same principles as those of the previous subsection. But before tackling this task we have to make some observations about tensor algebra in general.

We begin with linear maps of vector spaces. Let $\mathcal{V}$ and $\mathcal{W}$ be vector spaces; recall (Chapter 1, Note 2) that the set of linear maps from $\mathcal{V}$ to $\mathcal{W}$ may be made into a vector space in its own right. By choice of bases this space may be made isomorphic to the space $M_{m,n}(\mathbf{R})$ of $m \times n$ matrices; since the dimension of $M_{m,n}(\mathbf{R})$ is mn, so is the dimension of the space of linear maps $\mathcal{V} \to \mathcal{W}$. It is easy to spot a basis for $M_{m,n}(\mathbf{R})$: the so-called elementary matrices with just one non-zero entry, a 1 in some position. It is not quite so obvious how one describes a basis for linear maps. Observe, however, that it is possible to fashion a linear map $\mathcal{V} \to \mathcal{W}$ out of a fixed element w of $\mathcal{W}$ and a fixed element θ of $\mathcal{V}^*$, namely the linear map $v \mapsto \langle v, \theta \rangle w$. We shall denote this map $w \otimes \theta$ and call it the *tensor product* of w and θ.

Exercise 13. Show that if $\{f_\alpha\}$ is a basis for $\mathcal{W}$ and $\{\omega^a\}$ a basis for $\mathcal{V}^*$ then $\{f_\alpha \otimes \omega^a\}$ is a basis for the space of linear maps $\mathcal{V} \to \mathcal{W}$. Show that the matrices of these basis maps with respect to $\{e_a\}$, the basis of $\mathcal{V}$ dual to $\{\omega^a\}$, and $\{f_\alpha\}$, are the elementary matrices; that the components of a linear map with respect to this basis are just its matrix elements with respect to $\{e_a\}$ and $\{f_\alpha\}$; and that in particular, if $w = w^\alpha f_\alpha$ and $\theta = \theta_a \omega^a$ then $(w \otimes \theta)^\alpha_a = w^\alpha \theta_a$. □

Thus every linear map $\mathcal{V} \to \mathcal{W}$ is a linear combination of tensor products of elements of $\mathcal{W}$ and $\mathcal{V}^*$: we therefore call the space of all such linear combinations the *tensor product* of $\mathcal{W}$ and $\mathcal{V}^*$, written $\mathcal{W} \otimes \mathcal{V}^*$.

Exercise 14. Show that the tensor product is bilinear:

$$(a_1 w_1 + a_2 w_2) \otimes \theta = a_1 w_1 \otimes \theta + a_2 w_2 \otimes \theta$$
$$w \otimes (a_1 \theta_1 + a_2 \theta_2) = a_1 w \otimes \theta_1 + a_2 w \otimes \theta_2.$$ □

Exercise 15. Let $\acute{e}_a = A^b_a e_b$ be elements of a new basis for $\mathcal{V}$, with corresponding dual basis $\{\acute{\omega}^a\}$ of $\mathcal{V}^*$; let $\acute{f}_\alpha = B^\beta_\alpha f_\beta$ be elements of a new basis for $\mathcal{W}$. Show that if $\lambda \in \mathcal{W} \otimes \mathcal{V}^*$ with $\lambda = \lambda^\alpha_a f_\alpha \otimes \omega^a = \acute{\lambda}^\alpha_a \acute{f}_\alpha \otimes \acute{\omega}^a$ then $\acute{\lambda}^\alpha_a = (B^{-1})^\alpha_\beta \lambda^\beta_b A^b_a$. □

Another interpretation of the tensor product is possible. Beginning again with an element of $\mathcal{W} \otimes \mathcal{V}^*$ of the special form $w \otimes \theta$, we define a bilinear form on $\mathcal{W}^* \times \mathcal{V}$ by

$$(\phi, v) \mapsto \langle w, \phi \rangle \langle v, \theta \rangle.$$

This construction, extended by linearity to the whole of $\mathcal{W} \otimes \mathcal{V}^*$, may be used to associate a bilinear form on $\mathcal{W}^* \times \mathcal{V}$ with any element of $\mathcal{W} \otimes \mathcal{V}^*$. If $\lambda = \lambda^\alpha_a f_\alpha \otimes \omega^a$ is a general element of $\mathcal{W} \otimes \mathcal{V}^*$, and $\phi = \phi_\alpha \chi^\alpha \in \mathcal{W}^*$ (where $\{\chi^\alpha\}$ is the basis of $\mathcal{W}^*$ dual to $\{f_\alpha\}$) and $v = v^a e_a \in \mathcal{V}$, then the bilinear form defined by λ is given by

$$(\phi, v) \mapsto \phi_\alpha \lambda^\alpha_a v^a.$$

Therefore, shifting our viewpoint a little so as to separate the starred vector spaces from the unstarred ones, we redefine the tensor product $\mathcal{V} \otimes \mathcal{W}$ of two vector spaces $\mathcal{V}$ and $\mathcal{W}$ as the space of bilinear forms on $\mathcal{V}^* \times \mathcal{W}^*$. This is a vector space whose dimension is the product of the dimensions of $\mathcal{V}$ and $\mathcal{W}$.

Tensor product bundles. Suppose now that $\pi_1 : \mathcal{E}_1 \to \mathcal{M}$ and $\pi_2 : \mathcal{E}_2 \to \mathcal{M}$ are two vector bundles over the same base. We may define, in a manner similar to that used for the Whitney sum construction, their tensor product. It is a vector bundle with the same base whose fibres are the tensor products of the fibres of π_1 and π_2 over the same point of $\mathcal{M}$. The bundle space of the tensor product bundle is the union of the tensor products of the fibres $\bigcup_{x \in \mathcal{M}} \pi_1^{-1}(x) \otimes \pi_2^{-1}(x)$.

Exercise 16. Complete the construction, by defining the projection, and giving an atlas and a local trivialisation. □

Let $\{\sigma^1_a\}$ and $\{\sigma^2_\alpha\}$ be local bases of sections of π_1 and π_2. Then there is a corresponding local basis of sections of the tensor product bundle which we write $\{\sigma^1_a \otimes \sigma^2_\alpha\}$; any local section ρ of the tensor product bundle with the same domain may be written

$$\rho = \rho^{a\alpha} \sigma^1_a \otimes \sigma^2_\alpha,$$

where the coefficients $\rho^{a\alpha}$ are smooth local functions on $\mathcal{M}$.

This construction may be carried out repeatedly, with a single vector bundle and its dual as the initial component bundles. In this way we construct the tensor bundles corresponding to the original vector bundle. A particular case of the construction, starting from the tangent bundle, leads to the tensor bundles whose sections are the tensor fields described in Section 5 of Chapter 10.

4. The Frame Bundle

We return to consideration of bundles naturally associated with a manifold in the sense that the tangent and cotangent bundle are, rather than bundles in general. The ideas of this section do generalise to arbitrary vector bundles; but for the moment we concentrate on what is perhaps the most important case.

By a *linear frame* at a point x of a manifold $\mathcal{M}$ we mean a basis for the tangent space at x. The collection of all linear frames at all points of $\mathcal{M}$ is the bundle space of a fibre bundle whose base is $\mathcal{M}$ and whose projection is the map which sends each frame to the point in whose tangent space it lies. The differentiable structure, diffeomorphism of fibres, and local triviality all follow from the existence of local fields of linear frames, such as are provided by the coordinate vector fields in a coordinate patch. In fact, if $\{V_1, V_2, \ldots, V_m\}$ is a local field of linear frames on a neighbourhood $\mathcal{O}$ in $\mathcal{M}$, and $\{v_1, v_2, \ldots, v_m\}$ is a frame at $x \in \mathcal{O}$, then $v_a = \lambda_a^b V_{b\,x}$, where the numbers λ_a^b are the entries in a non-singular matrix λ. Thus relative to the local field, each linear frame determines an element of $GL(m,\mathbf{R})$; and each element of $GL(m,\mathbf{R})$ determines a linear frame. We therefore take $GL(m,\mathbf{R})$ as standard fibre for this fibre bundle. The bundle is called the *bundle of linear frames* or simply the *frame bundle* of $\mathcal{M}$, $\pi\colon \mathcal{L} \to \mathcal{M}$.

Exercise 17. Show, in detail, that the frame bundle is indeed a fibre bundle. □

Exercise 18. Show that the collection of bases of the fibres of any vector bundle may be made into a fibre bundle whose standard fibre is $GL(k,\mathbf{R})$, where k is the fibre dimension of the vector bundle. □

The frame bundle occupies a central position in the parade of tensor bundles (generated from the tangent and cotangent bundles) over $\mathcal{M}$, and thus in its "first order" differential geometry. The reason for this is that from a linear frame at x one may build a basis for $T_x^*\mathcal{M}$ (the dual basis to that of $T_x\mathcal{M}$ which is the linear frame), and bases for all the tensor product spaces (by taking tensor products). The frames are therefore skeletons not just for the tangent spaces to $\mathcal{M}$, but for the whole structure of tensor spaces; and so the frame bundle is the skeleton of the whole structure of tensors on $\mathcal{M}$.

The frame bundle will not in general be trivial. The necessary and sufficient condition for it to be so is the existence of a global field of linear frames on $\mathcal{M}$, that is, a global section of $\pi\colon \mathcal{L} \to \mathcal{M}$. For the existence of such a section allows one to fix a point in each fibre to correspond to the identity in $GL(m,\mathbf{R})$, and when this is done the identification of the whole fibre with $GL(m,\mathbf{R})$ follows. (The situation here is rather different from that occurring for vector bundles: each vector bundle has one global section, the zero section, whose value at each point of the base is the zero element of the fibre; in fact, in order for a vector bundle to be trivial there must be a global field of frames.)

The frame bundle is clearly not a vector bundle. Its standard fibre has instead the structure of a group. This group plays a significant role in the geometry of the frame bundle. It is therefore worth pausing here to clear up a tricky point about the way in which $GL(m,\mathbf{R})$ acts. We are interested mainly in the role played by $GL(m,\mathbf{R})$ in fixing the components of the elements of one basis $\{e'_a\}$ of an m-dimensional vector space $\mathcal{V}$ in terms of those of another $\{e_a\}$:

$$e'_a = \lambda^b_a e_b \qquad \text{where } \lambda = (\lambda^b_a) \in GL(m,\mathbf{R}).$$

Here (λ^b_a) represents a matrix with a labelling the columns and b the rows; matrix multiplication, in the usual row-into-column manner, is given by

$$(\lambda\mu)^b_a = \lambda^b_c \mu^c_a;$$

and with this definition $GL(m,\mathbf{R})$ acts on $\mathbf{R}^m$ to the left. The situation is different, however, in the case of the action of $GL(m,\mathbf{R})$ which is of most interest in the present circumstances, namely its action on bases of $\mathcal{V}$. We compute the effect of first applying λ to the basis $\{e_a\}$, and then applying μ to the result. We obtain the basis $\{e''_a\}$ where

$$e''_a = \mu^c_a e'_c = \mu^c_a \lambda^b_c e_b = (\lambda\mu)^b_a e_b.$$

Thus, in contrast to its action on $\mathbf{R}^m$, when $GL(m,\mathbf{R})$ acts in the natural way on the bases of $\mathcal{V}$ its action is to the right.

This right action of $GL(m,\mathbf{R})$ on bases of m-dimensional vector spaces generates a right action of $GL(m,\mathbf{R})$ on the frame bundle of the manifold $\mathcal{M}$. If $F \in \mathcal{L}$ we define $F\lambda$ as follows. Let F be the linear frame $\{v_1, v_2, \ldots, v_m\}$ at $x \in \mathcal{M}$; then $F\lambda$ is the linear frame $\{v'_1, v'_2, \ldots, v'_m\}$ at x, where $v'_a = \lambda^b_a v_b$.

Exercise 19. Show that this defines a right action of $GL(m,\mathbf{R})$ on $\mathcal{L}$ which is free (Chapter 12, Section 4), that the orbits of the action are just the fibres of $\pi: \mathcal{L} \to \mathcal{M}$, and that the action of $GL(m,\mathbf{R})$ on the fibres is simply transitive. □

The right action $R_\lambda: F \mapsto F\lambda$ of $GL(m,\mathbf{R})$ on $\mathcal{L}$ allows one to identify the Lie algebra of $GL(m,\mathbf{R})$ with certain vector fields on $\mathcal{L}$ in the manner of Section 11 of Chapter 12. The Lie algebra of $GL(m,\mathbf{R})$ is just $M_m(\mathbf{R})$, the space of $m \times m$ matrices. If $X \in M_m(\mathbf{R})$ the corresponding fundamental vector field $\tilde{X}$ on $\mathcal{L}$ is the generator of the one-parameter group $R_{\exp tX}$. It is tangent to the orbits of the action, that is, to the fibres: vector fields with this property are said to be *vertical*. The set of vectors $\{\, \tilde{X}_F \mid X \in M_m(\mathbf{R}) \,\}$ at a point $F \in \mathcal{L}$ spans the tangent space to the fibre there, since the action is transitive on the fibre. The map $X \to \tilde{X}$ is injective since the action is free, and is a Lie algebra homomorphism since the action is to the right.

As well as thinking of a linear frame at a point $x \in \mathcal{M}$ as a basis of $T_x\mathcal{M}$, one may think of it as a way of identifying $T_x\mathcal{M}$ with $\mathbf{R}^m$. Thus each linear frame at x defines a linear map $T_x\mathcal{M} \to \mathbf{R}^m$ in which each element of $T_x\mathcal{M}$ is mapped to its coordinates with respect to the given linear frame. If the map associated with the linear frame F is written Θ_F, then each component of Θ_F (pictured as a column vector of functions) is a linear form on $T_x\mathcal{M}$. That is, Θ_F is an $\mathbf{R}^m$-valued linear form. Given a local field of linear frames one has a locally defined $\mathbf{R}^m$-valued 1-form on $\mathcal{M}$, in the sense of Chapter 11, Section 5.

A local field of frames is merely a local section of $\pi: \mathcal{L} \to \mathcal{M}$. It is commonly the case that when the construction of some geometric object on $\mathcal{M}$ depends on the choice of a local section of some fibre bundle over $\mathcal{M}$ there is a "universal" object defined on the bundle space from which the geometric object in question may be derived by means of the section (the canonical 1-form on the cotangent bundle is an example). The frame bundle space carries a canonical $\mathbf{R}^m$-valued 1-form Θ defined as follows: for $F \in \mathcal{L}$ and $\varsigma \in T_F\mathcal{L}$, $\langle \varsigma, \Theta_F \rangle$ is the vector of coordinates of $\pi_*\varsigma \in T_{\pi(F)}\mathcal{M}$ with respect to the linear frame F.

Exercise 20. Show that if σ is the local section of $\pi: \mathcal{L} \to \mathcal{M}$ determined by a local field of linear frames then $\sigma^*\Theta$ is the local $\mathbf{R}^m$-valued 1-form on $\mathcal{M}$ described above. (It is the local basis of 1-forms dual to the local basis of vector fields defined by σ, but considered as an $\mathbf{R}^m$-valued 1-form.) □

Exercise 21. Show that $\Theta^b = x^b_c dx^c$ with respect to coordinates (x^a, x^b_c) on $\mathcal{L}$, where $x^b_c(F)$ are the entries in the matrix which represents F with respect to the coordinate frame $\{\partial_a\}$. □

The form Θ transforms in a straightforward way under the action of $GL(m,\mathbf{R})$. Notice that if $\varsigma \in T_F\mathcal{L}$ and $\varsigma' \in T_{F\lambda}\mathcal{L}$ where $\pi_*\varsigma' = \pi_*\varsigma$ then $\langle \varsigma', \Theta_{F\lambda} \rangle = \lambda^{-1}\langle \varsigma, \Theta_F \rangle$, since only the frame changes. Thus $\langle \varsigma, (R_\lambda{}^*\Theta)_F \rangle = \langle R_{\lambda*}\varsigma, \Theta_{F\lambda} \rangle = \lambda^{-1}\langle \varsigma, \Theta_F \rangle$ since $\pi \circ R_\lambda = \pi$. So one may write

$$R_\lambda{}^*\Theta = \lambda^{-1}\Theta,$$

where matrix multiplication of a vector is implied on the right hand side.

Exercise 22. Confirm that this result is consistent with the rule for pull-back maps, $(\phi \circ \psi)^* = \psi^* \circ \phi^*$. □

Exercise 23. Show that if $X \in M_m(\mathbf{R})$ then $\mathcal{L}_{\tilde{X}}\Theta = X\Theta$, where matrix multiplication is again implied on the right. □

Reconstructing the tensor bundles. We have derived the frame bundle from the tangent bundle; if one could set up the frame bundle first, it would be possible to reconstruct the tangent bundle, and by a similar construction the cotangent bundle and all the tensor bundles too. We describe how this is done.

We describe first how one might reconstruct a vector space from all its bases. A basis, or frame, $E = \{e_1, e_2, \ldots, e_m\}$ for the vector space and an element $\xi = (\xi^1, \xi^2, \ldots, \xi^m)$ of $\mathbf{R}^m$ together determine an element $\xi^a e_a$ of the vector space. But many different pairs (E, ξ) determine the same vector: in fact $(E\lambda, \lambda^{-1}\xi)$ determines one and the same element of the vector space whatever the choice of $\lambda \in GL(m,\mathbf{R})$.

Exercise 24. Show that if (E_1, ξ_1) and (E_2, ξ_2) determine the same element of the vector space then there is some $\lambda \in GL(m,\mathbf{R})$ such that $E_2 = E_1\lambda$ and $\xi_2 = \lambda^{-1}\xi_1$. □

We define a relation on the set of pairs $\{(E,\xi)\}$ by setting $(E_1, \xi_1) \sim (E_2, \xi_2)$ if $E_2 = E_1\lambda$ and $\xi_2 = \lambda^{-1}\xi_1$ for some $\lambda \in GL(m,\mathbf{R})$. This is an equivalence relation, and each equivalence class defines a unique element of the vector space (and conversely).

One may carry out this somewhat roundabout reconstruction of a vector space point by point over $\mathcal{M}$ to obtain the tangent bundle from the frame bundle. On the product manifold $\mathcal{L} \times \mathbf{R}^m$ one sets $(F_1, \xi_1) \sim (F_2, \xi_2)$ if $F_2 = F_1\lambda$ and $\xi_2 = \lambda^{-1}\xi_1$, as before. This relation is an equivalence relation, and $T\mathcal{M}$ is the set of equivalence classes.

An alternative way of describing the same construction is to observe that $(E, \xi) \mapsto (E\lambda, \lambda^{-1}\xi)$ defines a right action of $GL(m, \mathbf{R})$ on $\mathcal{L} \times \mathbf{R}^m$; $T\mathcal{M}$ is the space of orbits.

We mention this construction because it justifies the remark made earlier that the frame bundle is central to the whole tensor structure on the manifold: by modifying the construction a little, one may build all the tensor bundles in a similar way.

In fact let $\mathcal{V}$ be any vector space on which $GL(m, \mathbf{R})$ acts to the right: define a relation on $\mathcal{L} \times \mathcal{V}$ by $(F_1, v_1) \sim (F_2, v_2)$ if $F_2 = F_1\lambda$ and $v_2 = \rho_\lambda v_1$, where ρ is the action; this relation is an equivalence relation, and it may be shown that the equivalence classes are the points of the bundle space of a vector bundle over $\mathcal{M}$ with standard fibre $\mathcal{V}$ and structure group $GL(m, \mathbf{R})$. This bundle is called the bundle *associated* to the bundle of frames by the action ρ. Thus the tangent bundle is associated to the frame bundle by the action $\xi \mapsto \lambda^{-1}\xi$ on $\mathbf{R}^m$. The action of $GL(m, \mathbf{R})$ on $\mathbf{R}^m$ by the transpose (that is, $\rho_\lambda \xi = \lambda^T \xi$) is a right action: the associated bundle is the cotangent bundle. More generally the action ρ of $GL(m, \mathbf{R})$ on a tensor product space $\mathbf{R}^m \otimes \mathbf{R}^m \otimes \cdots \otimes \mathbf{R}^m$ by

$$\begin{aligned}\rho_\lambda(\xi_1 \otimes \xi_2 \otimes \cdots \otimes \xi_p \otimes \xi_{p+1} \otimes \cdots \otimes \xi_{p+q}) \\ = \lambda^{-1}\xi_1 \otimes \lambda^{-1}\xi_2 \otimes \cdots \otimes \lambda^{-1}\xi_p \otimes \lambda^T \xi_{p+1} \otimes \cdots \otimes \lambda^T \xi_{p+q}\end{aligned}$$

is a right action, and the associated bundle is a bundle of tensors of type (p, q).

5. Special Frames

Many special geometric structures define, and may be defined by, special kinds of linear frame. The most obvious case in point is a metric, where the special frames are the orthonormal frames. It is useful to look at this phenomenon in terms of the frame bundle.

First, however, we mention some more examples. A conformal structure defines a set of frames whose vectors are mutually orthogonal and all of the same length (with respect to one, and hence every, metric in the conformal class). A volume form determines the set of frames which bound parallelepipeds of unit volume, and an orientation determines the set of positively oriented frames. The most extreme case is that of a complete parallelism, when the manifold admits a global field of frames: the frame of this field at each point may be taken as the single special frame defined by this structure. Less obvious, perhaps, is the case of a distribution, where the special frames are those whose first k vectors belong to the distribution (k being its dimension).

Each of these structures defines a collection of special frames, and in turn is defined by its special frames. Thus if one knows which frames are orthonormal one may reconstruct the metric. But this raises the question, what conditions must a set of frames satisfy in order that it may be the set of orthonormal frames of a metric? The orthonormal frames of a metric of signature $(r, m-r)$ at a point are related one to another by the right action of $O(r, m-r)$, the appropriate orthogonal group. Conversely, for it to be possible to reconstruct the metric at a point, the special frames at that point must include with each frame all those obtained from it by the right action of $O(r, m-r)$ but no others. For then if $v, w \in T_x\mathcal{M}$ one may define $g_x(v, w)$ by

$$g_x(v,w) = v^1w^1 + v^2w^2 + \cdots + v^rw^r - v^{r+1}w^{r+1} - \cdots - v^mw^m$$

where $(v^1, v^2, \ldots, v^m)$ and $(w^1, w^2, \ldots, w^m)$ are the components of v and w with respect to any one of the special frames; the assumption about the action of $O(r, m-r)$ ensures that this definition is self-consistent.

In each of the cases described above there is a subgroup of $GL(m, \mathbf{R})$ which plays the same role as does $O(r, m-r)$ for a metric. For a conformal structure the group is $CO(r, m-r)$, the group of matrices which preserves the standard scalar product of signature $(r, m-r)$ up to a non-zero factor; for a volume form it is $SL(m, \mathbf{R})$, the group of matrices of determinant 1; for an orientation it is the group of matrices of positive determinant; for a complete parallelism it is the group consisting of the identity; for a vector field system it is the group of non-singular $m \times m$ matrices of the form

$$\begin{pmatrix} A & B \\ 0 & C \end{pmatrix}$$

where A is a $k \times k$ matrix. The intersections of some of these subgroups will also yield interesting geometric structures: thus $SO(r, m-r) = O(r, m-r) \cap SL(m, \mathbf{R})$ is the appropriate group for an orientable pseudo-Riemannian structure.

The group by itself is not enough: it is also necessary to ensure that the structure is smooth. This requires the existence of local fields of special frames.

One may describe the set of special frames of a geometric structure as a subset of the frame bundle. It should be clear from the discussion that this subset is required to be a submanifold which is the bundle space, in its own right, of a fibre bundle which shares its projection with the frame bundle but has for its structure group and standard fibre the subgroup of $GL(m, \mathbf{R})$ appropriate to the structure. Such a bundle is said to be a sub-bundle of the frame bundle, and a *reduction of the frame bundle* to the appropriate subgroup of $GL(m, \mathbf{R})$. This discussion leads to a definition which unifies many of the geometrical structures studied by differential geometric methods. Let G be a Lie subgroup of $GL(m, \mathbf{R})$. Then a *G-structure* on a manifold $\mathcal{M}$ (of dimension m) is a reduction of the frame bundle over $\mathcal{M}$ to the group G. Many geometric structures, in other words, are G-structures for suitable groups G.

It is not necessarily the case that for a given group G and given manifold $\mathcal{M}$ there is a G-structure over $\mathcal{M}$ (for example, a manifold need not necessarily be orientable). Nor will it necessarily be the case that the existence of a G-structure for some G permits the existence of a further reduction to smaller G (a pseudo-Riemannian manifold need not necessarily define a complete parallelism). Defining the structure in this way, however, allows one to see how the problem of the existence of a G-structure is a matter of the interplay between the topological properties of the manifold and the characteristics of the particular group G. It also allows a coherent approach to the study of the different geometric structures.

6. Principal Bundles

The frame bundle of a manifold, and its reductions, are paradigms of an important type of bundle, namely bundles whose standard fibres coincide with their structure groups. Such bundles are called *principal bundles*. We now point out some of the main properties of principal bundles, and give some further examples.

Let $\pi: \mathcal{P} \to \mathcal{M}$ be a principal bundle with structure group and standard fibre G; the action of G as structure group on itself as fibre is taken to be left multiplication. Then one may define a right action of G on $\mathcal{P}$ as follows. Take an open covering $\{\mathcal{O}_\alpha\}$ of $\mathcal{M}$ over which $\mathcal{P}$ is locally trivial and let $P \mapsto (x, \Psi_{\alpha,x}(P))$ be the trivialising map, where $P \in \pi^{-1}(\mathcal{O}_\alpha)$ and $x = \pi(P)$. Then the transition function $\Psi_{\alpha\beta}: \mathcal{O}_\alpha \cap \mathcal{O}_\beta \to G$ is given by

$$\Psi_{\alpha\beta}(x) = \Psi_{\alpha,x} \circ \Psi_{\beta,x}{}^{-1}.$$

This formula may be rewritten

$$\Psi_{\alpha,x} = \Psi_{\alpha\beta}(x)\Psi_{\beta,x},$$

where left multiplication by $\Psi_{\alpha\beta}(x) \in G$ is implied on the right hand side. Now for any $g \in G$ and any $P \in \pi^{-1}(\mathcal{O}_\alpha \cap \mathcal{O}_\beta)$

$$\Psi_{\alpha,x}{}^{-1}(\Psi_{\alpha,x}(P)g) = \Psi_{\alpha,x}{}^{-1}(\Psi_{\alpha\beta}(x)\Psi_{\beta,x}(P)g) = \Psi_{\beta,x}{}^{-1}(\Psi_{\beta,x}(P)g),$$

where $x = \pi(P)$, as usual. We may therefore, without danger of ambiguity, define an action of G on $\mathcal{P}$, which is clearly a right action, by

$$R_g P = \Psi_{\alpha,x}{}^{-1}(\Psi_{\alpha,x}(P)g).$$

Note that $\pi \circ R_g = \pi$, so the action preserves the fibres of $\pi: \mathcal{P} \to \mathcal{M}$. Note also that if $R_g P = P$ then $\Psi_{\alpha,x}(P)g = \Psi_{\alpha,x}(P)$ and so g is the identity: the action is free. Moreover, given any two points P and Q on the same fibre of π there is a unique $g \in G$ such that $Q = R_g P$, namely $g = \Psi_{\alpha,x}(P)^{-1}\Psi_{\alpha,x}(Q)$. Thus G acts simply transitively on the fibres, which are its orbits. The action of G on $\mathcal{P}$ is effectively just right multiplication of G on itself, transferred to $\mathcal{P}$ by the identification of the fibres with G; this is possible because the transition functions act by left multiplication, and therefore do not interfere with the right action.

Conversely, if $\mathcal{P}$ is a manifold on which a group G acts freely to the right, in such a way that the orbit space $\mathcal{M}$ may be made into a differentiable manifold, with smooth projection $\pi: \mathcal{P} \to \mathcal{M}$, and if π admits local sections, then $\pi: \mathcal{P} \to \mathcal{M}$ is a principal bundle with group G.

The action of G allows one to define fundamental vector fields $\tilde{X}$ on $\mathcal{P}$ corresponding to elements X of $\mathcal{G}$, the Lie algebra of G, in the usual way: each $\tilde{X}$ is vertical, that is, tangent to the fibres, and at each point the vector fields of this form span the tangent space to the fibre. The map $X \to \tilde{X}$ is an injective Lie algebra homomorphism.

The construction of associated bundles may be generalised to the case of a principal bundle as follows. If $\pi: \mathcal{P} \to \mathcal{M}$ is a principal bundle with group G, and $\mathcal{F}$ a manifold on which G acts to the right *via* ρ, then the orbit space of the manifold $\mathcal{P} \times \mathcal{F}$ under the right action $(P, z) \mapsto (R_g P, \rho_g z)$ is the bundle space of a fibre bundle over $\mathcal{M}$ with standard fibre $\mathcal{F}$ and structure group G. This bundle is said to be *associated* to $\pi: \mathcal{P} \to \mathcal{M}$ by the action of G on $\mathcal{F}$. The case of greatest interest occurs when $\mathcal{F}$ is a vector space and the action is a linear one (a representation of G on the vector space), for then the resulting bundle is a vector bundle.

Examples of principal bundles.

(1) Let H be a Lie group which acts transitively to the left on a manifold $\mathcal{M}$ and let G be the isotropy group of a point x of $\mathcal{M}$. The group G acts on H to the right by group multiplication: $R_g h = hg$. This action is clearly free. The orbit of any $h \in H$ consists of all those elements of H which map x to the same point as h does. The map $\pi: H \to \mathcal{M}$ by $\pi(h) = hx$ is surjective, because H acts transitively; π maps each G orbit in H to the same point of $\mathcal{M}$, and indeed the inverse image of a point of $\mathcal{M}$ is precisely a G orbit. Provided that the action of H on $\mathcal{M}$ admits local cross-sections, which is to say that there is a neighbourhood $\mathcal{O}$ of x in $\mathcal{M}$ and a smooth map $\sigma: \mathcal{O} \to H$ such that $\sigma(y)x = y$, then $\pi: H \to \mathcal{M}$ is locally trivial and is a principal bundle over $\mathcal{M}$ with group G.

(2) Real projective space $\mathbf{R}P^n$ is the set of lines through the origin in $\mathbf{R}^{n+1}$. It is a differentiable manifold. The multiplicative group of non-zero real numbers $\mathbf{R}^*$ acts to the right (or left, since it is commutative) on $\mathbf{R}^{n+1} - \{0\}$ by $R_t \xi = t\xi$. The action is evidently free, and its orbits are just the lines through the origin in $\mathbf{R}^{n+1}$ (note the necessity of removing the origin, which is a separate orbit under any linear action on a vector space). Local sections of the resulting projection $\pi: \mathbf{R}^{n+1} - \{0\} \to \mathbf{R}P^n$ may be constructed as follows. Consider the subset $\mathcal{O}_a$ of $\mathbf{R}P^n$ consisting of equivalence classes of points of $\mathbf{R}^{n+1} - \{0\}$ whose ath coordinates are non-zero. Then the map

$$[(\xi^0, \xi^1, \ldots, \xi^n)] \mapsto \left(\frac{\xi^0}{\xi^a}, \frac{\xi^1}{\xi^a}, \ldots, 1, \ldots, \frac{\xi^n}{\xi^a}\right)$$

is a local section of π over $\mathcal{O}_a$. The $\mathcal{O}_a$ evidently cover $\mathbf{R}P^n$. Thus $\mathbf{R}^{n+1} - \{0\}$ is the bundle space of a principal bundle over $\mathbf{R}P^n$ with group $\mathbf{R}^*$.

The interest of this example lies in the base rather than the bundle space: $\mathbf{R}P^n$ is constructed as the orbit space of an action on the familiar space $\mathbf{R}^{n+1} - \{0\}$.

Note that lines through the origin in $\mathbf{R}^{n+1}$ should be carefully distinguished from rays or half-lines. The multiplicative group of positive reals $\mathbf{R}^+$ also acts freely on $\mathbf{R}^{n+1} - \{0\}$ to the right. The orbits of this action are rays, and the orbit space may be taken to be S^n, the unit n-sphere. Thus $\mathbf{R}^{n+1} - \{0\}$ is a principal bundle over S^n with group $\mathbf{R}^+$, and is actually trivial. This leads however to an alternative construction of $\mathbf{R}P^n$: it is the orbit space of S^n under the right action of the two-element group $\{+1, -1\}$, where -1 acts by interchanging antipodal points.

For $\mathbf{R}P^2$ one has the familiar prescription: identify the antipodal points on S^2, or in other words take a hemisphere and identify diametrically opposite points on its boundary, to give a model of the projective plane as the "cross cap", a one-sided non-orientable surface with self-intersection.

(3) The n-dimensional torus T^n is the orbit space of the action of $\mathbf{Z}_n$, the n-fold product of the group of integers $\mathbf{Z}$, on $\mathbf{R}^n$, by $(\xi^1, \xi^2, \ldots, \xi^n) \mapsto (\xi^1 + m_1, \xi^2 + m_2, \ldots, \xi^n + m_n)$. Thus $\mathbf{R}^n$ is the bundle space of a principal bundle over T^n with group $\mathbf{Z}_n$. This is a simple case of the covering space construction described in Chapter 12, Section 10. Other examples are furnished by the coverings $SU(2) \to SO(3)$ and $SL(2, \mathbf{C}) \to L_+^\uparrow$.

Summary of Chapter 14

A fibration consists of a differentiable manifold $\mathcal{B}$ (the bundle space), a differentiable manifold $\mathcal{M}$ (the base), a surjective smooth map $\pi: \mathcal{B} \to \mathcal{M}$ (the projection) and a differentiable manifold $\mathcal{F}$ (the standard fibre) such that, for each $x \in \mathcal{M}$, $\pi^{-1}(x)$, the fibre over x, is diffeomorphic to $\mathcal{F}$; and $\mathcal{B}$ has a local product structure in the sense that for each point of $x \in \mathcal{M}$ there is a neighbourhood $\mathcal{O}$ and diffeomorphism $\psi: \pi^{-1}(\mathcal{O}) \to \mathcal{O} \times \mathcal{F}$ such that $\Pi_1 \circ \psi = \pi$ where Π_1 is projection onto the first factor. If $\{\mathcal{O}_\alpha\}$ is a covering of $\mathcal{M}$ by open sets over each of which $\pi^{-1}(\mathcal{O}_\alpha)$ is diffeomorphic to $\mathcal{O}_\alpha \times \mathcal{F}$ then there are diffeomorphisms $\Psi_{\alpha,x}: \pi^{-1}(x) \to \mathcal{F}$, and the transition functions $\Psi_{\alpha\beta}$ are defined by $\Psi_{\alpha\beta}(x) = \Psi_{\alpha,x} \circ \Psi_{\beta,x}{}^{-1}$ for $x \in \mathcal{O}_\alpha \cap \mathcal{O}_\beta$. The transition functions, which take their values in the group of diffeomorphisms of $\mathcal{F}$, record the changes in the way fibres are identified with $\mathcal{F}$ when the local product structure is changed. The transition functions satisfy $\Psi_{\beta\alpha}(x) = \Psi_{\alpha\beta}(x)^{-1}$ and $\Psi_{\alpha\beta}(x) \circ \Psi_{\beta\gamma}(x) = \Psi_{\alpha\gamma}(x)$ on appropriate domains. In cases of interest $\mathcal{F}$ usually has some additional structure: it may be a vector space, a Euclidean space or a Lie group, for example, and the diffeomorphisms of $\mathcal{F}$ which preserve this structure form a Lie group G. A fibration whose local product decompositions may be chosen so that the transition functions belong to a Lie group G is a fibre bundle with group G.

A fibre bundle is trivial if it is diffeomorphic to the product of base and standard fibre, with π corresponding to projection onto the first factor. Every fibre bundle is locally trivial, but not necessarily globally trivial: a tangent bundle $T\mathcal{M}$ is trivial if and only if there is a global basis of vector fields on $\mathcal{M}$, for example.

A smooth map $\sigma: \mathcal{M} \to \mathcal{B}$ such that $\pi \circ \sigma$ is the identity on $\mathcal{M}$ is called a cross-section of the bundle $\pi: \mathcal{B} \to \mathcal{M}$. A cross-section defines a field of quantities on $\mathcal{M}$: for example, in the case of the tangent bundle, a vector field.

A bundle whose standard fibre is $\mathbf{R}^k$ and whose group is $GL(k, \mathbf{R})$ is called a vector bundle of fibre dimension k. Cross-sections of a vector bundle may be linearly combined, with coefficients in $\mathcal{F}(\mathcal{M})$, by using the linear structure of the fibres pointwise. Moreover, the usual algebraic constructions involving vector spaces extend to vector bundles by applying them pointwise. Thus one may define the Whitney sum of two vector bundles, whose fibres are the direct sums of the fibres of the constituent bundles; the dual of a vector bundle, whose fibres are the vector spaces dual to the fibres of the original bundle; and the tensor product of two vector bundles, whose fibres are the tensor products of the fibres of the constituent bundles, or in other words the spaces of bilinear forms on their duals.

A principal bundle is a bundle whose group and standard fibre coincide. The group acts freely to the right on the bundle space in a natural way, and its orbits are just the fibres. The base space may therefore be regarded as the space of orbits under this action. A particular and important example is the frame bundle of a manifold, whose points consist of all bases for all the tangent spaces of the manifold. From a principal bundle and an action of its group G on a manifold $\mathcal{F}$ one may construct an associated bundle whose standard fibre is $\mathcal{F}$ and whose group is G. The tensor bundles over a diffentiable manifold may be constructed in this way from its frame bundle.

Many geometrical structures may be interpreted as reductions of the frame bundle to some Lie subgroup of $GL(m, \mathbf{R})$: for example, a pseudo-Riemannian structure corresponds to a reduction to some orthogonal group; a volume form to a reduction to $SL(m, \mathbf{R})$. This interpretation brings to the study of geometric structure a single coherent point of view.

15. CONNECTIONS REVISITED

In Chapter 11 we described how the notion of parallelism in an affine space or on a surface may be extended to apply to any differentiable manifold to give a theory of parallel translation of vectors, which in general is path-dependent. An associated idea is that of covariant differentiation, which generalises the directional derivative operator in an affine space, considered as an operator on vector fields. We used the word "connection" to stand for this collection of ideas.

In Chapter 13 we showed that a connection on a manifold has an alternative description in terms of a structure on its tangent bundle, namely, a distribution of horizontal subspaces, a curve in the tangent bundle having everywhere horizontal tangent vector if it represents a curve in the base with a parallel vector field along it.

In Chapter 14 we defined vector bundles. These spaces share some important properties with tangent bundles (which are themselves examples of vector bundles), namely linearity of the fibre, and the existence of local bases of sections. It is natural to ask whether the idea of a connection may be extended to vector bundles in general, so as to define notions of parallelism and of directional differentiation of (local) sections of a vector bundle. We shall show in this chapter how this may be done, first by adapting the rules of covariant differentiation on a manifold, and then, at a deeper level, by defining a structure not on the vector bundle itself but rather on a principal bundle with which it is associated.

1. Connections in Vector Bundles

It is in fact a very straightforward matter to define a connection in a vector bundle in terms of a covariant differentiation operator if one takes for guidance the rules of covariant differentiation of vector fields on a manifold given in Chapter 11, Section 2, namely

(1) $\nabla_{U+V}Z = \nabla_U Z + \nabla_V Z$
(2) $\nabla_{fU}Z = f\nabla_U Z$
(3) $\nabla_U(Y + Z) = \nabla_U Y + \nabla_U Z$
(4) $\nabla_U(fZ) = f\nabla_U Z + (Uf)Z$.

The vector fields here play two different roles: a vector field may be regarded either as an object to be differentiated (Y and Z) or as defining the direction in which differentiation takes place (U and V). In generalising covariant differentiation to an arbitrary vector bundle we replace the vector field as object to be differentiated by a section of the vector bundle, but retain the vector field defining the direction of differentiation. The reader is already familiar with this distinction: it arises in the definition of covariant differentiation of 1-forms and of tensor fields in general.

A *connection in a vector bundle* $\pi: \mathcal{E} \to \mathcal{M}$ is a rule which assigns to each vector field V on $\mathcal{M}$ and each section σ of π a new section $\nabla_V\sigma$, which satisfies, for any vector fields U, V, any sections ρ, σ, and any function f on $\mathcal{M}$, the rules

(1) $\nabla_{U+V}\sigma = \nabla_U\sigma + \nabla_V\sigma$

(2) $\nabla_{fU}\sigma = f\nabla_U\sigma$

(3) $\nabla_U(\sigma + \rho) = \nabla_U\sigma + \nabla_U\rho$

(4) $\nabla_U(f\sigma) = f\nabla_U\sigma + (Uf)\sigma$.

The operator ∇_U is called *covariant differentiation* along U.

Many of the properties of a connection in the earlier sense—which it will be natural to describe in the present context as a connection in the tangent bundle—are reproduced, *mutatis mutandis*, for a connection in a vector bundle, remembering always that one is dealing with sections of the vector bundle. In particular, the operation of covariant differentiation, though stated in terms of global vector fields and sections, is actually local. It follows from condition (4), taking f to be a suitable bump function, that if σ is zero on some open set in $\mathcal{M}$ then so is $\nabla_U\sigma$ on the same open set, for any U. In general, then, the value of $\nabla_U\sigma$ in a neighbourhood depends only on the value of σ in that neighbourhood; and so it makes sense to discuss the covariant derivative of a local section of π. Suppose that $\{\sigma_\alpha\}$ is a local basis of sections of π (where $\alpha = 1, 2, \ldots, k$, the fibre dimension). Then $\nabla_U\sigma_\alpha$ may be expressed as a linear combination of the σ_α with coefficients which are functions on the domain of the σ_α and which depend on U:

$$\nabla_U\sigma_\alpha = \omega_\alpha^\beta(U)\sigma_\beta.$$

It follows from conditions (1) and (2) that the ω_α^β are actually local 1-forms: we call them the *connection 1-forms* with respect to the local basis. Thus

$$\nabla_U\sigma_\alpha = \langle U, \omega_\alpha^\beta\rangle\sigma_\beta.$$

The covariant derivative of any local section σ (whose domain contains that of the σ_α) may be expressed in terms of the σ: if $\sigma = S^\alpha\sigma_\alpha$ then

$$\begin{aligned}\nabla_U\sigma &= S^\alpha\langle U, \omega_\alpha^\beta\rangle\sigma_\beta + (US^\alpha)\sigma_\alpha \\ &= \langle U, dS^\alpha + \omega_\beta^\alpha S^\beta\rangle\sigma_\alpha.\end{aligned}$$

The component expressions $dS^\alpha + \omega_\beta^\alpha S^\beta$, which may be called the components of the covariant differential of σ with respect to $\{\sigma_\alpha\}$, are local 1-forms. It follows that the value of $\nabla_U\sigma$ at a point depends on U only through its value at that point. It therefore makes sense to talk of the covariant derivative of a local section with respect to a tangent vector v at a point x in the domain of the section: $\nabla_v\sigma$ is an element of the fibre $\pi^{-1}(x)$.

Exercise 1. Let γ be a curve in $\mathcal{M}$ and σ a field of vectors of the type defined by $\pi: \mathcal{E} \to \mathcal{M}$ along γ, that is, a section of π over γ. Define parallelism of σ along γ; and show that given any element of $\pi^{-1}(\gamma(0))$ there is a unique parallel field along γ having that element as its value at $\gamma(0)$. □

Exercise 2. Show that the necessary and sufficient condition for the local basis of sections $\{\sigma_\alpha\}$ to be completely parallel (parallel along any curve) is that $\omega_\alpha^\beta = 0$; under these circumstances parallel translation is path-independent. □

Exercise 3. Show that if $\acute{\sigma}_\alpha = \lambda_\alpha^\beta \sigma_\beta$ defines another local basis of sections of π, where the λ_α^β are functions on the intersection of the domains of $\{\sigma_\alpha\}$ and $\{\acute{\sigma}_\alpha\}$ and are the elements of a non-singular matrix-valued function, then

$$\acute{\omega}_\beta^\alpha = (\lambda^{-1})_\gamma^\alpha d\lambda_\beta^\gamma + (\lambda^{-1})_\gamma^\alpha \omega_\delta^\gamma \lambda_\beta^\delta.$$

Rewrite this expression in terms of matrix-valued functions and 1-forms. □

Exercise 4. Show that if there is a local basis for which the matrix-valued 1-form ω satisfies $\omega = 0$ then for any other local basis with the same domain $\acute{\omega} = \lambda^{-1}d\lambda$. Show that, conversely, if the (matrix-valued) connection 1-form for some basis is given by $\lambda^{-1}d\lambda$ then there is a basis for which it is zero. Show that any two bases with the same domain for both of which the connection 1-form is zero are related by a constant matrix. □

Thus the connection 1-forms for a vector bundle transform in the manner typical of connection 1-forms.

Given a covering of $\mathcal{M}$ by open sets, over each of which the vector bundle is locally trivial, and for each open set a local basis of sections, any collection of matrix-valued 1-forms, one for each open set, satisfying the transformation rule of Exercise 3, defines a connection in the vector bundle.

If the vector bundle in question has some additional structure one is usually interested in a connection which respects it. The most obvious example of such additional structure is a *fibre metric*, that is, a section of the bundle of symmetric twice covariant tensors formed from the original vector bundle (Chapter 14, Exercise 12) which is non-singular in the usual sense. If ϕ is a fibre metric then for any two sections σ, ρ, $\phi(\sigma, \rho)$ is a smooth function on $\mathcal{M}$. The differential version of the condition that parallel translation preserve inner products, and therefore the additional condition to be satisfied by the connection, is that

$$V\bigl(\phi(\sigma,\rho)\bigr) = \phi(\nabla_V \sigma, \rho) + \phi(\sigma, \nabla_V \rho)$$

for any vector field V. A connection satisfying this condition is said to be a *metric connection*.

Exercise 5. Show that if $\pi\colon \mathcal{E} \to \mathcal{M}$ has a fibre metric then from any local basis of sections one may construct one which is orthonormal with respect to the fibre metric. Show that the matrix of connection 1-forms with respect to an orthonormal local basis of sections is skew-symmetric (due regard being paid to the signature of the metric). □

There is no analogue of the uniqueness of the Levi-Civita connection for a general vector bundle with fibre metric, because the concept of symmetry of a connection depends on the double role played by the vector fields in the tangent bundle case.

Curvature. There is no reason to suppose that covariant differentiation operators on a vector bundle satisfy $[\nabla_U, \nabla_V] = \nabla_{[U,V]}$, any more than that it should be true on the tangent bundle. The *curvature* of the connection, R, is defined by

$$R(U,V)\sigma = \nabla_U \nabla_V \sigma - \nabla_V \nabla_U \sigma - \nabla_{[U,V]}\sigma.$$

Exercise 6. Show that $R(U,V)\sigma$ is a section of the vector bundle; that it is linear in all its arguments, skew-symmetric in its vector field arguments, and satisfies $R(fU,V)\sigma = fR(U,V)\sigma$ and $R(U,V)f\sigma = fR(U,V)\sigma$ for any function f on $\mathcal{M}$. Deduce, by the usual methods, that for any $x \in \mathcal{M}$, $\bigl(R(U,V)\sigma\bigr)(x)$ depends on its arguments only through their values at x, and therefore defines a map $R_x\colon T_x\mathcal{M}\times T_x\mathcal{M}\times\pi^{-1}(x)\to\pi^{-1}(x)$ which is linear in all its arguments and skew-symmetric in the first two. In other words, R is a section of the tensor product of the bundle of 2-forms over $\mathcal{M}$ with the bundle of type $(1,1)$ tensors constructed out of the vector bundle. □

Given a basis of local sections $\{\sigma_\alpha\}$ one may write

$$R(U,V)\sigma_\alpha = \Omega^\beta_\alpha(U,V)\sigma_\beta$$

where the Ω^β_α are 2-forms, the *curvature 2-forms* of the connection with respect to the given basis. Taken together, the curvature 2-forms may be regarded as the entries in a matrix-valued 2-form. The definition of curvature may be used to obtain a formula for the curvature 2-forms in terms of the connection 1-forms, which generalises Cartan's second structure equation (Chapter 11, Section 4).

Exercise 7. Show that $\Omega = d\omega + \frac{1}{2}[\omega\wedge\omega]$. (The bracket of matrix-valued 1-forms is defined in Chapter 11, Exercise 36.) □

Exercise 8. Show that if $\{\acute{\sigma}_\alpha\}$ is another basis such that $\acute{\sigma}_\alpha = \lambda^\beta_\alpha\sigma_\beta$ then $\acute{\Omega} = \lambda^{-1}\Omega\lambda$. Show that if $\omega = \lambda^{-1}d\lambda$ then $\Omega = 0$. □

When the vector bundle has a fibre metric and an orthonormal basis is chosen, the curvature 2-form matrix, like the connection 1-form matrix, is skew-symmetric (as a matrix; the fact that it is skew-symmetric as a 2-form requires no comment). Orthonormal bases are related by orthogonal matrices, and skew-symmetry is the hallmark of the Lie algebra of the orthogonal group. More generally, when there is some additional geometric structure on the vector bundle which may be defined in terms of the reduction of the bundle of its frames from $GL(k,\mathbf{R})$ to a Lie subgroup G, in the sense of Chapter 14, Section 5, then the connection 1-form and curvature 2-form matrices with respect to a special frame field take their values in $\mathcal{G}$, the Lie algebra of G. This condition on the connection 1-form is necessary and sufficient for parallel translation to map special frames to special frames. Note that the results obtained so far here, and in Chapter 11, have a new aspect when considered in terms of Lie group theory: thus, for example, the transformation property of the curvature 2-form corresponds to the adjoint action of the Lie group on its algebra. This applies equally well when there is no special structure, the group G then being $GL(k,\mathbf{R})$ itself; however, in that case the Lie group aspect does not present itself so forcibly.

One interesting observation is that if $\omega = \lambda^{-1}d\lambda$ (which is formally the same as the formula for a left invariant 1-form on a matrix Lie group given in Chapter 12, Exercise 32) then $\Omega = 0$ and so $d\omega = -\frac{1}{2}[\omega\wedge\omega]$ (which is formally the same as the Maurer-Cartan equation for the group, Chapter 12, Exercise 31).

Exercise 9. Let $\{X_i\}$ be a basis for the Lie algebra $\mathcal{G}$, so that the X_i are constant matrices and $[X_i,X_j] = C^k_{ij}X_k$ where the C^k_{ij} are the structure constants. Let $\omega = \omega^i X_i$, where the ω^i are ordinary 1-forms. Show that the condition $d\omega = -\frac{1}{2}[\omega\wedge\omega]$ is equivalent to the condition $d\omega^k = -\frac{1}{2}C^k_{ij}\omega^i\wedge\omega^j$. □

Since ω is defined on the base manifold of the vector bundle, not on the group, these observations are at best merely suggestive. However, they do suggest a further elaboration of the significance of the vanishing of the curvature of a connection, which generalises the argument of Chapter 11, Section 6 so that it applies to special frames and G-structures as well as to vector bundles (see Section 5 of Chapter 14).

It is clear from Exercises 3, 4, 8 and 9 that if there is a completely parallel field of special frames then the curvature vanishes and the connection 1-form for any field of special frames takes the form $\lambda^{-1}d\lambda$ where the matrix-valued function λ takes its values in G; moreover if $\omega = \lambda^{-1}d\lambda$ for some special frame field and λ takes its values in G then there is a special frame field for which $\omega = 0$. The question of the converse arises. The argument of Chapter 11, Section 6 generalises easily to show that if $\Omega = 0$ then there is a completely parallel frame field. The additional ingredient is that this may be taken to be a special frame field. We consider the manifold $\mathcal{O} \times G$, where $\mathcal{O}$ is an open subset of $\mathcal{M}$ on which is defined a local field of special frames. Let ω be the corresponding connection 1-form, regarded as taking its values in $\mathcal{G}$; we assume that $\Omega = 0$ or equivalently $d\omega = -\frac{1}{2}[\omega \wedge \omega]$. Let X be the matrix-valued function on G such that $X(g)$ is g itself, considered as a matrix; thus the entries in X are the coordinate functions on $GL(k, \mathbf{R})$ restricted to G. Then $\theta = X^{-1}dX$ is a $\mathcal{G}$-valued left invariant 1-form on G, and it satisfies the Maurer-Cartan equation $d\theta = -\frac{1}{2}[\theta \wedge \theta]$. Consider now the $\mathcal{G}$-valued 1-form

$$\Theta = \theta - \omega$$

on $\mathcal{O} \times G$ (strictly speaking one should distinguish between θ, ω and their pull-backs to $\mathcal{O} \times G$). Then

$$\begin{aligned} d\Theta &= d\theta - d\omega = -\tfrac{1}{2}[\theta \wedge \theta] + \tfrac{1}{2}[\omega \wedge \omega] \\ &= -\tfrac{1}{2}[(\Theta + \omega) \wedge \theta] + \tfrac{1}{2}[\omega \wedge \omega] \\ &= -\tfrac{1}{2}[\Theta \wedge \theta] - \tfrac{1}{2}[\omega \wedge (\theta - \omega)] \\ &= -\tfrac{1}{2}[\Theta \wedge \theta] - \tfrac{1}{2}[\omega \wedge \Theta] \\ &= -\tfrac{1}{2}[\Theta \wedge (\theta + \omega)]. \end{aligned}$$

Thus Θ satisfies the conditions of Frobenius's theorem. An integral submanifold of Θ passing through (x, e) in $\mathcal{O} \times G$ is the graph of a function $\lambda: \mathcal{O} \to G$: on this submanifold, Θ vanishes, and so $\omega = \lambda^{-1}d\lambda$. Applying λ to the special frame field with which we started we obtain another special frame field, for which $\hat{\omega} = 0$. This special frame field is parallel, as required.

2. Connections in Principal Bundles

So far, we have described a connection in a vector bundle $\pi: \mathcal{E} \to \mathcal{M}$ entirely in terms of local structures on the base manifold $\mathcal{M}$, by expressing any section of π in terms of a basis of local sections. Now the collection of bases of fibres of $\mathcal{E}$, or frames, is a principal fibre bundle with group $GL(k, \mathbf{R})$ (where k is the fibre dimension of $\mathcal{E}$), and a basis of local sections of π is just a local section of this principal bundle. Moreover, if one has additional geometric structure of the kind discussed in Section 1 and in Chapter 14, Section 5 then the special frames constitute the bundle space of a reduction of the bundle of frames of $\mathcal{E}$ to the group G associated with the structure, and this reduced frame bundle is again a principal bundle, this time with group G.

A connection on the vector bundle may be described in terms of a structure on its frame bundle or, if the connection preserves extra structure, on the reduced frame bundle. In fact the definitions make sense for any principal fibre bundle, and serve to define a connection on any vector bundle associated with it. The advantage of this approach is that the connection is given by a global structure on the principal bundle; its representations in terms of local sections of the vector bundle are partial and local glimpses of this global structure. Moreover, all vector bundles associated with a given principal bundle are dealt with in one fell swoop: if the principal bundle is the bundle of frames of a manifold, so that it consists of all the bases of all its tangent spaces, then a connection in the principal bundle defines at once the connections (and covariant derivative operators) in all the tensor bundles.

The best place to start the description of this construction is with the notion of parallelism of frames. We shall use this to motivate the definition of a connection in a principal bundle.

Suppose that $\pi: \mathcal{E} \to \mathcal{M}$ is a vector bundle with connection, defined as in Section 1. Let γ be a curve in $\mathcal{M}$. A frame field along γ is parallel if the covariant derivatives of its component fields (which are sections of π over γ) are zero. Let $\{e_\alpha\}$ be a basis for $\pi^{-1}(\gamma(0))$. Then there is a unique parallel frame field along γ, consisting of component fields ε_α say, such that $\varepsilon_\alpha(0) = e_\alpha$. If $\acute{e}_\alpha = \lambda_\alpha^\beta e_\beta$ specifies another basis for $\pi^{-1}(\gamma(0))$, where $\lambda \in GL(k, \mathbf{R})$, then $\{\lambda_\alpha^\beta \varepsilon_\beta\}$ is the parallel frame field determined by $\{\acute{e}_\alpha\}$. If there is a G-structure under consideration, and the connection is compatible with it, then the same comments apply to the special frame fields, which are preserved (as a set) by parallel translation; the matrix λ corresponding to a change of frame must now belong to G.

In order to express these facts about a connection on $\pi: \mathcal{E} \to \mathcal{M}$ in terms of its frame bundle $\tau: \mathcal{L}(\mathcal{E}) \to \mathcal{M}$ we point out first of all that a curve in $\mathcal{M}$ with a frame field defined along it may be thought of as a curve in $\mathcal{L}(\mathcal{E})$; and conversely, any curve in $\mathcal{L}(\mathcal{E})$ which is nowhere tangent to the fibres defines a curve in $\mathcal{M}$, namely its projection, and a frame field along it. A curve in $\mathcal{L}(\mathcal{E})$ corresponds therefore to what is sometimes called a "moving frame" on $\mathcal{M}$. The curves in $\mathcal{L}(\mathcal{E})$ which correspond to curves in $\mathcal{M}$ and parallel frame fields along them are evidently of a special type: they are called, as in the case of the tangent bundle discussed in Chapter 13, Section 3, *horizontal* curves. Thus the observations of the previous paragraph may be restated as follows: given a curve γ in $\mathcal{M}$ and a point F in the fibre of $\mathcal{L}(\mathcal{E})$ over $\gamma(0)$ (that is, a frame at $\gamma(0)$) there is a unique horizontal curve $\hat{\gamma}$ in $\mathcal{L}(\mathcal{E})$ which projects onto γ and passes through F (so that $\hat{\gamma}(0) = F$). Moreover, for $\lambda \in GL(k, \mathbf{R})$ the horizontal curve over γ through $R_\lambda F$ (where R_λ represents the right action of λ on the frame bundle) is just $R_\lambda \hat{\gamma}$.

At each point $F \in \mathcal{L}(\mathcal{E})$ there is determined a subset of the tangent space $T_F \mathcal{L}(\mathcal{E})$ consisting of the vectors tangent to horizontal curves through F. It is a consequence of the axioms for a connection that this subset, together with the zero vector, is a subspace of $T_F \mathcal{L}(\mathcal{E})$. This subspace is complementary to the subspace consisting of the vectors tangent to the fibre. The latter subspace is usually called the ***vertical subspace*** of $T_F \mathcal{L}(\mathcal{E})$ and so the former is naturally called the ***horizontal subspace***.

Exercise 10. Let $\{\sigma_\alpha\}$ be a basis of local sections of π and suppose that the common domain $\mathcal{O}$ of the σ_α is also a coordinate neighbourhood on $\mathcal{M}$. Define local coordinates on $\tau^{-1}(\mathcal{O})$ in $\mathcal{L}(\mathcal{E})$ so that the coordinates of a point F are (x^a, x^β_α) where (x^a) are the coordinates of $\tau(F)$ and $\{x^\beta_\alpha \sigma_\beta(x)\}$ is the frame at x determined by F. Let (γ^a) be the coordinate presentation of the curve γ and let $\{\rho_\alpha\}$ be a frame along γ with $\rho_\alpha(t) = \rho^\beta_\alpha(t)\sigma_\beta(\gamma(t))$. Then the corresponding curve in $\mathcal{L}(\mathcal{E})$ has the coordinate presentation $(\gamma^a, \rho^\beta_\alpha)$. Finally, let $\omega^\alpha_\beta = \omega^\alpha_{\beta a} dx^a$ be the connection 1-forms corresponding to $\{\sigma_\alpha\}$. Show that the curve $(\gamma^a, \rho^\beta_\alpha)$ is horizontal if and only if $\dot\rho^\beta_\alpha + \omega^\beta_{\delta a}\rho^\delta_\alpha\dot\gamma^a = 0$. Deduce that a tangent vector $\xi^a\partial/\partial x^a + \eta^\beta_\alpha\partial/\partial x^\beta_\alpha$ at F is horizontal if and only if $\eta^\beta_\alpha + \omega^\beta_{\delta a}x^\delta_\alpha\xi^a = 0$. Confirm that the horizontal subset is a subspace, of dimension $\dim \mathcal{M}$, complementary to the vertical, and that $\tau_*: T_F\mathcal{L}(\mathcal{E}) \to T_{\tau(F)}\mathcal{M}$ is an isomorphism when restricted to the horizontal subspace at F. □

Exercise 11. Show that the fact that horizontal curves remain horizontal when acted on by R_λ implies that the horizontal subspace at $F\lambda$ is the image of the horizontal subspace at F under the action of $R_{\lambda*}$. □

Exercise 12. Make what modifications are necessary to the above description so that it will apply to a G-structure. □

Thus a connection on a vector bundle defines a collection of horizontal subspaces on its frame bundle, each one complementary to the vertical subspace at the same point, and mapping one to another under the right action of $GL(k, \mathbf{R})$. These subspaces form a smooth distribution on $\mathcal{L}(\mathcal{E})$.

Exercise 13. Show that if H_a is the horizontal vector field which projects onto $\partial/\partial x^a$ then $[H_a, H_b] = -\Omega^\beta_{\gamma ab}x^\gamma_\alpha\partial/\partial x^\beta_\alpha$ where Ω is the curvature 2-form (a matrix-valued form). Deduce that the distribution is integrable if and only if the curvature vanishes. □

Conversely, such a distribution of horizontal subspaces on $\mathcal{L}(\mathcal{E})$ determines a connection: for given a curve γ in $\mathcal{M}$ and a frame $\{e_\alpha\}$ at $\gamma(0)$ there will be a unique curve $\hat\gamma$, projecting onto γ and passing through the point corresponding to $\{e_\alpha\}$ in $\tau^{-1}(\gamma(0))$, which is horizontal in the sense that its tangent vector at each point lies in the horizontal subspace at that point; this curve determines a field of frames along γ, and the connection is that for which such a field of frames is parallel.

The definition of a connection in a principal bundle is just an adaptation of these ideas. A connection in the principal bundle $\pi: \mathcal{P} \to \mathcal{M}$ with group G is an assignment, to each point P of the bundle space $\mathcal{P}$, of a subspace $\mathcal{H}_P$ of $T_P\mathcal{P}$ which is complementary to the vertical subspace $\mathcal{V}_P$ of vectors tangent to the fibre, which defines a smooth distribution on $\mathcal{P}$, and which is invariant under the right action of G on $\mathcal{P}$ in the sense that $\mathcal{H}_{Pg} = R_{g*}\mathcal{H}_P$ for all $g \in G$. The subspace $\mathcal{H}_P$ is called the *horizontal subspace* of the connection at P.

Let a connection be given on $\pi: \mathcal{P} \to \mathcal{M}$: then to each tangent vector $v \in T_{\pi(P)}\mathcal{M}$ there corresponds a unique horizontal vector at P which projects onto v; this vector is denoted v^h_P and called the *horizontal lift* of v to P. Similarly, to each vector field V on $\mathcal{M}$ there corresponds a unique horizontal vector field on $\mathcal{P}$ which projects onto V, which is denoted V^h and called the *horizontal lift* of V to $\mathcal{P}$; and to each curve γ in $\mathcal{M}$ there is a unique horizontal curve passing through a preassigned point P in $\pi^{-1}(\gamma(0))$ (say) which projects onto γ, which is denoted γ^h and called the *horizontal lift* of γ through P. The horizontal lifts of curves define what passes for parallel translation in this context. Let γ be a curve in $\mathcal{M}$ joining points x and y. A map $\Upsilon_\gamma: \pi^{-1}(x) \to \pi^{-1}(y)$ is defined as follows: if $P \in \pi^{-1}(x)$ then $\Upsilon_\gamma(P)$ is the point where the horizontal lift of γ through P meets $\pi^{-1}(y)$. The map Υ_γ is called the *parallel translation* map from x to y along γ determined by the connection.

Exercise 14. Show that, for $g \in G$, $\Upsilon_\gamma \circ R_g = R_g \circ \Upsilon_\gamma$. □

Holonomy. When γ is a closed curve, beginning and ending at x, parallel translation along γ maps $\pi^{-1}(x)$ to itself. This map need not, and in general will not, be the identity. In fact it will be the identity for every closed curve γ if and only if the principal bundle is trivial and there is a global section which is parallel along every curve. For if parallel translation around closed curves always reduces to the identity then from a fixed point $P \in \mathcal{P}$ one may generate a global parallel section ρ of π by setting $\rho(y) = \Upsilon_\gamma(P)$ for any curve γ joining $\pi(P)$ to y; the choice of a different curve will merely give the same result. (This assumes, of course, that every point of $\mathcal{M}$ may be reached from $\pi(P)$ by a smooth curve.)

In general, however, parallel translation around a closed curve beginning and ending at x will define a non-trivial bijective map of $\pi^{-1}(x)$. If we fix a point $P \in \pi^{-1}(x)$ then we may write $\Upsilon_\gamma(P) = Ph$ for some suitable element $h \in G$.

Exercise 15. Show that if $Q = Pg$ then $\Upsilon_\gamma(Q) = Qg^{-1}hg$. □

Thus the action Υ_γ on the whole of $\pi^{-1}(x)$ is determined from its action on P simply by conjugation. If now δ is another closed curve beginning and ending at x then so is the composite curve $\delta \circ \gamma$ obtained by joining δ and γ end to end: if γ and δ are both defined on $[0,1]$, with $\gamma(0) = \gamma(1) = \delta(0) = \delta(1) = x$, then $\delta \circ \gamma: [0,1] \to \mathcal{M}$ is defined by

$$(\delta \circ \gamma)(t) = \begin{cases} \gamma(2t) & \text{for } 0 \le t \le \frac{1}{2} \\ \delta(2t-1) & \text{for } \frac{1}{2} \le t \le 1. \end{cases}$$

Exercise 16. Show that parallel transport is unaffected by a proper reparametrisation, and deduce that $\Upsilon_{\delta\circ\gamma} = \Upsilon_\delta \circ \Upsilon_\gamma$. □

Thus composition of curves gives rise to composition of parallel translation operations. Moreover, traversing a curve in the reverse direction replaces a parallel translation operator by its inverse. Thus the collection of maps of $\pi^{-1}(x)$ generated by parallel translation around closed curves beginning and ending at x has a group structure; it is called the *holonomy group* of the connection at x. By fixing a point $P \in \pi^{-1}(x)$ one may consider the holonomy group as a subgroup of G. To be precise, we define the holonomy group at P to be the subgroup of G consisting of those elements $h \in G$ such that $Ph = \Upsilon_\gamma(P)$ for some closed curve γ. That this is a subgroup follows from the fact that if $\Upsilon_\delta(P) = Pk$ then

$$\Upsilon_{\delta\circ\gamma}(P) = \Upsilon_\delta(Ph) = \Upsilon_\delta(P)h = Pkh.$$

Exercise 17. Show that if $Q = Pg$ then the holonomy group at Q is the conjugate by g of the holonomy group at P. □

Exercise 18. Show that if P and Q are points of $\mathcal{P}$ which may be joined by a horizontal curve then the holonomy groups at P and Q are the same. □

3. Connection and Curvature Forms

The horizontal subspaces of a connection on a principal bundle may be defined (as may any distribution) in terms of 1-forms. One has merely to find $\dim G$ linearly independent 1-forms whose zero set at each point is the horizontal subspace at that point. In order to do this one might choose a basis for the vertical vector fields, say $V_1, V_2, \ldots, V_k$, and define 1-forms $\omega^1, \omega^2, \ldots, \omega^k$ as follows: for $\varsigma \in T_P\mathcal{P}$, $\langle \varsigma, \omega^i_P \rangle$ is the ith coefficient of the vertical component of ς with respect to the basis of the vertical subspace of $T_P\mathcal{P}$ provided by the V_i. This is well-defined since the direct sum decomposition of $T_P\mathcal{P}$ into a horizontal and a vertical subspace picks out the vertical component uniquely. Moreover, ς is horizontal if and only if $\langle \varsigma, \omega^i_P \rangle = 0$, $i = 1, 2, \ldots, k$, as required.

One question is begged by this construction, however: how should one choose the vertical basis $\{V_i\}$; and indeed can one find a global basis or only local ones? Here the action of G on $\mathcal{P}$ provides the answer. As we have already pointed out, each element X of the Lie algebra $\mathcal{G}$ defines a vertical vector field $\tilde{X}$ on $\mathcal{P}$, whose integral curves are the orbits of the right action of the one-parameter group $\exp tX$; since G acts simply transitively on the fibres of $\mathcal{P}$ the vector fields $\tilde{X}_i$, for any basis $\{X_i\}$ of $\mathcal{G}$, form a global basis of vertical vector fields on $\mathcal{P}$. Thus the 1-forms $\{\omega^i\}$ determined in this manner by the basis $\{X_i\}$ of $\mathcal{G}$ are such that for $\varsigma \in T_P\mathcal{P}$ the vertical component of ς is the vertical vector at P determined by the element $\langle \varsigma, \omega^i_P \rangle X_i$ of $\mathcal{G}$. It is therefore convenient to regard the 1-forms ω^i as the elements of a $\mathcal{G}$-valued 1-form ω, defined in terms of the basis $\{X_i\}$ of $\mathcal{G}$ by $\omega = \omega^i X_i$. But now there is no longer any necessity to choose a basis for $\mathcal{G}$. We may define ω as follows: $\langle \varsigma, \omega_P \rangle$ is that element Z of $\mathcal{G}$ such that $\tilde{Z}_P$ is the vertical component of ς. It is clear that ω is well-defined, $\mathcal{G}$-valued, linear, and smooth; and that $\langle \varsigma, \omega_P \rangle$ vanishes if and only if ς is horizontal. Evidently, for $X \in \mathcal{G}$, $\langle \tilde{X}, \omega \rangle = X$. We call ω the *connection 1-form* determined by the connection.

How does ω transform under the right action of G on $\mathcal{P}$? To derive the transformation rule we apply the formula $R_{g*}\tilde{X} = \widetilde{\operatorname{ad} g^{-1} X}$ which we obtained in Chapter 12, Section 11. For any $X \in \mathcal{G}$ we have

$$\langle \tilde{X}, R_g{}^*\omega \rangle = \langle R_{g*}\tilde{X}, \omega \rangle = \langle \widetilde{\operatorname{ad} g^{-1} X}, \omega \rangle = \operatorname{ad} g^{-1} X = \operatorname{ad} g^{-1} \langle \tilde{X}, \omega \rangle;$$

while for any horizontal vector field H we have

$$\langle H, R_g{}^*\omega \rangle = \langle R_{g*}H, \omega \rangle = 0$$

since horizontal vector fields remain horizontal under the action of G. We may summarise these results in the equation

$$R_g{}^*\omega = \operatorname{ad} g^{-1}\omega,$$

where it must be remembered, when reading the right hand side, that ω takes its values in $\mathcal{G}$.

Exercise 19. Show that a $\mathcal{G}$-valued 1-form ω on $\mathcal{P}$ which satisfies $\langle \tilde{X}, \omega \rangle = X$ for $X \in \mathcal{G}$ and $R_g{}^*\omega = \operatorname{ad} g^{-1}\omega$ for $g \in G$ defines a connection on $\mathcal{P}$. □

Exercise 20. Suppose that $\pi: \mathcal{P} \to \mathcal{M}$ is a $GL(k, \mathbf{R})$ bundle; choose a local section σ of $\mathcal{P}$ over a coordinate patch $\mathcal{O}$ in $\mathcal{M}$, and take coordinates in $\pi^{-1}(\mathcal{O})$ as follows: the coordinates of P are (x^a, x^β_α) where (x^a) are the coordinates of $\pi(P)$ and (x^β_α) the elements of the matrix $g \in GL(k, \mathbf{R})$ such that $P = \sigma(\pi(P))g$. Suppose given a connection on $\mathcal{P}$, whose connection 1-form is ω. Show (by using the properties of ω given above) that the coordinate representation of ω may be written

$$\omega = x^{-1}dx + x^{-1}\omega_0 x = (\bar{x}^\beta_\gamma dx^\gamma_\alpha + \bar{x}^\beta_\gamma \omega^\gamma_\delta x^\delta_\alpha) E^\alpha_\beta,$$

where $x = (x^\beta_\alpha)$, $x^{-1} = (\bar{x}^\beta_\alpha)$, $\{E^\alpha_\beta\}$ is the standard (matrix) basis of $\mathcal{G} = M_k(\mathbf{R})$ and $\omega_0 = \omega^\beta_\alpha E^\alpha_\beta$ is a matrix-valued 1-form on $\mathcal{O}$ (pulled back to $\pi^{-1}(\mathcal{O})$) which depends on the choice of local section defining the coordinate system. Show that the horizontal subspaces are spanned by the vectors $\partial/\partial x^a - \omega^\beta_{\gamma a} x^\gamma_\alpha \partial/\partial x^\beta_\alpha$ (where $\omega^\beta_\alpha = \omega^\beta_{\alpha a} dx^a$). Show that if a new local section $\acute{\sigma} = \sigma\lambda$ is chosen, where λ is a $GL(k, \mathbf{R})$-valued function on $\mathcal{O}$, then the expression for ω in terms of the new coordinates determines a matrix-valued 1-form $\acute{\omega}_0$ related to ω_0 by $\acute{\omega}_0 = \lambda^{-1}d\lambda + \lambda^{-1}\omega_0\lambda$. □

In this way we reproduce the results of Exercise 3 in the new framework. Thus the connection 1-form on a frame bundle is a global object from which the various connection 1-forms on the base are obtained by taking local sections.

The bracket of two arbitrary horizontal vector fields will not in general be horizontal: it will be so only in the special case in which the connection is integrable, so that its holonomy group reduces to the identity and there is a global parallel section which trivialises the bundle. If V, W are vector fields on $\mathcal{M}$ with horizontal lifts V^h, W^h then in general $[V^h, W^h] \neq [V, W]^h$: but $[V, W]^h - [V^h, W^h]$ is vertical, since both of its terms project onto the same vector field on $\mathcal{M}$, namely $[V, W]$. Exercise 13 suggests that this vertical vector field should be the basis of a definition of the curvature of the connection. In fact, as we shall now show, this observation leads to a generalisation of Cartan's second structure equation.

We shall define a curvature 2-form Ω on the principal bundle $\mathcal{P}$ which, like the connection 1-form, is $\mathcal{G}$-valued, and which generalises the concept of a curvature 2-form on the base manifold in much the same way that the connection 1-form on $\mathcal{P}$ generalises the concept of a connection 1-form on the base. The starting point of the definition is the fact that the vertical component of $[V^h, W^h]$ is given by $\langle [V^h, W^h], \omega \rangle$; but since ω vanishes on horizontal vector fields

$$\langle [V^h, W^h], \omega \rangle = -d\omega(V^h, W^h).$$

This is an expression between horizontal vector fields only; to define a curvature 2-form we must give its value also when one (or both) of its arguments is vertical. In fact, we define Ω by the requirements that

$$\Omega(V^h, W^h) = d\omega(V^h, W^h) \qquad V, W \in \mathcal{X}(\mathcal{M})$$

$$\Omega(\tilde{X}, U) = 0 \qquad X \in \mathcal{G},\ U \in \mathcal{X}(\mathcal{P})$$

that is, that Ω agrees with $d\omega$ on horizontal vector fields but vanishes when one of its arguments is vertical.

Exercise 21. Show that Ω may be equivalently defined by

$$\Omega(U, V) = d\omega(\mathrm{h}(U), \mathrm{h}(V))$$

where $\mathrm{h}(U)$ is the horizontal component of U. □

We may express Ω more explicitly in terms of ω and $d\omega$ as follows. We have already observed that $d\omega(V^h, W^h) = \Omega(V^h, W^h)$. We evaluate $d\omega(\tilde{X}, W^h)$ and $d\omega(\tilde{X}, \tilde{Y})$, the other possibilities which have to be considered. For the first, note that $[\tilde{X}, W^h] = \mathcal{L}_{\tilde{X}} W^h$; since the one-parameter group generated by $\tilde{X}$ is the right action of a one-parameter subgroup of G and therefore maps horizontal vectors to horizontal vectors, it follows that $[\tilde{X}, W^h]$ is horizontal. Thus $d\omega(\tilde{X}, W^h) = -W^h\langle \tilde{X}, \omega\rangle$ since the other terms involve the pairing of ω with horizontal vector fields. But $\langle \tilde{X}, \omega\rangle$ is constant, and so

$$d\omega(\tilde{X}, W^h) = 0.$$

On the other hand

$$d\omega(\tilde{X}, \tilde{Y}) = -\langle [\tilde{X}, \tilde{Y}], \omega\rangle = -[X, Y]$$

since the other terms vanish because they again involve differentiating constants. Guided by the Cartan structure equations derived in Chapter 11, and by Exercise 7 of this chapter, we check that these results may be combined into the one formula

$$d\omega + \tfrac{1}{2}[\omega \wedge \omega] = \Omega.$$

When both arguments are horizontal this amounts to the definition of Ω. When one is vertical and one horizontal each term gives zero. When both are vertical the right hand side vanishes, and so does the left hand side, by virtue of the fact that $d\omega(\tilde{X}, \tilde{Y}) = -[X, Y]$. This establishes Cartan's structure equation in its general form.

Exercise 22. Show that for the frame bundle $\pi\colon \mathcal{L}(\mathcal{E}) \to \mathcal{M}$ with connection ω and curvature Ω, if σ is a local section then $\sigma^*\omega$ and $\sigma^*\Omega$ are the connection and curvature forms on $\mathcal{M}$ associated with the local frame field defined by σ, and $\sigma^*(d\omega + \frac{1}{2}[\omega \wedge \omega]) = \sigma^*\Omega$ is the Cartan structure equation for $\sigma^*\omega$ and $\sigma^*\Omega$. □

Exercise 23. Prove the Bianchi identities: $d\Omega(U^h, V^h, W^h) = 0$ for any $U, V, W \in \mathcal{X}(\mathcal{M})$. □

Finally, we establish the transformation properties of Ω under the right action of G. Note that $R_g{}^* d\omega = \operatorname{ad} g^{-1} d\omega$: this follows directly from the corresponding rule for ω. But R_{g*} preserves the decomposition of tangent spaces to $\mathcal{P}$ into vertical and horizontal subspaces. It follows that Ω satisfies the same rule, that is

$$R_g{}^*\Omega = \operatorname{ad} g^{-1}\Omega :$$

this is a direct consequence of the definition of Ω and the corresponding result for $d\omega$ when both arguments are horizontal; and both sides vanish when one argument is vertical.

Summary of Chapter 15

A connection in a vector bundle $\pi\colon \mathcal{E} \to \mathcal{M}$ is a rule which assigns to each vector field U on $\mathcal{M}$ and each section σ of π a new section $\nabla_U\sigma$, such that

$$\nabla_{U+V}\sigma = \nabla_U\sigma + \nabla_V\sigma \qquad \nabla_{fU}\sigma = f\nabla_U\sigma$$
$$\nabla_U(\sigma + \rho) = \nabla_U\sigma + \nabla_U\rho \qquad \nabla_U(f\sigma) = f\nabla_U\sigma + (Uf)\sigma.$$

If $\{\sigma_\alpha\}$ is a local basis of sections of π, the 1-forms ω_α^β defined by $\nabla_V \sigma_\alpha = \langle V, \omega_\alpha^\beta \rangle \sigma_\beta$ are the connection forms with respect to the local basis and the covariant derivative $\nabla_V(S^\alpha \sigma_\alpha)$ may be written $\langle V, dS^\alpha + \omega_\beta^\alpha S^\beta \rangle \sigma_\alpha$. Change of local basis to $\acute{\sigma}_\alpha$ gives a new set of connection 1-forms $\acute{\omega}_\alpha^\beta = (\lambda^{-1})_\gamma^\beta d\lambda_\alpha^\gamma + (\lambda^{-1})_\gamma^\beta \omega_\delta^\gamma \lambda_\alpha^\delta$. The connection 1-forms may be regarded as the entries in a matrix-valued 1-form, in which case this relation becomes $\acute{\omega} = \lambda^{-1} d\lambda + \lambda^{-1} \omega \lambda$. The curvature R of the connection is defined by $R(U, V) = [\nabla_U, \nabla_V] - \nabla_{[U,V]}$, and defines a matrix-valued 2-form Ω by $R(U, V)\sigma_\alpha = \Omega_\alpha^\beta(U, V)\sigma_\beta$. The connection and curvature forms are related by $\Omega = d\omega + \frac{1}{2}[\omega \wedge \omega]$, a generalisation of Cartan's second structure equation. If Ω vanishes then there is a matrix-valued function λ defined locally on $\mathcal{M}$ such that $\omega = \lambda^{-1} d\lambda$, and therefore a field of completely parallel frames. Similar results follow if the vector bundle has structure (for example, a metric); ω and Ω then take their values in the Lie algebra of the appropriate group.

A connection in a principal bundle $\pi: \mathcal{P} \to \mathcal{M}$ with group G is an assignment to each point $P \in \mathcal{P}$ of a subspace $\mathcal{H}_P$ of $T_P\mathcal{P}$ which is complementary to the vertical subspace, which defines a smooth distribution on $\mathcal{P}$, and which is invariant under the right action of G on $\mathcal{P}$. The subspace $\mathcal{H}_P$ is said to be horizontal. Given any curve γ in $\mathcal{M}$, and any two points x, y on it, a map $\Upsilon_\gamma: \pi^{-1}(x) \to \pi^{-1}(y)$ is defined as follows: $\Upsilon_\gamma(P)$ is the point where the unique horizontal curve through P above γ meets $\pi^{-1}(y)$. This is a general form of parallel translation. In particular when γ is a closed curve beginning and ending at x, Υ_γ maps $\pi^{-1}(x)$ to itself. For any $P \in \pi^{-1}(x)$ we may write $\Upsilon_\gamma(P) = Ph$ for some $h \in G$. The set of elements of G defined in this way by all closed curves at x is a subgroup of G called the holonomy group of P. By the right invariance of the horizontal distribution $\Upsilon_\gamma \circ R_g = R_g \circ \Upsilon_\gamma$; it follows that points on the same fibre of π have conjugate holonomy groups. If the holonomy group is the identity everywhere then the principal bundle is trivial and there is a global section which is parallel along every curve.

The connection form ω is defined as follows: ω vanishes on horizontal vectors; $\langle \tilde{X}, \omega \rangle = X$ for $X \in \mathcal{G}$. It is a $\mathcal{G}$-valued 1-form, and satisfies $R_g{}^*\omega = \operatorname{ad} g^{-1}\omega$. In the case of the bundle of frames of a vector bundle, the pull-back of ω by a local section gives the connection 1-forms of the connection with respect to the local field of frames corresponding to the cross-section. The curvature 2-form Ω is also a $\mathcal{G}$-valued form, and is defined by $\Omega(U, V) = d\omega(\mathrm{h}(U), \mathrm{h}(V))$, where $\mathrm{h}(U)$ is the horizontal component of U. It also satisfies $R_g{}^*\Omega = \operatorname{ad} g^{-1}\Omega$. Moreover, it satisfies the equation

$$\Omega = d\omega + \frac{1}{2}[\omega \wedge \omega],$$

which could be described as the apotheosis of Cartan's second structure equation.

Bibliography

Books mentioned in the text.

Arnold, V. I. (1973). *Ordinary differential equations.* Cambridge, Mass.: M. I. T. Press. (Note to Chapter 3.)

Bishop, R. L. & Goldberg, S. I. (1968). *Tensor analysis on manifolds.* New York: Macmillan. (Note 2 to Chapter 1.)

Brickell, F. & Clark, R. S. (1970). *Differentiable manifolds: an introduction.* New York: Van Nostrand. (Note 1 to Chapter 10.)

Cartan, É. (1936). *La topologie des éspaces representatifs des groupes de Lie.* Paris: Hermann. (Note 3 to Chapter 12.)

Coddington, E. A. & Levinson, N. (1955). *Theory of ordinary differential equations.* New York: McGraw-Hill. (Note to Chapter 3.)

de Rham, G. (1955). *Variétés différentiables.* Paris: Hermann. (Note 2 to Chapter 10.)

Flanders, H. (1963). *Differential forms.* New York: Academic Press. (Note 2 to Chapter 10.)

Fraenkel, A. A., Bar-Hillel, Y. & Levy, A. (1973). *Foundations of set theory.* 2nd revised edition. Amsterdam: North-Holland. (Note 1 to Chapter 1.)

Halmos, P. R. (1958). *Finite dimensional vector spaces.* Princeton, N. J.: Princeton University Press. (Note 2 to Chapter 1.)

Halmos, P. R. (1960). *Naive set theory.* Princeton, N. J.: Van Nostrand. (Note 1 to Chapter 1.)

Held, A. ed. (1980). *General relativity and gravitation: 100 years after the birth of Albert Einstein.* New York: Plenum Press. (Note to Chapter 11.)

Helgason, S. (1978). *Differential geometry, Lie groups and symmetric spaces.* New York: Academic Press. (Note 1 to Chapter 11.)

Kelley, J. L. (1955). *General topology.* New York: Van Nostrand. (Note 1 to Chapter 2 and Note 1 to Chapter 10.)

Kirillov, A. A. (1976). *Elements of the theory of representations.* Berlin: Springer. (Notes 1 and 3 to Chapter 1.)

Klein, F. (1939). *Elementary mathematics from an advanced standpoint. Part II, Geometry.* New York: Dover. Translated from the third German edition. (Note to Chapter 0.)

Lang, S. (1969). *Analysis II.* Reading, Mass.: Addison-Wesley. (Note to Chapter 3.)

Loomis, L. H. & Sternberg, S. (1968). *Advanced Calculus.* Reading, Mass.: Addison-Wesley. (Notes 1 and 2 to Chapter 1.)

MacLane, S. and Birkhoff, G. (1967). *Algebra.* New York: Macmillan. (Note 3 to Chapter 1 and Notes to Chapter 4.)

Montgomery, D. & Zippin, L. (1955). *Topological transformation groups.* New York: Interscience. (Note 1 to Chapter 12.)

Porteous, I. R. (1969). *Topological geometry.* London: Van Nostrand. (Note 1 to Chapter 1.)

Sanchez, D. A. (1968). *Ordinary differential equations and stability theory.* San Francisco: W. H. Freeman. (Note to Chapter 3.)

Schrodinger, E. (1954). *Space-time structure.* Cambridge: Cambridge University Press. (Note 2 to Chapter 11.)

Spivak, M. (1965). *Calculus on manifolds.* New York: Benjamin. (Note 2 to Chapter 2.)

Sternberg, S. (1964). *Lectures on differential geometry.* Englewood Cliffs, N. J.: Prentice-Hall. (Note 4 to Chapter 4.)

Warner, F. W. (1971). *Foundations of differentiable manifolds and Lie groups.* Glenview, Ill.: Scott, Foresman. (Note 1 to Chapter 10 and Note 2 to Chapter 12.)

Some books for further reading.

Abraham, R & Marsden, J. E. (1978). *Foundations of Mechanics.* Reading, Mass.: Benjamin.

Arnold, V. I. (1978). *Mathematical methods of classical mechanics.* Berlin: Springer.

Arnold, V. I. (1983). *Geometrical methods in the theory of ordinary differential equations.* Berlin: Springer.

Beem, J. K. & Ehrlich, P. E. (1981). *Global Lorentzian geometry.* New York: Marcel Dekker.

Bott, R. & Tu, L. (1982). *Differential forms and algebraic topology.* Berlin: Springer.

Bröcker, T.D. & Jänich, K. (1982). *Introduction to differential topology.* Cambridge: Cambridge University Press.

Caratheodory, C. (1965). *Calculus of variations and partial differential equations of the first order.* San Francisco: Holden-Day.

Cartan, É. (1951). *Leçons sur la géométrie des espaces de Riemann.* Paris: Gauthier-Villars.

Cartan, É. (1952). *Leçons sur les invariants intégraux.* Paris: Hermann.

Chevalley, C. (1946). *Theory of Lie groups, vol. 1.* Princeton, N. J.: Princeton University Press.

Chern, S. S. (1967). *Complex manifolds without potential theory.* New York: Van Nostrand.

Chillingworth, D. R. J. (1976). *Differential topology with a view to applications.* London: Pitman.

Choquet-Bruhat, Y., deWitt-Morette, C. & Dillard-Bleick, M. (1982) *Analysis, manifolds and physics.* 2nd edition. Amsterdam: North-Holland.

Dodson, C. T. J. (1980). *Categories, bundles and space-time topology.* Orpington, Kent: Shiva.

Edelen, D. G. B. (1985). *Applied exterior calculus.* New York: Wiley.

Eisenhart, L. P. (1949). *Riemannian geometry.* Princeton, N. J.: Princeton University Press.

Greub, W., Halperin, S. & Vanstone, R. (1972). *Connections, curvature, and cohomology.* 2vv. New York: Academic Press.

Griffiths, P. A. (1983). *Exterior differential systems and the calculus of variations.* Boston: Birkhauser.

Guillemin, V. W. & Sternberg, S. (1984). *Symplectic techniques in physics.* Cambridge: Cambridge University Press.

Hawking, S. W. & Ellis, G. F. R. (1973). *The large scale structure of space-time.* Cambridge: Cambridge University Press.

Hermann, R. (1977). *Differential geometry and the calculus of variations.* 2nd edition. Brookline, Mass.: Math Sci Press.

Hermann, R. (1975). *Gauge fields and Cartan-Ehresmann connections*. Brookline, Mass.: Math Sci Press.

Jacobson, N. (1962). *Lie algebras*. New York: Interscience.

Klingenberg, W. (1982). *Riemannian geometry*. Berlin: Walter de Gruyter.

Kobayashi, S. & Nomizu, K. (1963). *Foundations of differential geometry*. 2vv. New York: Interscience.

Lang, S. (1972). *Differential manifolds*. Reading, Mass.: Addison-Wesley.

Milnor, J. (1965). *Topology from the differentiable viewpoint*. Charlottesville, Va.: University of Virginia Press.

O'Neill, B. (1983). *Semi-Riemannian geometry with applications to relativity*. New York: Academic Press.

Penrose, R. & Rindler, W. (1984). *Spinors and space-time*. 2vv. Cambridge: Cambridge University Press.

Schouten, J. A. (1954). *Ricci-Calculus*. 2nd edition. Berlin: Springer.

Singer, I. M. & Thorpe, J. A. (1967). *Lecture notes on elementary topology and geometry*. Glenview, Ill.: Scott, Foresman.

Spivak, M. (1970). *A comprehensive introduction to differential geometry*. 5vv. Boston, Mass: Publish or Perish.

Wells, R. O. (1980). *Differential analysis on complex manifolds*. Berlin: Springer.

Weyl, H. (1946). *The classical groups*. Princeton, N. J.: Princeton University Press.

Willmore, T. J. (1982). *Total curvature in Riemannian geometry*. Chichester: Ellis Horwood.

Woodhouse, N. (1980). *Geometric quantization*. Oxford: Oxford University Press.

Yano, K. & Ishihara, S. (1973). *Tangent and cotangent bundles*. New York: Marcel Dekker.

Yano, K. & Kon, M. (1984). *Structures on manifolds*. Singapore: World Scientific.

Index

Lightning Source UK Ltd.
Milton Keynes UK
UKOW04f2153150215

246297UK00001B/8/P